Malleable Matter / Stretchable Space

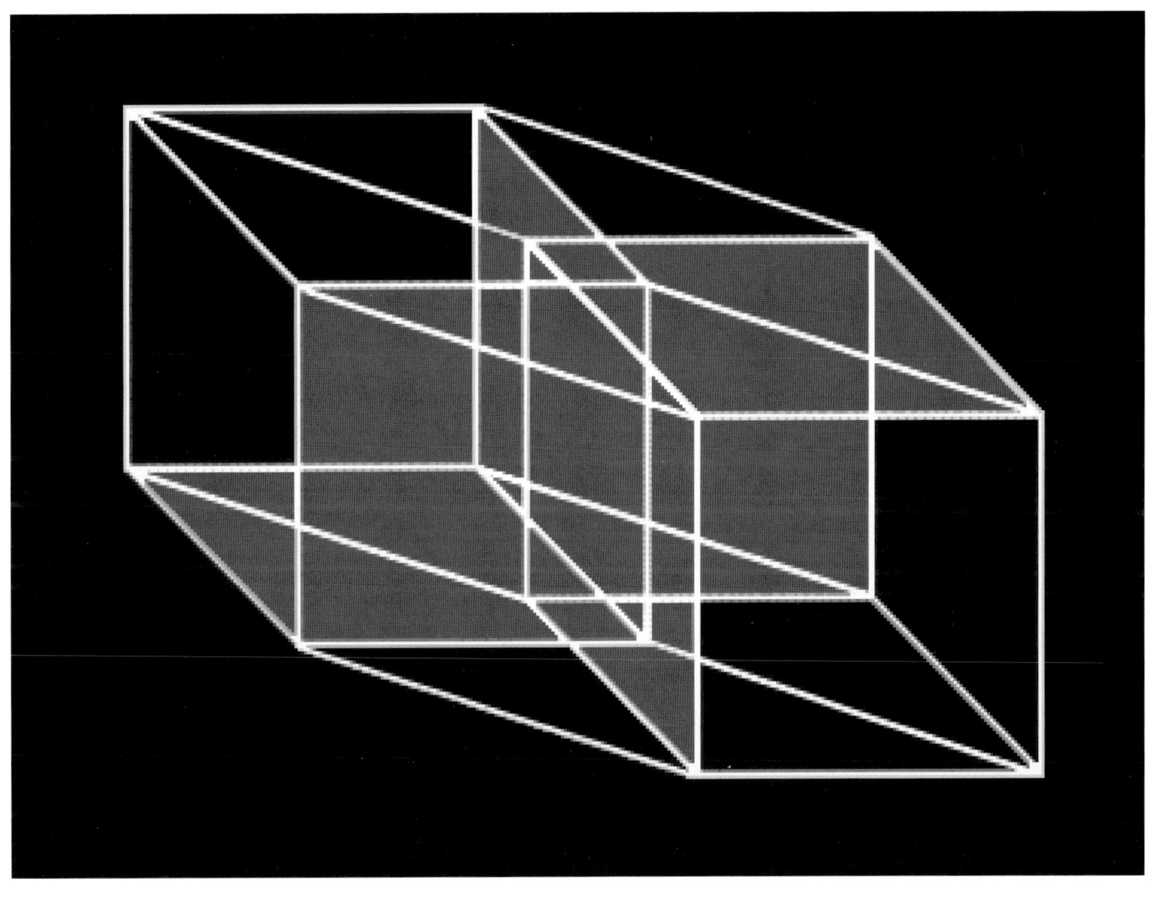

Interweaving Art, Math & Nature in n-Dimensions

Malleable Matter / Stretchable Space
Rochelle Newman

Mathematics Consultants

Judith F. Moran, Ph.D   Tom Hull, Ph.D

Photographer

Richard C. Newman

Illustrations, Drawings

Susan Libby

Gail Maciejewski

Computer Diagrams, Illustrations & Design

Claire Belanger, *Senior Illustrator/Designer*

Mary Calnan, *Assistant Illustrator*

pythagorean press

P.O. Box 5162
Bradford, MA
01835-0162

ISBN: 0-9614504-5-2

Printed in the United States of America
First Printing

About the contributors:
Rochelle Newman, author, is an artist and Professor Emerita of Northern Essex Community College., Haverhill, MA. She also has been Senior Lecturer at Bradford College, Haverhill, Massachusetts. She gives workshops and lectures on the interdisciplinary connections of art, math, & nature.

Richard C. Newman is an artist, Professor Emeritus of Bradford College, MA. He held the Kopf Chair in Art History and Theory and was the Creative Arts Division Chair for over twenty years. He has always been there to capture the ever elusive image. Cover photo: Slot Canyon, Page, Arizona

Susan Libby is a graphic artist and illustrator who took the concepts and translated them into beautiful images.

Gail Maciejewski is an artist who let her imagination take wing to play with a theme and variations.

Claire Belanger, is a recent graduate of the Graphic Design Program at Bradford College. Her skills and creativity are evident in the multitude of quality computer diagrams and illustrations in this book. Her assistant, Mary Calnan, is also a graduate of Bradford College, also working in the Graphic Design field.

Judith F. Moran, Ph.D., is a faculty member at Trinity College, Hartford, Conn.
Tom Hull, Ph.D, is a faculty member at Merrimack College, North Andover, MA.
Both of these persons are interested in the math and art connections.

The sole responsibility for this book, however, lies with Pythagorean Press. Our mathematics consultants were with us to correct our errors and make suggestions for the improvement of the mathematical content. It was the Press, however, that ultimately made the choices.

10 9 8 7 6 5 4 3 2 1

This book is dedicated to the memory of Johnny Appleseed, born John Chapman, who was more interested in the scattering of seeds than in the harvesting of apples.

Ideas are seeds that, when given fertile ground, flourish, bear fruit, and propagate the future.

1% of all sales is donated by Pythagorean Press to its Nurture Nature Fund for the support of environmental protection projects.

Acknowledgements

This book acknowledges the fact that powerful changes start with the acts of single individuals. We encourage each and every individual to reach out in positive ways to create new modes of thinking and new ways of acting that work for the betterment of this planet.

We are grateful for the examples set by the following persons:

Buckminster Fuller whose insight saw that the earth was a closed system and more could be done with less so that the bounty of the planet could be enjoyed by everyone.

Rebecca Hoffberger whose concern for creativity in all its manifestations gave form to the American Visionary Art Museum.

Maya Lin whose integrity as an artist and a human being is a model for all of us.

Dr. Penny Patterson who is best friend to Koko and Michael and is the founder of the Gorilla Foundation.

Anita Roddick whose business ethics and practices show that money and heart do not have to be antithetical.

Harvey Fite who showed that creation is a patient search for unity.

The professional and student artists who shared their works with us for your benefit.

In Memoriam: Cynthia Hastings, friend, colleague, and admirable human being passed away February, 1999. She did computer diagrams and illustrations for the book, *The Surface Plane,* also published by Pythagorean Press. Her work enhanced the quality of the book considerably. We will miss her presence.

Contents

Preface

Let us begin this book with an image, one that represents both an actuality and a metaphor and one that can symbolize a new paradigm for our new millenium. The image is that of the tree that stands erect in space with branches reaching up and out into the light of the sky while roots burrow deep and far into the dark for the nutrients of our planet. It is a fitting symbol for the aspirations and restrictions of humankind. The three major divisions of root, trunk, and bough connect to the imagination's experience of the underworld, earth, and heaven. In our human desire to link with the eternal reality of all things, the tree gives us the visual for our quest. It is the symbol of the psyche in the process of becoming an individual.

Throughout time and throughout cultures, the image and use of the tree bonds us together as one species. There is a fundamental connection between the inner life of a human and the vegetal life of the tree. A person strives throughout his/her lifetime to gain a sense of freedom and to reach maximum potential. The tree, with its roots and its many branches with their multitude of leaves, visually expresses this search. We speak of the tree as the axis of the Universe; the Tree of Knowledge; the Tree of Life, the Cosmic Tree, and the Family Tree.

Without the tree, the breath of life of this planet would be extinguished. Each and every tree revitalizes both the flesh and spirit of every individual. It is the essence of regeneration. Perhaps, because we have lost much of our ability to think in metaphors, we destroy individual trees, and forests of them, without believing that we are destroying our spiritual selves as well.

In Africa, Malay, Sumatra, India, and Persia, the tree is seen as the dwelling place of the godhead. Buddha was reputed to have been incarnated as a tree many times over. He found enlightenment when sitting under the bo-tree. Many gods and goddesses are identified with a particular species of tree. There is Athena and the olive; Artemis and the laurel. There is the banyan, the cypress, the oak, the date-palm, and the cedar. In Norse mythology, the center of the world was found in the ash tree. Because of the sacred connection, it was a sin in many places to even break off a twig from the protected trees.

Before agriculture became the way of life of civilizations, most of the earth was covered with trees. Today, less than one-third of the estimated original sixteen billion acres of forested land remains. Obviously, despite the fact that the sacred groves were protected by the priests and priestesses of old, many of the trees were taken and not many were replaced. Who protects them now? Is it the United States Con-

gress or the Forestry Service or the Parks Department? Is it the logging industry or the paper mills?

In pre-Christian Europe, forests were everywhere. They were thick, dark, and impenetrable. Think of the heroines and heroes of the fairy tales who had to first traverse the woods before encountering and overcoming the formidable evil.

First, it was the tree itself that became an object of worship. Then, people placed offerings at its base and it became an altar. Then, just the trunk became a pillar which could be erected anyplace for veneration. The pillars, in time, became columns made of stone and marble with capitals at the top suggesting their vegetal origins. Or they became totem poles joining man to his animal ancestry.

When we cut down a tree without replacing it, we are ignoring the needs of this planet. For the tree gives us oxygen to breathe; fruits and nuts to eat; medicines to cure; fuels to warm; habitats to shelter birds and animals; shade to cool; and roots to hold the soils that protect the earth from the ravages of erosion.

The tree has, because of its size, great advantages over the ground-hugging soft plants. Its photosynthetic abilities are impressive. A single tree, such as a maple, has several thousand green leaves which take carbon dioxide from the air, hydrogen from the water in the soil, combining with sunlight to give off oxygen. We humans, on the other hand, take in oxygen and give off carbon dioxide. An average tree in one year takes in twenty-six pounds of carbon dioxide and gives off enough oxygen to keep four people breathing for that year.

But that is one tree and this planet is home to many more persons than just four. If there were 240 billion trees in the United States, they could absorb almost all of the carbon dioxide this country produces each year. Carbon dioxide has become a major problem in this century because of the excessive burning of fossil fuels, the increase in industrial activity, the loss of plant life, and the continuing destruction of the world's forest, despite what we know of the greenhouse effect.

Therefore, it is imperative that we develop a new model of behavior for living on this planet. We need a worldview that is holistic and treats the planet as a single living organism, the Gaia principle. It is a view that is built on the values of conservation and recycling; of diversity and variety; of scaling down rather than increasing up; of caring rather than competing; of sharing rather than hoarding; of preserving rather than consuming. Let the Tree of Life become its icon.

"Loss of tree cover can drastically reduce a land's capacity to support life. Trees no longer pump nutrients out of the subsoil with their roots and return them to the topsoil with their fallen leaves, and even non-cultivated land declines in fertility. Wind and water erosion scrape away at the denuded topsoil. In effect, the land loses its third dimension when the protecting screen of treetops is destroyed. Life become flattened, desiccated."
David Rains Wallace
Biologist

The symbolic Tree of Life with its Axis Mundi (axis of the world) in which all creation begins. This axis stands at the center of the Universe and brings together the sky, the earth, and the underworld. Here the latter contains the chemical elements from which all things are made.

Introduction

This book is an exploration about space, which we cannot see, and matter, which we can. We humans physically know only the three spatial dimensions of this planet. With our bodies we understand the movements of forward/backward; left/right; and up/down. With our minds, however, we are capable of imagining and thinking about spatial dimensions beyond three. Physicists can explore them through experimentation, mathematicians can provide the tools for exploration, and artists can give them aesthetic form.

How do we experience dimensions? If each of us existed as a point (.) which has 0 dimensions, we would be isolated, alone, with no direction in life. We would always be at a standstill. If, however, we were granted one degree of freedom, we would be pushed to extend ourselves and become a line (----). If that line, however, were able to distinguish its head from its foot, then it could move slightly. It would always be following in the footsteps of another line. If now space could be granted two degrees of freedom, then our line could bend and move around on the plane and rub shoulders with others of its kind. It could slither and slide, but it could not take a leap of faith into the three dimensions with which we are so familiar. With each added dimension, we are granted a greater degree of freedom of movement; i.e., we have more space in which to move matter around.

In this century, scientists in many fields are investigating the notion that there are spatial dimensions beyond three that can explain the workings of the Universe. They reach out with their telescopes to probe the outer reaches of space. They perturb us with the audacious questions that they ask. They want to know "how?" And they want to know "why?" And "when?" "Who?" is the question they leave others to answer. "What is space?" they ask. "How large is space?" "How many dimensions does space have?" "Does it curve in upon itself or does it spread out to infinity?" "Does it have a beginning and does it have an end?" "What are the most basic rules that govern the building of matter?" "How are space and matter connected?"

Scientists dream of unlocking the secrets of the Universe by searching for the simplest explanation of complex phenomena. They believe that the more fundamental the theory, the more it is inclusive enough to answer the most difficult problems. Physicists, specifically, look at smaller and smaller components of matter trying to fathom the larger and larger questions. Scientists observe and then they try to prove, by repeated controlled experiments, that nature is comprehensive, comprehensible, logical, and yet always full of surprises.

In our three dimensional world, the basic forces of gravity, electromagnetism, the strong nuclear force, and the weak nuclear force are seemingly separated. When taken to the geometry of higher dimensions, however, these forces appear to be united into one grand unity. Some scientists claim that ten is the exact number of dimensions which bring all natural phenomena together in a simple and elegant way.

If we consider time a related dimension, then we inhabit a four-dimensional spacetime continuum. The world renowned physicist Albert Einstein postulated through his Theory of Relativity that space and time were linked into an undulating curved and seamless fabric. Matter influences space and space influences matter. Gravity is the result of their interaction. He showed how gravity is affected by warps in space/time.

The question is: how do we, who exist in a physical space of three dimensions, with time giving us a fourth dimension, which is not spatial, contemplate dimensions beyond these? Well, once upon a time long, long ago, all the people believed that the world was flat and that one would fall off it when one came to the edge. Then someone suggested that the world was really round and the people were horrified. How could it be possible to live on the surface of a sphere? It took them quite a while to come round, so to speak. Now there are those scientists talking of dimensions beyond three! How can one conceive of five or six or ten-dimensional space? It is mathematics that gives us the tool to intellectually examine what our senses cannot grasp.

These concepts are difficult to present to the layperson since what Einstein explored exists far beyond the limits of our solar system. Within our system, we assume flat planes and weak gravity. What we know of the limited space and short distances that we find in the Euclidean plane, does not necessarily hold true for extensive spaces and long distances. The structuring of Einstein's concepts are realized in the language of mathematics, geometry in particular. Believing his essential premises resides in the domain of faith until experiment proves what is first believed.

Perhaps, it is always faith that takes us to new places and facts that document our arrival there. But there are no absolute answers. There are only tentative solutions and more perplexing questions. What always remains is mystery, awe, and humility. But we are brave and optimistic creatures holding up the candle of reason to light the dark vastness of space. Who knows, but we might succeed in finding answers if we can only learn to be farsighted, patient enough, and to be kind to one another while travelling along the path on our spiritual quest.

> "Not only can space stretch and shrink, it can be bent and distorted."
> Paul Davies

revolution. A change in the way in which space and matter
scientific
is to restore to wholeness as many fragments as we
E=Mc²
equation. A science-educating person. A
teaches that our task in life
Truth does not lie in the mere accumulation of facts, but in an attitude of
beginning

are perceived
E = Mc²
encounter along our path of our life...
This is the art of being human
E = Mc²
A. E.
must be renewed from generation to generation
mind, and a mode of life and action, that
E = Mc²

1 A Matter of Perspective

The recurring content throughout this book is space, itself, and as the container of form. Mathematically it is an undefined term. For the visual artist, however, it is the essence of artworks. What is it? How is it perceived? How is it represented in n-dimensions?

Each age has a unique expression of space that sets it apart from other times. The worldview of Ancient Egypt is far different from that of Renaissance Rome. Its spatial vision used to express that worldview is different. Each age adds to the accumulated body of knowledge of previous times. What is found to be erroneous is removed when new facts displace the old, but no one age has a handle on the "Absolute Truth". Its particular ideas are framed by the entire cultural continuum.

So we must remember that any viewpoint is just a matter of perspective. It speaks of a particular vantage point and a specific place of reference. It speaks of a particular set of circumstances in a particular moment of time where we are receiving particular sensory information from the world.

Each of our senses gives us certain information about our three-dimensional environment. The overlapping of this information gives us knowledge. And yet, in the visual arts, touch and sight have come to predominate while taste, smell, movement, and sound are de-emphasized. These last two have become the domain of the performing arts. Of the two emphasized in the visual arts, sight still takes precedence despite the fact that is through the sense of touch that we first come to know the world.

A question, which would take many words to explore, would be to ask why sight has come to be considered our most prevalent information giver. Is it because it gives information over longer distances? With touch, objects must be very close in order to be experienced. Information is gained for only a small portion of an object. Each object has to be inspected one section at a time in order to gain a sense of the whole. Sight, however offers an overview of objective relationships. It affords us distance and a long rang view. But having too much distance and we lose touch.

In the Twentieth Century, tools have expanded our vision. We use cameras, radar, microscopes, telescopes, scanners, computers, satellites, space probes, space ships to see beyond, between, under, and through. Virtual Reality blurs the distinctions between what is there and what is perceived to be there. Sight and touch merge into an ambiguous relationship.

These facing pages show paintings by "naive" or academically untutored artists. They are self-taught and come to art via spiritual revelation.

This page:
Top right:
Painting by Ivan Rabuzin, Yugoslavia, born 1919. Paysage, 1975. Courtesy of the Musee D'Art Naif, France.

Bottom right:
Painting by Jean Klissac, Italy, born 1919. Le Rêve du Petit Patre au Pipeau, 1980. Courtesy of the Musee D'Art Naif, France.

If we accept that sight dominates, how does the artist choose which sight impression he or she should depict in the act of image-making in which the three-dimensional world must be translated down to the two dimensions of the Euclidean surface plane? What aspect of the world out there resonates so importantly inside the artist that he or she must put down those personal impressions in a more or less permanent form? How does one capture and contain what is essentially unable to be captured and contained--that elusiveness that we call life. Are there some elements in the world that seem more permanent than others, or invariant, to use a mathematical word? What elements remain constant while others change drastically?

How does an individual, or a culture, decide how the picture plane, the image surface, is to be treated? As an actual object, it is a physical and tactile thing on which to make marks. But if the marks are going to be used to define and represent objects, persons, and events, what criteria does one use in the depiction of these? What is the image going to present to the viewer? Will it be related to what is known, what is felt, what is thought, or what is seen?

When the choice is the latter, the artist must be selective about which visual pieces of information are going to be used to convey a particular spatial idea. Who decides what the answers are for the many, many questions that the artist must address in the course of creating any artwork? Many of the answers come from underlying assumptions that either the individual artist or the individual culture has accepted in a definition of their image-making.

On this page, there are four different photographic views of the same subject, an architectural object in space. The photographer has already set a viewpoint in each of these. If another visual artist wants to use these photographs as a jumping off point for another type of artwork, how would he/she decide which is the "correct' view for a particular drawing or painting? How "faithful" does the artist want to be to the original image? Does the artist want to combine parts of each image into a single unity? Does the artist want to ignore aspects of the image?

The answers to these questions go beyond mere representation to involve issue of size, proportion, composition, choice of materials, techniques of application, illusion of space, illusion of light, time of day, etc. What if the artist chooses to incorporate all four views into the same work? How would the images be organized so as to be comprehensible to the viewer?

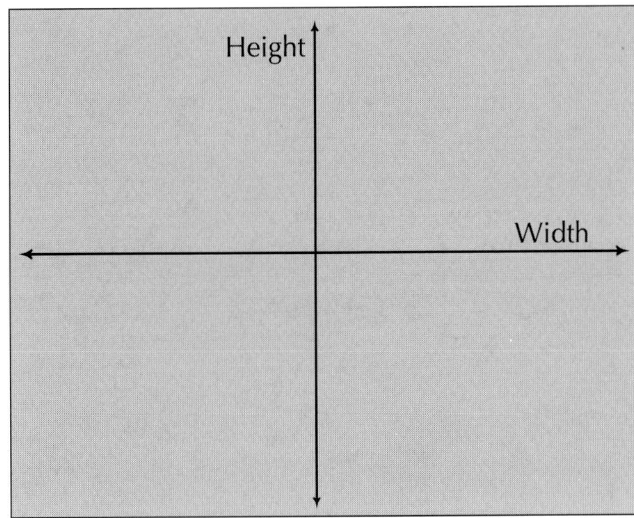

The Surface Plane

The Picture Plane

Given a bounded portion of the plane with a specific ratio of width to length, and a desire to place an image in that rectangle, an individual still has a great range of freedom. That surface is used to represent the space of three dimensions. Any system of representation that is chosen reflects a choice whether conscious or unconscious. Some of these choices are so embedded within a culture that they appear as absolute givens rather than relative values. The Twentieth Century, because of our global connectedness and cultural diversity, allows for a tremendous range of visual opportunity. But many times choice is more difficult to handle than the comfortable, but invisible, restrictions of societal traditions.

How must an object be presented to the viewer so that it looks "right?" If we are looking only for clear information, perhaps the architect's use of plan, side, and front elevation views will do. But, what if the desire is to include an aesthetic component--a desire to construct beautiful shapes, or textures, or patterns, etc.?

How does the artist create a feeling of depth on this two-dimensional plane? How is a feeling for volume achieved? How is the movement of air and wind suggested? How is optical realism achieved? One must remember, however, that "real" can be defined in a great many ways and that there are as many realities as there are people.

What we have to accept is that each cultural age sets a paradigm, a model, that frames a definition of what we believe to be true. All the "facts", then, relate to that definition. It is when new facts enter the picture, however, that a new model of the way things work has to be constructed. Therefore, Reality is greater than any particular model. But models definitely have their uses as we shall see throughout this text.

This page:
The picture plane.
Without images on it, we refer to it as the surface plane.
Usually the shape of the picture plane is rectangular. Circles and ellipses, however, are also well used format shapes.
Top left: Fig 1.1A
Length and width are the only two dimensions of the picture plane. Determining which ratio is pleasing is an important consideration since the internal divisions relate to the outer relationships.
Bottom right: Fig 1.1B
Illusions of three-dimensionality can be created on the two-dimensional surface using a variety of compositional devices which may include linear perspective.

Several Ways of Representing An Aspect of Reality On a 2-D Surface

As a mathematician would say, let "Reality" be an undefined term. Given a particular subject matter, be it plant, flower, or landscape, what kinds of choices can an artist make in representing that reality on a two-dimensional surface?

The camera frames a view and flattens the 3-D world onto the 2-D plane. The original color, in the photograph at the right, is translated here into various tones of gray. All volumes in 3-space become shapes in 2-space.

The materials are photographic paper, light, and the eye of the photographer.

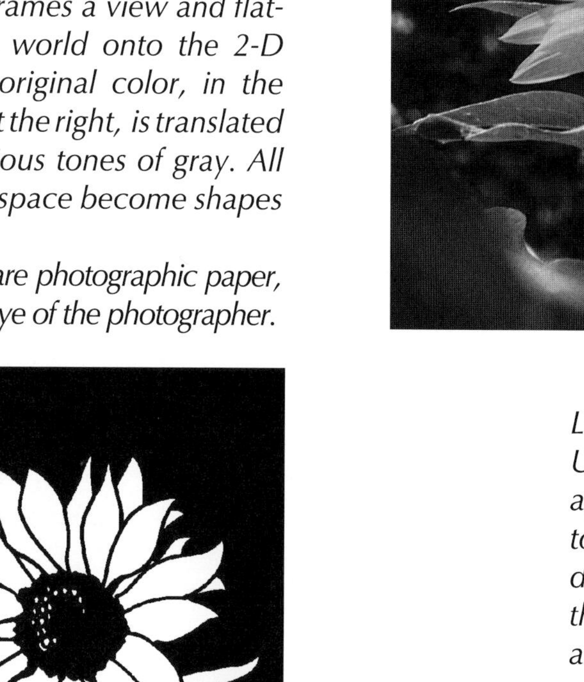

Line:
Using line only as in the image below, the edge, or contour, of the shape is given definition. The line separates the space into areas that read as figure and ground (object and background).

High Contrast:
From this photographic information, the graphic artist chooses another aspect of the image on which to focus. This time it is a high contrast image which ignores the subtleties of gray and reduces all tones to only contrasts of black and white. The tool is pen, the material is ink and the surface is bristol paper.

Point:
Now the artist, in the drawing at the right, using only black and white, but in the form of points (dots or pixels in the computer), creates the sense of tone change through the manipulation of the density of these dots. This technique is called stippling when done by hand. The tool is still pen, the material is still ink, and the surface is still bristol paper.

 Hatching and Cross-hatching:
Using parallel lines only, in the drawing on the left, some in one direction and others in an opposite direction, the artist also creates a sense of tone change by the overlapping of lines. This technique is called hatching and cross-hatching. Pen and ink on bristol paper.

Tone:
Now the artist changes tool from pen and ink to pencil in the drawing to the right. This gives a continuous surface that is "smudged" on the plane. Change in hand pressure allows the artist to control changes in tone. Notice how the pencil affects the "mood" of the drawing.

This page:
Top:
Painting by Jean-Baptiste Siméon Chardin. Still Life of Kitchen Utensils. *c. 1730-35. 12 1/2" x 15 1/4". Ashmolean Museum, University of Oxford, Oxford, England.*

Bottom:
Painting by Lubin Baugin. Still Life with a Checkerboard. *Allegory of the five senses. Circa 1630. Oil on panel. 55 x 73 cm. Paris, Louvre. Photographique de la Réunion des Musées Nationaux.*

The still life, as an aspect of visual reality, has had a long and honorable history in the annals of art. The objects of everyday use, whether humble or not, have been arranged and re-arranged in order to make compositions of personal meaning. In many works, the objects stand for qualities outside their original use and hint at the moral and spiritual values of an age.

On these facing pages are several representational paintings of still life arrangements. The two-dimensional surface is used to suggest an actual three-dimensional space. Though the works are of similar subject matter they differ in their degree of representing reality.

What do these works have in common? What aspects differ? How has the surface been treated? Is there an illusion of deep space or do the objects seem to stay on the surface? Set up a group of your favorite objects and try a drawing using one of the techniques discussed on the previous two pages.

Cézanne (1839-1906). whose painting is seen on this page, had a great influence on painters of the Twentieth Century. The strong sense of structure within his work directed them to make paintings that had their own internal logic which would not necessarily be based on the outward appearance of nature. The artistic movement called Cubism engendered artworks that utilized the interplay of angles and planes played out on the two-dimensional surface. As opposed to the unified sense of time, space, and place, in a perspective painting such as that of Baugin or Chardin, cubist works provided fragments of objects as seen from multiple viewpoints simultaneously. Theirs was not an illusion of three-dimensionality, but an affirmation of the integrity of the two-dimensional picture plane.

Juan Gris (1887-1927), along with the artists Pablo Picasso and Georges Braque, was considered a founding father of the Cubist movement. He was interested in abstract forms that would hint at recognizable objects such as bottles, guitars, and fruit bowls; however, underlying geometric relationships were extremely important to him. He often began his paintings with an underlying framework based on subdivisions using the Golden Ratio (1:1.618 approx). This ratio provides harmony between the parts and the whole because of the property of similarity and is connected to ratios that exist in the natural world. See the Design Appendix C for more information on this very special ratio.

This page:
Top:
Cézanne. Still Life with Peppermint Bottle, *Chester Dale Collection, © 1999 Board of Trustees, National Gallery of Art, Washington, c.1894, oil on canvas, .659x .821 (26x32 3/8); framed.*

Bottom:
Juan Gris (José Victoriano Gonzáles. Guitar and Flowers, *1912. Oil on canvas, 44 1/8 x 27 5/8. The Museum of Modern Art, New York. Bequest of Anna Erickson Levene, in memory of her husband Dr. Phoebus Aaron Theodor Levene. Photograph ©1999 The Museum of Modern Art, New York.*

Perceptual Cues for Representing 3-D Space on a 2-D Surface

In order to "read" an artwork in two-dimensions, we rely on cues obtained from actual experiences in the three-dimension world as viewed through our two eyes (binocular vision). Each eye receives slightly different information which the brain combines into one image. We also, however, receive monocular cues about the world. It is these pictorial spatial cues that help the viewer to understand images on the picture plane.

Essentially the artist is working with shapes that have line, texture, color, tone, and pattern. These shapes are interpreted by the viewer so as to stand for objects such as dogs, cats, flowers, landscape, etc.

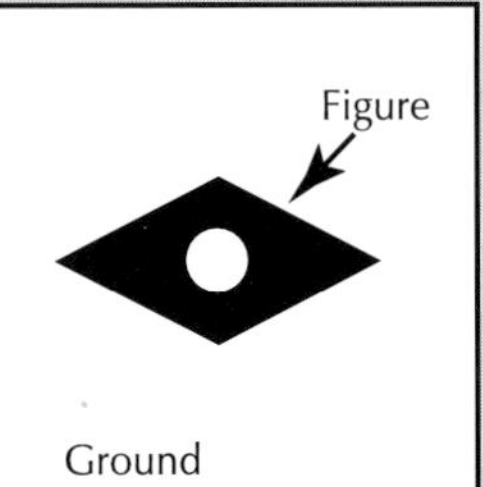

Figure/Ground
Sometimes called positive shape/negative space. In a representational work this term refers to the foreground (positive) and background (negative).

Foreshortening
When a figure, usually human or animal, is directly in front of the viewer and aspects of that figure are hidden, different parts may appear shorter or larger or distorted.

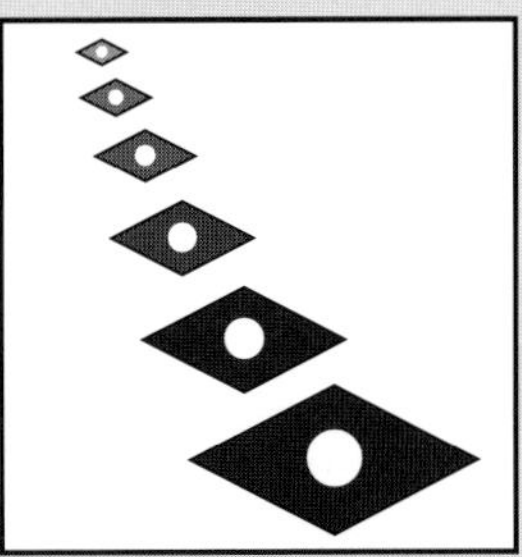

Gradient Change
A gradient is an orderly, gradual change in a particular visual element. This could be a size change. The larger the shape, the more it is perceived as being closer to the viewer while the smaller size is perceived as being further away. If there is a slow progression in the change of size, greater illusion of depth is created.

This could be a tone change. The dark tonal values appear to be closer to the viewer than the lightest values. This is sometimes referred to as aerial perspective in which not only do the tones change but the edges of form become less distinct.

This could be a textural change. More heavily textured areas appear closer to the viewer than do plain areas.

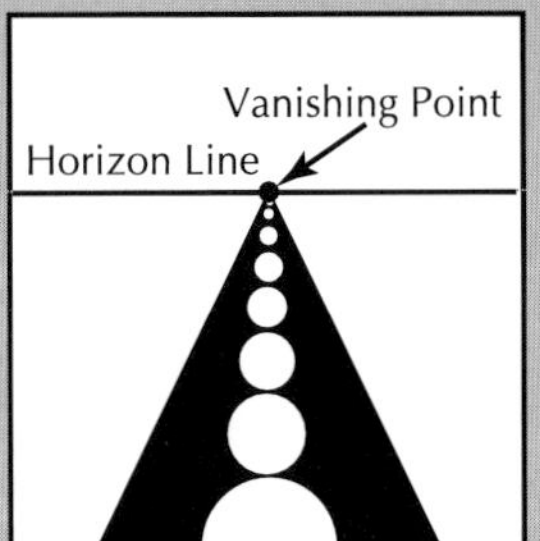

Linear Perspective
The apparent convergence of parallel lines as they recede to the horizon line and meet at a vanishing point. There are one-point, two-point, and three-point perspective views.

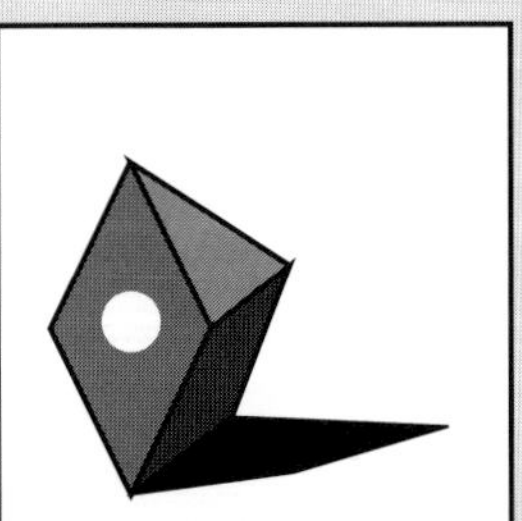

Modelling of Tone
The use of shading on an object and the shadows cast by that object give the viewer th sense of volume of three-dimensional objects.

Overlap
If one shape is placed over another shape so as to obscure part of it, we have a sense of one shape being in front or in back of the other.

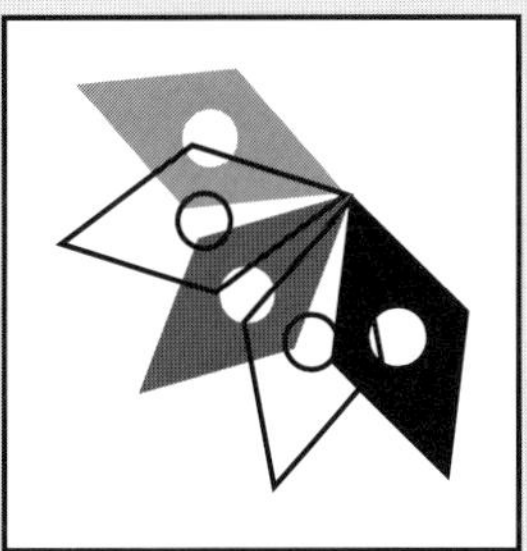

Transparency
When shapes overlap and both are read completely, the shapes are assumed to be transparent. This can occur through the manipulation of line, tone, color, or all three.

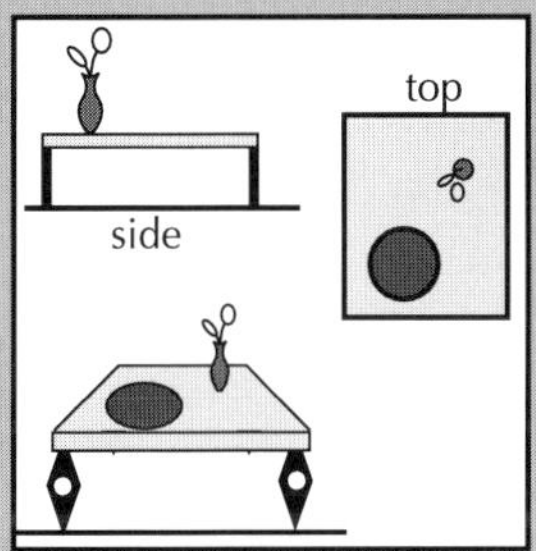

Multiple Perspectives
When several viewpoints are combined into a single image, there is a sense of movement through time and space. Interior and exterior scenes can be combined into one image. Look at the images of Ancient Egypt or Persian miniature paintings.

Vertical Location
The bottom of the picture plane is usually associated with elements that appear closer to the viewer. The higher up a shape is placed on the plane, the farther away it is perceived. If the size of the shape is also changed, greater illusion of depth is created.

Reverse Natural Perspective
Objects are made large, not for their relative position in space, but for their importance in the particular drama or scene presented. Objects are made narrow at the front and flare out at the back. This allows for the seeing of more of the object.

This page:
Black and white ink drawing interpretation of the miniature painting by Kamaluddin Bihzad. Yusuf Escapes from Zulaykha. *32 x 25 cm. 1488. Ink, color, gold on paper. Adab Farsi 22, folio 52v. Egyptian National Library.*

Our interest here is to show the complex use of layered planes and intricate patterning.

The drawing, and painting, show the fusion of inside and outside, pattern and shape within a strong geometrical structure. Notice that there is no single center of interest but, rather, many places to give your attention.

Try a drawing of your own using this work as a jumping off point. Link inside and outside, natural and geometric forms, texture and tone, color and pattern.

Perhaps the views of the castle on page four could become the basis for the invention of a black and white drawing that is similar to the one on this page.

On this page is a contemporary interpretation of a Persian painting done in the 15th Century. The original work has ink, color, and gold on paper while the above image is done in only black ink and white paper.

What objects have been used as the basis for this artwork? What are the similarities and what are the differences between this work and artworks on the previous pages in this chapter? How has this artist used spatial cues to make the work comprehensible to the viewer? What spatial cues were ignored? Which ones were emphasized? What do you like about this image? What do you dislike? Can you understand your response in either case? How does this work relate to those examples of "visionary" artists presented at the beginning of this chapter?

This page:
Student work. Cathy Flynn.
Imaginary Landscape. *Pen and ink on paper. Notice the use of stippling, hatching, and high contrast sections. Lay a piece of tracing paper over the drawing on this page and change aspects of it making your version quite different from what is presented here. Try a version where you substitute pencil tones for the high contrast black and white. Or consider adding color to the black and white version.*

In representational images, however, there must be a sense of accuracy in presenting what the objects are, where they exist in space, where the light source is located, where the shadows are cast. Without the appropriate skills and techniques, the two-dimensional illusion becomes ineffective.

In the Twentieth Century, there are artists who have chosen to assert the tension between the two-dimensional surface and the illusion of three dimensions depicted on it. The image is not perceived as a window onto an event, as in Renaissance perspective, but rather an object in its own right. These artists create ambiguous relationships between the figure and the ground. Research the superior graphic work of the Twentieth Century Dutch artist, M.C. Escher.

What spatial cues seem to be operating here in this student work? How is it like the image on the facing page? How does the artist take advantage of contradiction and ambiguity of shape or position or angle?

According to scholar Margaret Hagen in her book, *Varieties of Realism, Geometries of Representational Art,* there are four different types of geometry (metric, similarity, affine, and projective) that artists worldwide have used in representing three dimensions on the two-dimensional surface. All of them involve the concept of mapping.

Most laypersons are familiar with either two-dimensional roadmaps or three-dimensional globes on which areas of the earth are depicted. A mapmaker who constructs images of a portion of the earth's surface is involved with the concepts and problems of transferring information about a three-dimensional experience onto the plane roadmap or the surface of a sphere.

When the planet is represented on a globe, the north and south poles are the endpoints of an implied axis that runs through the center of the globe. Halfway between the poles is a great circle that is named the equator. The equator is divided into 360 equal segments called degrees. Each degree in a circle is divided into 60 minutes. The arc from either of the poles is divided into 90 degrees. Latitude lines run parallel to the equator while longitude lines run parallel to lines that are called meridians. These are great circles of the earth passing through both geophysical poles. Fig. 1.2 A.

On a globe both the parallels and the meridians have the same length. In other types of flat projections, parallels, meridians, or some other set of lines can have the same length, as on the spherical globe, but all other sets of lines will be either shorter or longer. The scale on a flat map cannot be true everywhere. Fig. 1.2 B.

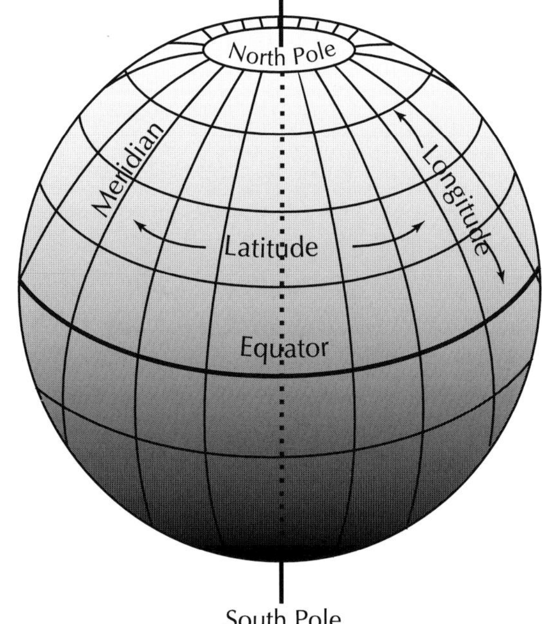

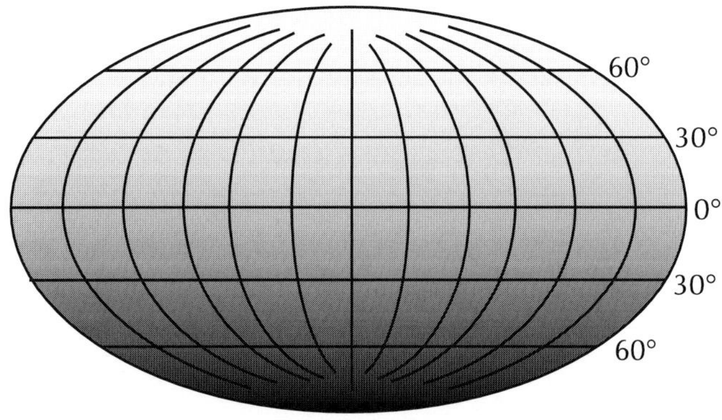

Mollweide Projection

The mapmaker constructs any of the various projections by first laying out the parallels or meridians with their true lengths. All other components relate to these. Mapping is related to correspondences between figures or events. A drawing is a form of projection . What the artist is doing in representational art is making a map of what is seen before him/her attempting to be faithful to the orginal objects so that the viewer is able to recognize them. But the creative process involves a great deal of selection. Every artist chooses what he or she thinks are the most salient features of an event and disregards what is unimportant. Sometimes, for the sake of effect, elements are left out or distorted or changed.

Geometry is involved with the description of objects in space and their positional relationships to each other. What are the characteristics of shape and what are the characteristics of position? These are the kind of questions that mathematicians might ask. The concept of congruence (having the same shape and size, not position) is of major importance in Euclidean geometry. In congruent figures, length is preserved. That is, the distance between two points of an original and the distance between two points in the image reproduction are equal. All angles remain unchanged. This is not so for other geometries.

It is possible to represent the three-dimensional sphere by using a two-dimensional image. Stereographic projection is a means by which the globe's surface is transferred onto the plane. For every point on the sphere, except the north pole, which has no image point, there is a corresponding, very specific, point on the plane. The equator projects to a circle. The upper half of the sphere projects outside this circle and the lower half of the sphere projects inside the circle. Distances become very distorted. Points that are very close to the north pole on the sphere become arbitrarily far apart in the projection. So distance is not preserved. A circle on the sphere, however, goes to a circle or line on the plane. The angle between two great circles on the sphere is equal to the angle between their images except if they meet at the north pole. Fig. 1.3.

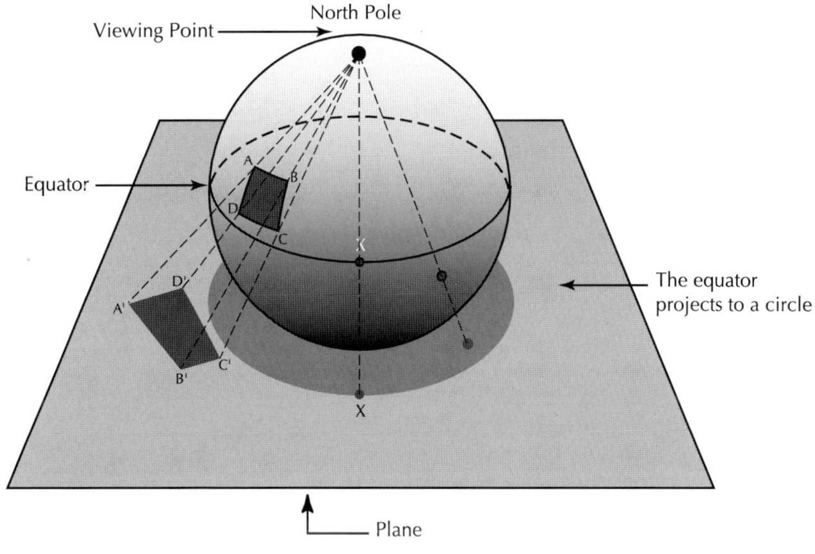

This page:
Student work: Cut black construction paper on a white bristol paper background.
Karen Landry
All three-dimensional objects have been simplified into two-dimensional shapes. The recognition of the objects requires a certain cultural familiarity with them.

smooth. It can be hard edged or blurred. Yet it still remains a shape.

The intent of the assignment for the student artworks on these facing pages was to have the viewer recognize objects from the studio environment solely through their contour shape. The goal, therefore, was to understand the relative distances between various parts of each object, not necessarily the relative size relationship between objects. The student artist had to try to keep as close as possible to the actual size of the original object. The figure/ground relationships (positive shape to negative space) were consciously determined in order to make an interesting composition through black and white contrast alone. The shapes had to be cut out with a scissors rather than being drawn with a pencil. The process of using a scissors demanded a greater precision in achieving each shape.Details were kept to a minimum. The information about the shapes was mapped in a series of rows one above the other, as is found in Egyptian wall painting. The suggestion of deep space was suppressed. Notice that none of the objects has been overlapped but instead are arranged across the surface plane.

The objects chosen by the artists may be familiar in the western culture of the Twentieth Century, but could be unrecognizable to those who had come before this era or those who might come after this time. How many of them are familiar to you ? Which can you not recognize?

Drawing an object from three-dimensional space onto the two-dimensional drawing surface is an act of projection. The contours of a three-dimensional volume become the edges of a two-dimensional shape.

Shape is a very strong element for the visual artist. It provides the essential skeleton on which to hang the flesh of an image. Even an ephemeral cast shadow becomes a shape for the artist to manipulate. The shape can be colored, textured, or patterned. It can be transparent or opaque. It can be rough or

Try doing a project like this your-self. Take one of the images present-ed here and do a variation using a contrast of color or texture or pat-tern. Or set up a group of objects that you find have interesting shape dif-ferences. Place a light source near them so that you can see the shadows they cast. What happens to the shad-ow shapes as you move the light source around?

Each artistic transformation leads to other ideas that can gener-ate further ideas. Try laying a grid over the image and shear the grid or warp it or fold it, etc.

Any profession has its own spe-cial group of tools. You could work with those of the health field, the sciences, exercise fitness, theater, music, or dance.

This page:
Objects in an art studio.
Student work.
Cut black construction pa-per shapes glued onto a white bristol paper back-ground. Notice that groups of objects are represented in bands across the surface. There is no sense of deep space nor cast shadows.

Top right:
David Romano.

Bottom left:
Tara Jamison

One of the categories of geometry that scholar Margaret Hagen explores in her previously mentioned book is that of Metric Geometry. Egyptian art is a very good example of this type. It is an art that fascinates people and has done so for a great many centuries. What is it that enthralls the viewer so? Is it the mystery associated with the pyramids? Is it the Egyptian particular attitude towards life-after-death? Is it the opulence of the gold and the jewels found in the tombs of the pharaohs? Is it the craftsmanship with which each object was lovingly executed? Is it because Western art took such a different turn beginning with the Renaissance? For whatever reasons, people have attempted to unravel the many secrets of Egyptian art.

Let us look at a painting on the wall of a tomb in order to begin to understand the special Egyptian way of rendering objects in space, their mapping. This Egyptian metric system of representation lasted for about 3,000 years. That is a very long time to define and refine a visual approach. No one system of representation can include everything that we know about the world; therefore, what is left out of an image is just as important as what is put into that image. What are some of the constraints of this method and what are some of its advantages? What was the purpose of the art? What was the intent of the artist?

According to Margaret Hagen, the "image plane", the surface on which the drawing or painting is made, was always set parallel to the most characteristic aspects of the objects represented like the drawings on the previous two pages. That characteristic view would be like a silhouette shadow cast upon a surface directly in front of the artist (thus, the viewer). Was there a quality about the light in Egypt that would preserve the shape of an object without distortion? Would there be something miraculous about a shadow for a culture that would not necessarily understand the cause and effect of light rays? Can we still retain the wonder of a shadow? For the Egyptian artist emphasis was on the essential quality of things rather than the transitory. And yet, a shadow is most transitory and mobile unless captured in a drawn image. Was the cast shadow the beginning of drawing? Shadows can be cast by moving light across cave walls or falling upon desert sands. Tracing the outlines of a shadow would be like capturing the essence of whatever form happened to be. Would that not be powerful magic?

In Ancient Egyptian artworks, regardless of the actual depth between objects, there is no sense of three-dimensionality. All of the objects are situated in a single frontal plane. Many times, the images are placed in rows with each row meant to be read separately and horizontally so as to be suggestive of activity. The complexity of a scene determined the number and type of vantage points from where the artist/viewer would be located. The choice for the artist would be between an elevation view (looking at something from a side) or a plan view (looking at something from above).

Each image might contain multiple viewing points but each object, aside from the human figure, was presented as seen from only one. Preserving the recognizability of each object was a primary concern for the artist. The metric system attempts to preserve the most constant elements of an object. Each image would have precise information about the relative size and shape of the parts of an object. There was no foreshortening across an object. Size differences between objects are not optical but hierarchical. Those objects that are most important are larger in size. Overlap is used to suggest that some objects are behind. There never is a sense of volume indicated nor a suggestion of cast shadows.

The Egyptian System

"Old Canon"
closed fist =one square unit
18 fists (squares) = 24 handbreadths
cubit = 4.5squares
24 handbreadths = 4 cubits
4 cubits = soles of feet to the forehead
Little Cubit = six handbreadths
(elbow to the tip of the thumb)
Royal Cubit = seven handbreadths
(elbow to the tip of the middle finger)

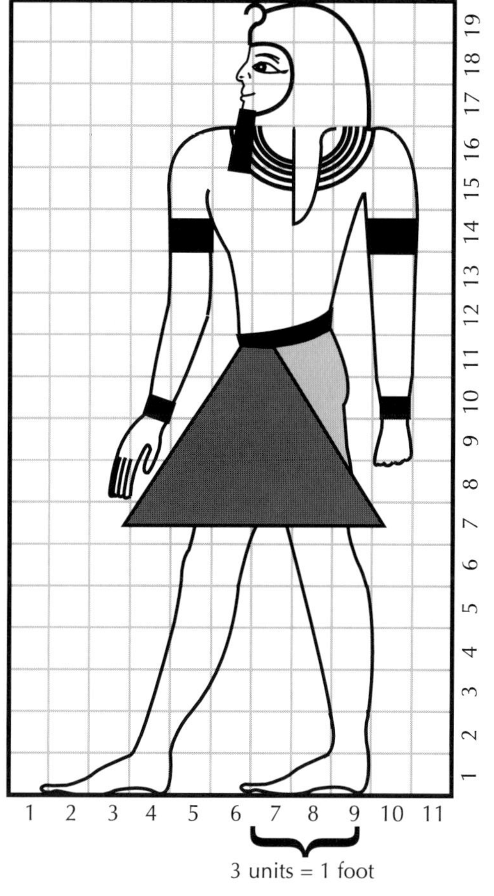

Upper left:
Fig. 1.4
The square grid was used by the Egyptian artist to establish a system for representing the human figure. The full length of the body measured 18 fists (square units) from the soles of the feet to the forehead. Notice that different views of the parts of the body are combined into a unified image. The head is seen in profile; the chest is seen frontally; the legs are shown in profile, etc. Each of these views gives the most visual information without distortion. It is the bringing together of these several views, however, that makes the image seem "strange" to those viewers used to other depiction systems.

Bottom left:
Stela of Imy
Egyptian (from Mesheikh), First Intermediate period, Dynasties 6-10. Painted limestone. H: 56 x 35.5 cm.
Museum Expedition
Courtesy, Museum of Fine Arts, Boston

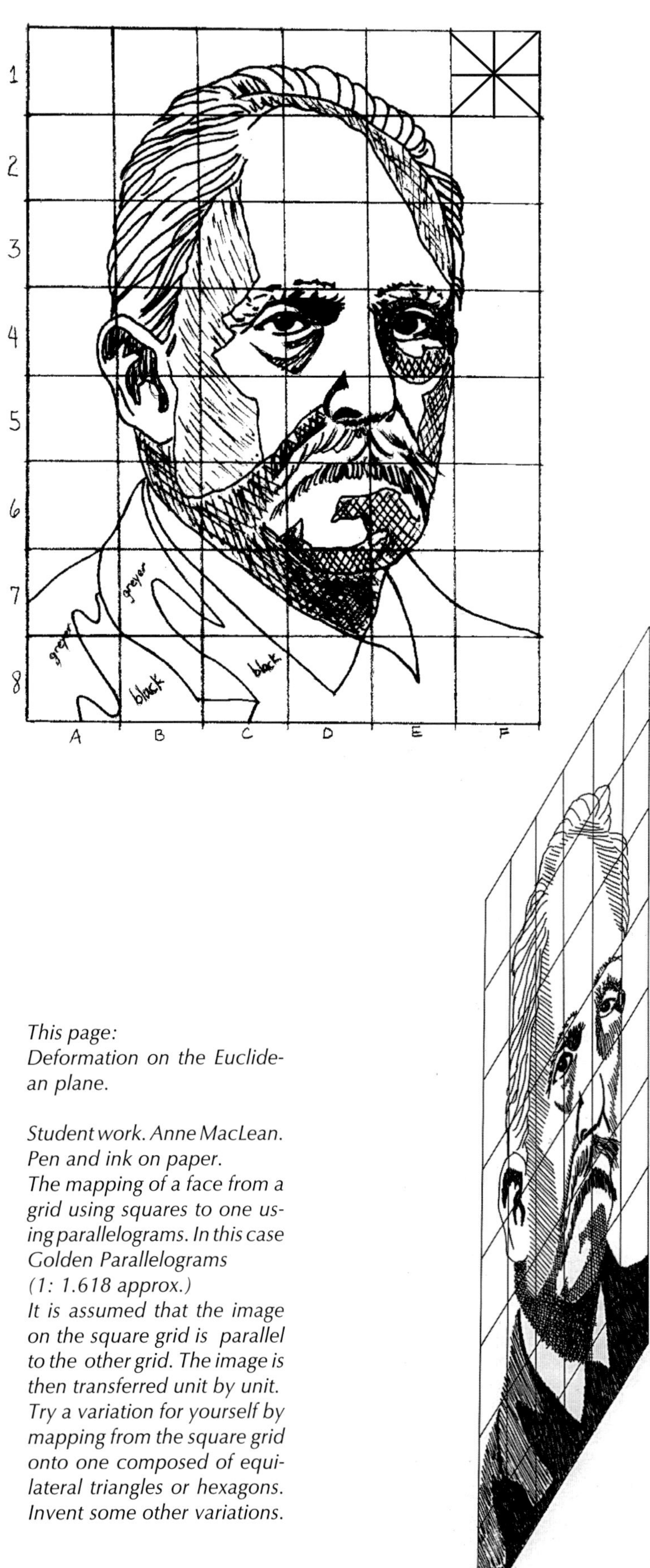

This page:
Deformation on the Euclide-
an plane.

Student work. Anne MacLean.
Pen and ink on paper.
The mapping of a face from a
grid using squares to one us-
ing parallelograms. In this case
Golden Parallelograms
(1: 1.618 approx.)
It is assumed that the image
on the square grid is parallel
to the other grid. The image is
then transferred unit by unit.
Try a variation for yourself by
mapping from the square grid
onto one composed of equi-
lateral triangles or hexagons.
Invent some other variations.

A grid can be used as the basis for a projection. What happens when the artist begins to stretch the grid as well as the imagination? The simple square grid can be transformed in a number of ways by pulling , skewing, or bending. The flat plane can be rolled into a cylinder or a cone or folded into a cube making it into a three-dimensional object.

In this student example, the artist used a photograph of a face and overlaid a piece of tracing paper marked with a grid on it. The grid is composed of 6 square units by 8 square units. Remember, the square can be of any size. This image was then transferred onto another grid that was also 6 units by 8 units but this time the unit is a Golden Parallelogram which is related to a Golden Rectangle. Instead of the square grid unit, which has a ratio of 1 to 1, the Golden Parallelogram unit has a ratio of 1: 1.618 approx. (See the Design Appendix.)

Each square unit of the original drawing was mapped (transferred) to a corresponding parallelogram unit. Even though the visual information is the same in both, the look of the final image is not. To obtain finer and finer detail within a particular unit, all the artist needs to do is use diagonals to find midpoints of each square which allows for the construction of smaller and smaller squares. See upper right section of image at top. Transfer what is found within the newly subdivided area.

Try a variation for yourself. Use colored pencils or inks or markers on bristol paper. Or try cut colored papers and tissues. Work from either image given here or find another photograph and start from the first step.

In what ways can you further change the grid? Notice that the number of units of each grid has not changed only the shape of the unit. The square has been stretched into a parallelogram but still, it contains the same area and, thus, the same amount of visual information.

Think about alternating black and white squares so that there are white lines on black squares and black lines on white squares. Do the same for the grid using the parallelograms. Change the angles in the parallelogram. Think about using a grid with curved rather than straight lines or a grid with jagged units. Turn your results into a picture puzzle with which others can play.

Using the square grid again, the artist can lay out a curved irregular shape so that it looks as if it were drawn in linear perspective as in Fig. 1.6B. (The following chapter will deal with the subject of linear perspective in more detail). First, the shape is drawn onto the square grid and then transferred to a grid where the units do not stay the same shape or size from row to row. But notice that the number of units in each grid remains the same.

What would happen if the grid was no longer flat but one that was on a sphere or a cylinder or any other irregular form?

This page:
Two grids with shapes drawn on them.

Fig. 1.6A
Grid which gives the illusion of a linear perspective space. The shapes are the same as those in the square grid.

Fig. 1.6B
Square grid with curved shapes.

Photocopy the square grid and transfer it unit by unit onto another perspective illusion grid of your own choosing. To help in transferring, remember to use diagonals to obtain midpoints.

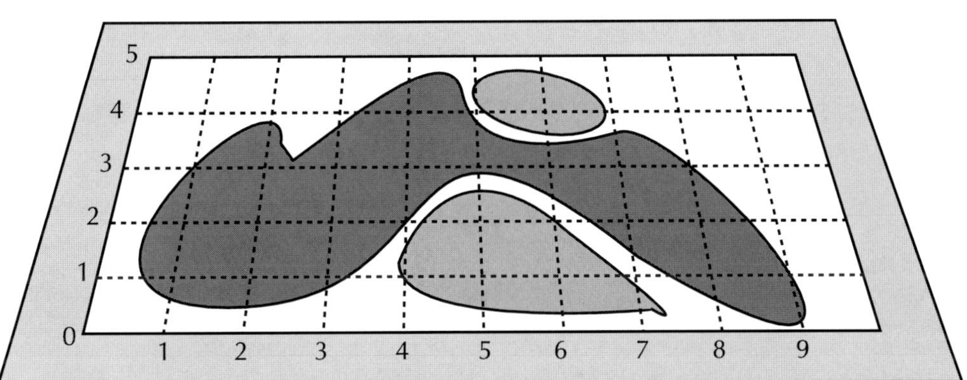

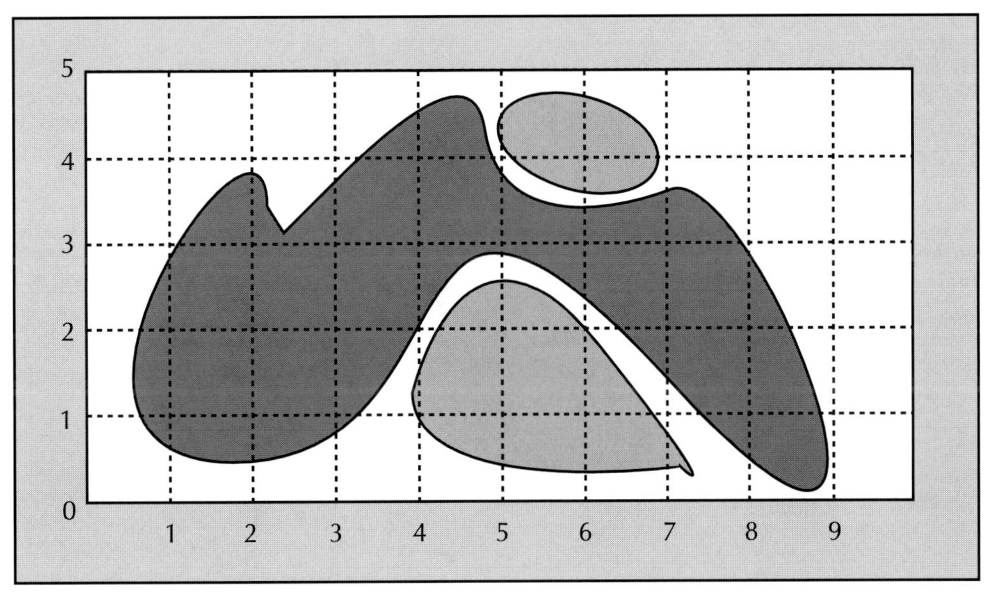

Procedure for Warping the Plane

Given a quadrilateral

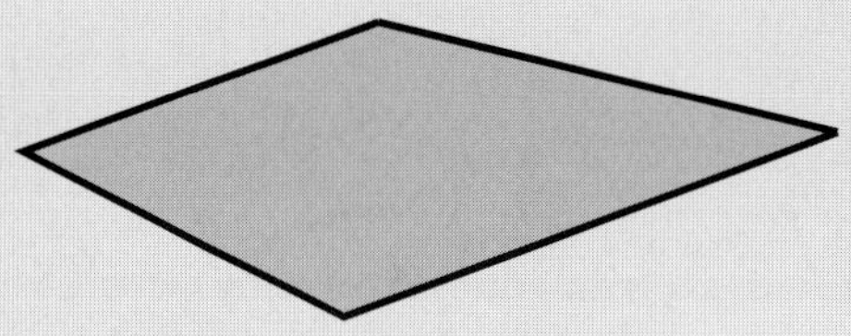

1. Develop a grid of *n* units by *n* units. In this example the grid is 7 units x 5 units.

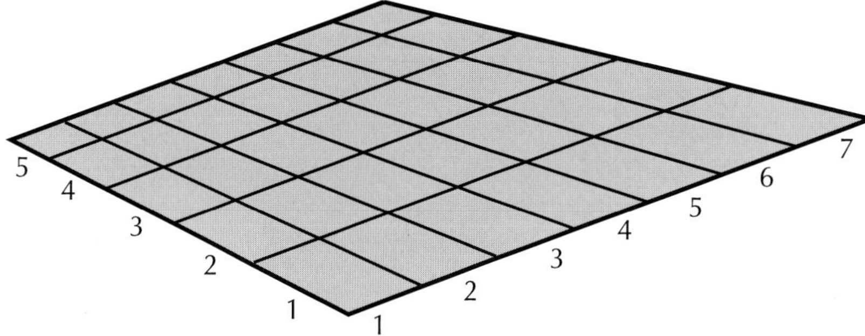

2. At each vertex of each grid unit, erect a vertical of any length. The lengths should vary from unit to unit in order for the plane to eventually appear to bend and fold. The height of each vertical is an arbitrary choice but will affect the suggestion of depth of the warping of the final surface.

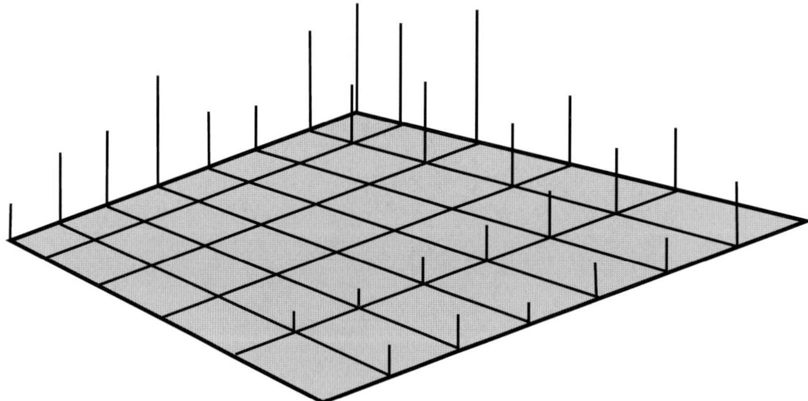

3. Connect the endpoints of each of the verticals through the use of diagonal line segments which are dashed lines in the diagram. As you move along, erase any line segments that would be hidden by the warped surface.

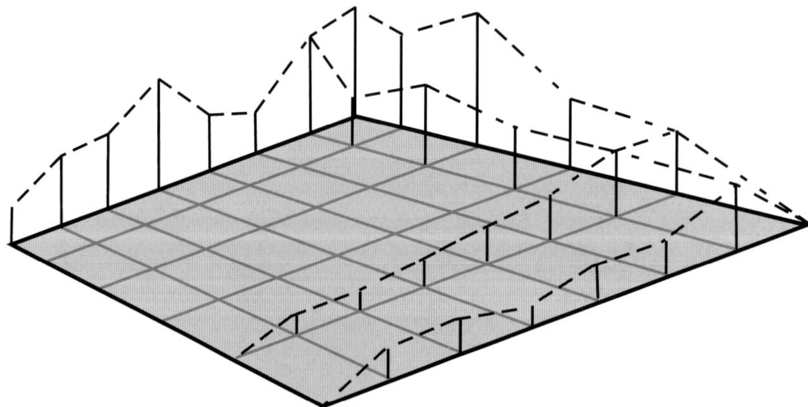

4. Continue connecting vertical to vertical until all are joined to each other.

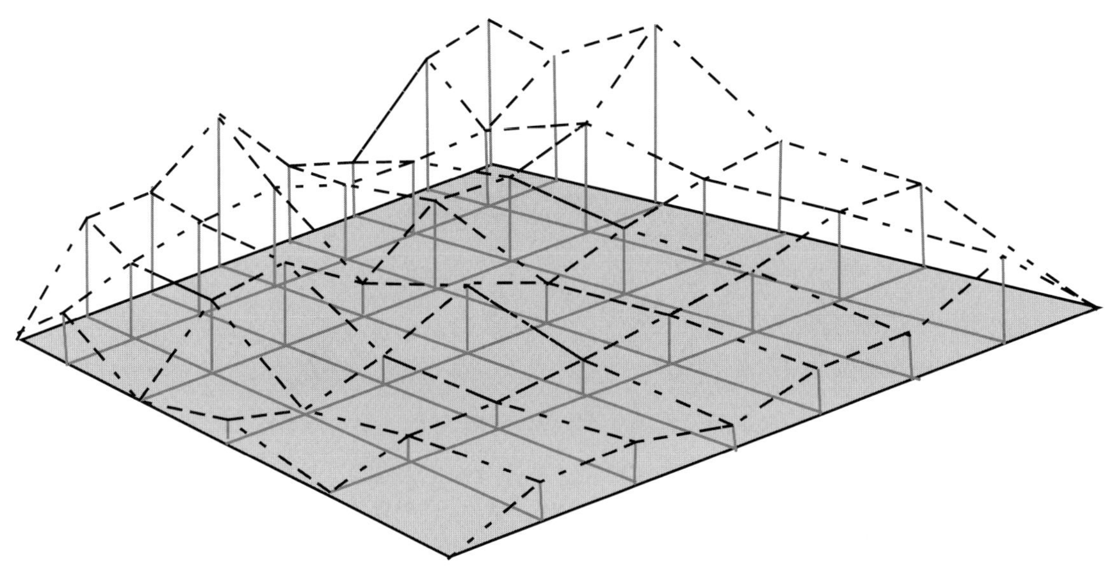

5. Remove all unecessary lines to achieve the illusion of a warped surface.

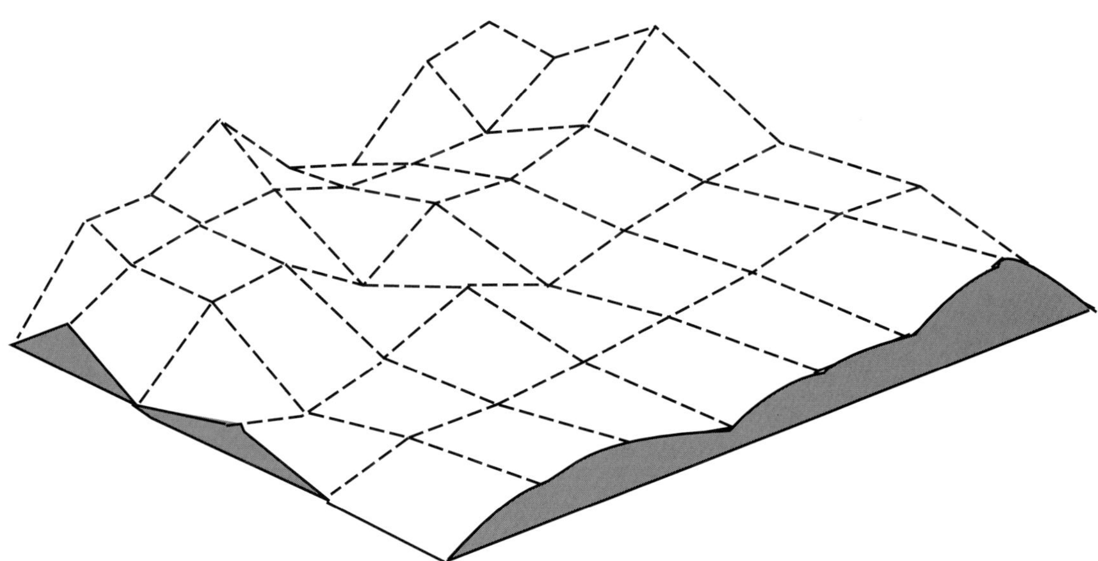

6. Now try a drawing of your own for transferring to your warped plane. First draw your image on a plane grid. Then copy it, unit by unit onto the warped grid. Do a still life or landscape or portrait. The greater the number of units within the grid, the more naturalisitc the image can be made to appear. (There are computer software packages that do this with a click of a mouse.)

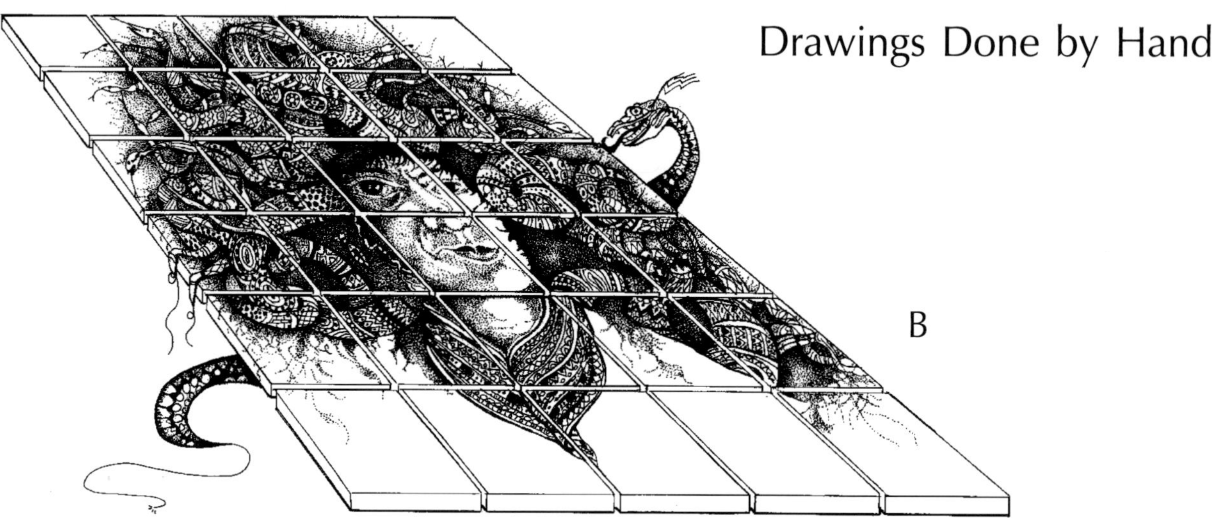

Drawings Done by Hand

B

A

This page:
Hand-drawn self-portrait variations of the artist, Gail Maciejewski, as Medusa. Pen and black ink on bristol drawing paper. The technique of stippling was used in all three images.
The straightforward plane drawing, A, is transferred to a skewed plane B, and then to an illusion warped plane, C.
The organization of the features of the human face is so powerfully imprinted on the human brain that images of the face can be manipulated to an extreme and still be recognizable as a human face.
These drawings are mappings of the three-dimensional face onto the two-dimensional plane, whether actual or illusional.

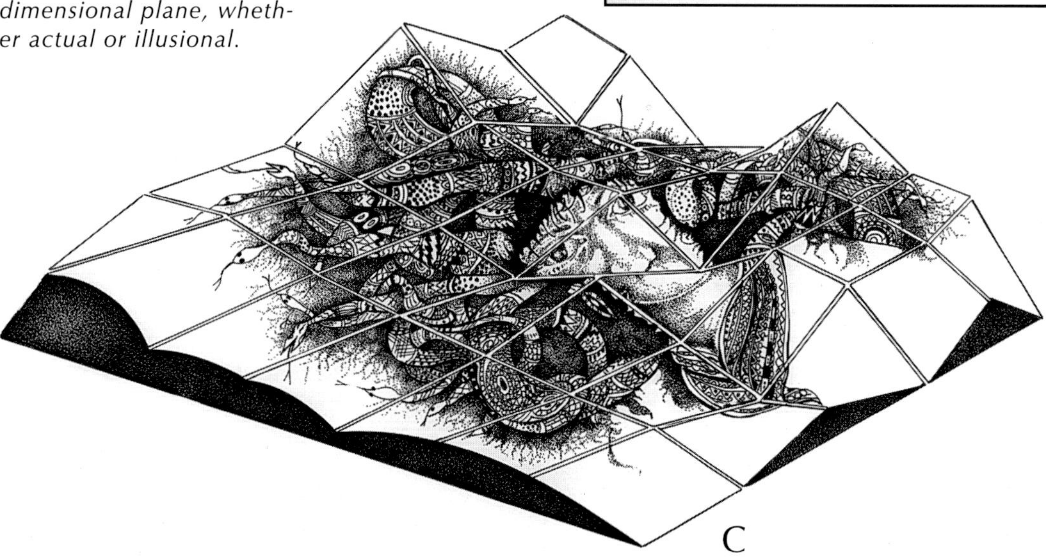

C

Computer Generated Transformations of the Drawings Done by Hand

D

This page:
Computer graphics programs enable an individual to transform a particular kind of picture plane into another type of picture plane by the alteration of the algorithms for the perspective grids that are inherent in the software. The images generated by the computer could then be manipulated by hand, completing the loop.

The first hand-drawing, A, was scanned into the computer and manipulated in the software program, Adobe Photoshop, using the Distort filters "Pinch", D, and "Ripple", E.

Thus, there is no need to choose between handwork and computer work, but the combination of both provides exciting possibilities for the artist.

E

The Mathematical Concept of Mapping

Mathematicians study many different "kinds" of geometries only one of which is the familiar Euclidean one. The different geometries are characterized by the ways that a shape can be transformed and still be "the same" in the context of that geometry. For example, in the portraits on the previous two pages, the viewer is still able to recognize the initial subject although a number of variations have taken place.

Each type of geometry is associated with a group of motions that leave intact the geometrical properties of figures in space. Those properties that remain the same after any of these motions are called the invariants. There are only three kinds of transformations that can take place in Euclidean geometry. They are the symmetry operations of translation, reflection, and rotation. These operations do not alter the shape or the size of the geometrical figure itself, but change the position and orientation of the figure in space. Fig. 1.7. (See Design Appendix on symmetry.)

The number and type of invariants differ from geometry to geometry. Euclidean, or metric geometry, contains the most invariants while topology, viewed as geometry is the broadest sense, contains the fewest invariants.

The Euclidean plane is a special kind of surface. The properties of congruence (staying the same shape and size) and the preserving of distance between points are the most important aspects of this geometry. Position and orientation in the plane do not alter these properties.

Mathematician Felix Klein, 1849-1925, looked at a more general way of studying geometry. At age 23, he proposed a new view that was based on looking for invariants. This has come to be known as the "Erlanger programme". He introduced the idea of classifying the various branches of geometry according to their types of transformation. In the chart on the facing page, a portion of the square grid on the Euclidean plane is shown in relation to a particular transformation which puts it into a different category of geometry. In Euclidean geometry, with each transformation, the invariants get fewer. The motions that can be applied to a geometric object so that the end result is the "same" as the original are the rigid motions of reflection, rotation, and translation. This kind of sameness of figures is called congruence.

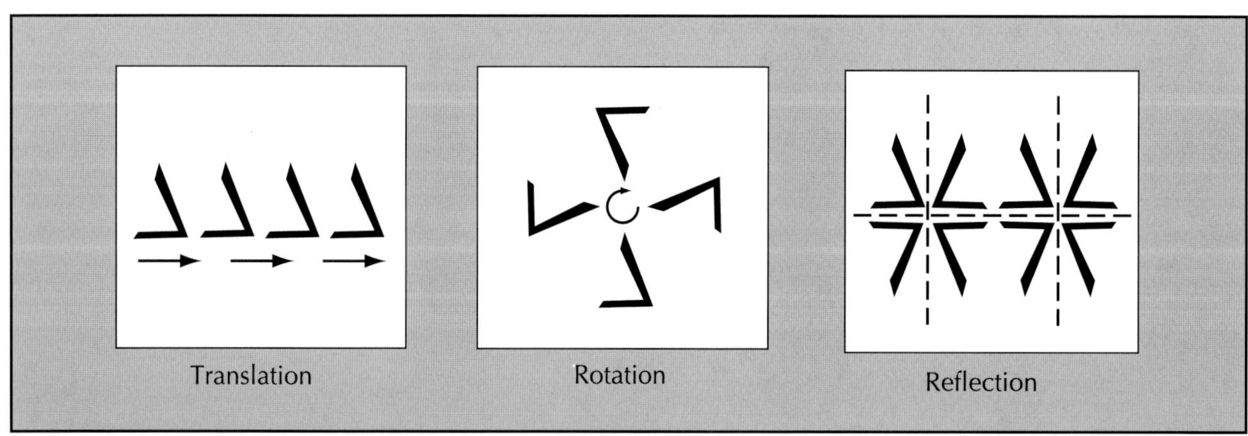

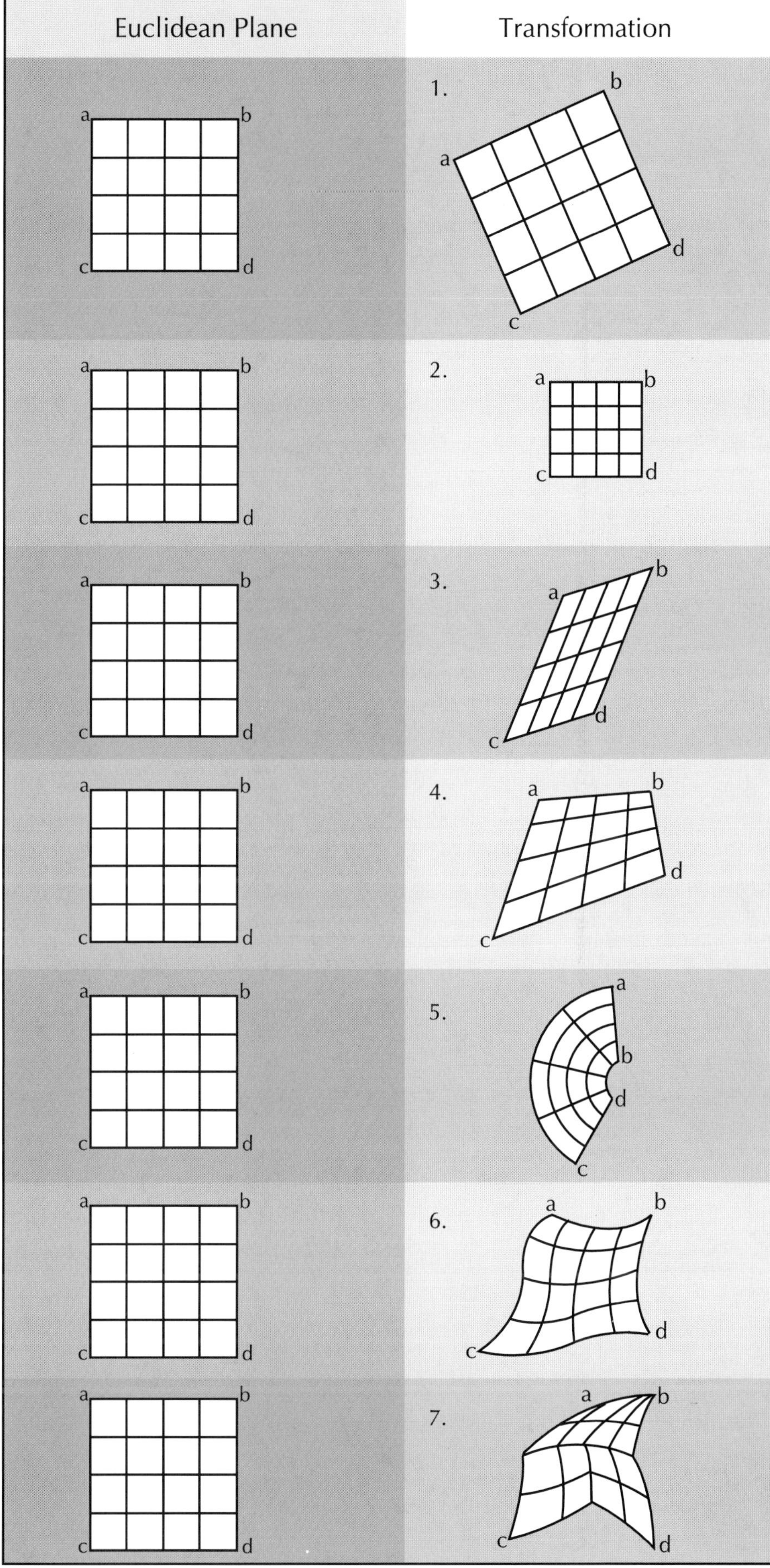

Invariants
The Erlanger Program of Felix Klein, 1872.

1. Position and orientation of the grid in the plane changes while the size and shape and angles remain the same. (These are called the invariants, objects subjected to rigid motions).
This is the concept of congruence.

2. Grid changes size, position, orientation but maintains the same shape. It shrinks or expands. Length is no longer a constant but the ratio of lengths is maintained.
This is the concept of similarity.

3. The grid is sheared or skewed. Parallel lines, with their length ratios, stay parallel but the Euclidean concept of angle and length are no longer invariant. Grid changes shape, orientation, position, and angle.
This is the concept of affinity.

4. The grid has been projected in the plane. Lines which are parallel in the original object may not be parallel in its image. In order to do this, there has to be a line at infinity. Grid changes shape, size, orientation, position, and angle. All quadrilaterals are considered equivalent.
This is the concept of perspectivity.

5. The square grid is not a circular polar grid. The Euclidean concepts of angle and circle are retained. In order to do this there has to be a point at infinity.
This is the concept of inversion.

6. The square is mapped onto a manifold in 3-D space. This is the concept of differential geometry.

7. All simple closed curves, such as squares and circles are considered equivalent. The only recognized properties are those that are preserved under continuous mappings. The grid can change shape size position and angle. *This is the realm of topology.*

Fooling the Viewer (Trompe l' Oeil) Images

In contrast to the flat silhouette images of the Egyptians, there is a genre of painting called "Fool-the-Eye" (Trompe l'oeil) that creates illusions of three dimensions that trick the viewer into believing that what is directly in front of him/her is physically real and could be lifted up off the surface of the plane. In order to be effective, works made this way display many of the pictorial spatial cues discussed previously. Can you identify them by looking at the paintings presented here?

There is a tension between the actual physical object that is the painting and the illusion of matter and space that is not. It is a game played from artist to viewer. The artist has

This page:
Top:
William Harnett. My Gems. *1888. Oil on panel, 46 x 35.5. cm, Washington, National Gallery of Art, gift of the Avalon Foundation.*
Bottom right:
William Harnett. Music and Good Luck. *1888. Oil on canvas, 40 x 30". Metropolitan Museum of Art, New York (purchased by the Catharine Lorillard Wolfe Fund, 1963).*

to create an initial impression of surprise through a dialogue with the viewer. Both parties know the duplicity of the game and both delight in it. The viewer takes pleasure in the skills of the artist as he/she is able to create an effective illusion. The artist takes pleasure in the portrayal of acquired skills needed to fool the audience. Many of these devices can be seen in the fresco works that are found in Pompeii, Italy of the first century B.C.

There are, however, other layers of meaning that may be hidden from the viewer unless he/she has spent time with many artworks. There is a concern for the composition; the symbolism of the various objects represented, both personal and cultural; and the color structure. The objects depicted are at the same time both clear and hidden in their meanings; clear, because they are recognizable, and hidden, because they may suggest ideas about morality, life, truth, and beauty.

Notice that the subject matter, technique, and materials of these three paintings are similar. How deep does the space seem to you? How big do you think the actual objects are? What makes you believe that the objects have volume? Do you recognize which ones? Why are they brought together in a single composition? What are the differences in composition between the three paintings? How does the composition of each affect your perception of each? What do you think is the symbolic meaning (content) of each painting as opposed to its subject matter? Set up a still life and attempt to play the game.

Definition of a picture by perceptual psychologist, James Gibson:

"...is a surface treated so as to contain the same kind of invariant information as the section of the world it depicts."

This page:
William Harnett.
Old Models. *1892. The Museum of Fine Arts, Boston, The Hayden Collection, 39.761.*

29

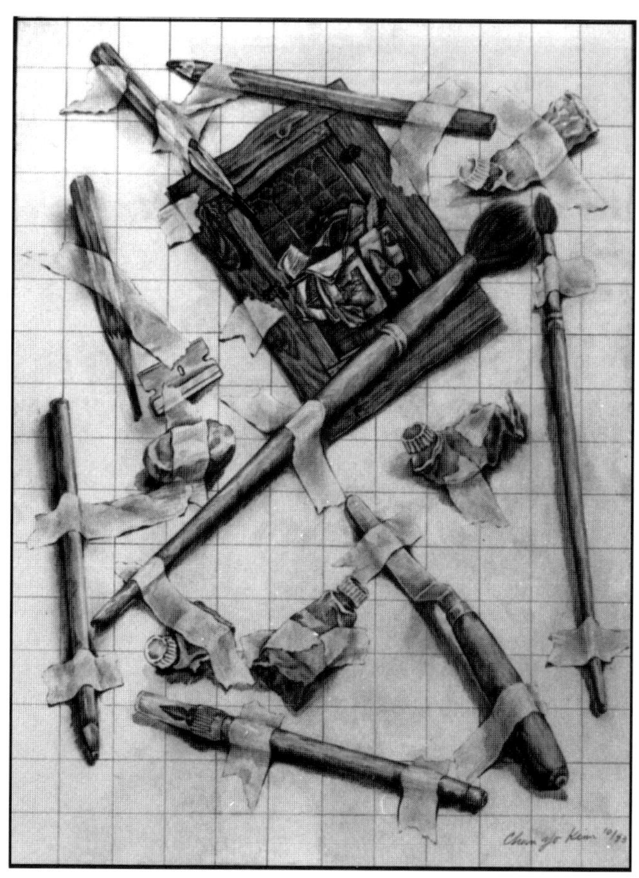

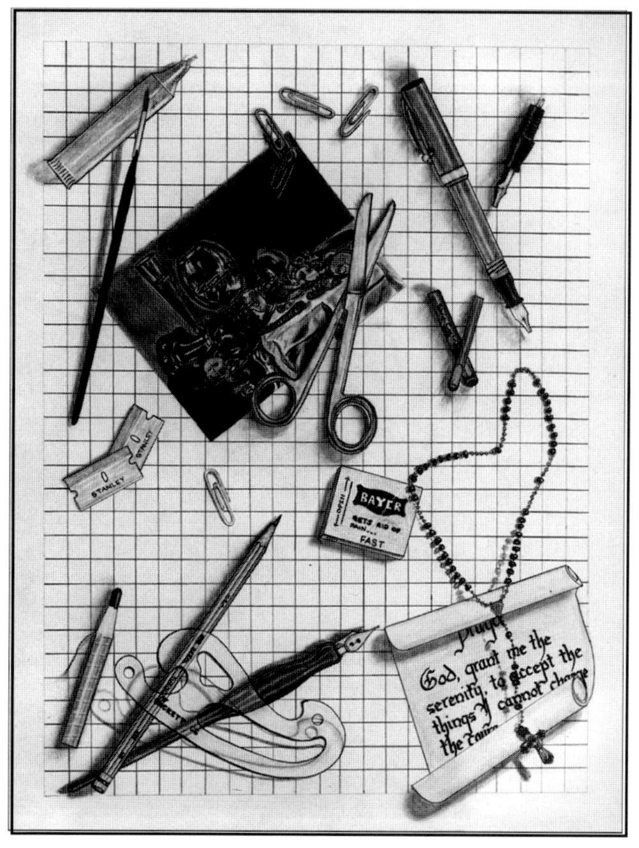

This page:
Student works. Trompe l'oeil drawings of objects within their own studio environments. Pen and ink and pencil on bristol paper. These are not computer generated.

Top right: Chun Yo Kin.

Lower left: Mary Calnan

There are "rules" for playing this "fool-the-eye" game that make the images work. On these facing pages, in these compositions, the student artists were asked to build drawings, using tone only, of objects that would be used by the artists themselves in and around their personal studio spaces. There are some elements that are common to all four images presented here because of the limits set for the assignment. Can you ascertain what these objects might be? And why? What objects might you choose for your set-up?

You will notice that each drawing utilizes a square grid as a background layer. This device is used to remind the viewer that the drawing is on a plane surface. It also establishes a way for the artist to check the actual size of each object to see how the parts of the composition relate to the whole. For this game to be effective, it is necessary for the image to be life-size. The work is also best seen hung in a setting where the objects usually would be found naturally.

Every object within the image must be seen in its entirety. No part of it must be cut off by the edge of the drawn or painted surface. The objects are commonplace and suggest the real physical properties of shape, size, mass, and position. Though visual, they activate the sense of touch and sometimes imply sound as do objects in the Harnett paintings on the previous pages. The sense of depth must necessarily be shallow because of the use of the spatial cues of overlap, transparency, and tone gradient. Many of the objects depicted are relatively flat such as letters, sheet music, watches, ribbons, fabric, etc.

The techniques for rendering the objects should be unobtrusive and almost invisible so that the texture of the paint or pencil does not become obvious and thus part of the subject matter and content of the image. The gestures of the artist's hand must be removed. The objects must exist and look like they were always there and not created by the artist.

Usually human figures are not found within the trompe l'oeil images since we tend to associate life with movement. If, however, you look at the sculptural works of artist Duane Hanson, you will see three-dimensional portraits of common people that have the feeling of sculptural trompe-l'oeil. He takes casts from actual persons and translates them into fiberglass and dresses them with wigs and actual clothing.

This page:
Student works. Trompe l'oeil drawings.
Top left: Brenda Dorney.

Bottom right: Bob Dumas.

Drawing Systems Using Parallel Lines

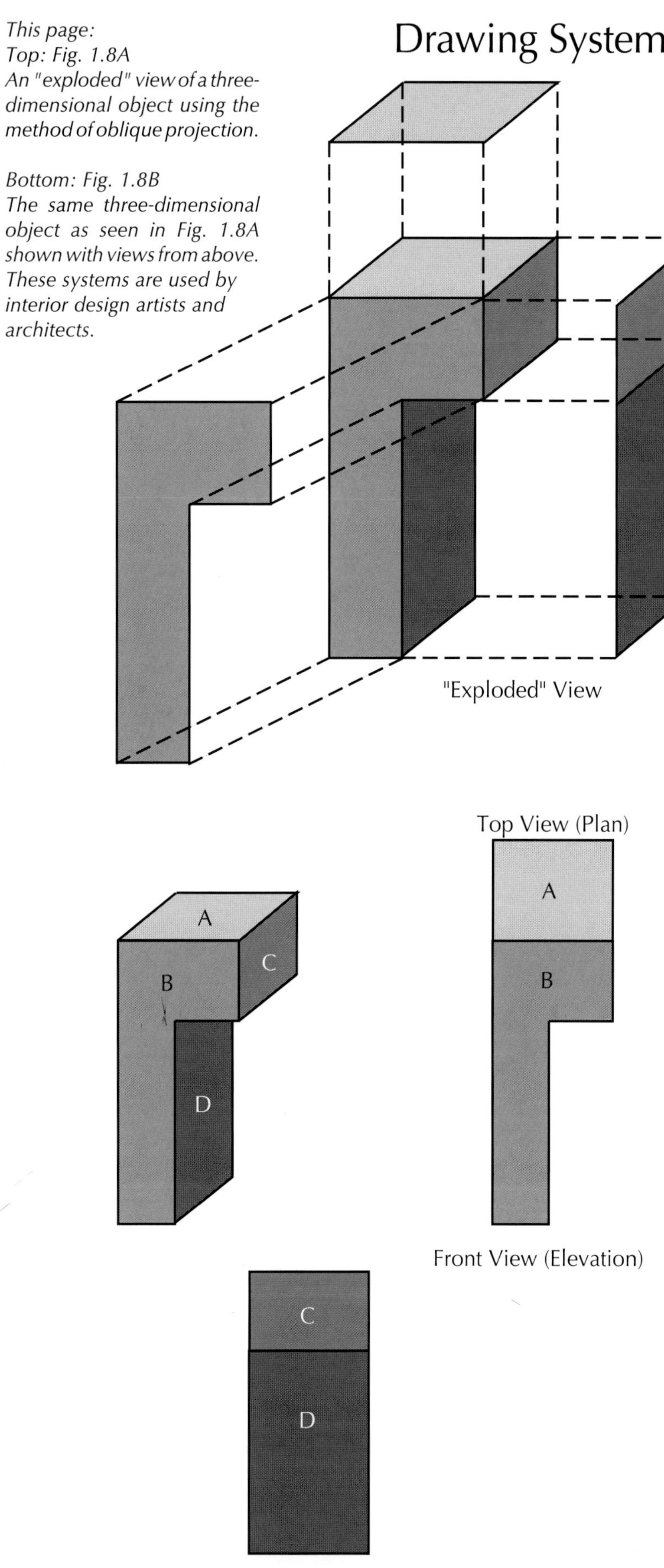

This chapter has been exploring ways of representing three-dimensional objects on the two-dimensional plane. In technical drawing, objects need to be presented in the most direct and complete manner. An orthographic projection, which consists of a plan view and two elevation drawings, Fig. 1.8B, is a most common solution.

For the layperson however, this type of drawing may be difficult to "read" since the several plane views have to combine in the viewer's mind's eye in order to be interpreted as three-dimensional.

There is no distortion of shape in these drawings and one axis must always remain parallel to an imagined picture plane. All measurements that are used are either actual or to scale. These are called metric projection drawings and they retain length, breadth, and height dimensions (the three axes perpendicular to each other) in actual measureable form.

This type of image-making is used extensively in the international fields of architecture, design, and engineering because once an artist is versed in drawing this way, information about forms is clear and concise and can be communicated to others who are also fluent in this system. Fig. 1.8A and 1. 8B.

In the next chapter, we shall look at a drawing system in which some of the parallel edges of the actual three-dimensional objects become lines in an image that converge to a vanishing point on a horizon line. There are no vanishing points, nor a horizon line, in the system depicted on this page.

There are two basic systems called planometric and oblique projections. Each has its own subcategories as you can see in the chart below.

In the planometric system, each member of the group depicts an object that is tilted in relation to the imagined image plane. Thus, the proportions and the size of the angles of each object vary with the number of possible positions that the object can take. Neither of these systems uses converging diagonals but always parallel pairs of lines and edges.

The second system is called oblique in which one face is parallel to the viewer. Fig. 1.9. This face could be either the plan view (looking down from above) or the elevation view (looking from the side). In our diagrams, we shall show the latter type.

We shall confine our discussion of these types of drawing to the simple form of the cube. This is a most basic building block for solids whether physical or drawn. The cube can be stacked or slid or skewed in order to build more complex forms. It also relates to the other basic blocks of the cylinder, cone, and sphere. In Fig. 1.9, each of the three visible faces of the cube has a 10 x 10 unit grid on the surfaces related to the decimal system. This can make it easier to plot changes in shape on the faces.

This page:
Fig. 1.9. Illusion cube using the oblique system of drawing and an elevation view. The three faces have been given a grid.

Using diagonals and midpoints, three three visible faces of the cube can be manipulated so as to be able to obtain circles and ellipses which become the groundwork for drawing more complex forms.

Chart showing a comparison of systems of drawing.

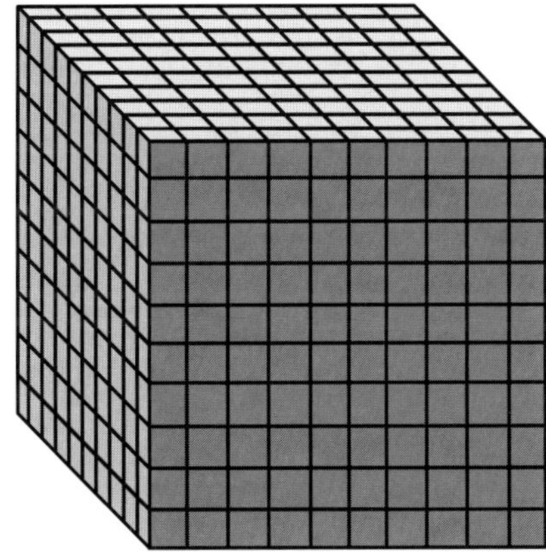

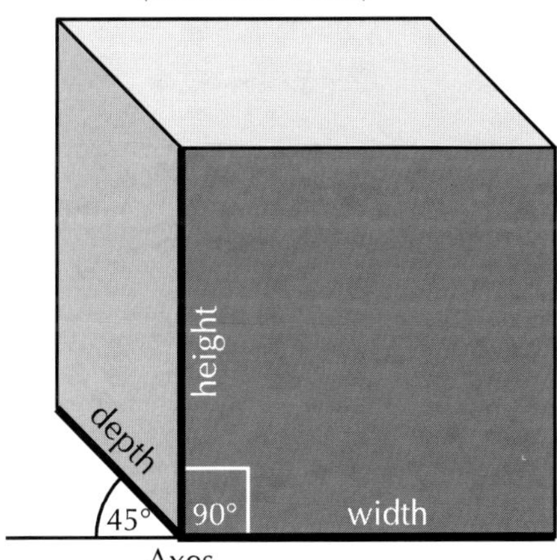

<table>
<tr><td colspan="2" align="center">Axonometric Projection Systems</td></tr>
<tr><td colspan="2" align="center">Axes =Height, width and depth</td></tr>
<tr><td>
Planometric System
(no face is parallel to the viewer)
Isometric
All axes have equal lengths. Angles a = b.
Dimetric
Two axes have same lengths.
Trimetric
All three axes have different lengths. Three different angles for the three different faces.
</td><td>
Oblique System
(one face parallel to the viewer)
Cavalier
Has true length in the axes.
Cabinet
Has half length in the axes.

For more information refer to a book on drafting.
</td></tr>
</table>

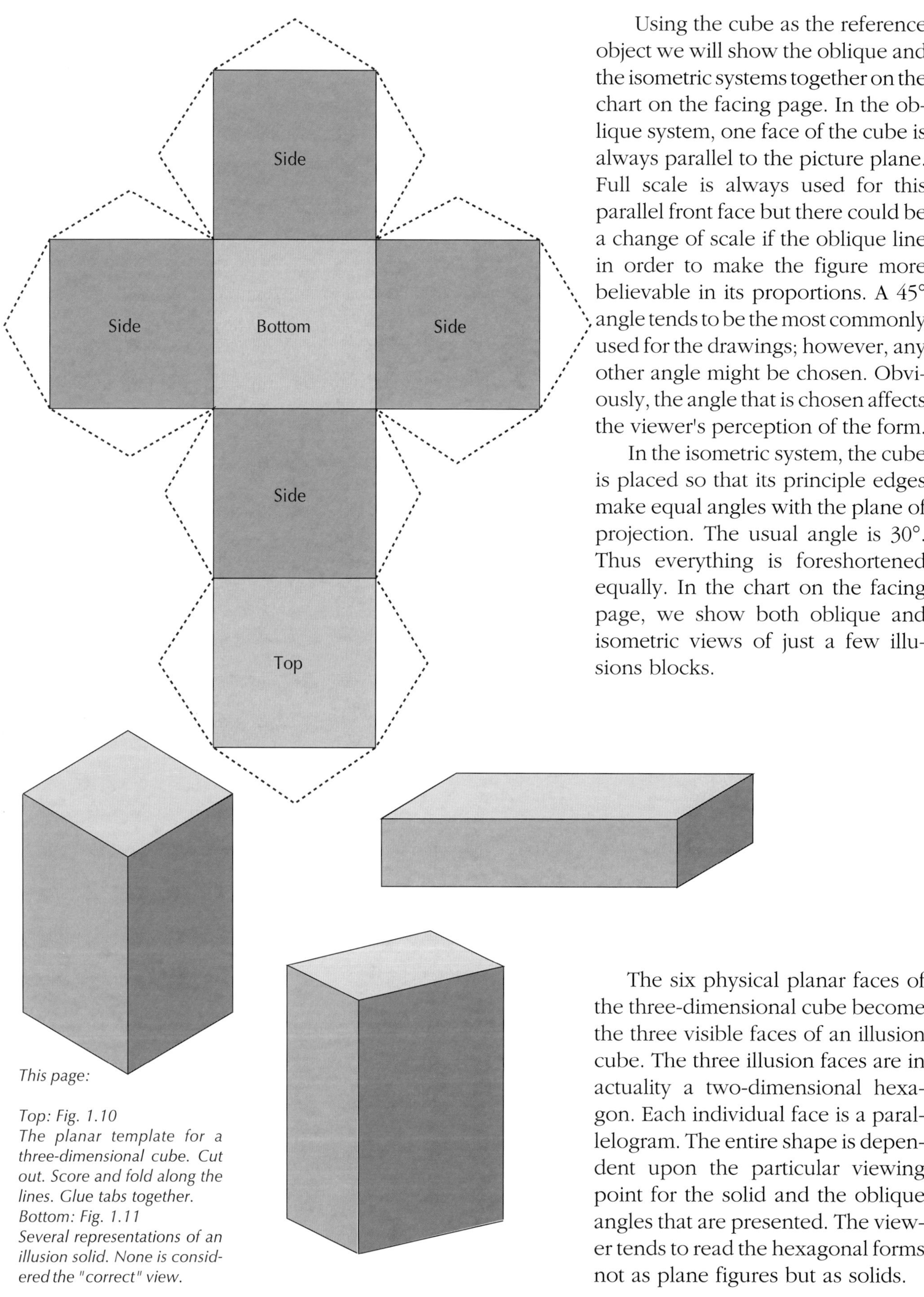

Using the cube as the reference object we will show the oblique and the isometric systems together on the chart on the facing page. In the oblique system, one face of the cube is always parallel to the picture plane. Full scale is always used for this parallel front face but there could be a change of scale if the oblique line in order to make the figure more believable in its proportions. A 45° angle tends to be the most commonly used for the drawings; however, any other angle might be chosen. Obviously, the angle that is chosen affects the viewer's perception of the form.

In the isometric system, the cube is placed so that its principle edges make equal angles with the plane of projection. The usual angle is 30°. Thus everything is foreshortened equally. In the chart on the facing page, we show both oblique and isometric views of just a few illusions blocks.

The six physical planar faces of the three-dimensional cube become the three visible faces of an illusion cube. The three illusion faces are in actuality a two-dimensional hexagon. Each individual face is a parallelogram. The entire shape is dependent upon the particular viewing point for the solid and the oblique angles that are presented. The viewer tends to read the hexagonal forms not as plane figures but as solids.

This page:

Top: Fig. 1.10
The planar template for a three-dimensional cube. Cut out. Score and fold along the lines. Glue tabs together.
Bottom: Fig. 1.11
Several representations of an illusion solid. None is considered the "correct" view.

Ilusion Cubes
(There are three visible faces; A,B,C)

Oblique System *(Elevation View)*

One face, A, is always parallel to the viewer. Lengths of the three axes are not equal. To prevent visual distortion, the length of one axis requires shortening by a particular ratio related to the 90° angle.

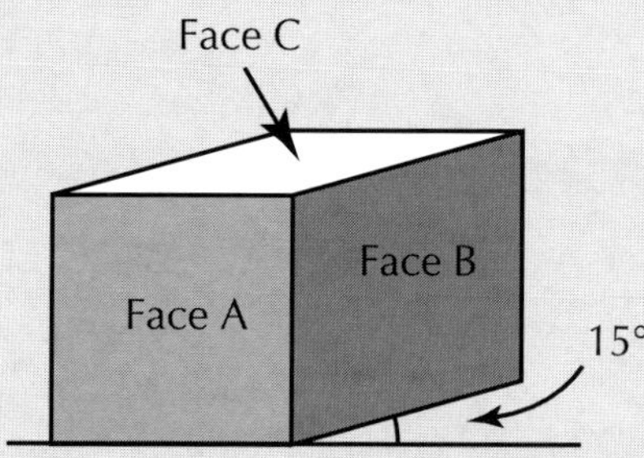

$$\frac{30}{90} = \frac{1}{3}$$

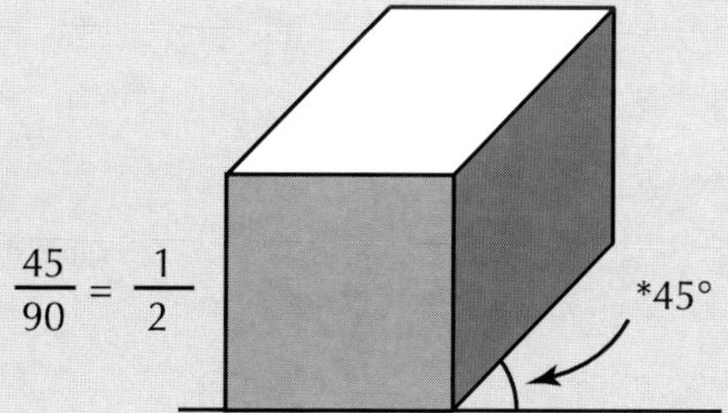

$$\frac{45}{90} = \frac{1}{2}$$

* This one is most frequently used angle.
45/90=1/2. Thus, the length of the oblique line must be reduced by 50%.

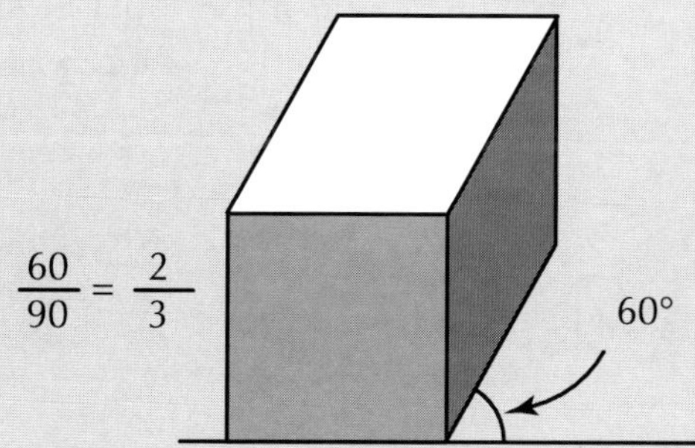

$$\frac{60}{90} = \frac{2}{3}$$

Isometric System *(Elevation View)*

One edge is always parallel to the viewer. Lengths of the three axes are equal, however measurements along the isometric axes are 0.816 of the actual size.

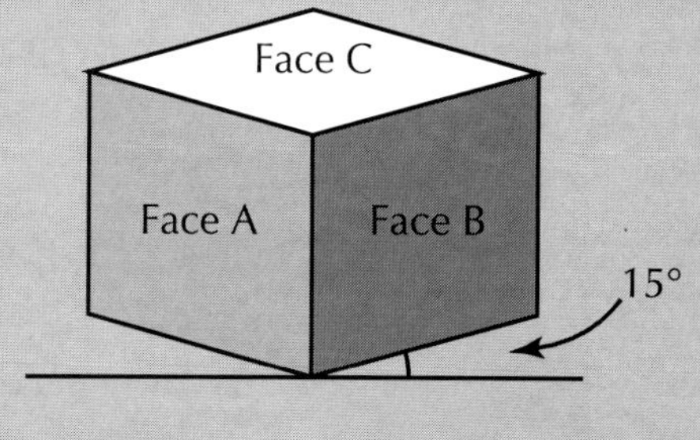

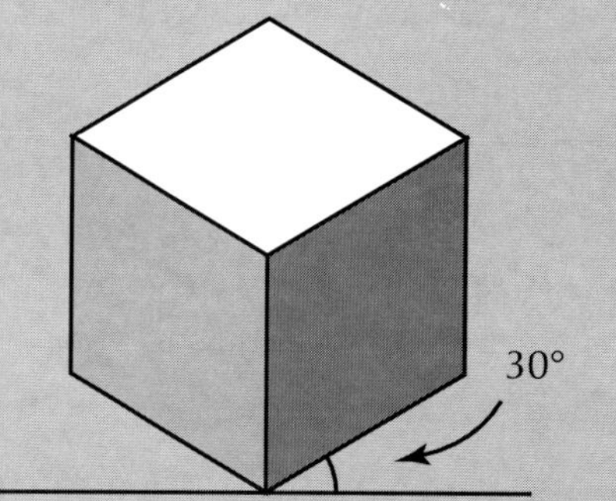

* This one is most frequently used angle.

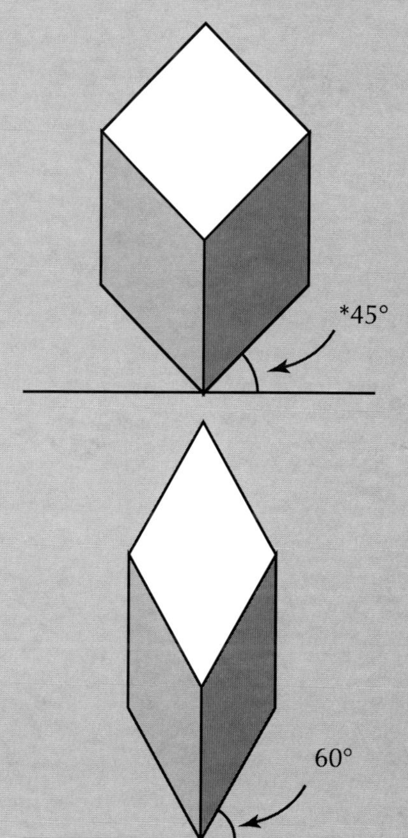

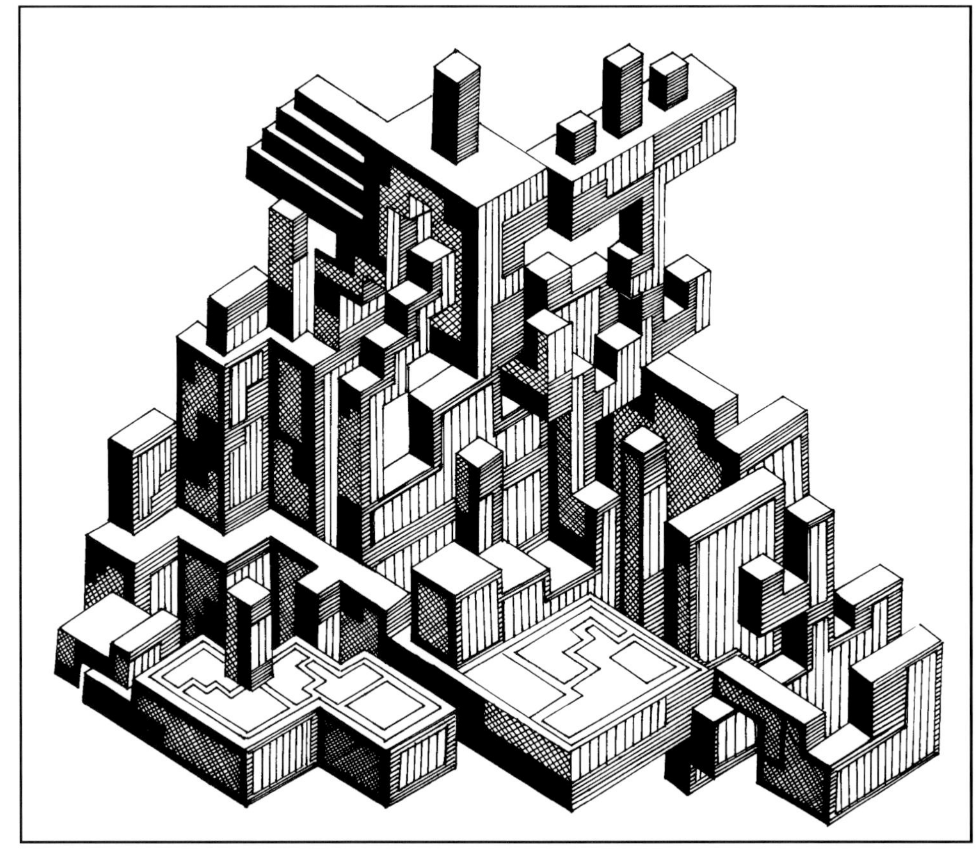

Facing pages:
Fantasy Cityscapes.
Student works using the iso-
metric system of projection
and an elevation view.
Pen and ink on bristol paper.

Top: Jay Maquire

Bottom: Joe Hastings

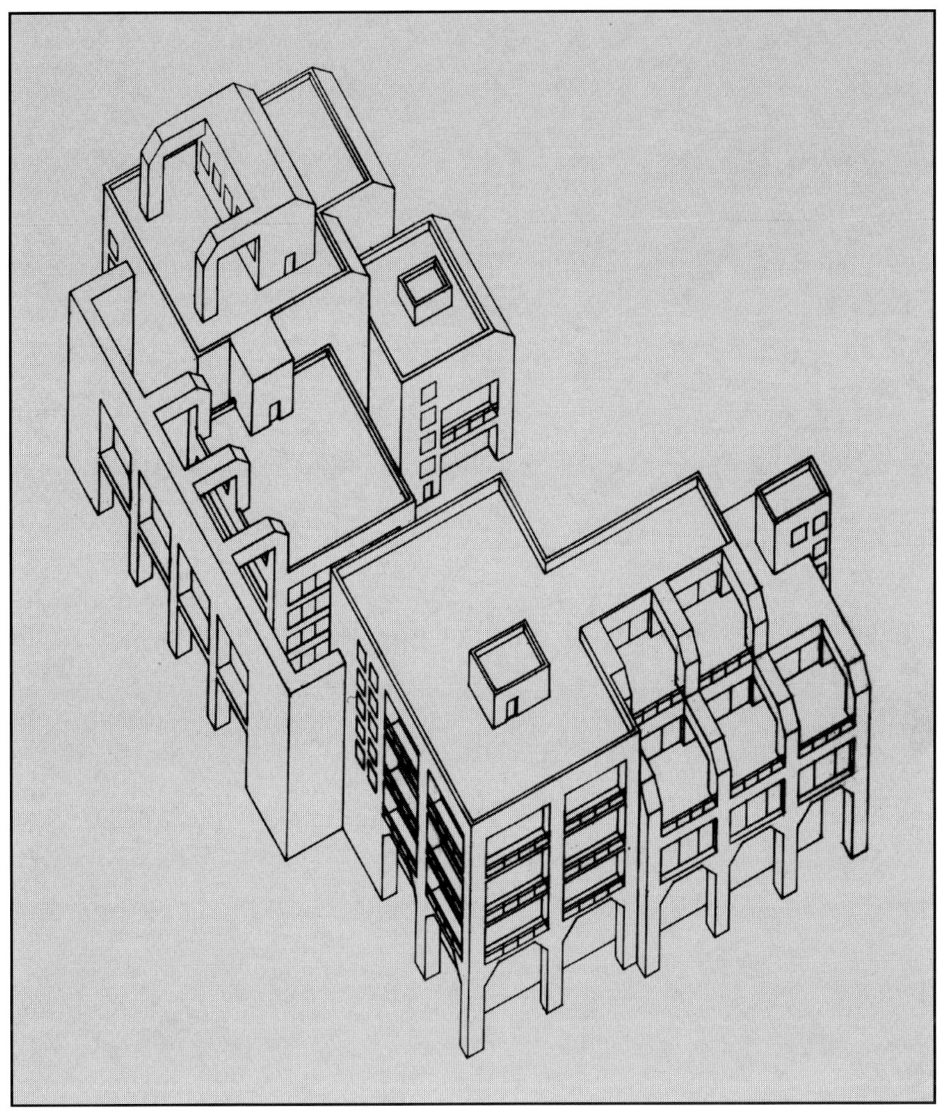

In these drawings, since the planes are tilted, there is a sense that the viewer is looking down upon a scene from a viewing distance that is somewhat removed from the subject. There is a mood of impartiality because of this. In addition, lines or edges that are parallel remain so whether nearer or further away from the viewer. Parallel lines never come together in the distance.Objects, therefore, do not get smaller as they move away from the viewer. Usually in the works of this kind, the sense of space is much shallower in feel since the element of modelling with light and shadow is eliminated. Usually, all edges have a consistent crispness. There is a confirmation of the flatness of the two-dimensional surface. The sense of depth is diminished while the sense of pattern across the surface is increased. This approach to the surface plane affected such artists as Vincent Van Gogh and Paul Cézanne.This system of representation is found in the artworks of China, Japan, and Persia. The human being is not placed within the event and thus is not the measure of all things as would become the prevailing attitude in the Renaissance as we shall see in the next chapter.

This page:
Robert Dumas. Pen and ink on bristol paper. This system contains no vanishing points. The diagonal lines suggest movement in space. Parallel lines remain parallel throughout the image.

We opened this chapter with a group of paintings of the landscape created by self-tutored "naive" artists whose visions interpret the landscape in very idiosyncratic ways. We end this chapter with a look at the paintings of a contemporary Twentieth Century artist, Sarah Supplee, who took on the task of translating the immensity of the landscape in a most naturalistic style. The artist, born in 1941 and died in 1998, professionally trained, painted and draw the landscape in a personally defined way despite the seeming objectivity of the subject matter and technique.

Her sketchbook, much of the time, was a camera which allowed her to record quickly the mercurial changes of light and shadow in the landscape of summer, her preferred painting season. Her works are meticulously crafted. The building up of the images in either pastel or oil paint becomes a form of meditation. Each minute movement of the brush, small in itself, gathers momentum in the large created vistas of solitude. The landscape is devoid of persons or the detritus of human habitation, and yet, hints of civilization peek out among the grasses, shrubs, and trees. There is a pervasive sense of green lushness and warm light. Throughout the history of art, spirit is always represented as a manifestation of light.

"In the summer I am a hunter. I listen for weather forecasts noting the type of light that may be likely. Canadian air brings the clear golden light of low humidity. Rain may result in morning mists, thunder showers in dynamic clouds. Haze is good for going into stream beds to photograph plants and rocks close-up. The hot humid days of August give good imagery for pastels."

Sarah Supplee

This page: painting by Sarah Supplee. Where the Heron Fishes, *68" x 68". Oil on canvas. 1996.*

This page:
Landscape paintings by the Twentieth Century artist, Sarah Supplee. Courtesy of the artist.

Top:
Bruegel Vista, 23" x 24", oil on canvas 1991.

Bottom:
Morning River, 48" x 53" oil on canvas, 1993.

The artist would split her time between her winter studio in an urban environment, Cambridge, Massachusetts, and her summer studio in a landscape setting, Hillsdale, New York in the Hudson River valley. Her works are heir to a long tradition of landscape painting in this country. The art of the 19th Century American painter, Frederic Church, resonated within her. But her artistic roots also reached across the ocean to Dutch, Flemish, and French artists, as it did her American predecessors.

In fact, Sarah Supplee spent her early years on a small farm in the state of Maryland. Without being conscious of it, she forged a lasting bond with rolling hills, pastures, and streams. Even in her more abstract period, between 1959-1972, she drew her inspiration from the light and open space of the landscape. But her return to painterly realms allowed her to connect more directly with qualities in the landscape that are both beautiful and threatening at the same time. Her audience, too, can more directly reach out to her content through the more accessible subject matter.

Remembrance of past things fuse with visceral feelings in the present. She was attracted to developed lawns and fields where the shape of the earth and the growing plants had developed individual characteristics. The afternoon light had a special sculptured quality with color that is full and rich. The tactile qualities of grass, plant life, water, and soil, are evoked.

Her two working situations allowed her to take advantage of their differences. Summer was her preparatory time. With her camera and her sketch tools she hunted the land in order to capture a special moment in time that is affected by weather changes. The sun and the moisture dictated where she would go to search out her prey. The weatherman was her constant companion in her forays. She stalked among three or four sites moving through morning to afternoon, watching the land as the subtle movements of the weather softened or accentuated the contours of the land. She will go out again and again through the month of November.

Winter allowed her to withdraw into her studio to take the summer inspiration and turn it into actual pastel drawings and oil paintings. She sifted through her photographs evaluating content, color, and composition before she moved onto paper or canvas. Since each large oil painting might take a minimum of three months of steady, long hours of work, she had to choose carefully. To quote the artist, "I build the painting slowly, watching it. I try to let the paint breathe and to allow the painting to make its own statement. The image has three realities: the original site as seen, the photo perception of a moment in time and the canvas." Her pastels allowed her to explore ideas and images more quickly due to their smaller size and more immediate in material. Yet, these, too, took intense concentration and long hours.

It is not easy to live in the latter part of the Twentieth Century and be a landscape painter. Our civilization has an electronic urban mindset that pushes nature off to the side. It was one of the missions of this artist to remind each of us that we really live in the natural world whether we realize it or not. Our future may depend upon this realization.

Problems, Projects, and Play

1. Find a photograph of a face, human or other animal, overlay a square grid on it. Trace the general structure of the face. Now develop another grid structure. Transfer the drawing from the square grid onto the new grid.

2. Locate technical engineering books that describe in detail orthographic, oblique, and axonometric drawing systems. Take a rectilinear object, such as a computer, and first do a free hand line drawing of it. Then, using one of these types of engineering systems, do another line drawing. Compare the results by matting or mounting the two drawings together.

3. Develop a comparative chart showing three different drawing systems using a 10 x 10 x 10 unit cube. Include the rules of the systems and make the chart artistic.

4. With both drawing and words, analyze a painting or drawing that is a Persian miniature or a Japanese interior. What similarities do you find? What are the differences? Which one do you prefer, and why?

5. In writing, using photocoies for documentation, compare and contrast at least four different still life paintings from four different historical periods within the Western Art tradition.

6. Find photographic examples of rock paintings from the Old Stone Age peoples and wall paintings of Ancient Egypt. In a written and visual essay, compare and contrast each cultural representation. What do these images say to you about each culture?

7. Explore in writing and images, American Indian Art, Eskimo Art, or the art of the Northwest Coast Indians.

8. Using the seven types of mapping, take any object and show how it would look in any three of the systems.

9. Research systems of mapmaking. In a written and visual essay, explore the differences between any two systems.

10. Find a reproduction of a painting or drawing that you like. On a tracing paper overlay, analyze the different perceptual spatial cues that the artist used.

11. Look up Trompe l'oeil painting. In writing, attempt to answer the question: Why would this type of image making develop in Western Art and not in the art of non-western cultures?

12. Write an essay on the history of map making. Give visual examples of various cultures.

13. Research the works of two naive artists. Compare and contrast their treatment of the picture plane with examples of two different Renaissance artists.
Do a small colored pencil drawing based on one of the four works.

14. Research the photographic work of Twentieth Century arist, David Hockney. In a written essay, explain how he represents space and time in a single work. Using a polairoid camera, try to do a similar work.

15. Research the paintings of Juan Gris, or Georges Braque and do an analysis of one work of each. How is the picture plane treated? How are objects in space represented? Do a small colored pencil drawing based on a single work of either of these two artists.

16. In drawing, painting, photography, or digital imaging, do a visual essay combining multiple viewpoints of a single subject into a single unified image.

17. Research "Trompe l'oeil" paintings. Based on your understanding, create a small painting.

18. In this chapter many of the artworks include tree elements. Research the subject of trees froma botanical viewpoint and from a symbolic attitude. Do an artwork, in any material and technique, based on your findings.

19. Do a photographic essay in which you take at least 10 different views of the same object or subject matter. Mat or mount your results.

20. Use the above photographic essay as the basis for a drawing in black and white, tone, color, pattern or texture.

21. Using a found photograph, do two drawings of the same subject using two different types of grids.

22 Do a drawing of an imaginary landscape using one of the systems of representation found in this chapter. Be sure to include a tree.

23. Using the chart of illusion cubes from this chapter as a basis for a drawing or painting, construct a work that uses a combination of cubes. Work with black line only. Then do a variation in three changes of tone, or a monochromatic scheme, or a primary system of yellow, red, and blue in appropriate values.

24. Take the image of the Egyptian figure on page 19 and place it on a grid of one of the following:
    a. a trapezoidal unit
    b. a Golden Rectangle unit (ratio
       of 1:1.618 approx.
Keep the number of units constant.

25. Research the wall paintings of Ancient Egypt. Using a group of contemporary objects, scenes, and or events, do a pencil drawing in the style of the Egyptian work.

26. Research the wall paintings of Ancient Egypt. Find an example that you like. In writing, analyze the various objects and their viewpoints. Do a variation drawing in black and white or color. Substitute contemporary objects for the ones used in the Egyptian image.

27. Use the construction techniques found on pages 22 and 23. Build a pliable plane structure on which you then put a drawing of a person, still life, or landscape. Use materials and techniques of your choice. Think about using the computer to generate your image.

28. Do a tone drawing in which the objects are built within the oblique system of projection.

29. Do a tone drawing in which the objects are built within the isometric system of projection.

30. Find a reproduction of Persian miniature painting that you like. Using collage elements as well as drawing materials, create an interpretation of that miniature.

31. Go into your own "studio" environment. Set up a still life arrangement based on what you understand from looking at the student examples on pages 30 and 31. Do a pencil drawing only of these objects. Or choose to use pen and ink.

32. Look up the Cubist paintings of Juan Gris, Georges Braques, and Pablo Picasso. Set up a group of objects and attempt to interpret them in a manner similar to one of these artists. Use materials of your choice.

33. Using the information on pages 6 and 7, try a group of drawings using the different approaches suggested. Invent at least one new approach and offer an example of your own using it.

34. Do a painting, drawing, collage, or relief painting based on the work on page 39.

35 Even though orthographic representation is a functional type of drawing, create a composition in which you make the functional much more artful.

36. Take a plane drawing or a photograph and transfer it onto a pyramid, cylinder, or cone. For a cyclinder, begin with a grid on a rectangle. For the cone, use a polar grid.

Further Reading

Bihalijhi-Merin, Oto. *Modern Primitives*. New York: W. H. Smith Publishers, Inc. (no copyright date.)

Cooper, Douglas. *Drawing and Perceiving,* Second Edition. New York: Van Nostrand & Reinhold. 1992.

Hagen, Margaret A. *Varieties of Realism*. Cambridge: Cambridge Unversity Press. 1986.

Milman, Miriam. *Trompe l'oeil Painting*. New York: Rizzoli International Publications, Inc. 1982.

Rénard, Hêlene. *The Dream and the Naifs*. New York:Vilo, Inc. Publishers. 1983.

2 A Single Viewpoint

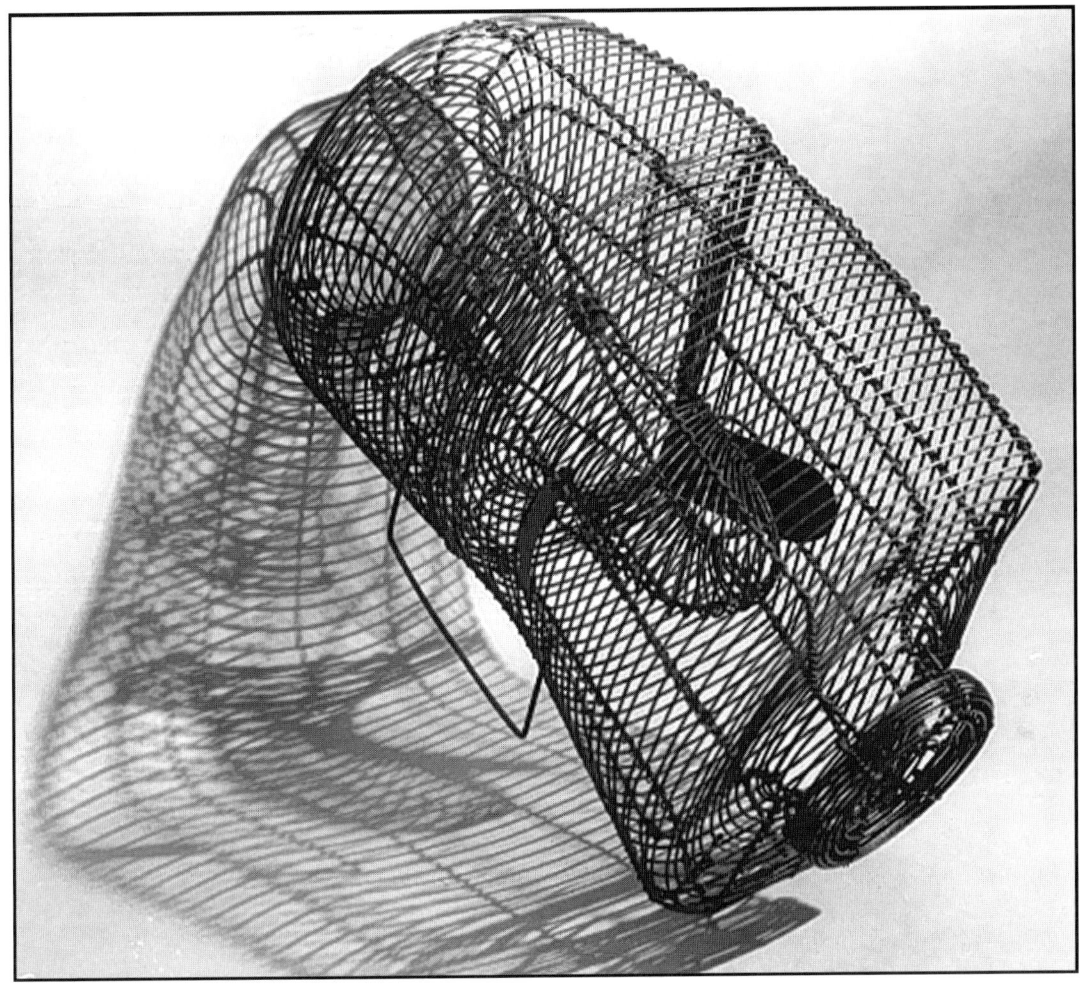

We humans are all familiar with shadows. Where there is light, there are shadows. When the sun shines and we come between the sunlight and a surface, we cast a shadow. How magical! As children we played the game of hide-and-seek with our close companion. We would try to run from it but it dogged our heels. We tried to sneak around the corner from it but it crept up right behind us. It is a dark silhouette-shape that stretches and bounces and shrinks everywhere we go. We are forever attached to it,though it is alwasy aloof from us. Yet, on a cloudy day, our second darker self disappears, not to reappear until the sun comes out once again.

This shadow is a projection of a three-dimensional object onto a two-dimensional surface. This surface could be planar or spherical or curved or crumpled. Despite the alterations to its shape as it projects onto the surface, the shadow contains all the visual information of the original form though it is topologically transformed. In Chapter 5, we shall see what that means. In Chapter 4, we will look at four-dimensional objects that cast "shadows" in both three-and two-dimensional space. Meanwhile, on the following two pages, we offer you a photo essay of shadows.

This page:
An object and its shadow. A not-so-familiar object, an old-fashioned mousetrap. Can you tell which is the actual physical object and what is the shadow? Where does the light seem to be coming from and what kind of surface is the shadow cast upon?
How would you try to draw this object? It has a very strong contour shape despite the extreme busyness of the interior form.

I have a little shadow
Who goes in and out with me.
What can be the use of him is
more than I can see.
        Robert Louis Stevenson

These facing pages:
Photographic exploration of shadow compositions. These images are two-dimensional projections of three-dimensional objects What characteristics of the objects allow for the recognition of them in the images?.

Think of all the shadows as simply shapes with which to work. Use high contrast, stippling, hatching or a combination of techniques with the materials of pen and ink.

Try doing a drawing that is based on one of these images.

Is a shadow considered a geometric object? It is not rigid. It cannot be picked up and moved to another position while maintaining its shape. It ever changes with the direction and angle of the light; with the texture of the surface it is cast upon; with the shape of the surface it covers. It is as elusive as a butterfly. What is the net for catching shadows? It is the lens of a camera. This is the mechanism for holding the shapes and tones of light onto a special piece of plane, photographic paper.

What about the act of drawing or painting? How do these capture shadows? What does an artist do when he or she sees objects out there in three-space and wants to transfer and translate that three-dimensional world accurately and in proportion, onto the two-dimensional piece of paper? By necessity, the three-dimensional objects of experience become the two-dimensional shapes of art.

The artist stands at a particular distance from what he or she is looking at. This is called a station point. An artwork could contain more than one station point and in many non-western cultures, it does.

The artist looks at an object or an event through a cone of vision. The apex of that cone is at the eye of the artist. Light strikes the object/event and it bounces off the object back onto the eye of the beholder, the artist. The points of light from the three-dimensional object must be placed onto a surface that is set between the artist and the object. See the chart opposite.

The painters of the Italian Renaissance of the 14th and 15th centuries took this particular problem very seriously. The images on a surface may be considered a projection of the three-dimensional world, the center of projection being found at the eye of the artist. In this process, lengths and angles are distorted depending upon the relative posi-

tions of the various objects chosen to be depicted. And yet, the drawing can be "read" and the objects identified. How does this happen?

Let us consider the subject of a pictorial drawing system in which there is a vanishing point on the horizon line and where some lines, which are parallel in three dimensions, become oblique converging lines in two dimensions. These meet at the vanishing point on the horizon.

Geometers became interested in properties of geometric figures in the plane and in three-space, especially those properties that remain unchanged when mapping from three-space into a plane or from one plane to another plane.

Artists needed the information of projective geometry in order for them to depict the space of three dimensions more accurately. Thus artists became mathematicians as they struggled with these involved geometrical problems.

But why did mathematicians bother with making rigorous axioms and theorems for projective geometry? The reason has to do with consistency. Mathematicians and artists needed to know if projective geometry was consistent with the "real world" geometry of Euclid. This is not trivial. In Euclidean geometry the concept of infinity does not really exist, except as a nebulous "forever." In projective geometry we decide that infinity exists at the points on the horizon line. This changes everything. A mathematician might ask, "Why should the geometric theorems that hold in Euclidean also hold in projective geometry/"An artist might ask, "How do I know if there are certain objects that exist in three-space that look completely different when I try to draw them in perspective?" To answer them one would need to write axioms for projective geometry and show that these do not violate the axioms of Euclidean geometry when we go from one to the other. That is what mathematicians did.

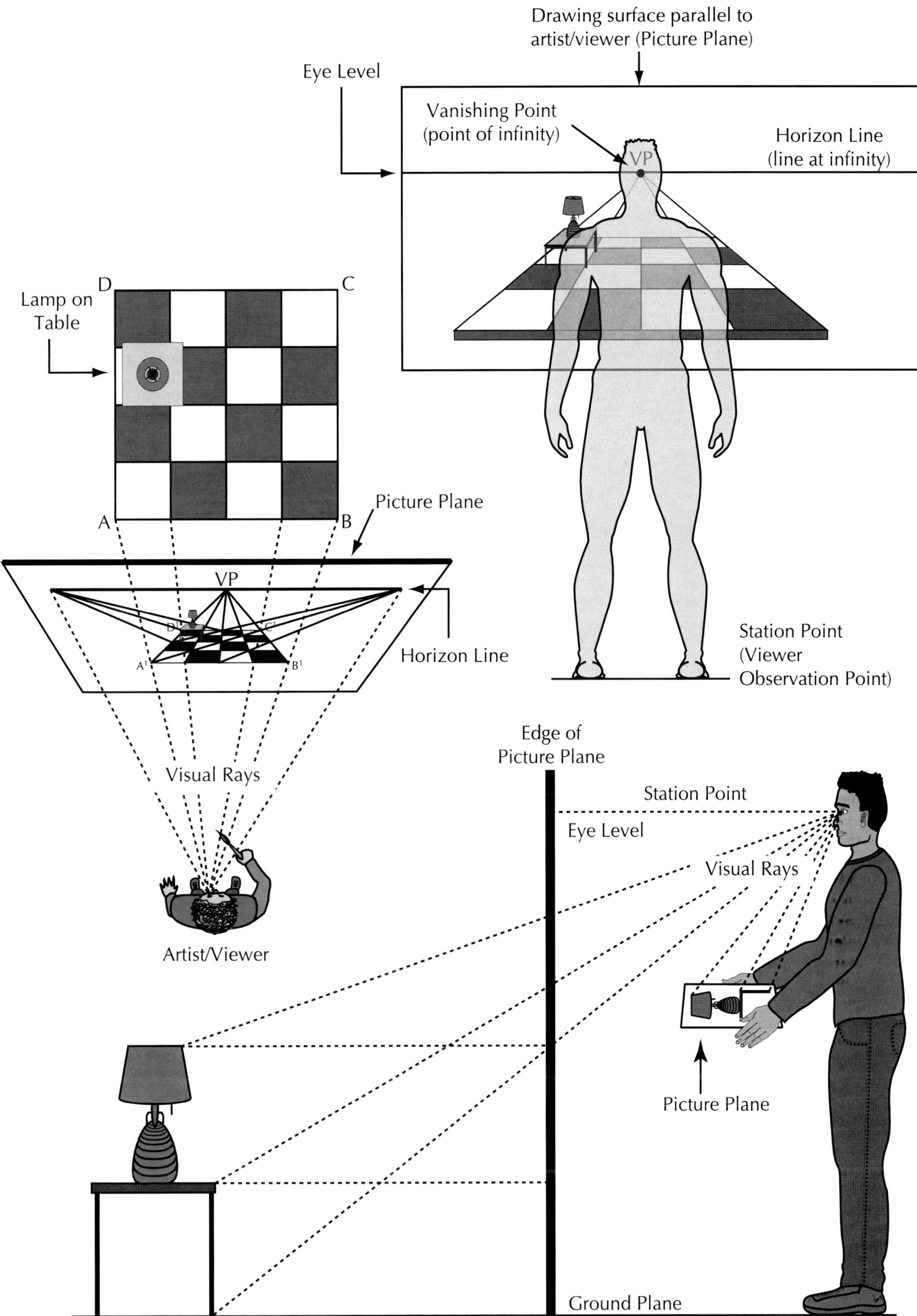
Drawing surface parallel to
artist/viewer (Picture Plane)
Eye Level
Vanishing Point
(point of infinity)
VP
Horizon Line
(line at infinity)
Lamp on
Table
D
C
A
B
Picture Plane
VP
D¹
C¹
A¹
B¹
Horizon Line
Station Point
(Viewer
Observation Point)
Visual Rays
Artist/Viewer
Edge of
Picture Plane
Station Point
Eye Level
Visual Rays
Picture Plane
Ground Plane

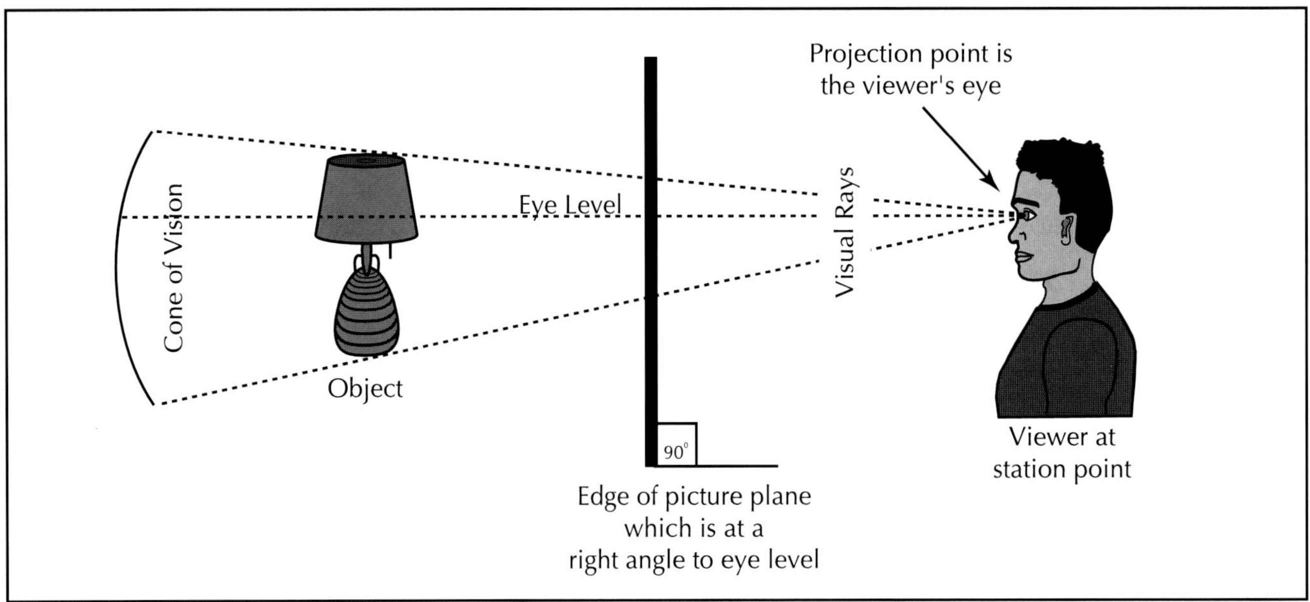

This page:
Above: Fig. 2.1
The cone of vision.
The cone of vision is a right circular cone generated by the rotation of a right triangle. See chart pg. 54.
What is framed by the cone of vision is limited in scope to 60°-80°, the maximum area. The closer the viewer is to the object, the more the object fills the field of vision. The further away the viewer is located, more of the object can be seen.
Light travels in straight rays from an object and reflects back to the eye. A single centric ray reflects from the point on which the eye is focused. This ray is right-angled. The section, or transparent glass plane, intercepts these rays as they pass through it. The section contains a scale image of the original object.
Despite the fact that we exist in a three-dimensional space, we receive much of our information about that world through visual sensations which are captured in two-dimensions by the retina of the eye.
The field of normal binocular vision (using two eyes) is approx. 214° horizontally and 135° vertically which produces a width to height ratio of 1: 1.58, which is close to a Golden Rectangle.

Projective geometry is related to the science of optics and that science is related to vision. Each of our eyes has a cone of vision which extends infinitely until stopped by objects in its path. The apex of that cone is the lens of each eye. Planes, sometimes called sections, can intersect this plane. From each point of an object, a ray of light travels in a line to one eye of the viewer/artist. A transparent bounded portion of a plane, like a glass window pane, is interposed between the object and the artist. All of the converging lines are called a projection and can be marked on the glass. These points can then be mapped onto a drawing surface. Fig.2.1. A drawing surface usually, however, is not transparent. Some means, therefore, has to be developed to enable the artist to draw on the opaque surface. This is where mathematics with its geometric theorems enters the picture, so to speak.

The mathematics of perspective was developed by the investigations of a Seventeenth Century self-educated mathematician, Gérard Désargues (1593-1662).While artists used the ideas of perspective in the 15th Century, it was the mathematicians of the 18th Century that studied it as a special discipline. According to current scholarly research, evidence that has turned up points to the fact that Désargues was both financially well-off and mathematically sophisticated. His interest in perspective moved beyond the practical concerns of the artist and into the generalized theorectical focus of the mathematician. He investigated what was not changed within a geometrical figure, invariants, while the artist was involved with the how-to-change an object and still make it recognizable to the viewer.He published a book on the subject in 1639, but he was a man whose ideas about projective geometry were far too early for his time. In fact, his °writings on the subject remained in obscurity for two centuries. Despite his friendship with, and support by, René Descartes, his work was not acceptable to many of the mathematicians of his era.

It was in the nineteenth century that this particular branch of mathematics surfaced again. In the intervening centuries, the study of geometry was eclipsed by algebra. It was rediscovered by geometers more through good fortune and happenstance than by deliberate consideration. Jean Poncelet, however, recognized that here was indeed a new branch of geometry so very different from that of Euclid.

Greek geometry concerned itself with the size and shapes of rigid figures in space. Projective geometry looks at a more basic aspect of figures, those properties which remain unchanged under a projection (that which the eye sees with the cone of vision).

Preserving the distance between points is not a property of geometric figures in projective geometry while it is in Euclidean. Neither is the property of preserving congruent angles. So figures can change shape and still be considered equivalent. For projective geometry, the goal involves understanding the relationship between an actual object and its projected image on the section whether it be a canvas or a wall or a piece of glass. The idea of two lines being parallel is not preserved. Where once there were parallel lines, there are now lines that converge to a point at infinity and a line at infinity is defined. This is called the horizon line. All points at infinity must lie on a single line. It is the concepts of a line and a point at infinity that makes projective geometry an important tool. You will become aware of this when you work with the checkerboard floor in perspective drawing. Later on in this chapter you will learn how to play the "Perspective Game".

Imagine a cast shadow of a circle which stretches to become any one of these other figures. These are transformations. The properties of shape, size, and angle lose their meaning here, but something basic remains. In a perspective drawing or painting, shapes and sizes and angles are altered and yet the viewer is able to recognize the source of the forms. There are aspects of the figure that are invariant.

In Euclidean geometry, parallel lines never meet, no matter what. Désargues, however, showed a way in which parallel lines, and planes, can meet at infinity. Further on in this chapter you will see that certain sets of lines meet at infinity at the horizon line in the picture plane. This had important repercussions for the image-making aspects of painting.

By generalizing, as mathematicians do, Désargues used the triangle to stand for any projection. The concepts apply to other polygons composed of triangles. All quadrilaterals are projectively equivalent to a fixed square. As you also shall see further on, it is the square, the square grid, and the cube that are essential elements in any perspective drawing.

For the mathematically inclined, do more in-depth research into the subject of projective geometry. What makes it more general than Euclidean? What other areas of life, besides art, does this geometry investigate? Does it have any relation to the hardware or the software of computers?

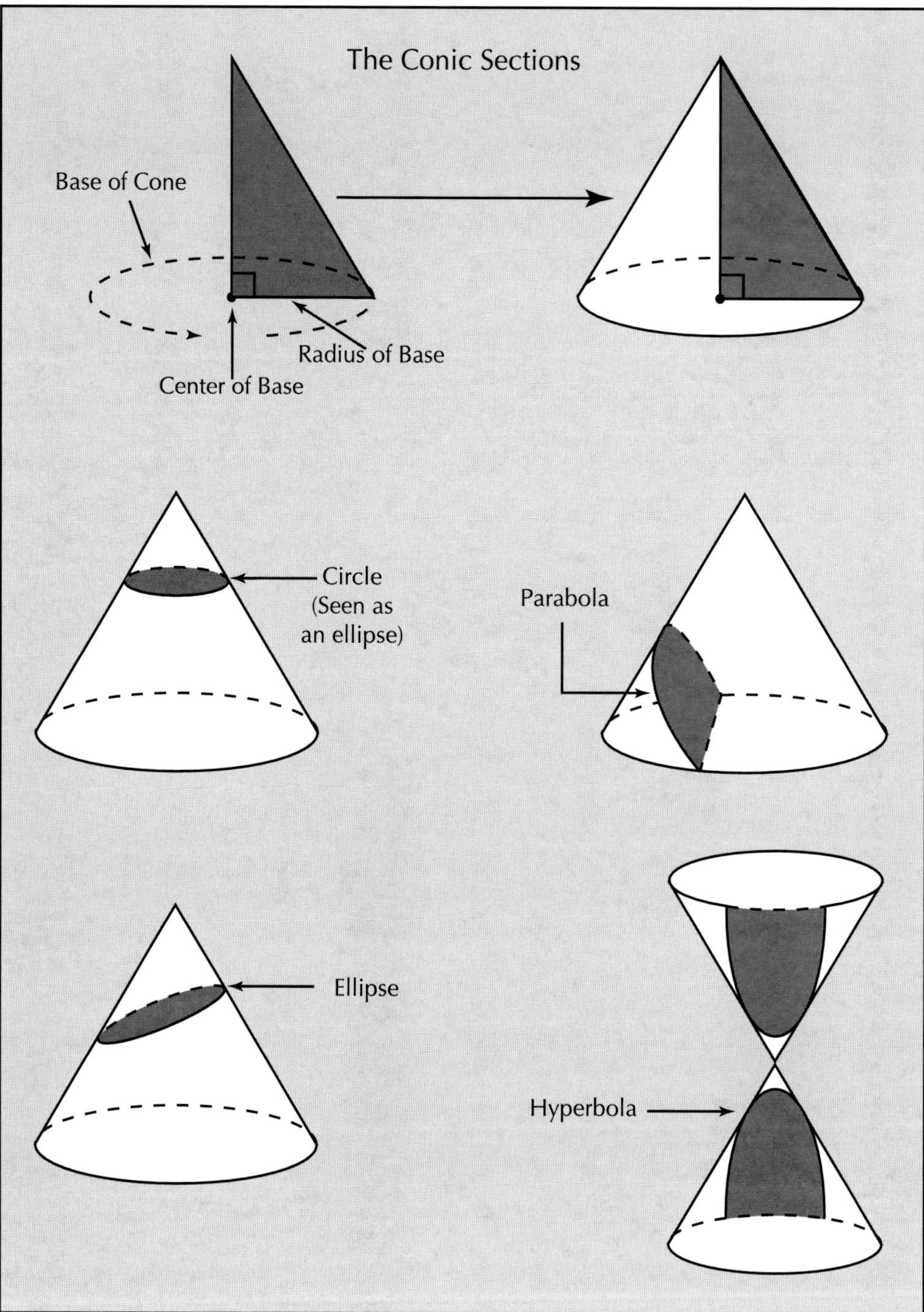

This page:
The conic sections:
*Conic sections are meaning-
ful in Projective Geometry.
The figures shown in the chart
can also be described by algebraic equations involving two
variables, x and y.
Geometrically, the conic sections are derived from a plane
slicing a right circular cone.
  The Right Circular Cone is
generated by rotating a right
triangle a full 360°.*

The conic sections are:
*a. The circle,
a horizontal slice of the cone.
$x^2 + y^2 = r^2$.*

*b. The ellipse,
an oblique slice of the cone.
$x^2/a^2 + y^2/b^2 = r^2$.*

*c. The parabola,
a parallel, to one side, slice of
the cone.
$y = x^2$.*

*d. The hyperbola,
a vertical slice of a double
cone.
$xy = r^2$.*

*These two-dimensional figures
give rise to a set of three-dimensional, forms.
The circle becomes the sphere.
The ellipse becomes the ellipsoid.
The parabola becomes the
paraboloid.
The hyperbola becomes the
hyperboloid.*

On pages 56 and 57, we show you Désargues theorem for triangles in the plane. But how does this theorem show how perspective functions? The vertex of the cone of vision seen in the diagram on page 52 is considered a center of perspectivity. The cone of vision has become the concept of the abstract ideal cone. Cross-sections of that cone give us the figures of the circle, the ellipse, the parabola, and the hyperbola. These are all related to each other and are unified into aspects of the single cone. It was Désargues who first theorized the connection in his book entitled "Rough Draft on Conics" which was intended for an audience of learned persons. On this page we give you a chart of the conic sections.

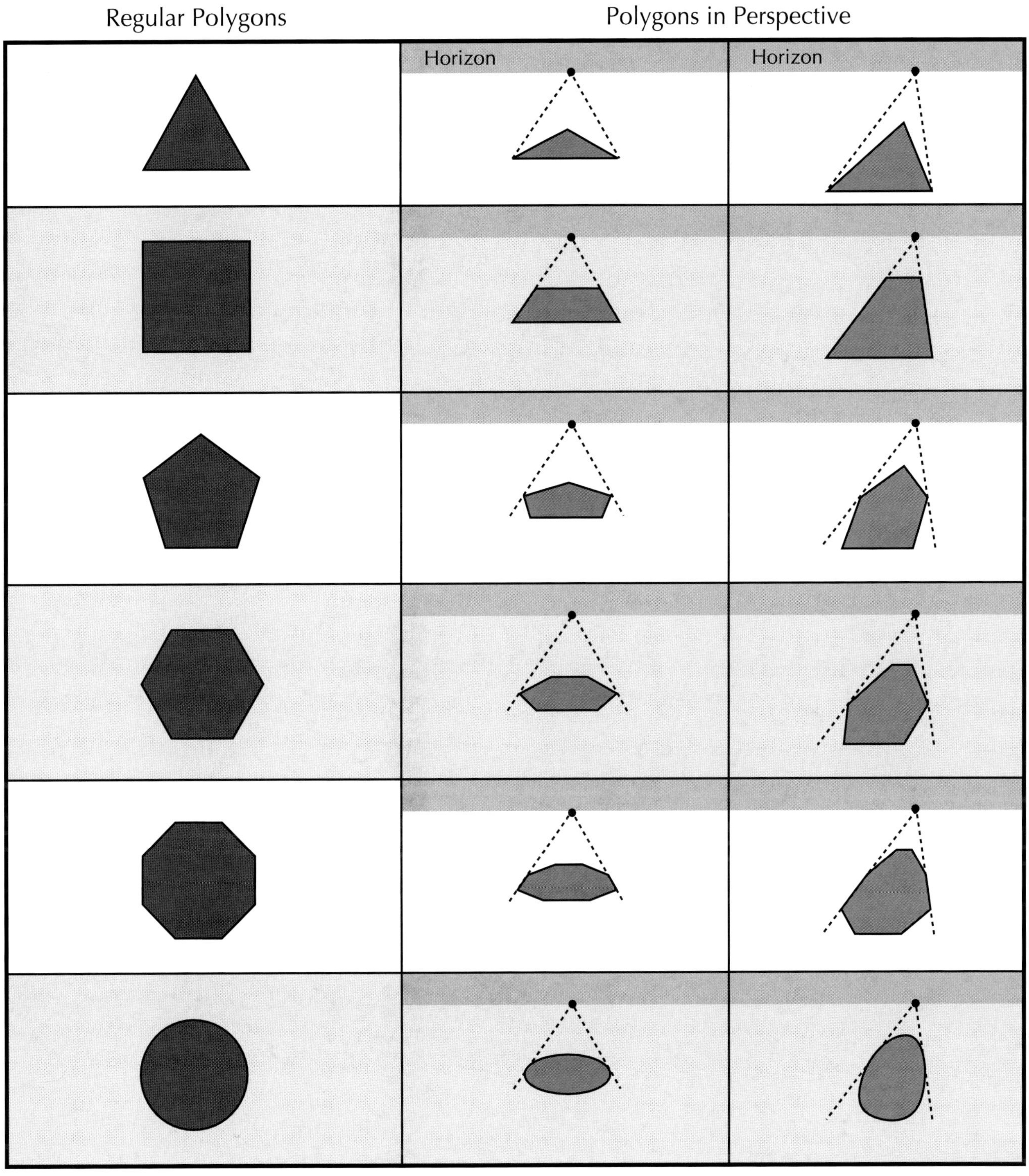

Regular Polygons   Polygons in Perspective

Geometric objects are transformed. Size, shape, and angle can change. All Euclidean planes extend to infinity in the projective plane. There is a line at infinity. In a projective plane there are only points and lines.

Cut polygons out of paper. Shine a light source on them and see what kinds of shadows these cast. On paper, draw the shape of these new polygons. Move the light source around to another position. Draw the shapes again. Do this a number of times. What conclusions can you come to with this experiment? Take these practical experiments and turn them into an interesting art experience by adding other shapes, textures, patterns, colors, etc.

Désargues Theorem in the Plane

> Let ABC and A'B'C' be two triangles in the plane such that lines AA', BB', and CC' intersect at a common point O.
>
> Let the lines AB and A'B' intersect at the point Q.
> Let the lines CB and C'B' intersect at the point P.
> Let the lines AC and A'C' intersect at the point R.
> Then the points, P, R, and Q are always collinear.
>
> This is true for *any* two triangles ABC and A'B'C'.

The facing page: Diagrams showing Désargues Theorem in the Plane.

Point O is the center of perspectivity, or the vantage point. It is also considered a point at infinity.

The axioms of projective geometry are:
1. There exists at least one point and one line.
2. Any two lines meet at one and only one point.
3. There is exactly one line through any two distinct points.
4. Any two distinct lines intersect in a point.
5. There are at least three points on any line.
6. There exist four points, no three of which are collinear.

Perspective drawing is a two-dimensional rendering of objects from three-dimensional space onto a particular surface. The study of perspective is an aspect of Projective Geometry. Désargues Theorem is a theorem of this geometry. Though Désargues Theorem works in three-dimensions as well as two-dimensions, it is the latter that concerns us here.

His fundamental theorem looks at an important property that unites two sections of the same projection of a triangle. A triangle may be thought of as either a set of three non-collinear points or as a set of three non-concurrent lines. This involves the principle of duality which allows for the symmetry of lines and points. Thus, another ways to state the theorem is to say: Any two triangles that are in perspective from a point are also in perspective from a line.

The two triangles ABC, A'B'C' can be considered projections of each other. The theorem states, and is shown in the diagrams on the facing page, that the triangles ABC and A'B'C' are two triangles in the plane such that lines AA', BB', and CC' intersect at a common point O. There are three pairs of corresponding sides in these two triangles. Each pair of corresponding sides for the two triangles will meet in a point. These three points will lie on a straight line.

Désargues Theorem states that in a plane if two triangles, ABC and A'B'C' are placed such that lines joining corresponding vertices meet in a point, the corresponding sides when extended, will meet in three points which all lie on the same line. The theorem in three-dimensions involves two given triangles lying in different non-parallel planes.

Point O can be thought of as a perspective point, or a vantage point. The points P, Q, And R are vanishing points that exist on the horizon line. The line that P, Q, and R determine is the horizon line that the two triangles would determine if they were to be drawn in perspective.

56

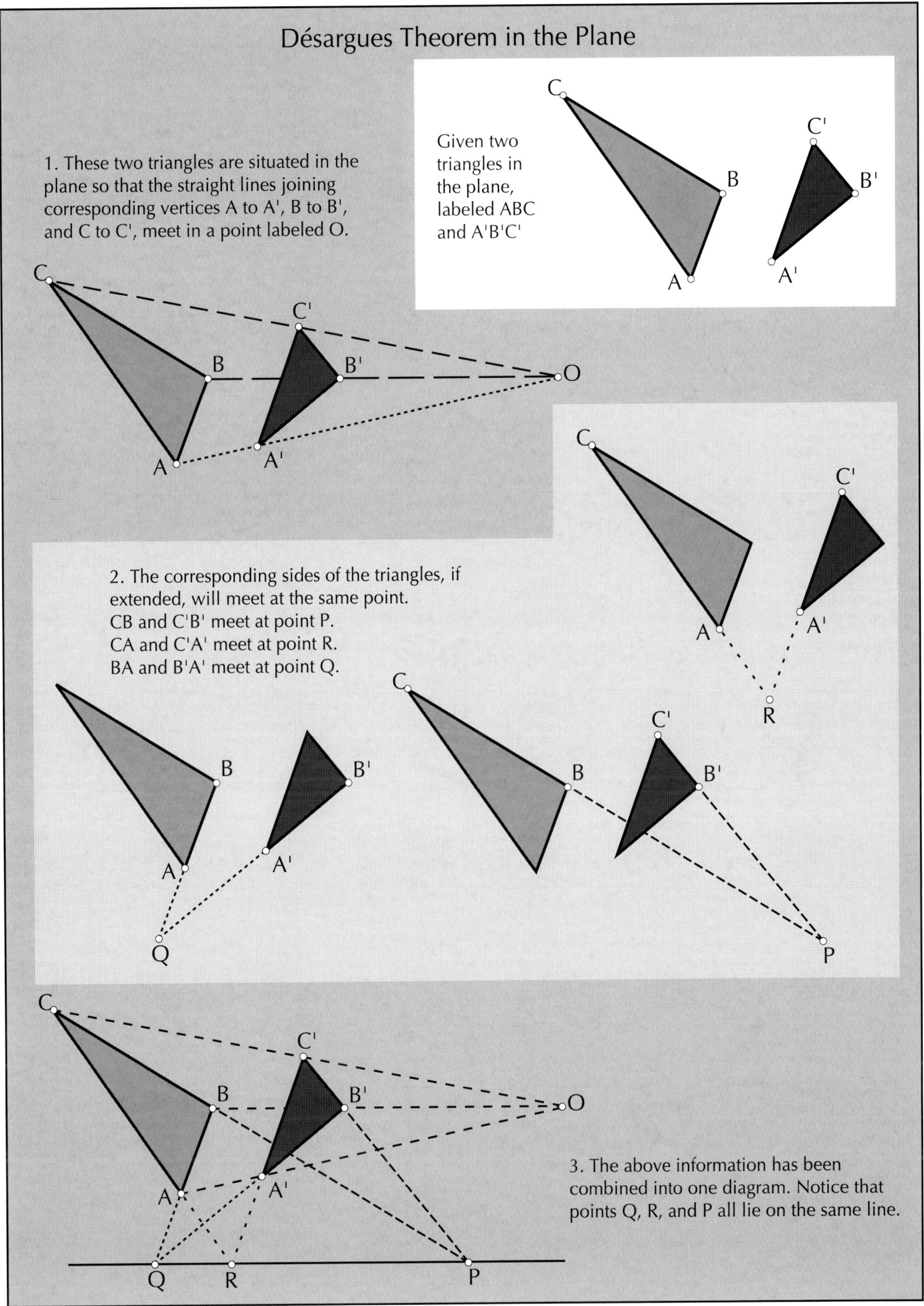

Désargues Theorem in the Plane

Given two
triangles in
the plane,
labeled ABC
and A'B'C'

1. These two triangles are situated in the
plane so that the straight lines joining
corresponding vertices A to A', B to B',
and C to C', meet in a point labeled O.

2. The corresponding sides of the triangles, if
extended, will meet at the same point.
CB and C'B' meet at point P.
CA and C'A' meet at point R.
BA and B'A' meet at point Q.

3. The above information has been
combined into one diagram. Notice that
points Q, R, and P all lie on the same line.

Before we examine the rules of the perspective game, let us look at a cube as presented in several views in order to recognize that what we see in two-dimensions is based upon what we know and experience from three-dimensions.

What we are using here is a particular drawing system. The cube we are looking at in Fig. 2.2 is placed directly in front of the viewer and is parallel to the viewer. It has been rotated a number of times and what is seen in the diagram are four different views of the same cube. The viewer perceives a cube but in actuality what is seen is two-dimensional and may be squares, trapezoids, or any other manner of quadrilateral. It should be obvious that there are many, rotations of this figure just in this position. Tilting the figure would change the rotation. Knowing about the symmetry of this figure would help also (Newman and Fowler).

Notice that the vertical lines all stay vertical as do the horizontals. What changes are the lines which in the actual cube are perpendicular to the verticals. These lines converge to an implied vanishing point on an implied horizon line. Since the cube can be rotated 360 ° before it returns to its initial position, it could be rotated only one degree each time. It could also be rotated to the left instead of to the right. It could rotate about a horizontal axis in the same manner or it could rotate about a diagonal one.

Construct a paper cube and insert a thin rod through the center for an axis. Set it up so that you can rotate the figure. Add strong lighting so that the cube casts shadows. Then try drawing it from a range of rotations using pencil or pen and the techniques suggested in the previous chapter. Make the drawing visually interesting as well as accurately drawn in terms of the perspective views. The cube is an essential building block of the perspective game.

This page:
Fig. 2.2
Four perspective views of a cube that has been rotated about a vertical axis. A face, ABCD, is parallel to the viewer.

A. Looking into the cube, its front face is transparent so that you can see into it.

B. Rotating the cube slightly to the right.
C. Rotating the cube still further to the right.

D. Looking at the back closed face which is not only a square.

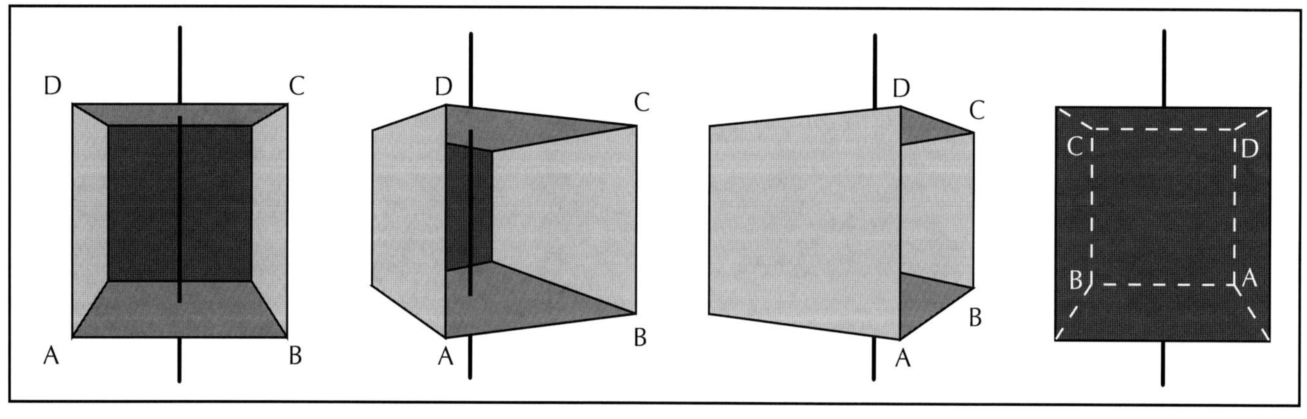

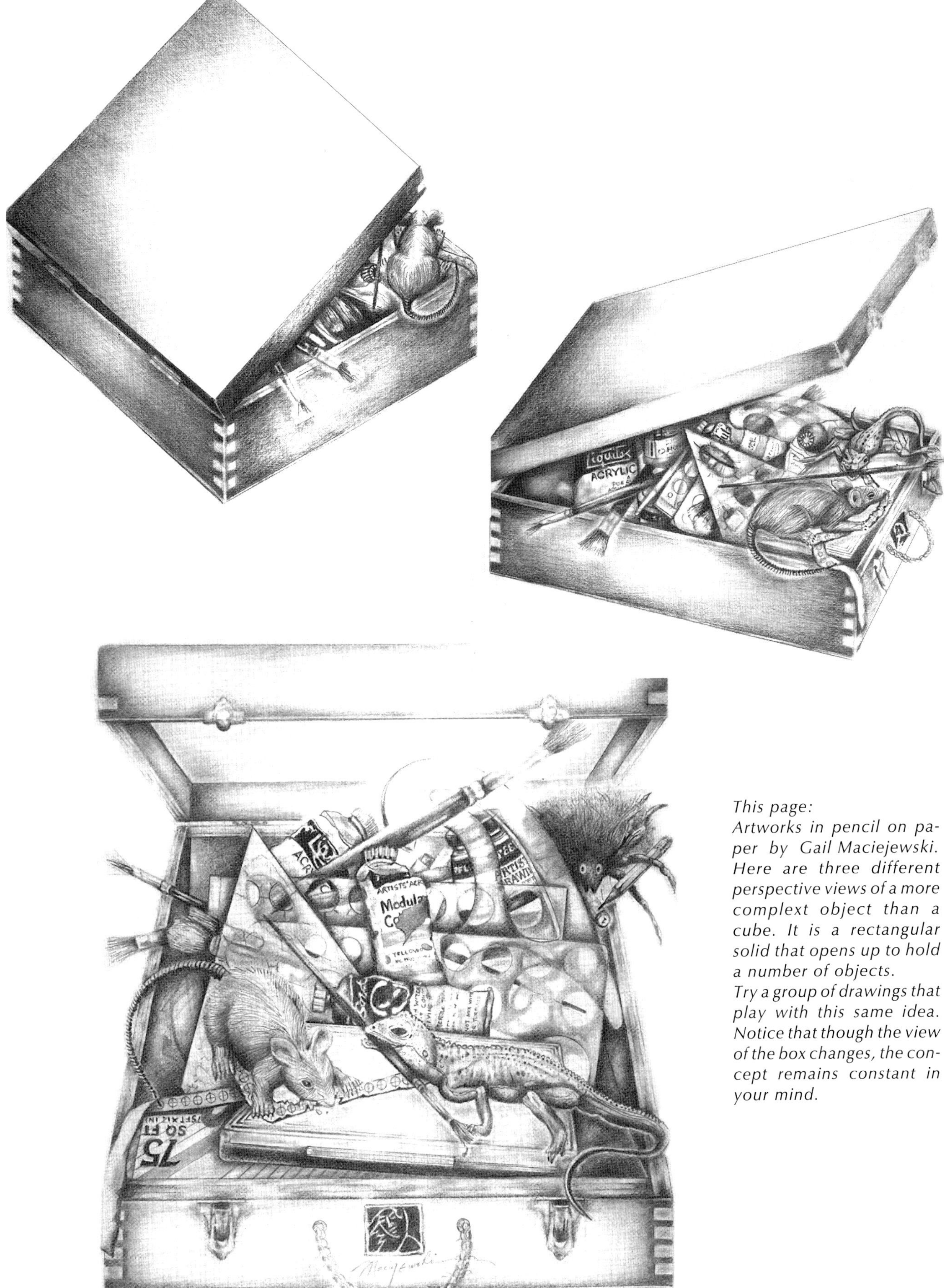

This page:
Artworks in pencil on paper by Gail Maciejewski. Here are three different perspective views of a more complex object than a cube. It is a rectangular solid that opens up to hold a number of objects.
Try a group of drawings that play with this same idea. Notice that though the view of the box changes, the concept remains constant in your mind.

Renaissance Optical Perspective and the Image Box

Art has multiple functions only one of which is to attempt to present what is seen of three-dimensional form and space onto a two-dimensional surface. There is a kind of prestidigitation in the skills required. Both the artist and viewer get pleasure from the experience. There is delight in knowing that a flat surface can also function as a deep space. A sheet of paper, a length of canvas, a section of wall, all offer a surface upon which to depict the events of three dimensions. Depending upon culture, time, or place, that depiction might range from extremely shallow to very deep space.

If one images a two-dimensional representation of a box with a transparent side parallel to the viewer, objects within that box can be arranged so that they appear to be situated in an actual space. The depth of the implied space can be controlled depending upon the intent of the artist.

If the threshold of the box is given some visual importance, the connection between the viewer and the viewed space is heightened. Renaissance perspective is one of the major tools for developing a sense of deep space under very specific viewing conditions.

The system for representing this illusion is sometimes known as central perspective, optical perspective, vanishing point perspective, but it is most often called Renaissance perspective because of the culture and time of its discovery. It has been the system most widely used in Western Art since the 14th and 15th centuries in Italy. It had its development in the city of Florence though information relating to it had been around much longer. The Renaissance looked back to the ideas of Classical Greece and Rome, ideas that had been dormant for many centuries. It was the reverence for the learning of the past with a sense of optimism for the present and a belief in the future that allowed Florence to take the lead in bringing about a rebirth of intellectual curiosity about the natural world. Despite its small size, this city took the initiative in the arts. To be an educated person meant to be learned in the wisdom, words, and ways of antiquity. Artists such as Piero della Francesca, Paolo Uccello, Filippo Brunelleschi, Giotto, Massaccio, Donatello, and Alberti gave visual life to an intellectual structure.

Some art historians believe that it was Alberti who was considered the father of modern perspective while others believe that it was discovered in 1425 by the architect/sculptor Brunelleschi. But it was Uccello who was in love with it and Leonardo da Vinci of Italy and Albrecht Dürer of the North country that pushed its use to the limits. Despite, or perhaps because of, the highly emotional aspect of the temperament of Italian artists, they respected reason and order. They sought harmony through the use of proportion within all things. It was a belief that even though trouble and travail exist, the world is a place that makes sense.

It was an interesting place to observe, to investigate, and to ask questions about. But it would be a world of ideal beauty, one in which nature was improved upon. Mathematics was seen as the key to unlocking the secrets of nature and the understanding of space. These values spread to the rest of Europe and began the shaping of our modern world, for good and for ill.

On this page are two variations of the same subject matter. Both were created by Renaissance artists; however, one is from the southern Renaissance and the other is Northern. Notice the difference in how the pictorial elements are used in both these images. Both make use of one-point perspective but the station point of the artist/viewer is shifted from center to left. In each work, there is an architectural threshold that invites you into the study space of St. Jerome. There is a back wall that keeps your eye from moving too far into the space of the work.

There is a checkerboard floor in one image; a common device that establishes units of measure for building all elements within the space. There is the singular figure of Jerome working at a desk. There are archways and windows; table and chairs; instruments of the scholar, and animals from the natural world set comfortably in the human environment. Each object that is used has both a literal meaning and a symbolic one, the latter may be hidden from the viewer unless he or she is versed in the symbolism of the times. Can you offer conjectures?

This page:
Two images of St. Jerome.

Top left:
Antonello da Messina. St. Jerome in His Study. c. 1475. Reproduced by courtesy of the Trustees, The National Gallery, London.

Top right:
Albrecht Dürer. St. Jerome in his Cell (engraving). 9.75" x 7.5". 1514. British Museum, London.
Dürer studied the principles of Italian perspective and wrote, in German, his own treatise on the subject.

This page:
Piero della Francesca. The Flagellation of Christ, ca. 1458 (Urbino, Galleria Nazionale dell March, Palazzo Ducale.(Copyright 1990 Archivi/Alinari/Art Resource, NY. All Rights Reserved.) Piero dell Francesca was both an artist and a mathematician. He wrote a well known text on the subject of perspective.

Place a piece of tracing paper over the above image. Can you locate and place the vanishing point, or points, and the horizon line? Can you estimate the height of each figure?

Though these Renaissance artworks depict objects and events with great reality, they have been built with a system which is based upon a great many assumptions. In fact, it is the connection between the mathematics of conic sections and optic reality that allows for the rationalization of visual data into a coherent and logical visual system. Because it plays by the "rules", we will introduce linear perspective as a game. In the next chapter, we will look more deeply into the concept of games, the Euclidean and the non-Euclidean ones.

Before learning about the game, try an analysis of the above painting. Lay a piece of tracing paper over this image. With a straightedge and pencil, draw in the diagonal lines that appear to converge at a vanishing point which is found behind one of the figures in the background. Which one? What do you notice about the vertical and horizontal compared to the diagonal ones. How tall, in your estimation, are each of the three gentleman at the front right or the four men in the background at the left? How large are the pavement squares found in the left middle section?

> *I define beauty to be a harmony of all the parts...fitted together with such proportion and connection, that nothing could be added, diminished, or altered, but for the worse.*
> Leon Battista Alberti, 1404–1472
> Renaissance Architect

This page:
Jan van Eyck. The Madonna with the Chancellor Rolin. c. 1435. Painting on panel. 26 x 24 1/2" (66 x 62 cm.) Louvre, Paris.
What do you notice about the elements of the composition in this painting in relation to the two works of St. Jerome found on page 61? What are the similarities? What are the differences? Can you locate the vanishing point within the work?

As the distance of the observer from the picture plane changes, the size of the objects change also., but they retain their proportions.
When viewing the painting, the distance from the picture plane to the viewer should be the same as it was for the artist in order for the image to appear at its most realistic.

The Eye of the Beholder

When you begin to play the perspective game, you will notice that the eye of the beholder controls all things as they appear at any given moment.The beholder stands still directly in front of an object/event in real physical space. Standing too far right or left would create too much distortion within the image. In the photograph below, however, you will notice that the viewer/photographer is standing a little to the left of the center.

This object/event is also without movement. The immediate is frozen in space and time. Nothing is in flux. The stationary eye flattens out all the elements. Objects appear to be two-dimensional when they are more than sixteen feet from the viewer. In a perspective image, the light is also held constant. This type of representation obviously leaves out a great deal of experience and, yet, it held the attention of artists and viewers for about 500 years. Why? What is it that we like about this particular system? The perspective game is based on a very obvious perceptual experience. Objects that are situated further away from the viewer appear smaller. Each of us knows, howevr, that objects do not acually change size just because they are moved further away from us. This size change is due to the fact that light rays travel to the eyes in a visual cone. The visual angle that is formed in this cone becomes narrower as the viewer moves away from the object. When objects are seen over a great distance, the angle becomes so small that all objects appear as a dot on a far off horizon line where the sky meets the earth. This dot is called the vanishing point.

When you read the rules of this perspective game, you will see that it uses two types of picture planes. The first type is the opaque drawing or painting surface that will act as the gameboard for the artist. It might have a lightly drawn grid on it in order to facilitate the transfer of the visual information that is found on the secound picture plane. This is situated between the artist/viewer and the object/event. It is an imaginary transparent plane that is set in front of, and parallel to, the object/event that is to be depicted. Think of it as a glass window that is flat, fixed, and immovable. It too can have a grid drawn upon its surface.

While keeping the head stationary, and with the assumption that there exists a single unblinking eye in the middle of the forehead, the artist traces the outlines of what is seen in the object/event onto the surface of the glass. In actuality, the glass is only a device that enables the artist to transfer the three dimensions of the physical world onto the plane in a scaled version. It is the drawing paper or canvas or wall on which this image is then actually constructed.

Try setting up a situation for yourself. Use a group of still life objects, a piece of window glass,or clear acetate paper, a permanent-black marker, pencil, and a drawing or painting surface.

> *Now it was they who, through illusion, could create environments more complete than architects could bring about in fact, and could populate them with more awesome images of human action than sculptors could easily realize. From this moment, therefore, the painter became the most powerful artistic force in the visual arts of Western civilization.*
>
> *Vincent Scully*
> *Architectural Historian*

This page: Photograph of the same garden arcade as in the previous image but from a different viewpoint. This format is vertical while the one on the previous page is horizontal. With enough slight shifts in viewpoint and a speed for showing these shifts, you end up with the motion picture, movies. If the movies can indicate movement and sound more effectively than the still camera, what should be the artistic task of the photographer?

The Body as the Basic Unit

The proportional relationships in a perspective image are predicated upon a given human figure with specific size dimensions. Therefore, we are taking the time to first develop a schematic human that has harmony of proportions based on a height of 7.5 heads, the head being the basi unit of measure. Classical Greek art emphasized an ideal where the body was 7.5 heads high and the navel was considered the Golden Cut of the entire body. See Design Appendix for further information.

One could also use a body that was 8 heads high or one that was five. The figure then becomes the module for all the elements in the perspective game which will be played out on the following pages. Essentially a figure has a front, back and side view. Each of these should be in proportion to the other.

We shall do the same for our schematic figure found on the facing page, Fig. 2.5 Any real world measurements can be substituted as you

shall see further on.

The head itself can be subdivided by the Golden Ratio so that its features are in harmony with the rest of the body. The Golden Rectangle, ratio of 1: 1.618 approx., can be the enclosing frame for the ideal face on the ideal head on the ideal body in an ideal world. The proportions of the head are very closely related to those of the hand. Fig. 2.4

Using the human body as the basis for developing a logical system of measurement has a long history. It can be seen in the figure analysis of Egyptian tomb painting and sculpture, architectural design in ancient Rome as well as in Twentieth Century architecture. Le Corbusier (Charles Edouard Jeanneret, 1887-1965), the international architect and painter brought the Golden Ratio into the Twentieth Century as the basis for design. He incorporated these proportions into most all of the buildings he designed using a system he developed called " Le Modulor".

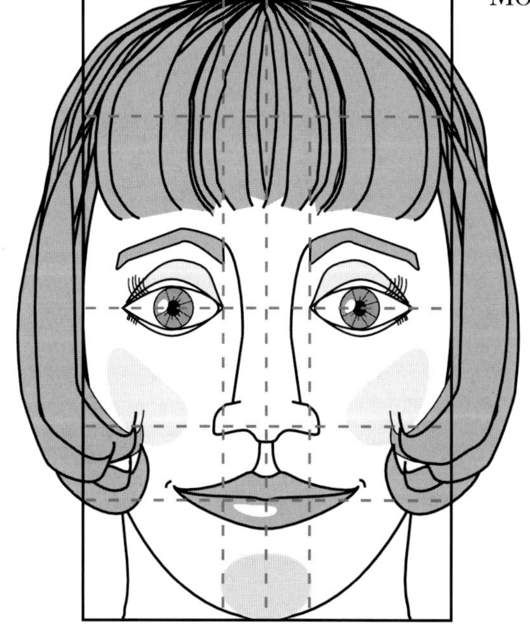

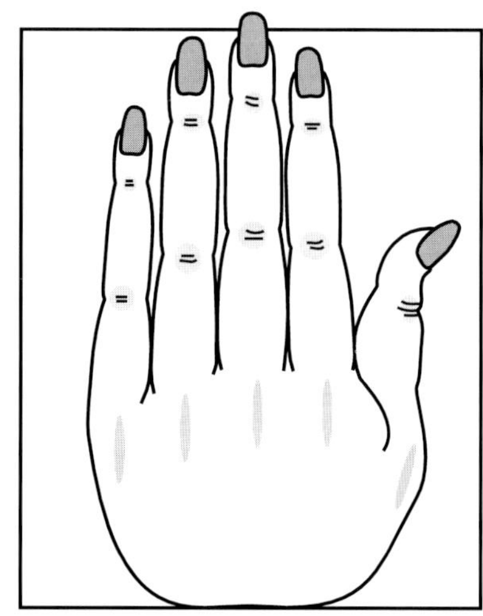

This system is based on the distance from the groundline to the navel (113 cm) of a 6 foot tall human (183 cm). When the arm is raised, it reaches to 226 cm. Within Le Corbusier's system, the height of a room would be twice the height of this figure. If the 113 cm were divided by the Golden Ratio, one would obtain 70 cm which would be used for the height of a table. The height of a chair would be 43 cm. Other divisions would become the measure for a countertop or cabinet placement for the sizes for window openings.

Below we show a human figure, 7.5 heads in height, 3 heads in width, and 1 head in depth as seen in the front, side, and back views of Fig. 2.5

It is also divided into thirds, a braccio in Italian units, in keeping with the way Alberti, the Renaissance, architect, developed figures for perspective images. The braccio becomes the unit length measure used for establishing the divisions for a checkerboard floor grid. It also aids in establishing the scale for the architectual elements within a particular image.

This page: Fig. 2.5
A schematic rendering of the human figure in front, side, and back views. Key divisions are indicated.
The Greek Canon of proportion for the body proposed by the sculptor Polyclitus state:
a. The figure is 7. 5 heads
b. The foot is 3 times the palm of the hand
c. The lower leg is 6 times the palm of the hand
d. The knee to the center of the abdomen is 6 times the palm of the hand.

Measure your own palm and see if this system holds true.

Byzantine art, on the other hand, utilized a body which was 9 heads tall.

Given a human figure that is 8 heads tall, develop a schematic diagram with all the components indicated.

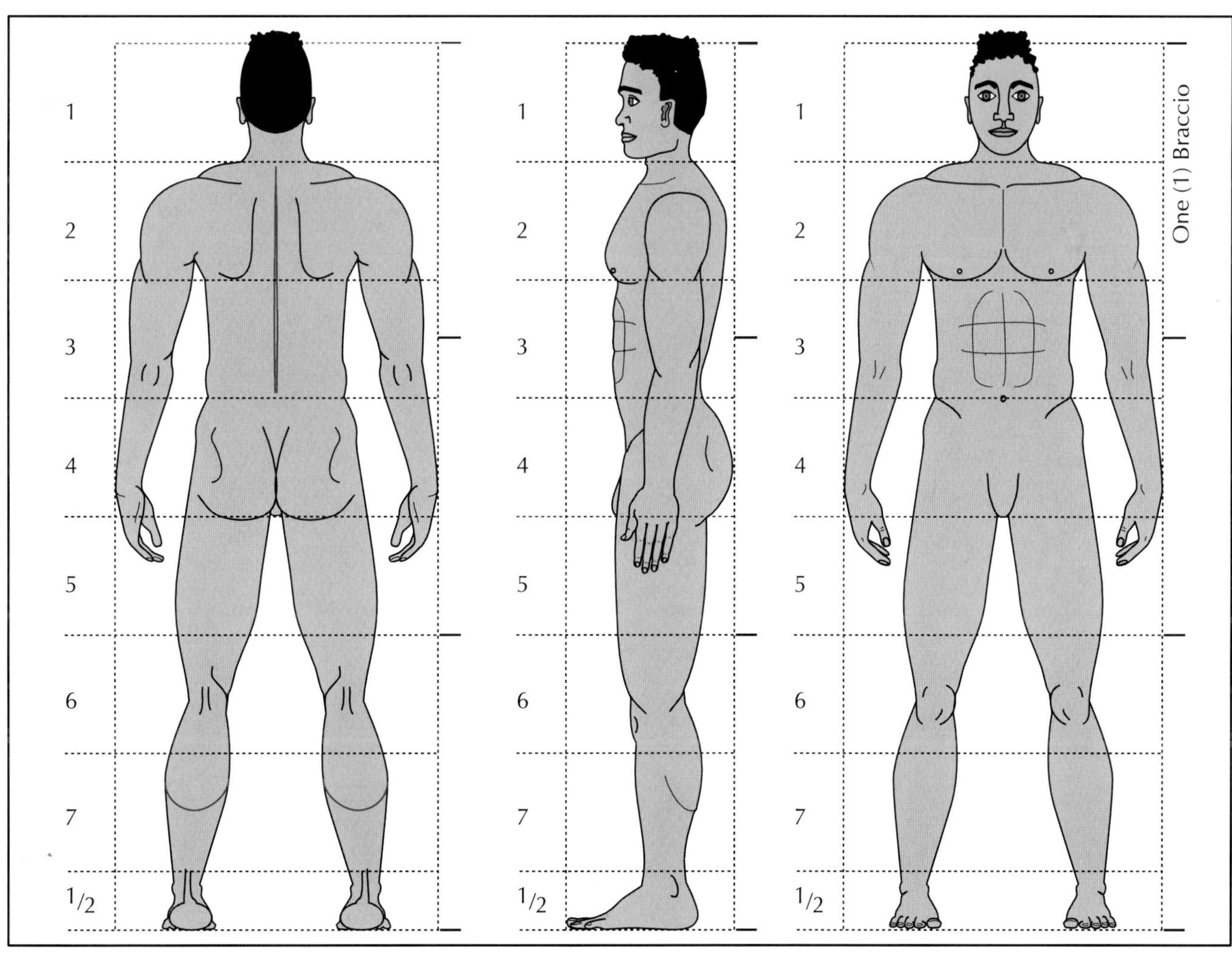

Rectangles and Rectangular Solids for Floor Plans and Rooms

This page:
Chart of Rectangular Solids

The proportions of rooms should have a direct relationship to the proportions of the human body. Here are some floor plans and their resultant rectangular solids that can be used for designing rooms.
We begin with the unit square (1) and generate the other rectangles from it. The width of each rectangle is always unit length 1.
Fibonacci numbers have a direct connection to the Golden Ratio especially after the fifteenth term. See Design Appendix.
The square and the cube. The ratio is 1:1.

The double square, ratio 1:2.

The 2:3 rectangle is a pair of Fibonacci numbers.

The 3:4 rectangle.

The 3:5 rectangle is also a pair of Fibonacci numbers.

Notice the use of the isometric projection system for representing the several solids.

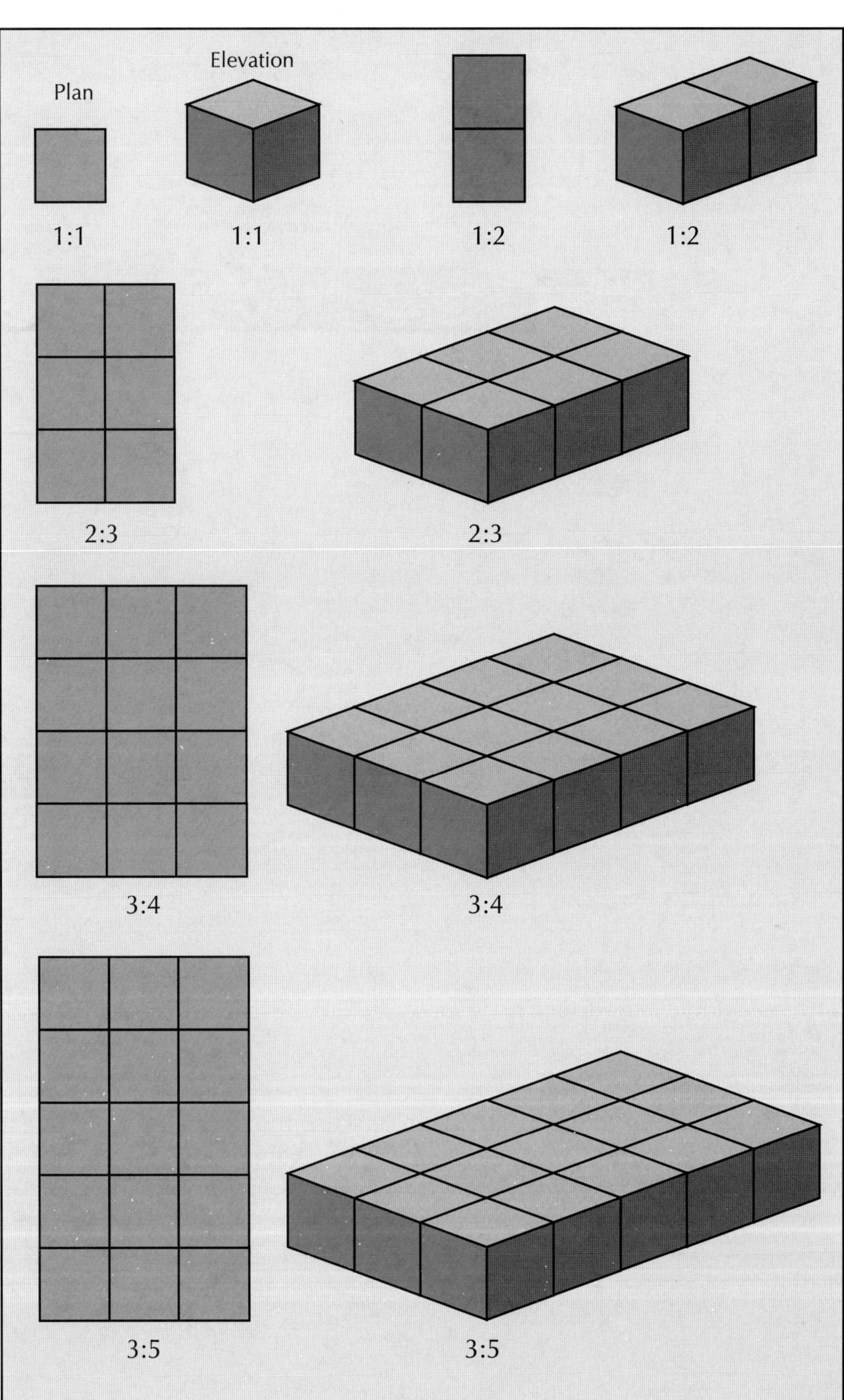

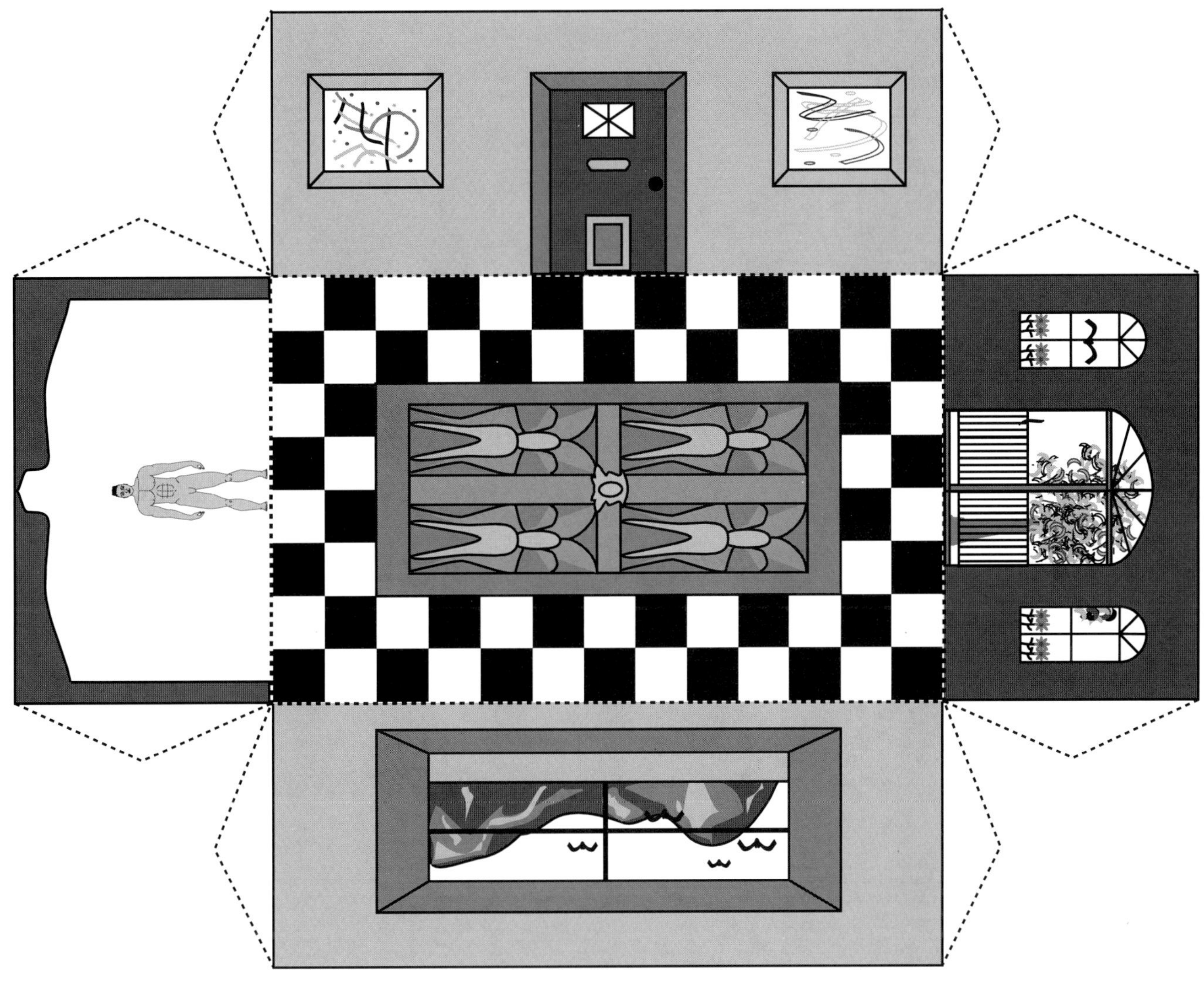

The Golden Room

In way of an introduction to the perspective game, we offer a template to build a small three-dimensional paper model room which has harmonic proportions related to the Golden Rectangle.

We are continuing to use whole numbers from the Fibonacci Sequence. The "Golden Room", as we shall call it, is 8 units wide, 5 units hight and 13 units deep. In this room, we indicate a checkerboard floor, windows, and doors that will relate to the perspective rendering of the same space.

The human figure who is three braccia high, 69" in this case, standing in the archway, determines all the proportional relationships within this particular room. Try an analysis of the rooms in your house.

Turn the above into a mini doll house. Adhere a photocopy of the template to illustration or fomecore board.Add other rooms, put a roof on, make a garden terrace, etc. Build furniture by using simple combinations of cubes and defining the limits of the object within the cube or groups of cubes.

This page: Fig. 2.6
Above is a template for a "Golden Room".
Score lines and fold up into a rectangular solid. We have left off the "ceiling" rectangle so that youcan see into the room. Add a celing and build a second floor onto this one. Construct a roof based on the same proportional system used elsewhere.

The Renaissance Perspective Game

Object of the Game:

a. For the artist.
   To make a convincing illusion of a three- dimensional object/event on a two-dimensional surface so that the illusion becomes virtual reality.

b. For the viewer.
   To believe that the reconstruction on the two-dimensional surface is the three-dimensional reality. To obtain the effect that the artist wishes to create, the viewer should be stationed in the same position, and at the same eye level, as that of the artist when he or she created the work.
   There are no winners, or losers, in the game, only participants.

Number of Players:

The game is played by one or two people at any given time. Player #1, the artist, begins the game and Player #2, the viewer, completes the game. At the onset, Player #1 and Player #2 are essentially one and the same, the artist/viewer.

Gameboard:

A bounded portion of the Euclidean plane. This surface is flat. It can be a piece of paper or canvas, a section of wall, or a board. What do you think would happen if the surface were curved, crumpled or twisted in a variety of ways?

Gamepieces:

Drawing or painting tools such as pencils, brushes, pens, inks, paints, etc.

Game Rules:

First, we shall give a written synopsis of the rules . Then the game shall be played out visually with a comparison between an isometric view of an interior and a Renaissance perspective view of the same interior.

1. There is a perceived 3-d object/event that is to be transferred onto the two-dimensional surface of the gameboard. This surface is considered to be parallel to the event. The object of the game is to construct an illusion of a three-dimensional architectural space on the gameboard. Various objects will be placed within that space.

2. Between the object/ event and the player is an imagined transparent surface on which the light rays that fall on the object/ event in 3-space become the points marked on this transparent surface. This set of lines is called a projection. The accumulation of points becomes the image. The transparent surface is at right angles to the ground plane on which the player is standing. The information obtained from this surface is what is marked on the gameboard. To facilitate the process a square grid is drawn lightly on both the transparent plane and the gameboard.

3. In the actual event, parallel lines never meet. On the gameboard, all horizontal lines and all vertical lines remain horizontal and vertical. All lines that appear to move away from the viewer, however, converge to a single vanishing point on the horizon line on the gameboard. This point is dependent upon the fixed position of the eye of the player. If the object has one face parallel to the player, there is a need for only one vanishing point.

4. The player has a given height which determines all the measurements on a particular gameboard. The height of the player could vary from game to game. All other human figures within the space are in relation to the player. The units of measure placed along the baseline and along the two sides of the gameboard are related to the height of the player. They are determined at the beginning of the game and influence the relative size of all objects on the gameboard. They are considered to be the "true" measurements. The size of each object diminishes in exact proportion to its distance from the player. The player selects the station point which is the place and distance from which the viewer sees the event. This station point determines the eye level which, in turn, determines the placement of the horizon line which is always at eye level. The eye level would shift if the player were seated, or kneeling, or laying down, or climbing a ladder. Eye level is perpendicular from the eye of the player to the gameboard. It defines the principle vanishing point. The center of vision should be close to the center of interest of the image.

The player is standing at a particular distance from the event. The farther away from the event, the more of it is seen. The closer to the event, less is seen. The player is considered to be standing still in a fixed position in relation to all the objects/events that are also considered to be standing still. The light and atmosphere as well as the shadows, are also considered fixed. Nothing is in flux.

5. The player is considered to have monocular vision (one eye) like the lens of a 35mm camera. In fact, we humans have binocular vision (two eyes) which means that each eye has its own cone of vision and that these two cones overlap to give a sense of depth. In this game, the head is held immobile with no ability for peripheral vision.

6. The cube is taken to be the basic building block for all objects drawn on the gameboard. It becomes the unit measure of length, area, and volume within the image. It can be stacked, subdivided, and it can provide the structure in which to build more complex forms such as pyramid, cones, cylinders, and spheres.

The face of the cube is considered to be parallel to the picture plane, and thus, the viewer. There is no distortion or foreshortening of objects when they are parallel to the picture plane.

The size of the objects within the space are relative to the size of the player and they diminish in size in exact proportion to their distance from the player. The more distant an object, the smaller it will appear. Renaissance perspective used the inverse size-to-distance law. If the distance to an object is doubled, its represented size is one-half the actual size. If the distance is tripled, its represented size is one-third. If the distance is quadrupled, the represented size is one-fourth, etc. A checkerboard floor which has units that are one-third the height of the player becomes a grid for measuring objects.

7. All the objects maintain a uniform clarity of form throughout the image regardless of whether or not they are close to the player. This does not hold true for seeing objects in the natural world since sight is influenced by the atmosphere and air surrounding them. In the actual event, light may vary as it falls on objects. On the gameboard, through the use of tonality and color, there is a uniform light consistently applied to all objects. This helps to emphasize their depth and solidity. There is a judicious use of cast and reflected shadows since one of the goals of the artwork is to provide a sense of three-dimensional solidity everywhere.

Playing The Perspective Game

Oblique Projection Box

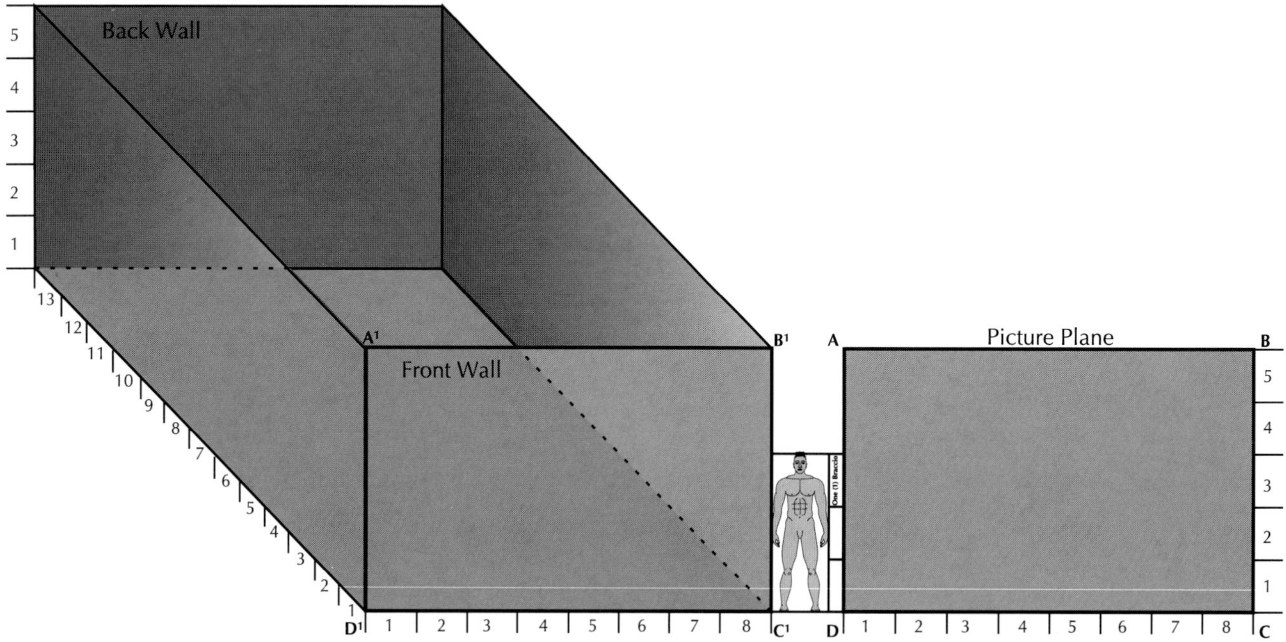

The Oblique View

1. Given a 5 unit x 8 unit x 13 unit oblique projection "box".
For the box dimensions, we have used numbers in the Fibonacci Sequence. These are whole numbers that relate to the Golden Ratio.
Each unit will be used as a braccio. Since our figure is 69" tall, one braccio equals 23". given this information what are the dimensions of this room in inches and in feet?

2. The oblique length of the box has been adjusted so that it appears "correct" to the eye. It has been shortened by one-half the length of the oblique line because of the use of the 45° angle in constructing the box (45/90° = 1/2).

If the box had contained a 60° angle, the length of the oblique line would have to be reduced by 2/3 (60/90° = 2/3).

Renaissance Perspective View

1. Given the face of the 5 unit x 8 unit perspective "window", the picture plane.
It is the same as the face of the oblique projection box. Rectangle ABCD is equivalent to Rectangle A'B'C'D'.

The figure between the two systems represented here is 3 braccia high. One braccio is the unit measure.
Units have been marked along both the baseline and heightline.

Setting the Horizon Line and the Vanishing Point

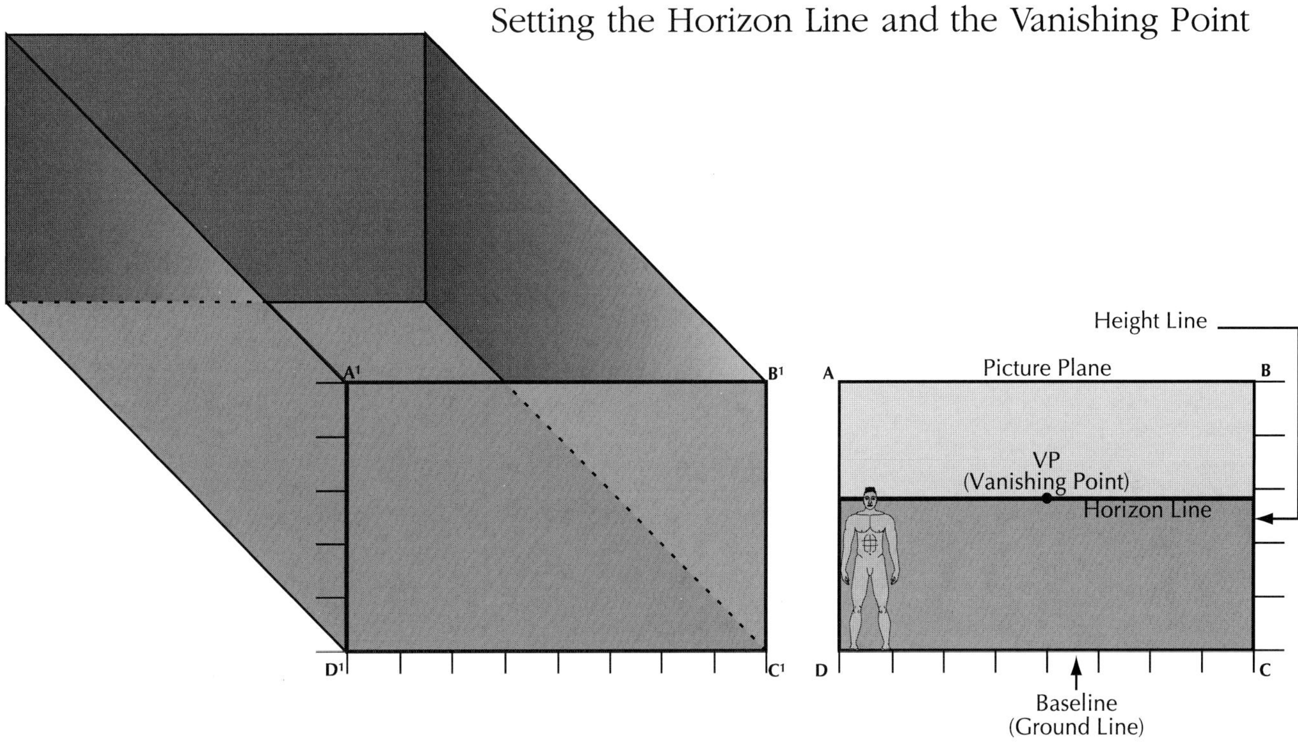

The Oblique View

2. There is no horizon line and no vanishing point when using the oblique projection box.

Renaissance Perspective View

2 The rectangle is called the Picture Plane and is considered to be parallel to the artist/viewer.

The person is 3 units tall. One unit becomes the measuring unit for the Baseline and the Heightline. The Baseline is divided into 8 units. It is sometimes called the groundline. The Heightline is divided into 5 units. HL = Horizon Line. It is at the eye level of the person. The upper portion of the Picture Plane is considered the "sky" area and the lower portion the "ground". The horizon line is located where the sky and the ground meet.

VP = Vanishing Point. In this case, it is located at the midpoint of the horizon line. It could be located left or right of center. Try developing another diagram where the VP is moved. Then follow the rules of the game using this new gameboard. Compare results from both of these versions.

Establishing the Grid

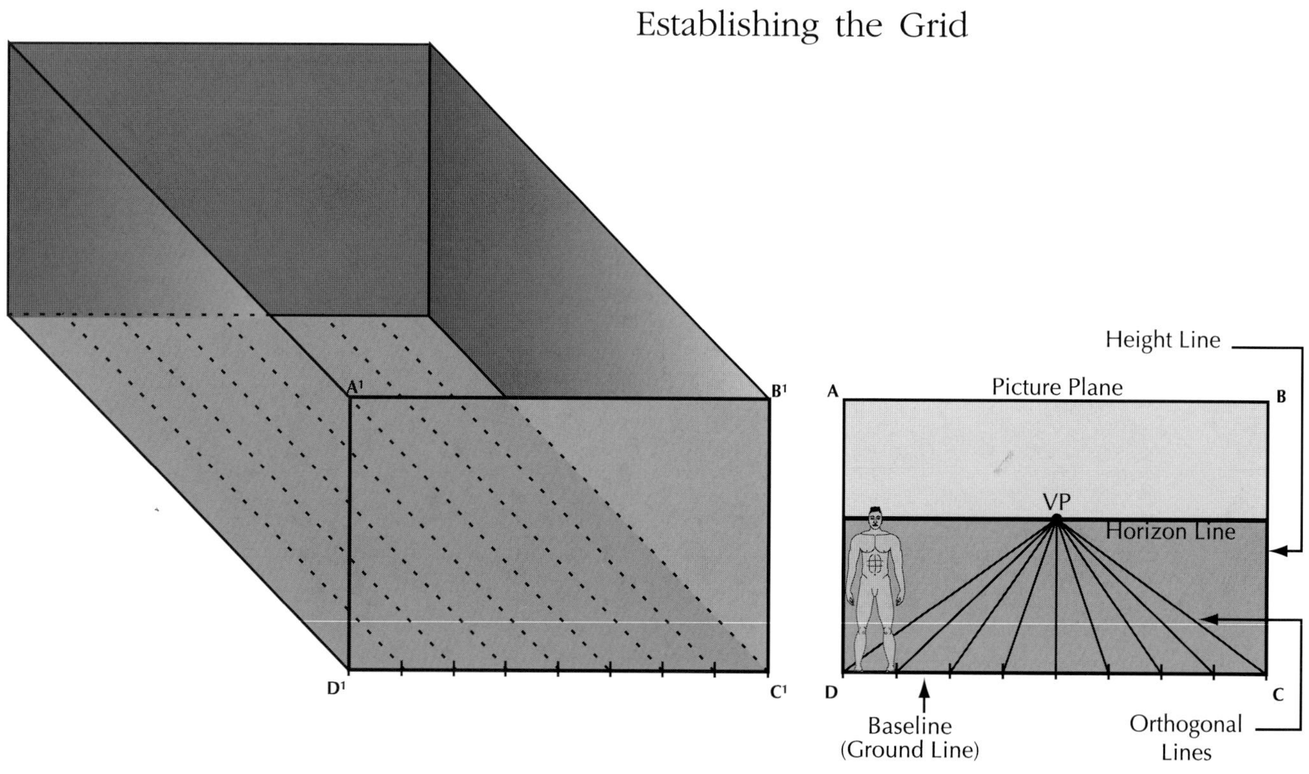

The Oblique View

3. Some of the walls of the room are to be considered transparent.
All measurements are true.
Nothing diminishes in size. There are no proportional changes.
All horizontal lines stay horizontal.
All vertical lines stay vertical.
All parallel lines stay parallel.
There are no converging diagonals.

Renaissance Perspective View

3. All measurements are in relation to the standing person.
All horizontal lines stay horizontal.
All vertical lines stay vertical.
All lines that are perpendicular to the player, converge on the horizon line. They are called orthogonal lines.
The baseline and the heightline are used to established the "true" measurements needed for the floor grid. The baseline may need to extend beyond the frame of the rectangle in order to obtain additional placement points for the floor grid.

Artists understood that in order to create a sense of depth, one or more points at infinity, which is the horizon line, had to be established. These points all had to lie on a single line. Mathematicians stated that each line in the plane had an 'ideal' point at infinity.
Any two parallel lines are assumed to intersect at their common point at infinity, the vanishing point.

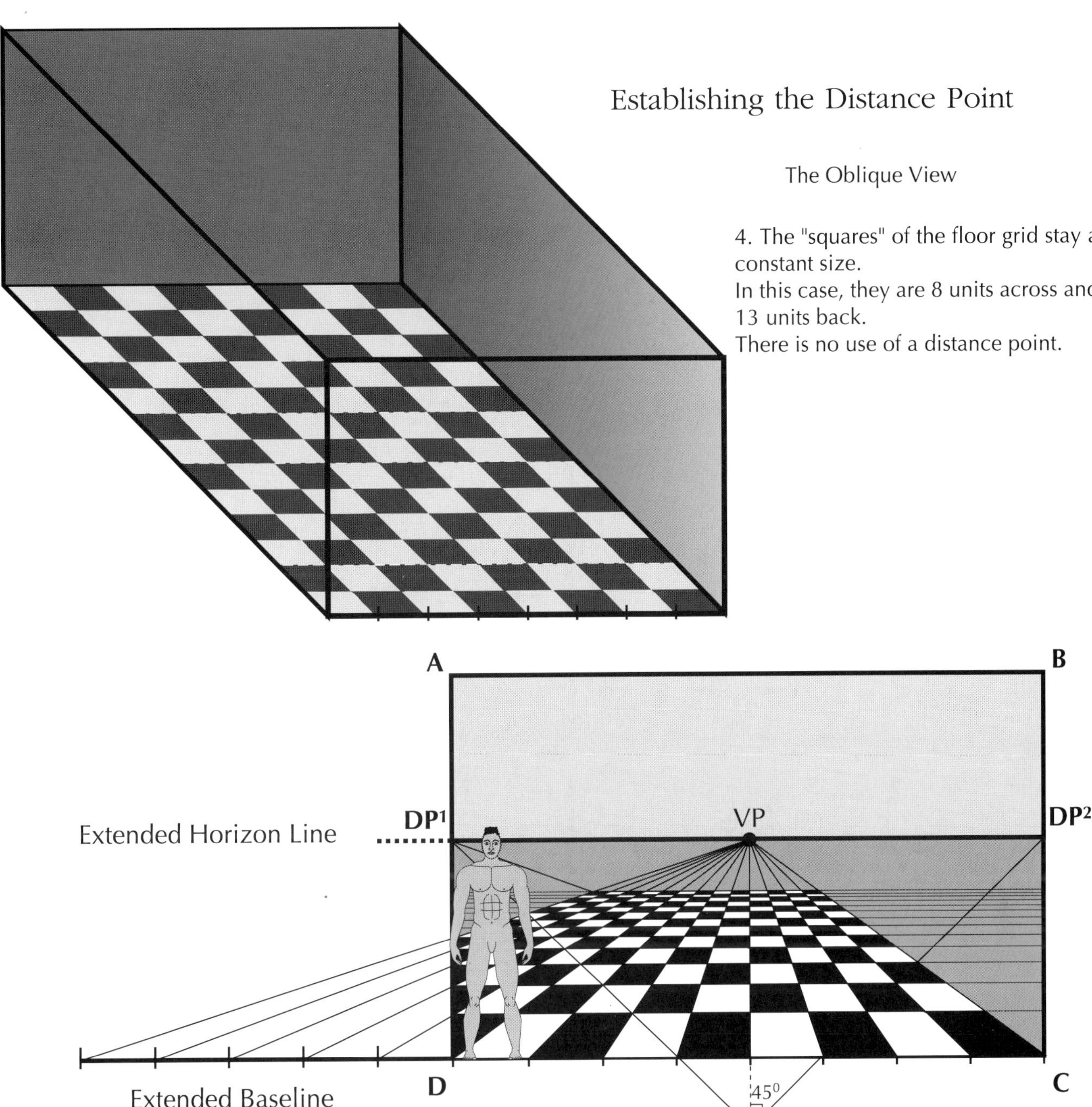

Establishing the Distance Point

The Oblique View

4. The "squares" of the floor grid stay a constant size.
In this case, they are 8 units across and 13 units back.
There is no use of a distance point.

Renaissance Perspective View

4. After establishing the horizon line, the VP, the baseline and the heightline divisions, and the orthogonal lines that converge at the vanishing point, the two Distance Points (DP¹ and DP²) need to be established. There are several ways to do this. Our method is as follows:
a. DP¹ and DP² are placed on extensions of the horizon line equidistant from the vanishing point. These are related to the Station Point (SP) of the artist/viewer.
The SP can be moved closer or further away from the picture plane. DP¹: VP is the same distance as VP: DP². SP:VP is the same distance as the VP is to the distance points.
Regardless of how far the SP or the DP are, the required angle is always 45°.

b. Where line segments E DP¹ and E DP² cross, the orthogonal lines become the points that establish the floor blocks.

c. Since the depth of our perspective room requires 13 units deep, additional points have to be marked on one side or the other along the extended baseline. Here we have extended the baseline to the left of the picture plane.

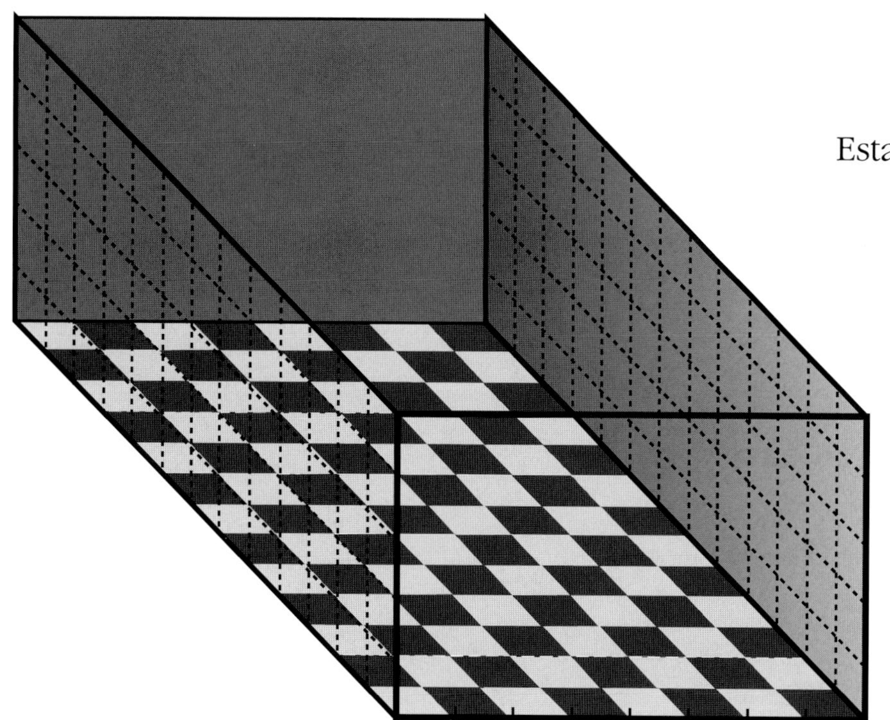

Establishing the Room Walls

The Oblique View

5. The walls and ceiling have already been established at the beginning.The checkerboard floor is in place. Equidistant lines are marked on the side walls.

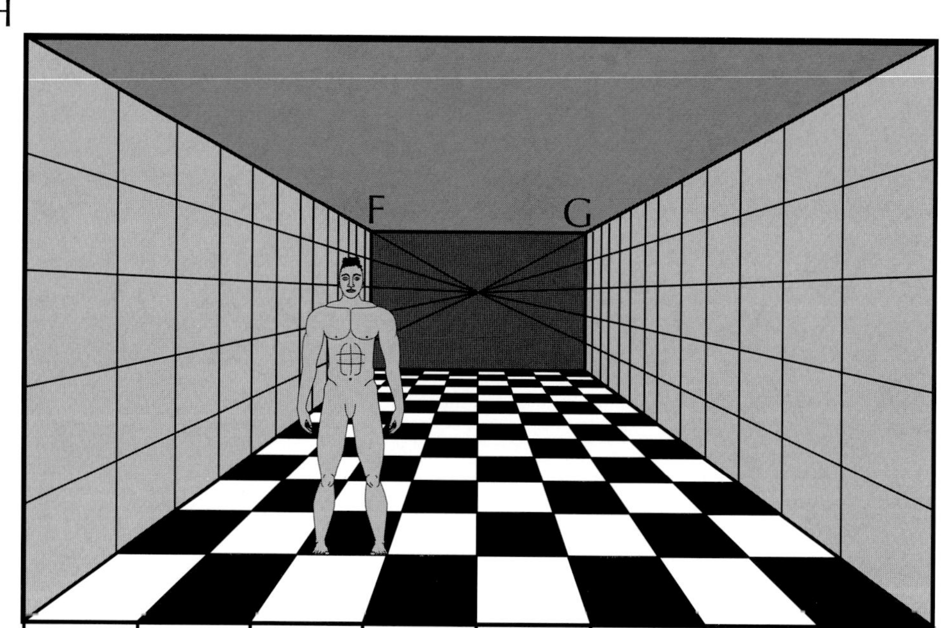

Renaissance Perspective View

5. To establish the side walls, back wall, and ceiling perpendicular and horizontal lines have to be placed accordingly.
a. Along the points on the heightlines of both sides, draw diagonal lines to the VP. These establish the horizontal lines for the wall grid.
b. Along the points where the floor grid meets the walls, erect perpendicular lines. These establilsh the vertical lines for the wall grid which, in turn, allows for the accurate development of wall elements, furniture placement, etc.

c. On the floor grid, count 13 units back to locate the back wall. Count 5 units up and draw in the horizontal line to mark the top of the back wall.
d. At the top two corners of the back wall, F and G, draw lines through these to the VP and to the top corners of the picture plane, HI, in order to set the tops of the side walls.

Establishing Placement of Doors and Windows

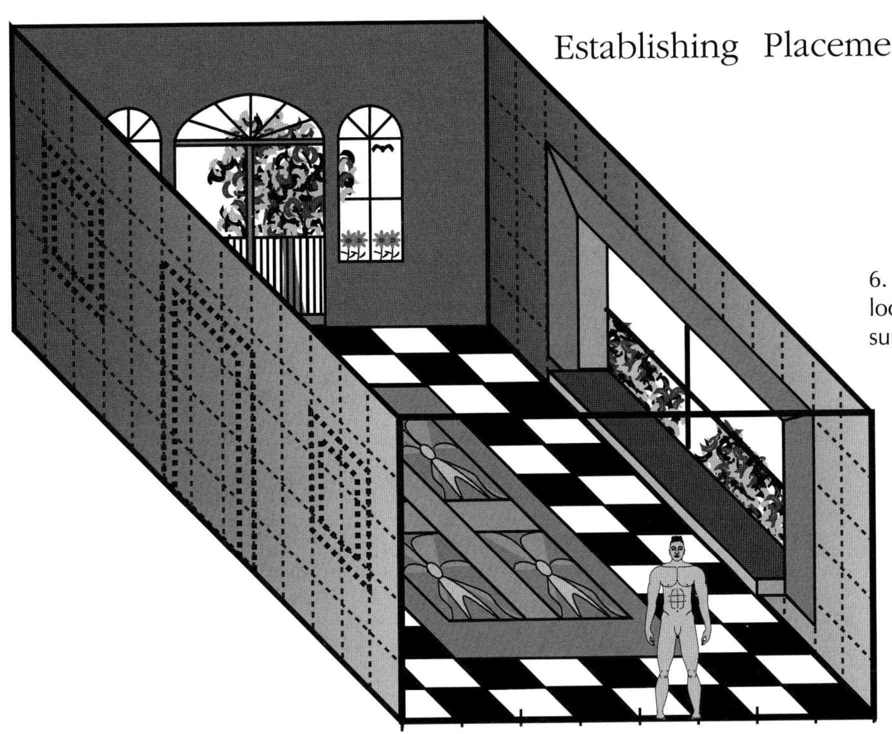

The Oblique View

6. The doors, windows, etc. can now be located and establshed with true measurements.

Renaissance Perspective View

6. To establish the accurate lengths and widths within the room space, you need to use the grid units on the floor grid within the space, as well as the units along both the baseline and the height-line.

a. Ascertain real room measurements. Then translate those measurements into units or parts of units.

b. Determine the location for these elements and lightly mark them on the wall grids.

In our diagram, there are the following elements:

1. On back wall, there are two windows and a door leading out onto a garden are.

2. On the right side wall, there is a picture window and a window seat.

3. On the left side wall, there are paintings on either side of the door.

4. There is a light fixture located in the center of the ceiling.

Placing Objects Within the Room

The Oblique View

6. To place objects within this room, locate their positions and determine their size and shape and set within a basic cube unit or units.

Renaissance Perspective View

7. To establish the accurate lengths and widths of objects within the room space, you need to use the units on the floor grid within the spaces as well as the units along the side and back walls.

a. Ascertain the real room measurements of objects that you would like to include within your perspective room. Translate those measurements into basic cubes, stacks of cubes- Develop furniture elements within these cube units.

b. Determine the location for these cubes and lightly mark them on the wall and floor grids.

In our diagram there are the following object:

    1. An oriental rug on the floor.

    2. A ceiling light fixture.

    c. A couch, side tables, lamps, a book shelf, a table on the back left wall.

The Completed Room

Renaissance Perspective View

An archway defines the completed room. Photocopy this image, or use a tracing paper overlay, and place additional objects within the space. Develop an elaborate tiling pattern within the black and white floor area. Look at paintings throughout history for inspiration. Give your image color.

Turn the image into a surreal space by placing out-of-context objects of various sizes within this "normal space". Research the paintings of the Twentieth Century Belgian artist Renee Magritte in order to see how to do this well.

This page:
Student one-point perspective drawings using a high contrast approach. The drawings were first done in pencil. They were then transferred onto scratchboard on which the white areas were removed from the black surface.

Top right: Joe Hastings

Middle left: Dionisio Genao

Lower right: Mary Calnan

Are you able to locate the central vanishing point in each of these drawings?

When a person or an object is seen in a foreshortened position, the drawing needs to show that the portion of the object that is closer to the viewer appears larger than the parts that are further away and these diminish proportionately. Both the diagram in Fig. 2.7 and the imaginative drawing on the following page, show a reclining figure in a foreshortened position.

In the diagram, the unforeshortened profile figure is placed horizontally with the feet touching the edge of the gameboard. (The figure could be of any height or width.) This figure is set within a rectangle which is subdivided vertically into any number of units that mark key points on the body. In this diagram, we have maintained the 7.5 heads that were set previously. We have also divided the figure horizontally into fourths. The feet touch one edge of the rectangle and the head touches the other. Notice that in the diagram key points are found at the neck, the hands, the feet, etc.

From the vertical division marks on the rectangle of the reclining figure and the figure itself, draw lines that meet at the vanishing point indicated with the picture plane rectangle. The points where these lines intersect the edge of the picture plane determine horizontal lines which give some necessary divisions of the foreshortened figure.

Notice that the feet fill the unit entirely. Other parts of the body, however, occupy less area within a unit. In order to locate the correct place for the foreshortened section of the figure, first ascertain which unit you are working within. Next, determine how far up within that unit a section of the body lies. Example: The ankle lies within the half units but takes up less than half the length of the unit. Therefore, within the first unit of the foreshortened figure, locate a point less than halfway up to set the position of the ankle.

This page: Fig. 2.7
Schematic diagram showing a figure in a foreshortened position. The portion of the figure closest to the artist/ viewer appears to be larger than the rest of the figure.

Find a friend who will stretch out on the floor or a bed and attempt to draw the fore- shortened view. Imagine a rectangle surrounding the body in order to "eyeball" where the various body parts would set within it.
Use a collection fo objects instead of a person and attempt the same kind of drawing.

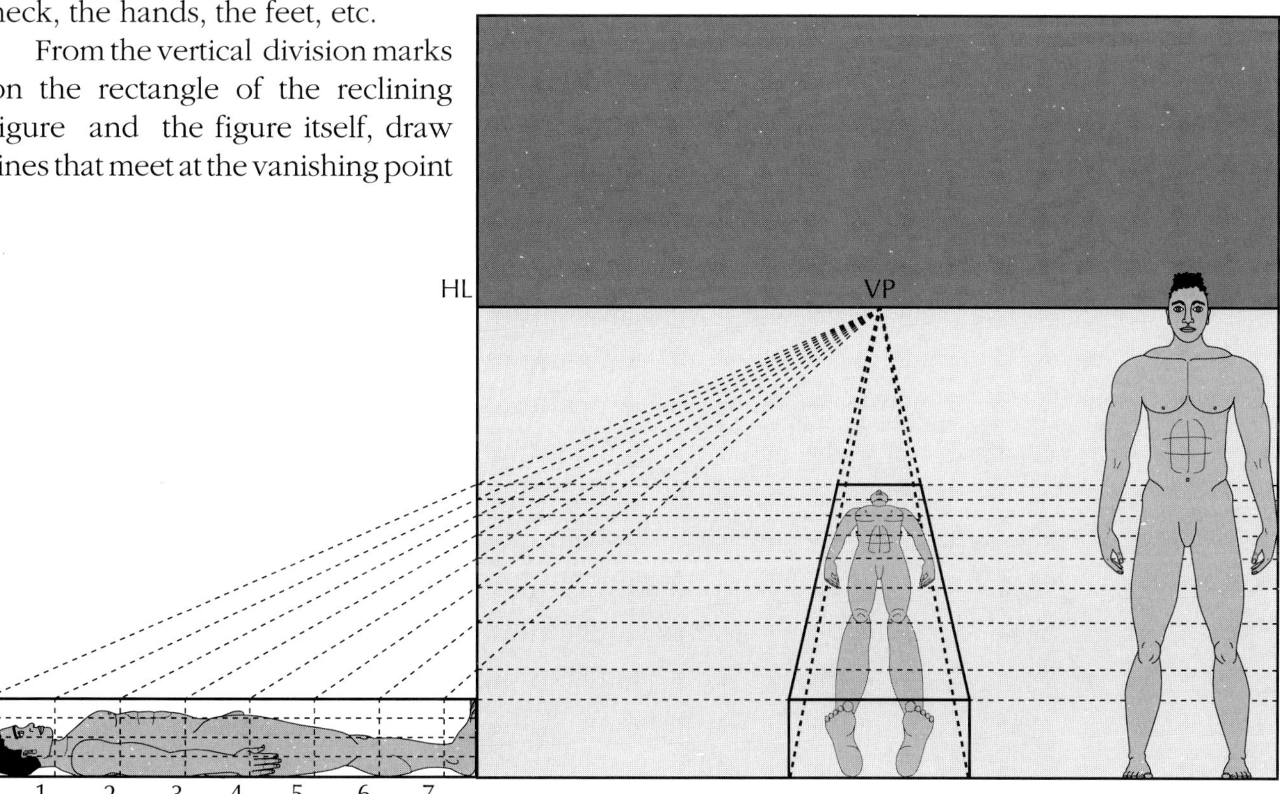

This page:
Right:
One-point perspective work by the student artist Bob Dumas. Monastery Interior. Pencil on illustration board.

Bottom:
Gail Maciejewski. Medusa No Longer Needs Her Wingtips.
This work is from her series of self-portraits using the myth of Medusa as the starting point. Pen and ink on bristol paper. This drawing is based loosely on an architectural engraving of Dutch 16th Century artist Jan Vredeman de Vries.

Try a drawing of your own using a self-portrait in a foreshortened view. Place the figure in an architectural setting that you have first researched from a period of art history.

Although it is a small country, Holland has produced some of the quintessential artists in the history of western art. Think of M.C. Escher, Piet Mondrian, Vincent Van Gogh, Jacob Ruisdael, Vermeer, and Rembrandt.

Vermeer, whose life remains a mystery and whose works lay unrecognized for almost two hundred years, is an artist who used light and perspective to construct paintings of quiet domestic interiors. The one or two participants within the painted space imply meaning beyond that of cool light, and cool color upon patterned and textured surface. Vermeer presents the viewer with moments of personal intimacy and serene silences.

There seem to be no existing drawings to tell us if the artist used preliminary studies in the construction of his paintings, or if he was involved with the use of lenses and the science of optics, or the mathematics of perspective. The thorough research by scholar Philip Steadman suggests that Vermeer may have used a device called a "camera obscura".

The few Vermeer paintings that are in existence are held together by a strong sense of an underlying architecture. George Bouleau in his book, *The Painter's Secret Geometry,* discusses in some detail the use of Golden Ratio subdivisions within the painting shown here. For further information on this ratio, see the Design Appendix.

We have looked most closely at one-point perspective within the Renaissance perspective system; however, an object, such as a cube, may be placed within the image space so that only one set of parallel edges is parallel to the picture plane, in which case two vanishing points are required. If one of the three sets of parallel edges of a cube are parallel to the picture plane, the drawing requires the use of three vanishing points. An image may also contain a number of objects each with different vanishing points. So the game can get more and more complex. The trick is to maintain a sense of visual unity and clarity. On this and the following pages, we show photographs of each of these three systems.

Attempt a drawing of an imaginary space in which you include all three types of perspective. Include areas of high contrast, tone changes, and color. Look at the works of lTwentieth Century Dutch artist, M.C. Escher, to see how he manipulates perspectives for very unusual intents. Do an analysis of one of his works to see if you can figure out his perspective techniques.

These next four pages: Photographs and diagrams of one-point, two-point, and three-point perspective, respectively.

Above: Photograph of a bridge, with shadows, that shows one-point perspective. Can you locate the horizon and the vanishing point?

Right: Fig. 2.8. Diagram of one-point perspective.

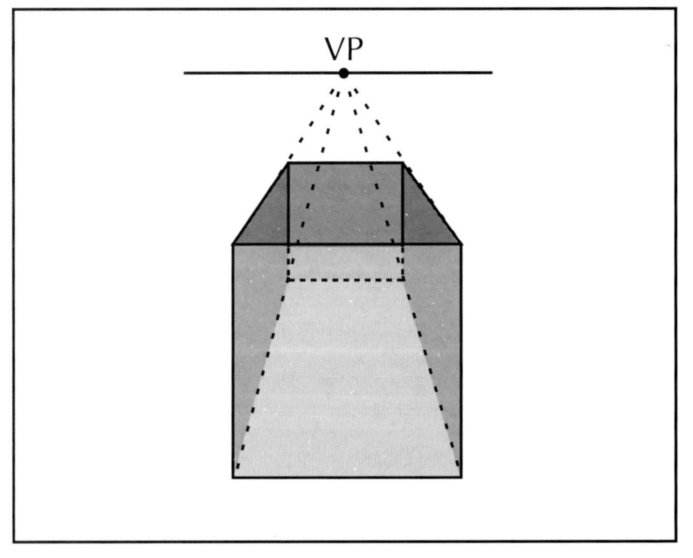

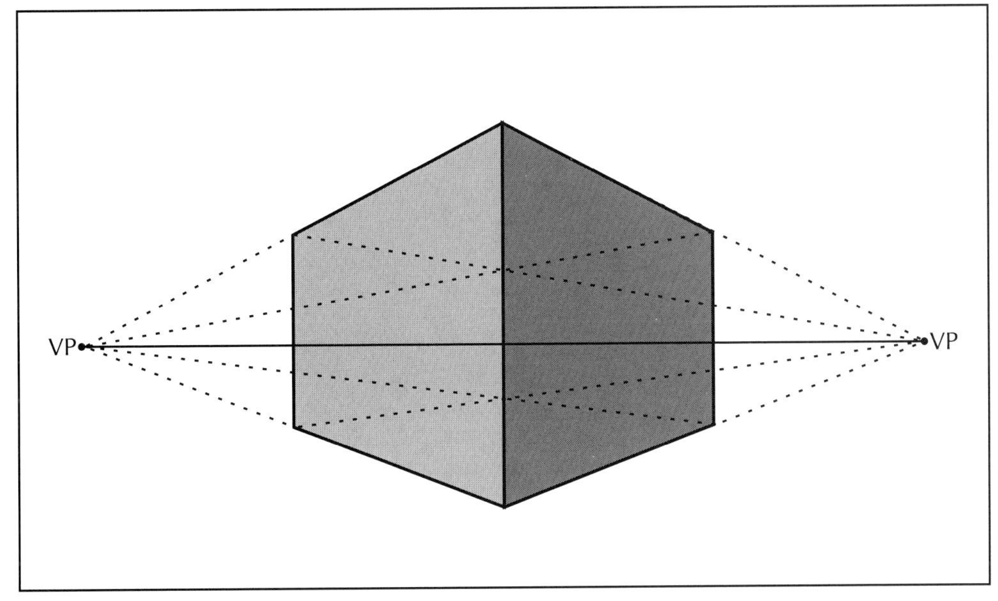

This page:
Above: Fig. 2.9
Diagram of two point perspective. Looking at the edge of the block requires the use of two vanishing points, both of which are located on the same horizon line.

Below:
Photograph of a parking garage showing two-point perspective.

Above:
Photograph of a house show-
ing three-point perspective as
seen from below looking up
at the structure. A vantage
point is assumed to be far be-
low or far above the viewer.
There are no parallel lines. All
lines converge to one of the
three vanishing points. Two of
them are located on the hori-
zon line and one is located
above or below that horizon
line.

Right: Fig. 2.10
Diagram showing three-point
perspective.

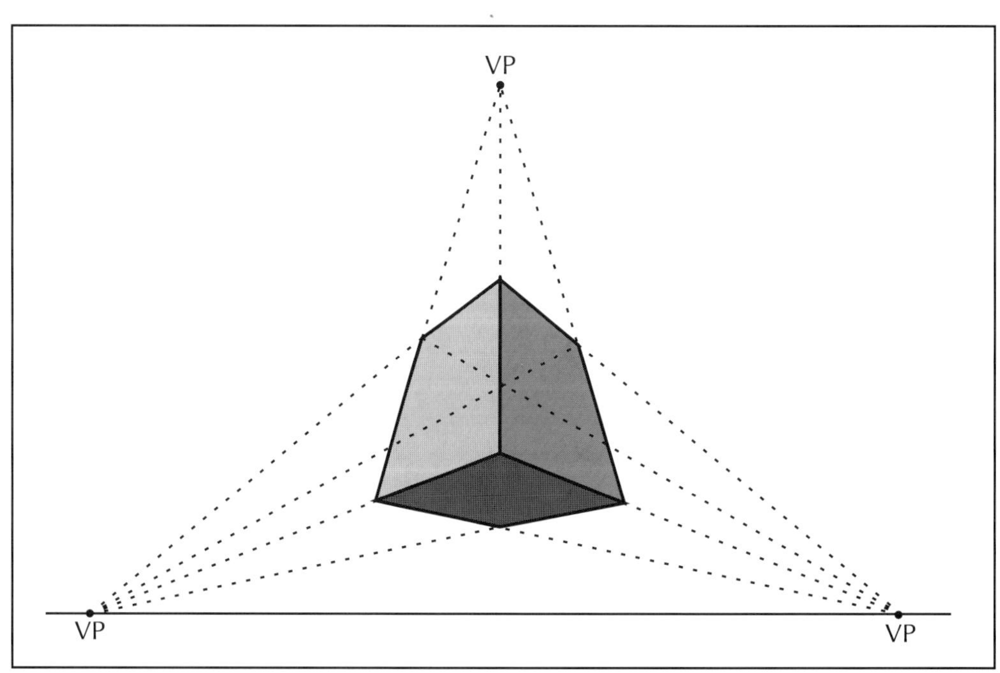

It is obvious that both the world of nature and the world of human-made forms is filled with complexity. To render either of these, one must first see the underlying essential structures.

In looking at the photograph on this page, how would you break down the image in order to create an interesting composition for drawing or painting? How would you go about rendering the spiral staircase? First consider doing a high contrast drawing in pen and ink. Then develop a pencil drawing using a range of tones. Taking the process a step further, convert the tones into a monochromatic colored pencil drawing or an acrylic painting.

Photocopy the above image, adhere the copy to a piece of drawing paper larger than the original and using your imagination, continue the drawing by inventing the forms that extend beyond what you can see in the photo.

This page:
Photograph of a spiral staircase showing three-point perspective.
Where is the viewer located? Where is the horizon line located? Where are the three vanishing points?

In any perspective system complex objects can be developed through the use of the unit cube, its diagonals and arcs of circles. The basic forms of the pyramid, the cone, the cylinder, and the sphere can be constructed within a basic cube.

Begin with any face of the unit cube in one, two, or three-point perspective. Draw in the two major diagonals. These determine the center, and so the midpoints of the sides. When a circle is inscribed within the square face, it can be seen that the center of the circle and the center of the square coincide. The circle is tangent to the four sides of the square at its midpoints. Fig. 2.11

You will notice that, depending upon the perspective system, the square face of the cube will take on other quadrilateral shapes. Any circle seen obliquely is an ellipse. It has a major and a minor axis. The major axis of the ellipse, indicating the circle, always is located in front of the center of the circle. The edges of the ellipse touch the quadrilateral (square in perspective) at its perspective midpoint on each side, but these points are not the endpoints of the major axis. The distance between the perspective center and the major axis of the ellipse depends upon the size of the circle and the distance between the front and back sides of the quadrilateral. The minor axis will always pass through the perspective midpoint and the center of the circle. If the square contains concentric circles, the minor axes of all the concentric circles coincide and pass through the perspective center as well. Fig. 2.12

It should be obvious by now that, in terms of perspective, there is more to it than meets the eye.

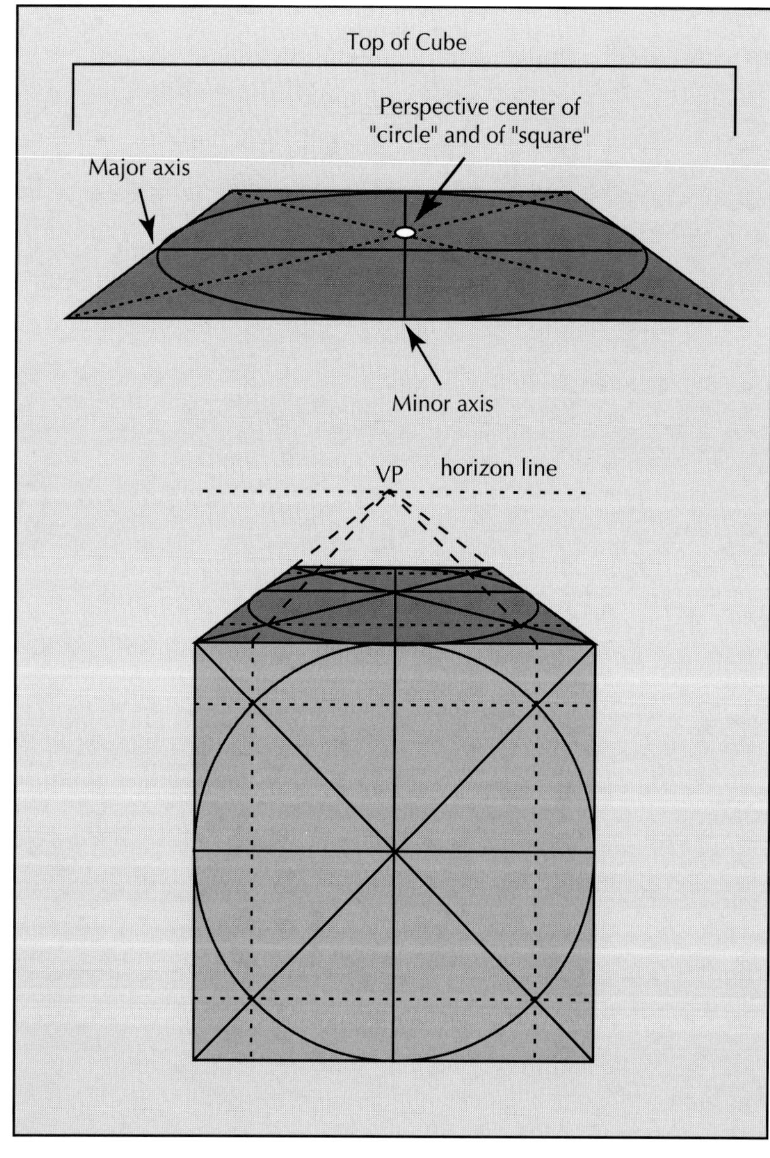

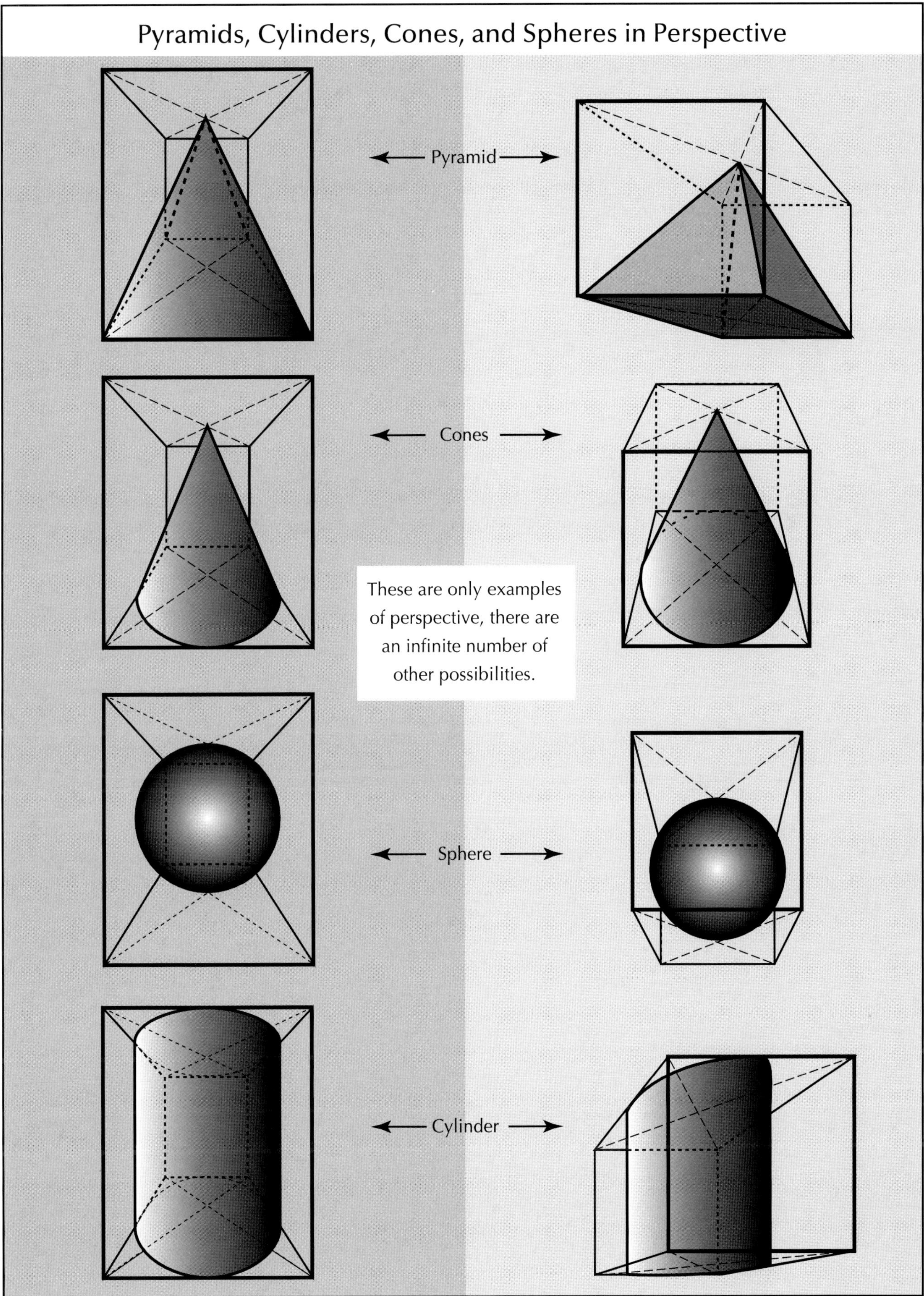
Pyramids, Cylinders, Cones, and Spheres in Perspective
Pyramid
Cones
These are only examples
of perspective, there are
an infinite number of
other possibilities.
Sphere
Cylinder

Problems, Projects and Play

1. Find information on Euclidean and Projective geometries. In a written and visual essay, attempt to show their connections and their differences.

2. Look up the work of two of the following Renaissance artists and in a written and visual essay show their studies of perspective. Where are they similar and where are they different?

    Alberti
    della Francesca
    Uccello
    da Vinci
    Massaccio
    Mantegna
    Giotto
    Dürer

3. Find a reproduction of da Vinci's painting, "The Last Supper". With a tracing overlay, do an analysis of the perspective involved.

4. Find a reproduction of Tintoretto's painting, "The Feast in the House of Levi". With a tracing overlay, do an analysis of the perspective involved. In writing, compare its use in this work and in the work previous.

5. Research the perspective in the paintings of Vermeer. Notice which ones seem to be of the same room. With tracing overlays on at least two work, analyze and compare their use of perspective. How does Vermeer's work differ from that of da Vinci and Tintoretto?

6. Using the schematic diagram found on page 53, make a small model reconstructing this image in actual three-dimensions.

7. Research the life and times of Gérard Désargues. Write a brief essay on how the times influenced the mathematics of that period in history. What were the other mathematicians exploring?

8. Find reproductions of Persian miniature painting and a Renaissance painting. With visuals, analyze the perspective systems in each. In writing, explain the advantages and disadvantages of both. If each explores an aspect of reality, how can one choose between them?

9. Compare and contrast, in writing, a "naive" work from Chapter 1 and a Renaissance work from this chapter. What do they have in common? How do they differ? Which one do you prefer and why?

10. Look through magazines and find photographic examples of one, two and three-point perspective images. Do tracing overlays to indicate where the various perspective points lie.

11. Set up a group of still ife objects so that all their faces are parallel to the viewer. Light the set-up in a dramatic way. Do a one-point perspective drawing of it using lines, hatching, and cross-hatching only.

12. Set up a group of rectangular solids. Have some objects with faces parallel to the viewer and some with edges parallel to the viewer. Do a pencil drawing of this set-up using gradation of tones only.

13. Do a colored pencil drawing of a three-point perspective view of an imaginary city seen from either the vantage point of a bird or a mouse.

14. Do a photo essay of shadows. Mat, mount, and frame your results.

15. Taking the above project as your starting place, do a high contrast drawing of the shadows using brush and ink only or cut out dark construction paper.

16. Develop a set of paper dolls using the information found on pages 66 and 67. Make them out of sturdy cardboard, fomecore, or plywood. Develop a wardrobe of contemporary clothing for these doll figures.

17. Using the information on page 68, construct a one-room doll-house based on one of the floor plans and rectangular solids found in the chart. Furnish it with persons of the correct proportions.

18. Walking around your own neighborhood, do a photographic essay in which you give examples of one, two, and three-point perspective. In captions, document the history of these structures.

19. Using a found photograph, do two drawings of the same subject using two different types of grids.

20. Do a drawing or painting of a foreshortened view of a male or female figure. Use pencil, or pen and ink or watercolor paint.

21. Do a drawing, painting, or collage of an interior space that includes a number of figures, male and female, sitting, standing, bending, lying down, foreshortened, etc.

22. Do a drawing or painting or collage which uses only pyramids, cylinders, spheres, and cones in a particular perspective system. Choose a favorite drawing material and a favorite technique.

23. As Vermeer did, do a painting of yourself as the artist in your studio. Consider the room; persons with you; the objects you would include; the viewpoint you would take; the light source for the space, and the color structure of the work.

24. Given the information about body measurements found in this chapter, and the following recipe, build an ideal couple in an ideal room with ideal objects. First make cardboard templates of the pieces required. Then make them in the dough. Bake and decorate. You can purchase edible colors for mixing into the dough or frosting recipes.

     1 stick of butter
     1/2 cup white sugar
     1/2 cup packed brown sugar
      2 eggs
     1/8 tsp salt
     1 tsp. cinnamon
     1 tsp. ginger
     1 tsp. cardamom
     1/2 tsp. cloves
     1/2 tsp. nutmeg
     2 1/2 cups white flour

Cream the butter and sugars until the mixture is soft and fluffy. Add the eggs and continue to beat until mixed well. Add all the spices and the flour. Beat until the mixture is very stiff and thoroughly mixed.

Turn out onto a floured board and knead in additional flour so that the dough is not sticky but workable.

Wrap in waxed paper and refrigerate for 24 hours. Roll out dough so that it is 1/4" in thickness. Lay tem-plates onto the dough. Trace around them with a pointed knife. Cut out the figures, walls, etc.

Place on greased and floured cookie sheets. Bake at 350° until the edges start to brown and the cookies feel set (about 15 minutes).

Cool before decorating.

Think about doing a self-portrait variation using this material or a playful variation of a perspective painting.

25. Do a collage/drawing that uses a mixed-perspective system.

26. Do variation of Project #24 using the "new" plastic clay.

27. Research Renaissance architecture and painting. See how the painters of the period represented the architectural concepts within their paintings.

28. As with Project #27, research contemporary architecture and painting and see how the artists of this century represent architecture in their works.

29. Do a collage/drawing that is a variation of Projects #27 or Project # 28.

Further Reading

Auvil, Kenneth, W. *Perspective Drawing*. Mountain View: Mayfield Publishing Co.1990.

Cole, Alison. *Perspective*. New York: Dorling Kindersley. 1992.

Courant, Richard and Herbert Robbins.*What is Mathematics?* New York: Oxford University Press. 1941.

deVries, Jan Vredeman. *Perspective*. New York: Dover Publications, Inc. 1968.

Devlin, Keith. *Mathematics The Science of Patterns*. New York: Scientific American Library. 1996.

Field, J.V. *The Invention of Infinity*. New York: Oxford University Press. 1997.

Gombrich, E.H. *Shadows*. London: National Gallery Publications, Ltd. 1995.

Hall, E.T. *The Hidden Dimension*. New York: Doubleday, Inc. 1966.

James, Jane H. *Perspective drawing*. Englewood Cliffs:Prentice Hall. 1981.

Montague, John. *Basic Perspective Drawing, Second Edition*. New York: Van Nostrand Reinhold. 1993.

Newman, Rochelle and Martha Boles. *Universal Patterns,* Second Revised Edition. Bradford: Pythagorean Press. 1992.

3 *Euclid and Beyond*

> *Summing up the formal characteristics of play we might call it a free activity standing quite consciously outside "ordinary life" as being "not serious", but at the same time absorbing the player intensely and utterly. It is an activity connected with no material interest, and no profit can be gained by it. It proceeds within its own proper boundaries of time and space according to fixed rules and in an orderly manner.*
>
> Johan Huizinga
> Philosopher

Human beings like to play games. Look at all the kinds of games there are and where they are played. There are card games, board games, rope games, string games, physical stamina games, riddles, and puzzles. They are played with sand and stones in the arid environment of Africa and with bones and feathers in the igloos of Eskimos in frigid climes. They are played in the royal courts of kings and in the back alleys of slums. They are played by the young and by the old. They are played by both men and women competition or in collaboration. The form and the rules of each game reflect the values of a society. Consider the very many video and computer games that are now available in our own Twentieth Century. What do they say about our culture?

The first written account in European literature on games was written by the Spanish King Alfonso X in 1283. Since he also wrote books on other learned subjects, we can assume that he considered games to be an important part of the cultural heritage.

Games and their boards can be traced back to the Near East and the cultures of the Egyptians, Babylonians, and Sumerians. Games have their roots in ancient rituals and rites that are connected to mazes and labyrinths. Think of the story of the minotaur, Theseus, Ariadne, and the ball of string. Can these games be connected back even further to Old Stone Age cutlures? These same roots connect art to the past and part to the notion of games.

Commercial sports are not the same as games since the competitive aspect takes the activity away from the idea of pure play. And it is play that we are concerned with here. To play is a voluntary act. It is a choice to play a game and never a chore. It is a temporary activity that interrupts the routine of everyday life. It takes place in a clearly defined arena which separates the game from secular tasks. It brings with it a sense of limits and boundaries. It has a clearly defined form. It involves risks, chance, choice, and responsibility. Play has no reason for exisiting except for its own sake. The skills obtained in the act of playing, however, may be the same skills that are needed for adult survival. Watch the activities of kittens and puppies and see how their playing

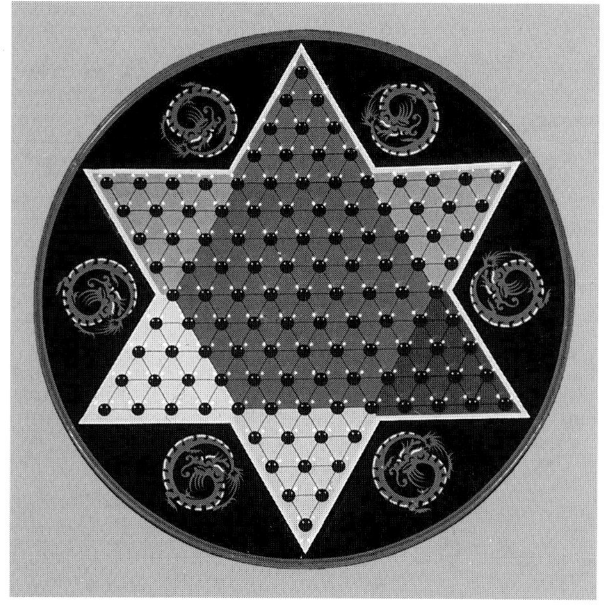

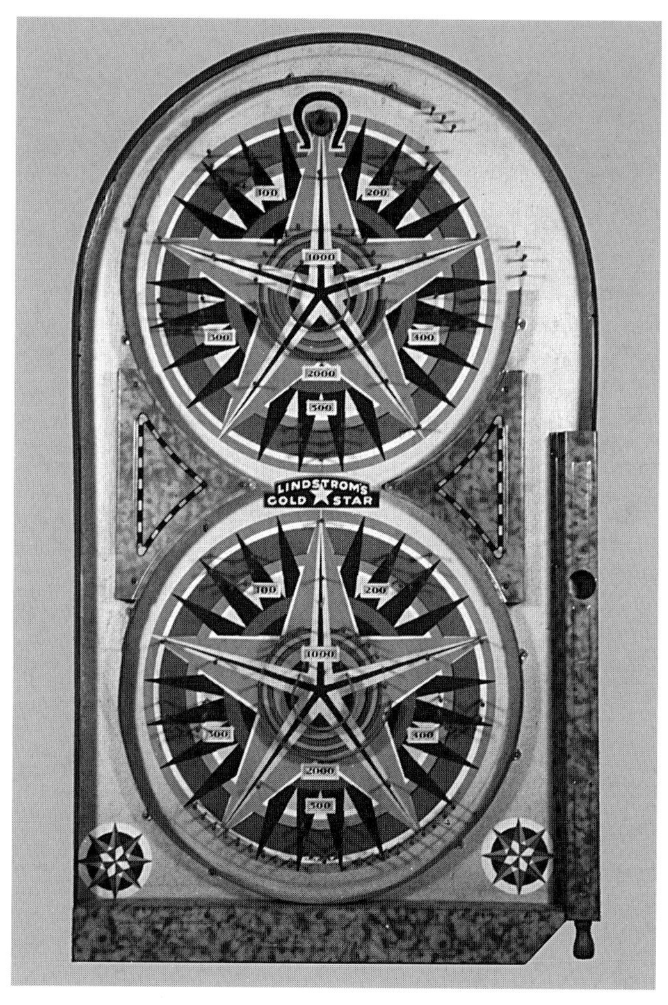

suggests adult behavior. Play is a special activity that is independent of the real world. It captivates its participants and its audience. Both take play extremely seriously with intensity and excitement because, above all, it is fun to do.

Our perspective game from the last chapter has many of the components of play and ties the notion of game and art together. Both art and mathematics can be likened to games since they both are autonomous rule-governed activities that use the manipulation of symbols, with order, clarity, and objectivity.

Mathematics is a system of relationships; art is a system of relationships. Both art and the mathematics of geometry are systems of relation-ships of objects in space, both of two and three dimensions.

In this chapter, we would like to articulate more of the general aspects of game strategy and to look at the geometry of Euclid as a very particular game. We shall then examine variations that were generated by deviations from the original rules. This is the way in which new games arise.

Geometry is about the transformations of objects in space. Some aspects of the objects stay invariant, immune to change, while others are transformed drastically. In Euclidean geometry, all objects are considered to be rigid, changing only their positions in space. In projective geometry, the transformations can distort the objects in very specific ways.

Left:
A game of cards which uses a circular format. Notice the division of four within the face cards as well as the use of the symmetry operation of rotation. See the Design Appendix for more information on symmetry.

Below:
The Japanese game of "Go" which has connections to the familiar western game of chess. The game uses only black and white "stones" for playing pieces on a gridded wooden board.

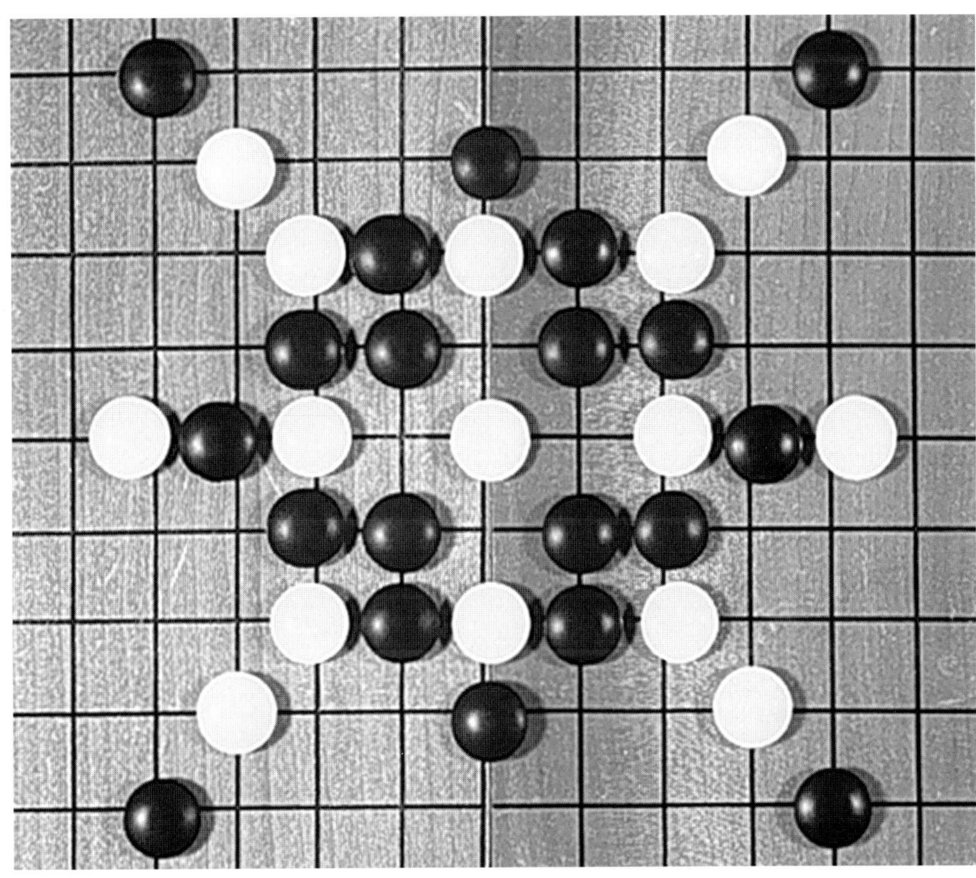

Games of Strategy

> The human being is one player in a huge game, the outcome of which is, for him, uncertain. He has to make full use of his capabilities to hold his own as a player and not become a plaything of chance.
>
> Manfred Eigen/ Ruthild Winkler

This page:
A game of darts and their cast shadows. Notice the variety of shapes that the shadows take. Where do you think the light is coming from?

Many people take games very seriously. Game theory is a branch of applied mathematics. It has only been within this century that a theory of games has been formulated by mathematicians. Understanding the structure of games allows for the application of that structure to widely differering areas of investigation such as economics, business, and military strategy.

There are both finite and infinite games. Chess is considered a finite game with optimal strategies that are limited in number. The game ends after a finite number of moves has been made. Games can also be related to networks and topology, the subject of Chapter 5.

A game involves a series of decision steps that are made by the players. There should be enough variation within the game structure so that the outcome is not predictable and, therefore, not boring. The game itself is called a system. Chance cannot completely determine the outcome of the game as in the cardgame Blackjack; however, chance and choice are complementary components of a game.

A game of strategy involves a set of rules which specify exactly what player is allowed to do under all possible circumstances. If the player is to receive any information, the rules determine what information is required and how it is to be acquired. If there is an element of chance, such as the roll of the dice, the rules define the ways in which chance is used. There are usually game pieces which stand for the different players.

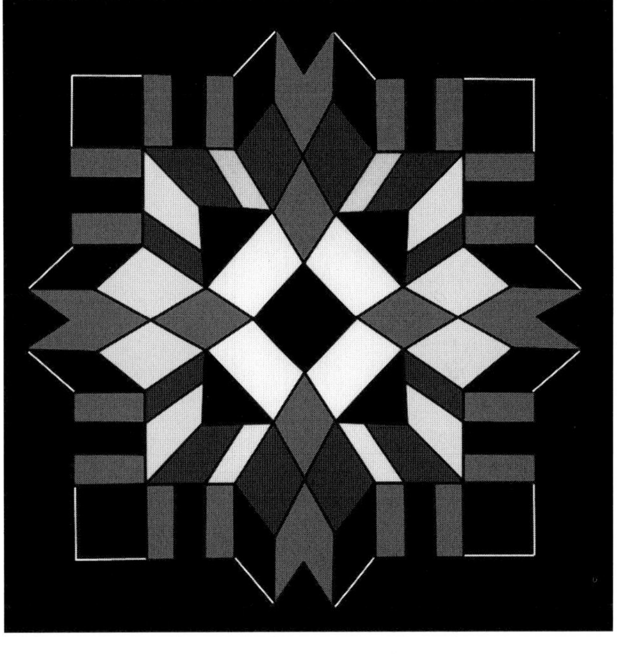

This page:
Student works. Gameboards using color structures, wood, illustration board, and acrylic paints.
Top left:
Kristen Francis. Puzzle with removable wooden pieces.

Center right:
Chris Bouressa. Painted square structure.

Bottom:
Peter Sheridan. Game utilizing concentric circles.

This page:
*Student work.
Marsha Johnson. The game-
board of the two-dimensional surface is used to play a
game with a variety of perspective systems.*

One can discriminate among difficulties, puzzles and problems. A difficulty is a problem with minimum definition. A puzzle, on the other hand, is a game in which there is a set of givens and a set of procedural constraints all precisely stated. A problem is a difficulty upon which we attempt to impose a puzzle form.

Jerome Bruner
Twentieth Century Philosopher

The Game of Chess

Chess, one of the most beloved and most played games, is an essential archetype for all games. It was developed in the 7th century in India as a war strategy game. It is a model for the battlefield in which the opponents need to have foresight, a sense of the totality of the playing field, the positions of the various plays, and a strong notion of which moves would produce winning results. Today, computer games function in somewhat the same way emphasizing speed, movement, strategy, and an overall sense of the game.

Chess travelled westward first to Persia and then to Europe via the Arab occupation of Spain during the Middle Ages. It was adopted by royalty and the game pieces of Indian elephants, chariots, and majarajas were replaced by castles, kings, and bishops. It became the passion of Medieval Europe and there are many references to the game in literature. There is a Chinese version, choo-hong-ki, and a Japanese version, Shogi. All are played with the same sense of excitement and urgency.

But chess is not the oldest of board games. The Sumerians of Ur, circa 4,500 BC., created a racing game on a board. And in the far east, people developed a game called "Go". It is similar to chess in that it is a game of battle but it is more severe and has a more rigorous logic to it. You might like to find copies of the rules for all of these chess variations and compare and contrast them. Then, try to develop a variation based on Twentieth Century objects and values.

Mathematics, too, is in a very real sense a game. It has its board, its rules,its intent, its passion, and its fun. Let us give you an abbreviated version of the rules for chess and you can use them as a basis for comparision when we soon look at Euclid's game and other geometry games later.

Gamepieces:

Board, 8 units x 8 units
King (2), one black, one white
Queen (2) one black, one white
Bishop (4) two black, two white
Knight (4) two black, two white
Castle-Rook (4) two black, two white
Pawn-foot soldier (16) eight black, eight white

Game Moves:

1. The game is played by two opponents who move their pieces about a square board which has 32 black squares and 32 white ones.
2. The player who has the white pieces starts the game.
3. The two players alternate turns.
4. A move is the translation operation of transferring one of the pieces from one square to another square. The square can either be vacant or occupied by the opponent's piece.
5. No piece, except the Knight or Rook when castling, can cross a square occupied by another piece.
6. A captured piece must be taken in the same move as when it is played.
7. Each of the various pieces can move in only prescribed ways. The outcome is always unpredictable because there is such a huge number of possible variations with the relatively few play structures. (How has the computer affected the playing of the game of chess?)

The RNA Bead Game

Rules and Play

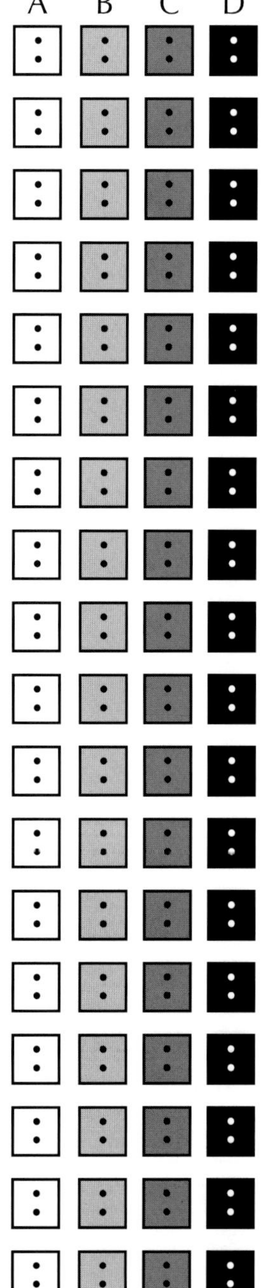

This page:
Fig. 3.1.
A set of square "beads" for use in the RNA Bead Game. There are four groups with 20 beads of each tone: black, white, light gray, dark gray, for a total of 80 beads.
You will need needle, thread, scissors, and beads. A set of paper ones is given below.

Nature also plays by rules. Scientists believe that they starts out with simplicity and build to complexity until the complexity hides the essentials. The outcome is a result of the "sensitive dependence on initial conditions". The authors Eigen and Winkler demonstrate this in their book with their RNA (ribonucleic acid) molecules game which they feel mirrors reality. Perhaps you would like to do more detailed research into the RNA molecules to see how the following game relates to the process of natural selection in evolution. You would have to do research into the work of Darwin.

Can one apply a game structure and game rules to the DNA molecules? Try the following game and see.

This game, using beads to represent the four building blocks of RNA, relates to the process of natural evolution. Given a random group, the objects is to quickly build a beaded structure that is folded and characterized by a maximum number of particular pairs of beads called complementary pairs. RNA's are made up of about 80 building blocks.

The rules of the game closely mirror aspects of reality.

At the beginning of the game, each player has a string of 80 randomly strung bead. Each bead presents a nucleotide. The tetrahedral die has four toned faces corresponding to the beads A, B, C, D. The die is used to roll mutations according to three basic rules which must be adhered to faithfully.

Points:

Every white-dark gray pair = 1 point.
Every black-light gray pair = 2 points.

Materials:

1. 4 sets of twenty toned "beads"" Black (D), White (A), Light gray (B), and Dark gray (C). There is a total of 80 beads. Fig. 3.1.

The four tones represent the four basic building blocks of the RNA molecules. Two pairs of the tones, white-dark gray and black-light gray, will be considered complementary pairs that stand for the nucleotides.

2. A pair of tetrahedra dice, Fig. 3.2. Each face of the tetrahedron corresponds to one of the four bead tones.

3. A length of string and a needle.

4. Remove type of adhesive tape so that beads can be constantly joined and rejoined.

Number of Players::

In this case, l. Number of players could increase to as many as desired. Then, the object of the game is to see who is the first to build a folded structure that has the most complementary pair of beads.

a. Thread the 80 beads in a random arrangement, Fig. 3.3. Then, fold the string-chain experimentally to see which structures contain the most complementary pairs of beads.

One can lay out the beads in a hairpin grouping or a cloverleaf formation.The hairpin shape is the simplest arrangement because it has only one loop. The cloverleaf, however, is more advantageous since more base pairs can form from the chance sequence. Cloverleafs with three or four loops are more conducive to solving the problem of pairing. Because of the relatively high number of chance complementary pairs these structures offer, they are most likely to provide an optimal beginning siutation.

b. Notice which beads form complementary pairs.

c. Roll the tetrahedra die. Which tone shows up?

In each round of play, the die is used to "mutate" in relation to the color rolled, a bead located in a certain position. Mutations are considered "selected" if they can form a complementary base pair according to the rules.

Five beads in a loop cannot undergo pair-formation. Because the formation of base pairs between different sections of a sequence can be accomplished only by folding the chain on a plane, loops will inevitably result (Steric Rule).

In the folded arrangement, if two beads of complementary pairs (white-dark gray or black-light gray) are oppsite each other (Rule of Complementarity) and simultaneiously, if there is an unbroken sequence of at least four white-dark pairs; two white-dark gray pairs, and one black-light gray pair; two black-light gray pairs, these beads are considered a pair and are linked together (Rule of Cooperativity). These are considered to be stable base pairs. Since they are "selected", they are no longer subject to the roll of the tetrahedra die.

If you are intrigued with this game, research the one called "life" developed by mathematician John Conaway. See how humble beginnings beget complex middles and ends. How does this compare to real life as played out every day by all living things on this planet?

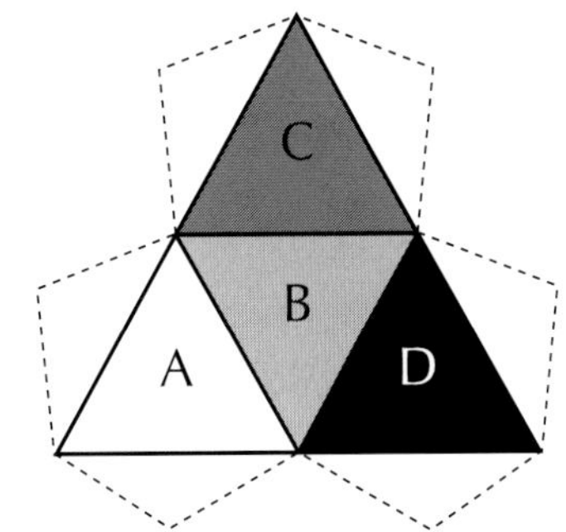

Tetrahedral die

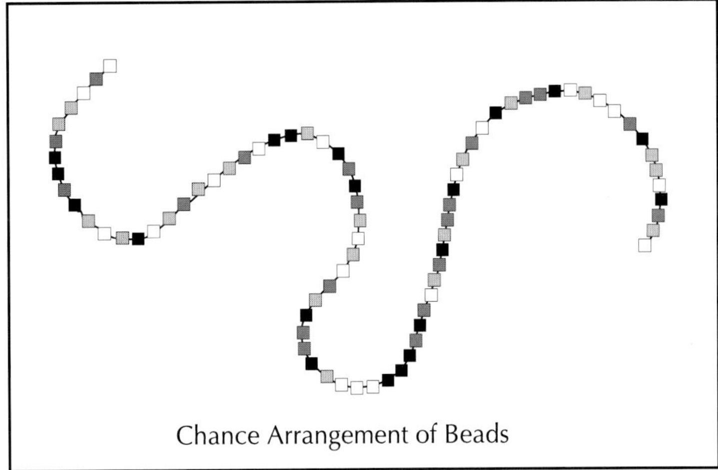

Chance Arrangement of Beads

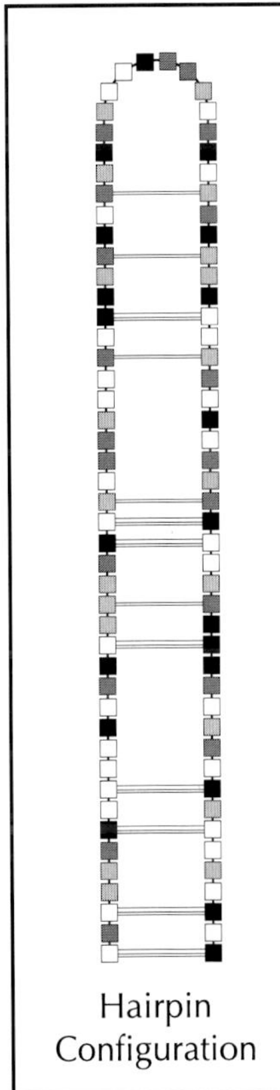

Hairpin Configuration

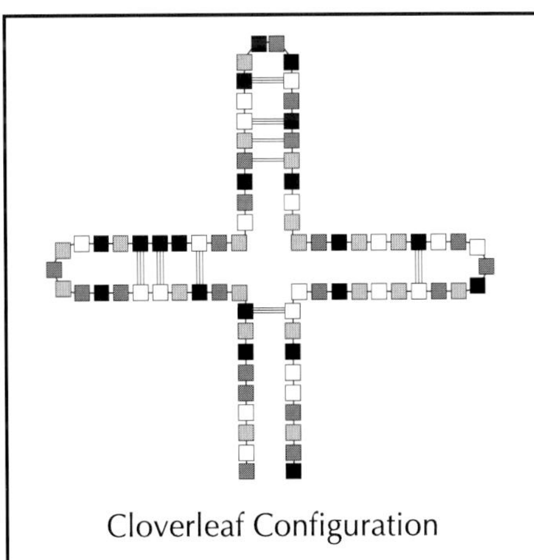

Cloverleaf Configuration

This page:
Top left: Fig. 3.2.
A template for a tetrahedra die. Make two of these.
Center: Fig. 3.3 A chance arrangement of beads as well as beads arranged in hairpin and cloverleaf configurations which offer the optimal pattern for 80 beads.

Try a game using other than 80 beads.

Euclid's Game

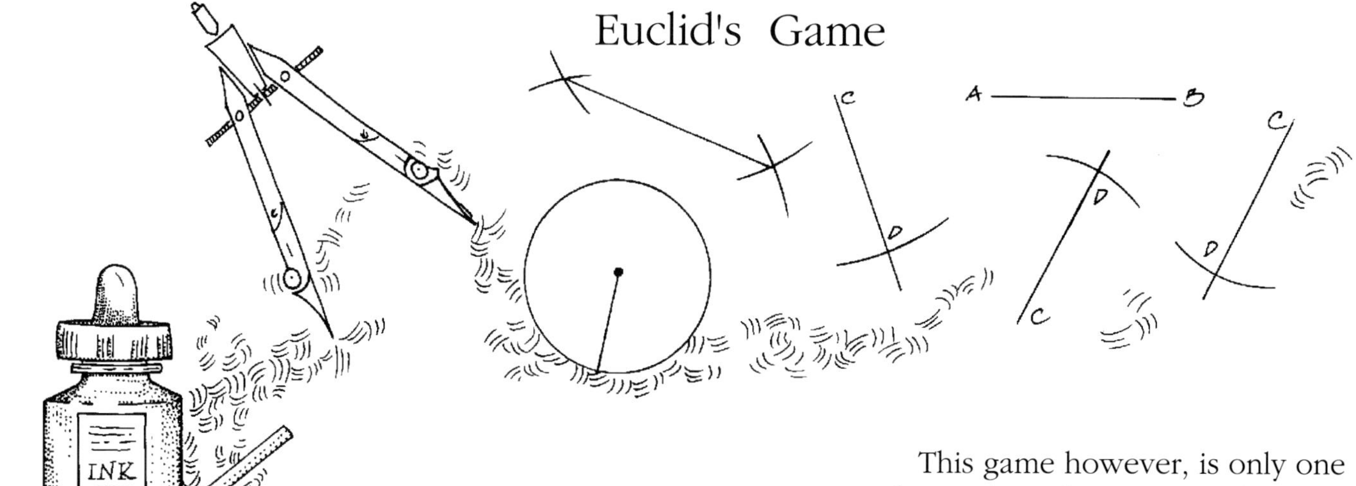

Many, many, many students over time have been playing Euclid's game, *The Elements*, without ever realizing it was a game with its own set of consistent rules and an order by which the game was meant to be played. It can be played on a flat surface, the plane, or in 3-space.

It is a game of geometry that pertains to properties of figures and the positions of these figures in the plane and in 3-space. The ancient thinkers of Greece were interested in understanding the principles that govern the forces and objects of the natural world. To that end, they tried to systematically codify aspects of it. Euclid's game was considered to be the model for what actually happens in physical space. It works quite well on the surface of the earth where there is no noticeable curvature and no strong gravitational force.

Euclid attempted to order what was known about space, and figures in space, by developing statements that held to a logical progression. For a very long time, it was thought that his was the only game in town. It has survived through time and translation for twenty-two centuries. It is the standard by which mathematicians judge other geometric games. And to his credit, it was the very first game.

This game however, is only one of a number of geometries that deal with different kinds of spaces whether real or theoretical. In general, each type of geometry is connected to a group of motions, or one-to-one mappings, of the space onto itself which leave the geometrical properties of the figures intact. Euclidean geometry is involved with the properties that preserve distance within the figure and that keep the figures congruent, identical in size and shape, and rigid. Position in space was not an issue. Euclid also looked at similar figures, those having the same shape, but not the same size, and equivalent figures, those having the same area but not the same shape. Artists are also interested in these properties. The figures Euclid chose to look at were basic. He examined the properties of points, lines, triangles, rectangles and circles, in two dimensions and analogous forms, such as the cube, tetrahedron, octahedron, dodecahedron, and icosahedron from three dimensions.

But above all, he was concerned with what he could explain and prove about these figures beyond a shadow of a doubt. This is the idea of the geometrical proof that is such a bugaboo for many persons. But it was the only method that would give his game validity.

This page:
The tools of Euclidean geometry. A bounded portion of the plane, such as a piece of paper, is used as the gameboard. The compass is used for measuring. When the setting of it is not changed, congruent lengths are obtained. The measure of a figure has nothing to do with its position in the plane.
Some compasses have means for inserting the different drawing instruments of pencil, marker, pen, etc.
A straightedge, which has no particular units marked on it, is used for obtaining straight lines.
An eraser is an aid in the thinking process.

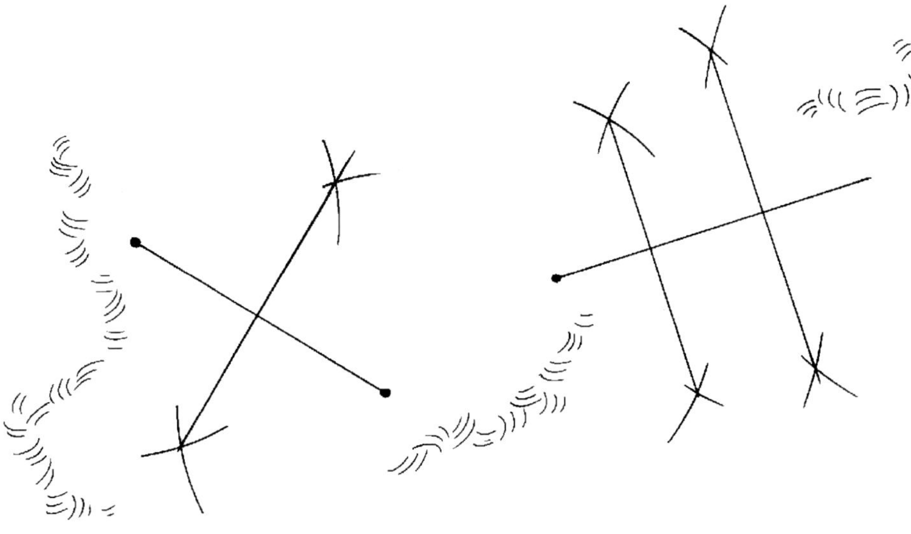

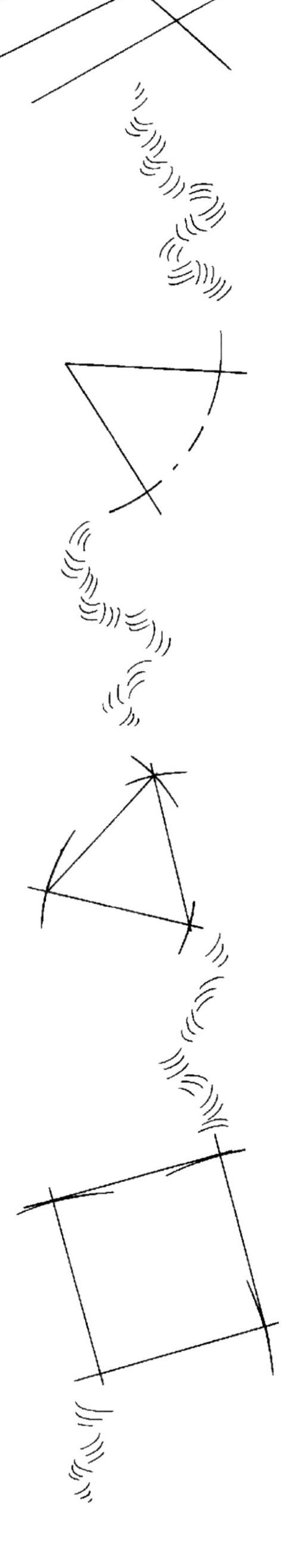

To play the game was to use the logic of deductive reasoning which calls for developing conclusions based on very few premises and previously established results or theorems. Each of the parties involved in the game must agree on the meaning of the symbols and the words that are used. But some terms must be left as undefined in order to begin the game. Euclid established axioms, sometimes called postulates, which are statements that are accepted as true without requiring verification. These he based on our intuitive grasp of figures in space.

Theorems develop from the axioms. They are mathematical statements that must be correct under every conceivable situation or else they lose their validity. All subsequent statements must be proved though the rigors of logical argument. Even the most "obvious" statements must be subjected to the process. In any theorem, the conclusions must be distinguished from the assumptions. A proof has to be general so that it covers all the particular examples. If a particular case is counter to the statement, then the proof is considered false.

The more one probes into searching for the essence of "Reality" and "Truth", the more one finds questions and not answers. For even the simplest of facts has layers of meanings, assumptions, and associations.

For instance, each of us takes a meassuring device, such as the foot ruler, for granted. Its purpose seems clear, its function is useful. Everybody understands it. But when one starts to ask questions such as" What is length? What is distance? Why this length? Why this tool? What kind of surface am I measuring? What, in fact, is measure? All at once, the solidity of absolute knowing slips away. If, however, each time we performed the act of measuring, we asked these questions, we would not be able to get on with the practical tasks of daily living.

But these are the deep questions that scholars, such as Euclid, ask so that what we eventually know about our world becomes more and more complete. Will we ever know everything? Will we want to know everything? Are some things unknowable? Or are there only more and more questions to ask with more and more answers to examine?

The thinkers of Classical Greece had a special affinity for the subject of geometry over and above its everyday function. They looked beyond the practical problems of geometry to the abstract questions of pure logical through the reasoning process of deduction. For the Greeks, geometry was not involved with numbers but shapes as pobserved in the physical three-dimensional world.

Despite the fact that *The Elements* of Euclid is one of the most well known and used book of western civilization, very little is known of his personal life. He lived during the reign of Ptolemy I, a general appointed by Alexander the Great, third century B.C. He probably received his mathematical education from persons who had studied with Plato.The question is, who taught Plato? And who taught the teacher of Plato, and on and on.....

Euclid's book has been the cornerstone of the foundation of mathematics with its concern for logic and exact reasoning. His achievement is recognized for the great skill he had in organizing the many bits and pieces into a coherent system of the then known mathematics.

> The Elements began with very simple concepts, definitions, and so forth, and gradually built up a vast body of results organized in such a way that any given result depended only on foregoing results.
>
> Douglas Hofstadter

Object of the Game:

To build a system of statements (Theorems), that are logical, consistent and which are deduced from the fewest possible definitions and statements that stand independent from one another (Axioms). The mathematical reasoning is called deduction. It can be applied to all types of games and situations.

Number of Players:

The game can be played by an individual, with a partner, in small groups, in communites, between communities, and between nations.

Gameboard:

An infinite flat surface (plane) considered to be uniform everywhere, or three-dimensional space, also to be considered uniform everywhere.

Gamepieces:

Geometric objects: points, triangles, rectangles, surfaces. An unmarked straightedge, compass.

Euclid's Game

Rules:

A. Definitions of terms

There are 23 defined terms in his game. They are about ideal objects that do not exist in the physical world but can be intuited by reference to that world. Mathematicians have since come to grips with the fact that some terms must always remain undefined within any system. For example, the terms space, plane, and point.

B. Axioms (sometimes called Postulates):

These relate to the terms. Each one needs to stand alone. The purpose of a system that uses axioms is to provide a way to free the mind from its dependence upon intuition. The system must be built on an internal logic. (It may be intuition, however, that gives the seeds for the concept of a new system).

Five Postulates

These are specific geometry assumptions. They are accepted without further justification.

1. There is exactly one straight line connecting any two distinct points.

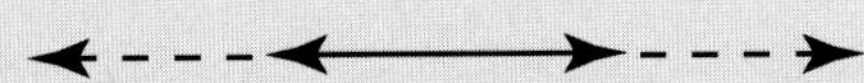

2. Every straight line can be extended indefinitely.

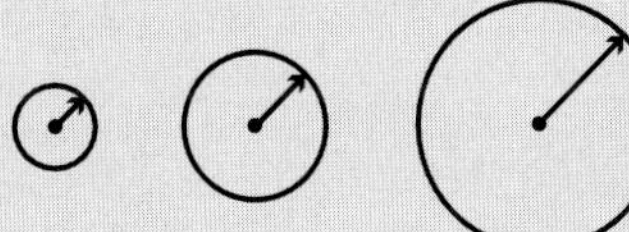

3. It is possible to construct a circle with any given center and any given radius.

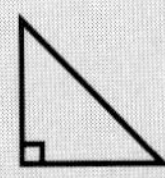 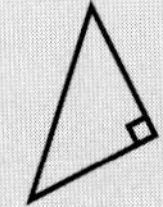 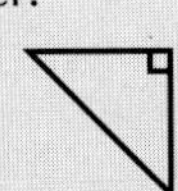

4. All right angles are equal to each other.

*5. Through any point not lying on a given straight line there is one, and only one, line parallel to the given line.

Five Common Notions

These can apply to other mathematics and to the other sciences as well.

1. Things which are equal to the same thing are also equal to each other.

2. If equals are added to equals, the whole are equal.

3. If equals are subtracted from equals, the remainders are equal.

4. Things which coincide with one another are equal to one another.

5. The whole is greater than any of the parts.

C. Therorems (sometimes called Propositions):

Book 1 of *The Elements* has 48 Theorems which deal with properties of straight lines, triangles, and parallelograms. From the above, Euclid derived 465 Theorems. These are statements that are logically deduced from previously accepted or established statements and definitions. These should be consistent and without contraditions.

Example: Given any straight line on which there are three points, a, b, c, there is always one point that lies between the other two points.

Game Moves:

Use the game pieces of Euclid plus a compass, straightedge, pencil, pen, eraser, etc., and the game rules to make all the right moves to win in the game of art. Artists of all persuasions need to use space, points, lines, angles, circles, arcs, triangles, rectangles, etc., to their collages, drawings, paintings, sculptures, etc. Artists also need to subdivide lines and areas. The surface of the two-dimensional drawing page, the volume of the three-dimensional space, becomes the arena of creation. The compass is the primary tool of action. You will notice that, because one point is fixed, and the other point rotates around, what you are doing is constructing a circle, or arcs, parts of a circle, which you can then use to mark off points. You use the straightedge to connect points and to make straight lines.

You, as an artist, are concerned with the "quality" of a line or a shape which has no concern for the mathematician. These are issues of the artist. The physical lines and shapes that an artist constructs are just mere approximations of these pure abstract elements which reside in the realm of the mind.

Below:
Fig. 3.3
The compass and its relation to
circles, arc, and diameters.

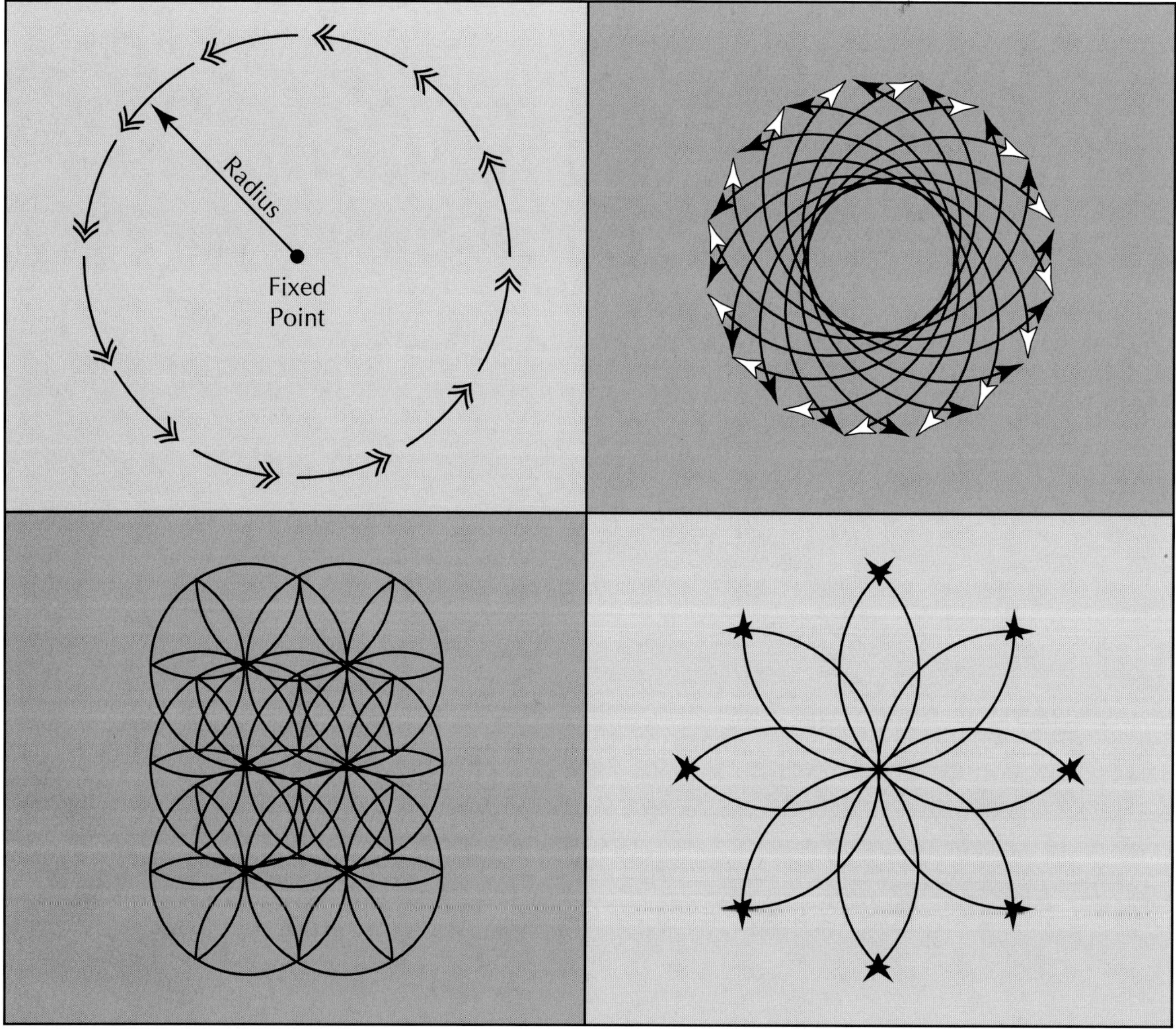

There are some basic constructions that allow you to play the game artfully. On the following several pages, we will show them to you with procedures for doing them yourself. The words used will be in the language of mathematics which, when you are familar with it, is more precise than common talk. Here we summarize the constructions using the math words and in parentheses we use common talk so that you can see the difference. The procedures will also have diagrams so that the combination of words and images should make the constructions accessible to you even if you find the words unfamiliar.

This page:
How would you go about proving that these constructions work? This is a math concern and some familiarity with the process of proof is required.

Constructions Useful to the Artist/Designer
1. To Construct a Line Segment Congruent to a Given Line Segment. (To make a copy of a line)
2. To Bisect a Given Line Segment. (To cut a line in half)
3. To Divide a Line Segment into Congruent Segments Numbering the Powers of Two. (To keep cutting a line in half, then half again, etc.)
4. To Construct an Angle Congruent to a Given Angle. (To make a copy of an angle)
5. To Bisect an Angle. (To cut an angle in half)
6. To Divide a Line Segment Into a Given Number of Congruent Segments. (To cut a line into as many pieces as you like)
7. To Construct a Perpendicular to a Line Through a Point on the Line. (To make a right angle on a line)
8. To Construct Perpendicular to a Line Through a Point Not on the Line. (To make a right angle from a point)
9. To Construct a Line Parallel to a Given Line. (To construct a pair of parallel lines)

Given a Line Segment, Construct a Line Segment Congruent to It

Given line segment FG   F——————G

1. First, construct what is given (FG) anywhere in the plane. The example here will differ from what you might construct. Work from your figure.

F————————————G

2. Draw another line segment anywhere in the plane other than where you drew the first line segment. This should be longer than the given. On this segment, choose a point to be designated as one endpoint. Label this point H. (You could have chosen either endpoint).

3. Go back to the given line segment . Place the metal tip of the compass on F and the pencil tip on G, respectively, in order to measure the length. Without changing the setting of the compass, place the metal tip on H and with the pencil tip, cut an arc that intersects the line segment. Label the point of intersection J.

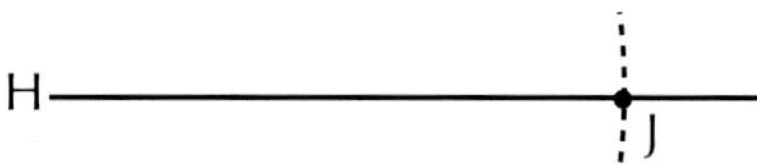

Now line segment HJ is congruent to line segment FG.

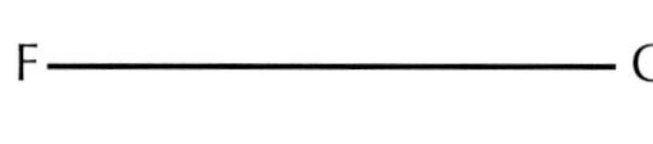

Given an Angle, Construct an Angle Congruent to It

Given angle S 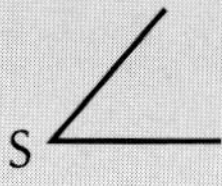

1. First, construct the given angle S anywhere in the plane.

2. Draw another line segment anywhere in the plane. Label one endpoint V.

3. Go back to the given and place the metal tip of the compass on S. With the pencil tip, cut an arc that intersects both sides of angle S. Label the points of intersection T and U. Without changing the setting, place the metal tip on V and cut an arc that intersects the line segment. Label the point of intersection W.

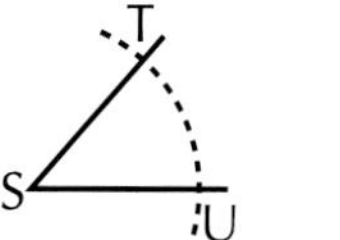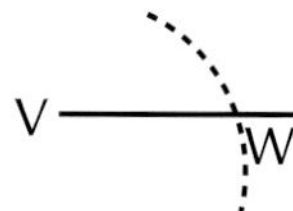

4. Place the metal tip of the compass on U and adjust the setting so that the pencil tip measures UT. Without changing the setting of the compass, place the metal tip on W, and cut an arc that intersects the line segment through W. Label this point of intersection X.

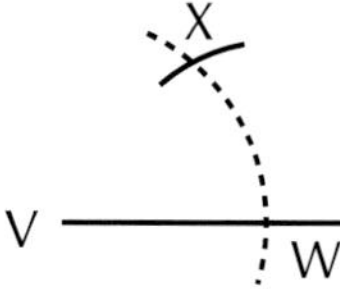

5. Draw ray VX.

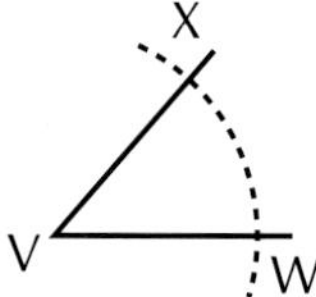

Now angle V is congruent to angle S.

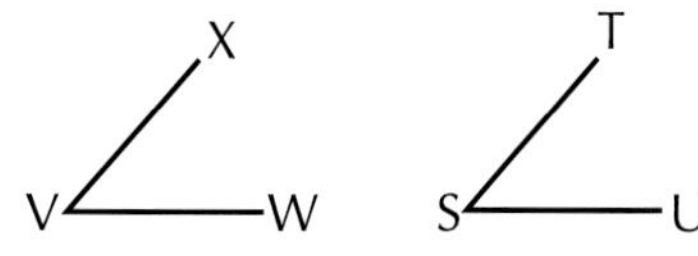

Bisect a Given Line Segment

Given line segment AB   A———————B

1. First, construct what is given (AB) anywhere in the plane. Open the compass to a distance that is more than half the length of AB. Place the metal tip of the compass on A and cut an arc that goes through line segment AB.

2. Without changing the compass setting, place the metal tip on B and cut an arc that goes through line segment AB. Label the points of intersection C and D.

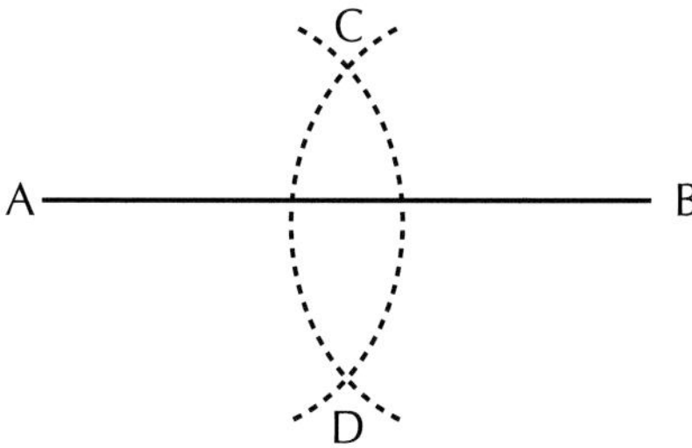

3. Draw line segment CD. Label the point of intersection E.

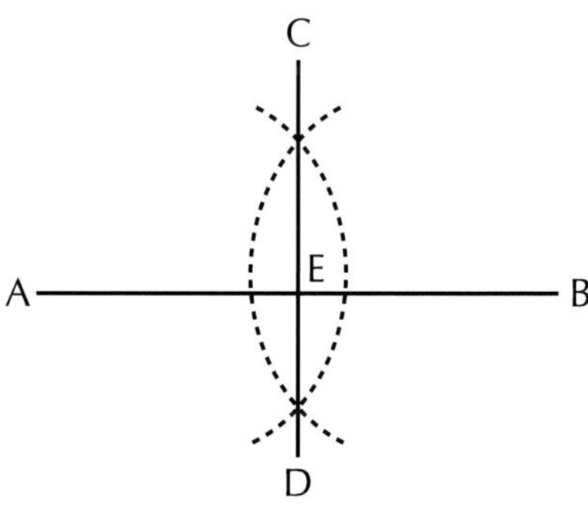

Now line segment AB is bisected.

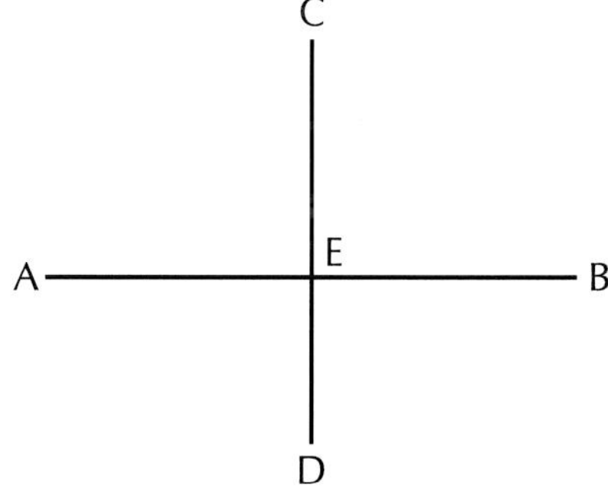

Bisect a Given Line Segment into Congruent Segments Numbering Powers of Two

Given line segment AB   A———————B

1. First, construct what is given (AB) anywhere in the plane. Bisect line segment AB as in the previous construction. Label the point of intersection C.

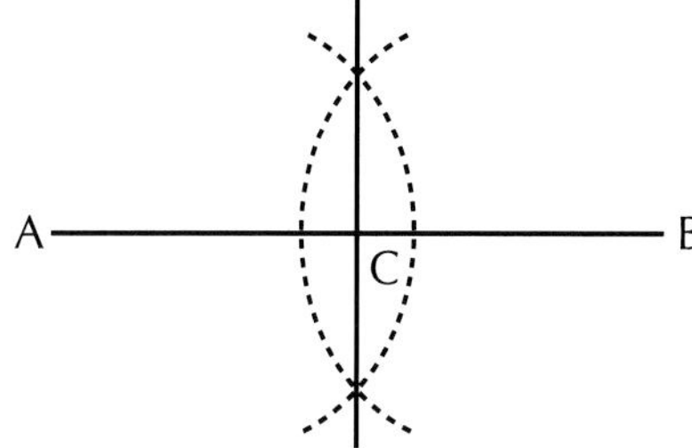

2. Bisect line segments AC and CB. Label the points of intersection D and E.

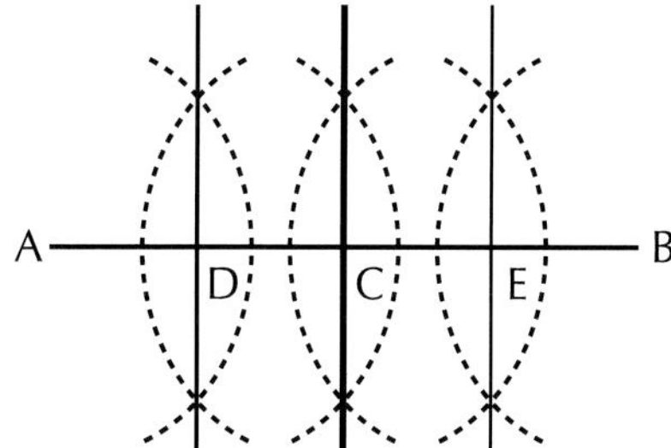

3. Bisect line segments AD, DC, CE, EB. Label the points of intersection F, G, H, I. You can continue to bisect these line segments as many times as you choose.

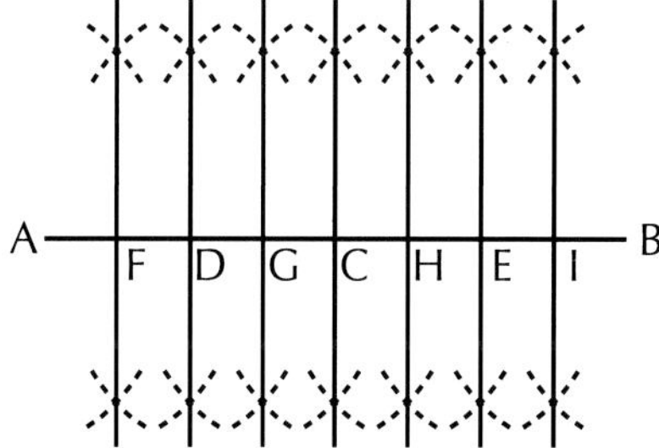

Now line segment AB is divided into a group of congruent segments.

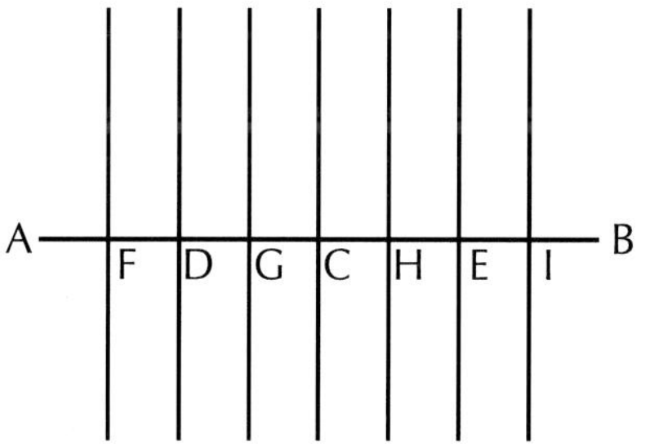

Divide a Given a Line Segment into a Given Number of Congruent Segments
(We have chosen to use three divisions; however, any other number may be used.)

Given line segment FG   A———————B

1. First, construct what is given (AB) anywhere in the plane. Draw a ray from A to form an acute angle with line segment AB.

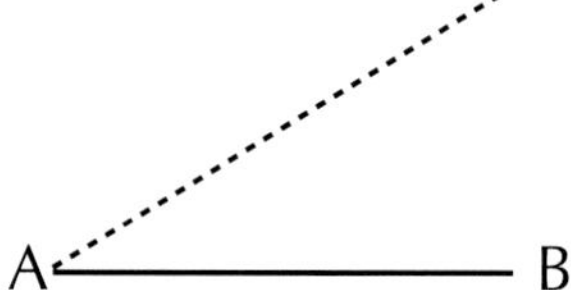

2. Use an arbitrary length, mark off a given number of congruent segments on this ray (in this case 3) by first placing the metal tip on A and cutting an arc on the ray. Label the point of intersection C. Repeat this step without changing the compass setting. Each time, place the metal tip on the point of the most recently cut arc. Label the points of intersection C, D, and E, respectively.

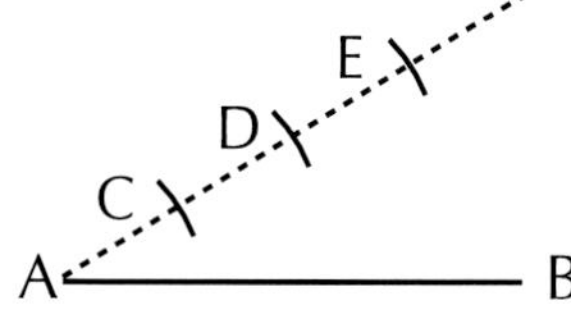

3. Draw line segment EB.

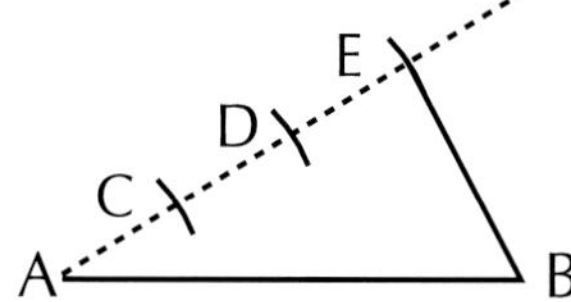

4. Place the metal tip on E and cut an arc that intersects both EA and EB. Label the points of intersection R and S.

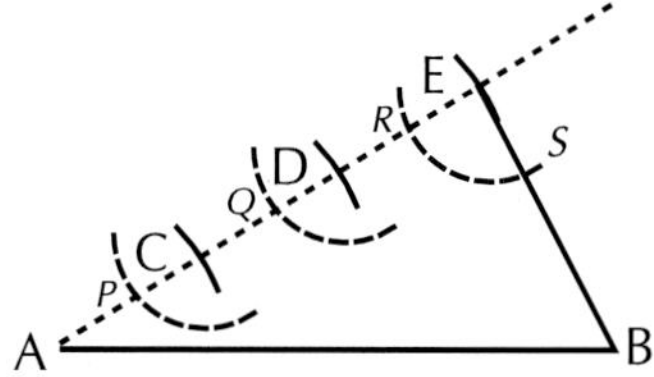

5. Without changing the compass setting, cut arcs at D and C as you did in the previous step. Label the points of intersection Q and P. Place the metal tip on R and the pencil tip on S. Without changing the compass setting, put the metal tip on Q and cut an arc that intersects the arc through Q. Repeat at point P. Label the points of intersection T and U.

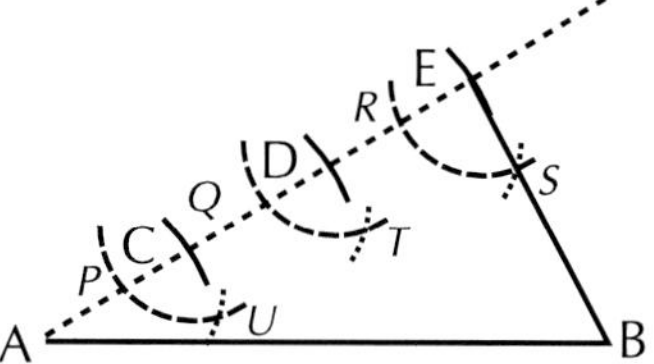

6. Draw line segments through points D and T, C and U extending the line segments until they touch AB. Label the points of intersection F, B, and G.

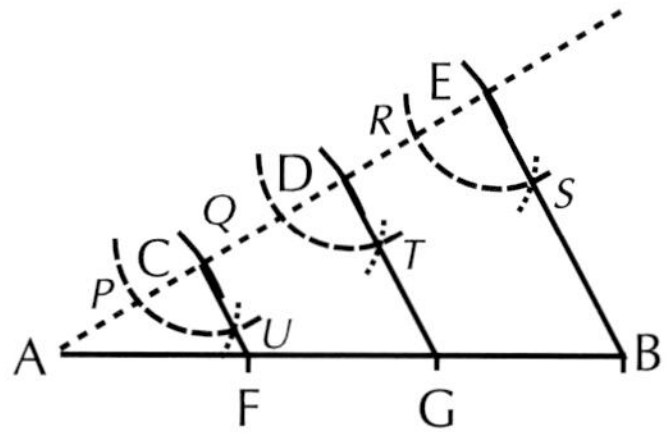

Now line segment AB is divided into a given number of congruent segments.

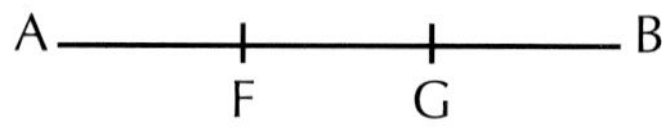

Construct a Perpendicular to a Line Through a Point on the Line

Given point A on line segment *l*.
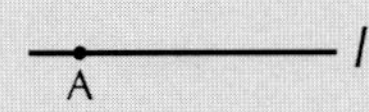

1. First, construct what is given (A*l*) anywhere in the plane. Place the metal tip of the compass on point A and cut arcs that intersect line *l* on both sides of A. Label the points of intersection B and C.

2. Open the compass to more than half the distance from B to C. Place the metal tip of the compass on B and cut an arc on one side of line *l*. (It could be either above or below the line). Without changing the compass setting, place the metal tip on point C and cut an arc that intersects the one drawn in the previous step. Label the point of intersection D.

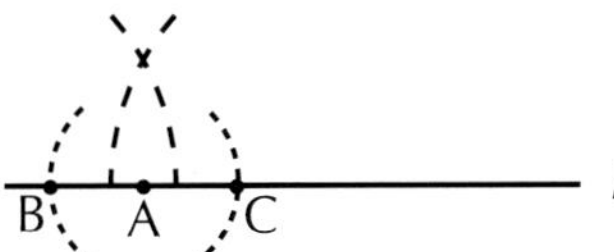

3. Draw line segment AD.

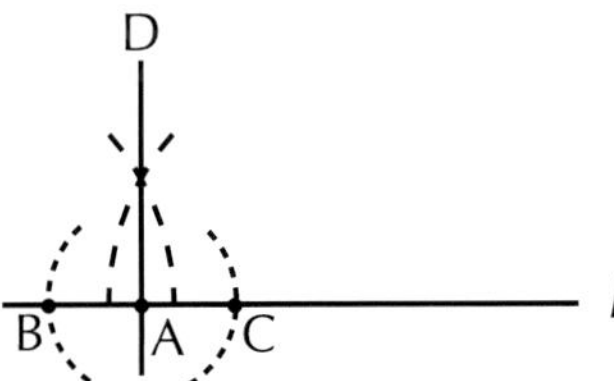

Now line segment AD is perpendicular to line segment l.

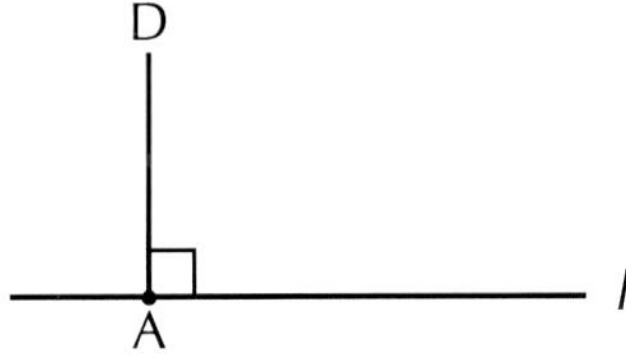

Construct a Perpendicular to a Line Through a Point Not on the Line

Given line segment *l* and point A which is not on line *l*.
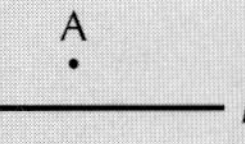

1. First, construct what is given (*l*) anywhere in the plane. Open the compass to a distance more than half from point A to line *l*. Place the metal tip of the compass on point A. Cut two arcs that intersect the given line *l*. Label the points of intersection B and C.

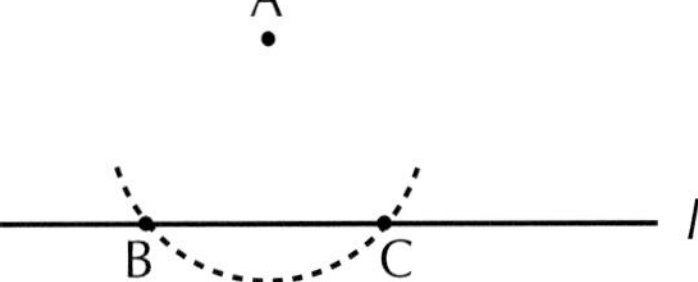

2. Place the metal tip of the compass on B and cut an arc on the side of line *l* which is opposite A.

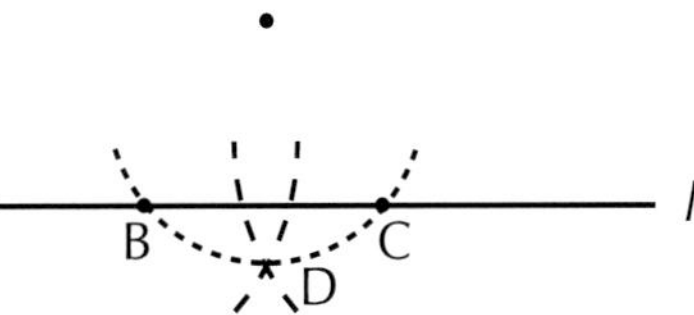

3. Without changing the compass setting, place the metal tip on C and cut an arc that intersects the previously drawn one. Label the point of intersection D. Draw line segment AD.

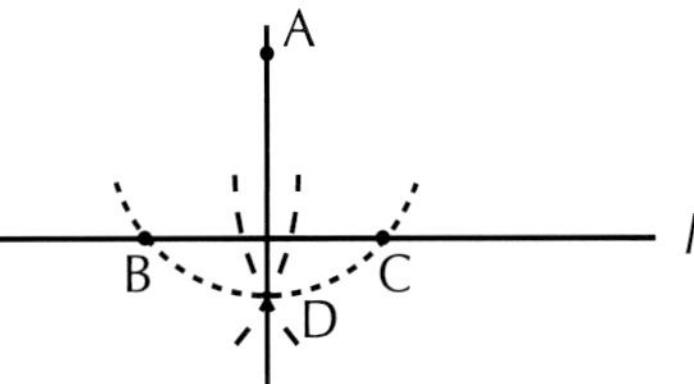

Now line segment AD is perpendicular to line segment l.

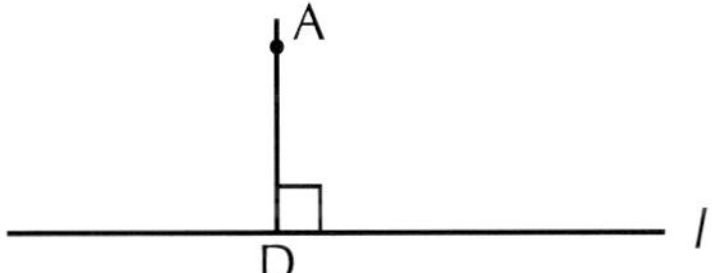

*Construct a Line Parallel to a Given Line Through a Given Point

Given line segment *l* and point A which is not on line *l*.

1. First, construct what is given anywhere in the plane. Draw a line through point A that intersects line *l*. Label the point of intersection B.

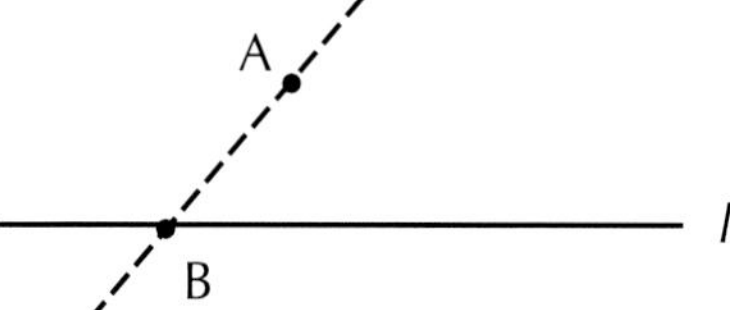

2. Place the metal tip of the compass on B. Cut an arc that intersects both AB and line *l*. Label the points of intersection C and D.

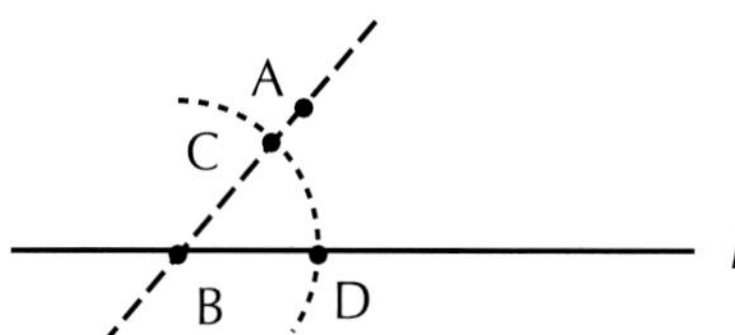

3. Without changing the compass setting, place the metal tip on A. Cut an arc that is the same as the arc drawn in the previous step. Label the point of intersection E.

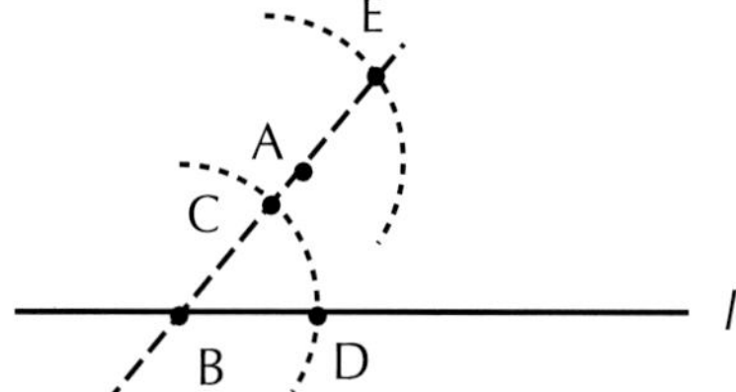

4. Place the metal tip on C and the pencil tip on D. Without changing the compass setting, place the metal tip on E. Cut an arc that intersects the arc previously drawn. Label the point of intersection F.

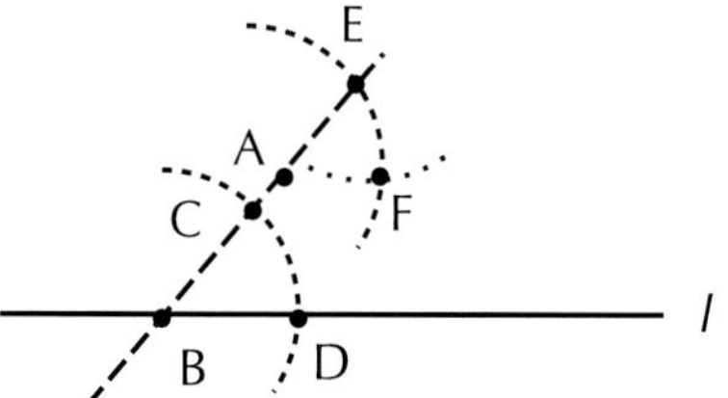

5. Draw line segment AF.

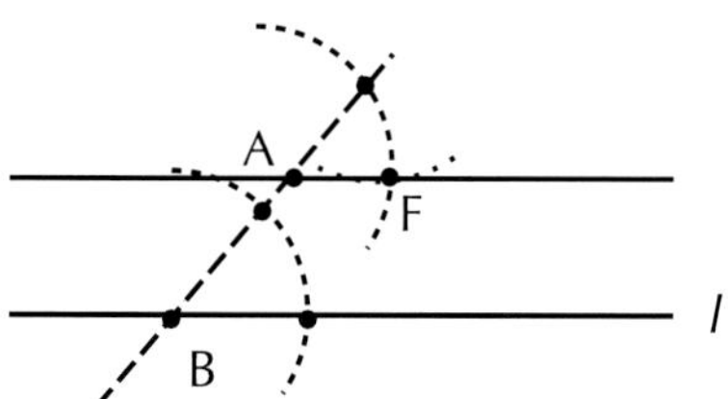

Now line segment AF is parallel to line segment l.

A ———————————— F

———————————— *l*

* *This is the construction that relates to the 5th postulate of Euclid.*

Other Geometry Games

For a long time, everyone played by Euclid's rules. Then some players began asking questions about the validity of one particular aspect of the game, Postulate #5, which dealt with parallel lines as you have just constructed on the previous page. Even Euclid was somewhat reluctant about it. This axiom did not seem so obvious. It was this questioning that changed the shape of the gameboard and the rules that accompanied it.

Parallel lines are central to Euclidean geometry. The definition of parallel lines does not necessarily assert that there are any parallel lines in space. Postulate #5, however, implies the existence of these lines. To geometers it, therefore, sounded more like a theorem than an axiom, and one that could never be proven through experiment.

Up until the nineteenth century, mathematicians did not doubt that this postulate held true for physical space as they knew it. What niggled at them was the sense that this postulate was not so clearcut as the other four. All of the other axioms deal with finite portions of lines or with plane figures that extend finitely. Early on, mathematicians sought to replace Postulate #5 with an equivalent one which would allow for the derivation of the same theorems.

Poseidonios, circa. 135-51 BC, questioned this postulate. Proclus, 4th Century AD, also felt uncomfortable with it. In the 17th Century, a Jesuit priest, Saccheri (1667-1733) tried to, based on the other Euclidean axioms, provide a proof for this postulate. Lambert, 1728-1777, also worked on this proof to no avail. Ultimately, math-

ematicians came to the conclusion that this postulate could not be proved because it stood outside the domain of the others. It became evident that the inability to find a proof for Postulate #5 was not due to lack of trying but because this postulate could stand separate from the four others. While appearing as a flaw in thinking, Euclid's postulate provided a way to broaden the base of geometry.

Three persons in different countries, Gauss in Germany, Bolyai in Hungary, and Lobachevsky in Russia, all came to the same conclusion. This postulate could be removed from the game and be replaced with another stating that, "through a point not on a line there exists more than one parallel to the line". Thus, it became possible to construct a system of geometry that is consistent, free from internal contradictions, and which still deals with points, lines, planes, etc. by deductions containing the first four postulates of Euclid, plus a new one.

Once it was established that other geometric systems were equally as valid as the Euclidean model, the question arose as to which system more accurately described the physical world.

The answer to this really depends upon whether one is examining the "local" conditions on this planet or those conditions at a greater distance than a few million miles. What happens on the moon? What happens in other parts of this solar system? Or this particular galaxy? How about this universe? Or the Universe that contains all the other universes? Questions, questions, questions.....
and then more questions.....

"In mathematics the goal is always to give an ironclad proof for some unobvious statement."

Douglas R. Hofstadter
mathematician

In formal systems, a theorem does not need to be a statement--rather it could be a string of symbols. Instead of being proven, theorems are produced according to the rules of the system.

"A good proof well explained shows not only that something is true but also why it has to be true."

Peter Cromwell
mathematician

A mathematician states, "All artists are liars." Is this statement true or false? Can you prove it one way or the other? This is called the Epimenides Paradox and led the 20th Century mathematician Kurt Gödels to examine the logic within any given system.

> The fact gradually forced on the mathematicians is that geometry is not the truth about physical space but the study of possible spaces. Several of these mathematically constructed spaces, differing sharply from one another, could fit physical space equally well as far as experience could decide.
>
> Morris Klein
> Mathematical Philosopher

This page: Fig. 3.4 Alternatives to Postulate #5 given line l.

A. No lines parallel to line l (Elliptical geometry)

B. More than one line parallel to line l. (Hyperbolic geometry)

C. Hyperbolic postulate: through a point not on a given line, there are many lines not meeting the given line. At least two such lines exist.

The truth in mathematics is decided by proof through deduction from some initial sets of assumptions called axioms. They take the form: If A, then B.

Now there are two ways in which this postulate could fail to be true.

A. It might be that there are no lines parallel to *l* through the point P. (Elliptic-Riemannian geometry)

B. It might be that there is more than one line which passes though point P parallel to line *l*. (Hyperbolic geometry)

Are there geometries in which the straight line is not infinite but finite and closed? Consider the sphere in which a straight line is in actuality a great circle of the sphere. Hyperbolic geometry what you get if you replace Euclid's postulate with alternative B as in Fig. 3.4. Non-euclidean geometries lead to some surprising results. For example, most persons recall from their high school geometry courses, that the sum of the degree measure of the angles of a triangle is 180°. But in Non-euclidean geometry this theorem is not true. Hyperbolic geometry includes triangles with fewer degrees than 180° and the larger the triangle, the less the angle sum. Elliptical geometry contains triangles with greater than 180°. For small triangles, the sum of the angles is not much greater than 180° which is why we can approximate small regions of a sphere as a flat map.

Karl Friedrich Gauss (1777-1855) worked at this postulate and came to the conclusion that the axiom could be replaced, but he did not present his findings publically. He felt that, despite his solid reputation, he would be ridiculed by the intellectuals of his time because Euclid's ideas were so deeply entrenched in the cultural ground of western civilization. His work on this subject was found after his death in 1855. Can you think of any analogous situations today?

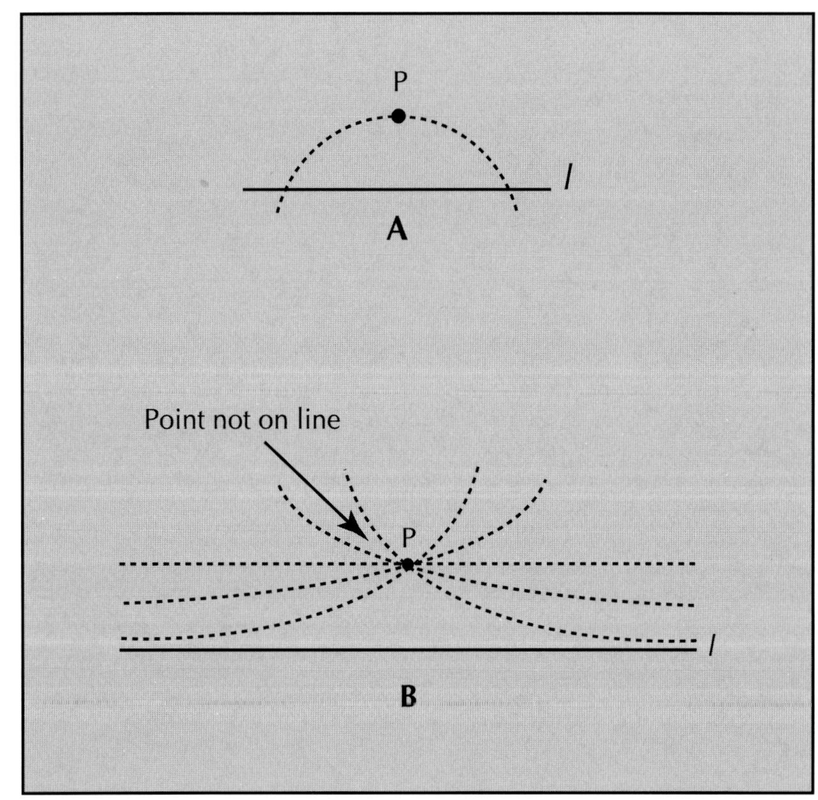

It was left to the Russian, Lobachevsky (1793-1856) and Bolyai (1775-1856) in Hungary to publish their deductions which were similar to those of Gauss. It took a long time to convince the mathematical community that these geometries were not just games of the mind, but could reflect aspects of the real world. Albert Einstein's Theory of Relativity makes use of the theorems of Non-euclidean geometry.

George Bernard Riemann (1826-1866), a student of Gauss, took the quest even further. He was interested in the geometry of curved spaces in arbitrarily many dimensions. By considering the surface of a sphere as a space by itself rather than part of three-dimensional Euclidean geometry, one could build a system that would coincide with that particular space. Riemann was also involved with topology, which shall be discussed in Chapter 5.

The rectangular coordinates of our well-known Cartesian grid could no longer be used to describe the surface of a sphere. Rather what was needed was the notion of arcs of great circles that would function as lines in this geometry. Triangles formed by the arcs of great circles would have angle sums of more than 180°. One could then study other types of surfaces as well. Hyperbolic geometry describes a surface of constant negative curvature. What does such a surface look like? We can only visualize a finite piece of such a surface since the hyperbolic plane would not even "fit" into Euclidean three-dimensional space. Thus we have to use schematic diagrams to suggest what happens on that plane.

In a hilly landscape, one notices

that there are peaks and valleys, no two having exactly the same height or depth. This is a surface whose curvature varies from one location to another like in our physical environment. whose curvature varies from one location to another, like in our physical environment.

What about a piece of crumpled paper? What geometric description would work well here? Interestingly enough, the crumpled paper is still considered flat because it is composed of many planar pieces. It is fractal-like and can be thought of as exisiting somewhere between two and three dimensions.

This page:
Fig. 3.5. Three different surfaces are represented here.

A. The Euclidean plane has zero curvature. We can describe the plane as a sphere of infinite radius so that the curvature is zero. Parallel never meet.

B. A sphere has positive curvature and parallel lines always meet. This includes parallel lines that are arcs whose centers coincide with the center of the sphere.

C. A "saddle" shape which has negative curvature. There are an infinite number of lines parallel to a given line that go through a fixed point.

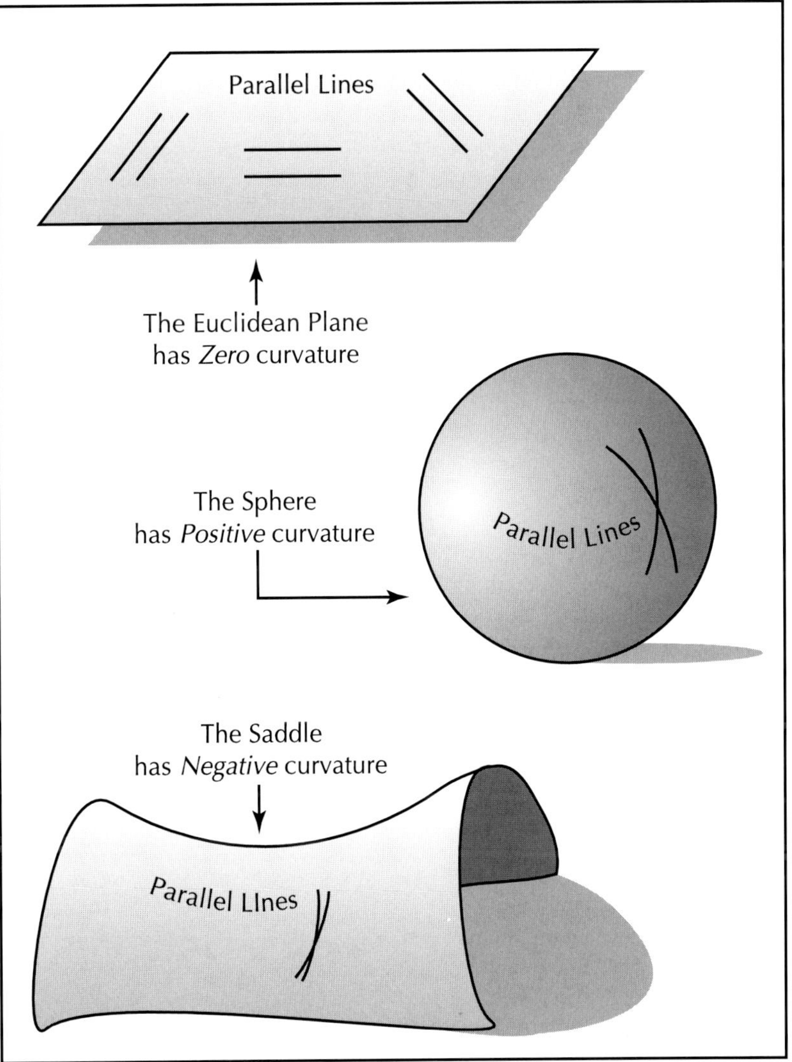

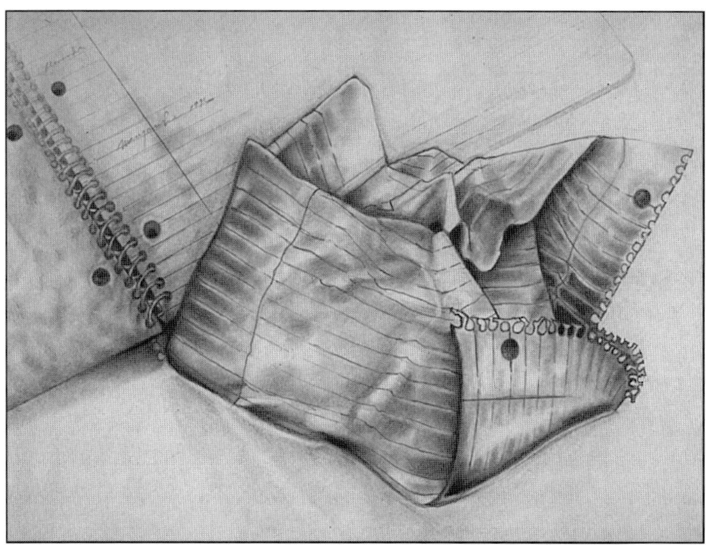

This page:
Artistic representation of a crumpled piece of paper by Gail Maciejewski. The drawings use pencil, pen and ink on bristol paper.
The two-dimensional surface acts as a space for the illusion of a three-dimensional object.

Top left: Notebook and crumpled notebook paper.

Center: A skull image emerging on the crumpled notebook paper surface.

Bottom right:
The unfolded crumpled notebook paper revealing the two sides simultaneously.

A sheet of paper is considered a flat surface and, thus, is a two-dimensional Euclidean plane. Crumple the paper and it now has a rough surface. It is no longer two-dimensional, but not quite three-dimensional. It can be described as having a fractal dimension somewhere between 2.1 and 2.9

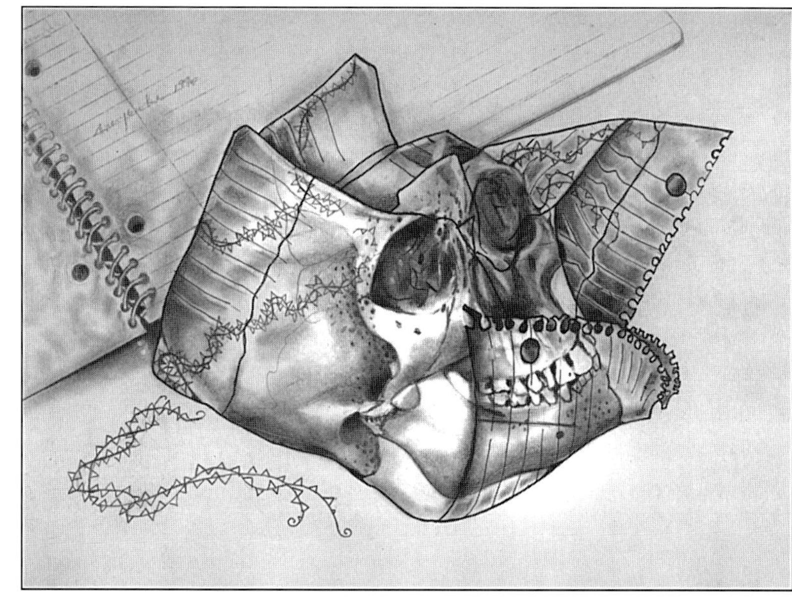

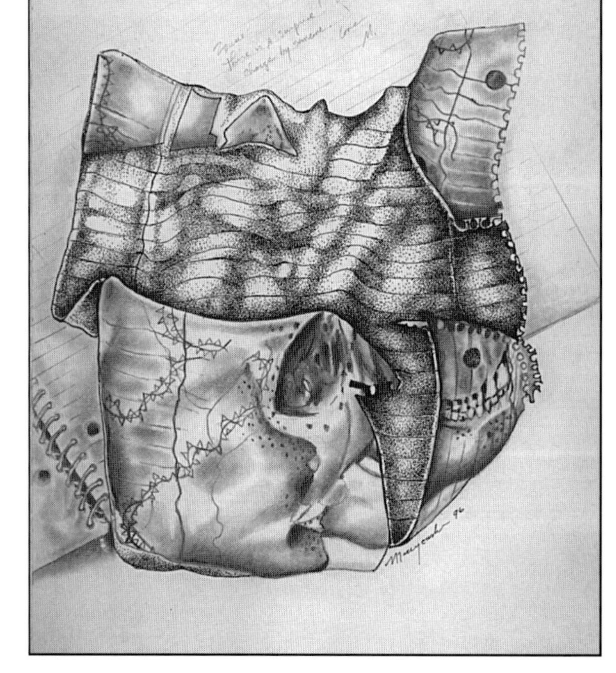

Comparison Between Euclidean & Non-Euclidean Geometries

	Euclidean	Non-Euclidean	
		Hyperbolic	**Elliptic**
Plane	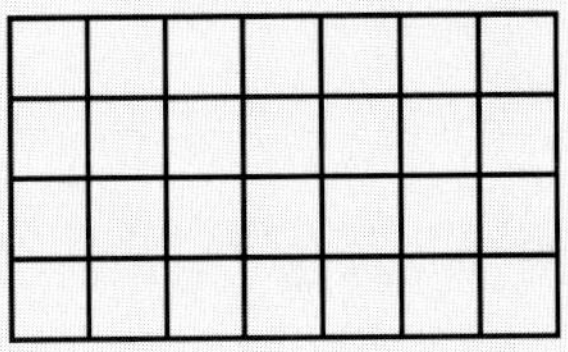 Geometry on a plane. A plane has zero curvature.	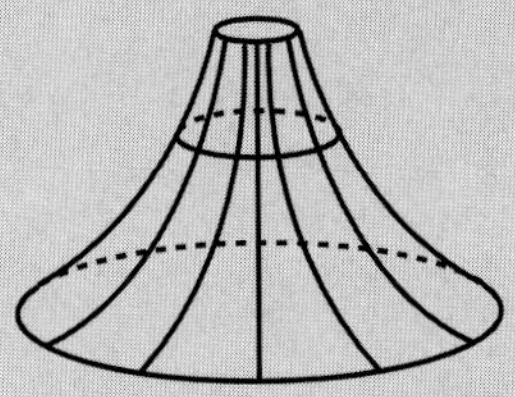 Geometry on a surface like the pseudosphere.	 Geometry on a sphere. A sphere has positive curvature. Space is finite but unbounded, it curves back on itself.
Point	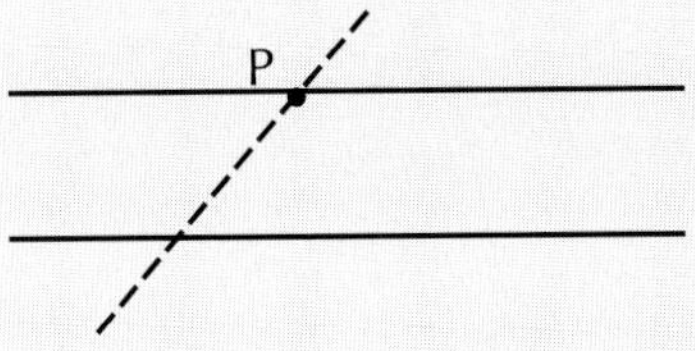 Given point A not on line *l*, exactly one parallel line can be drawn through point A.	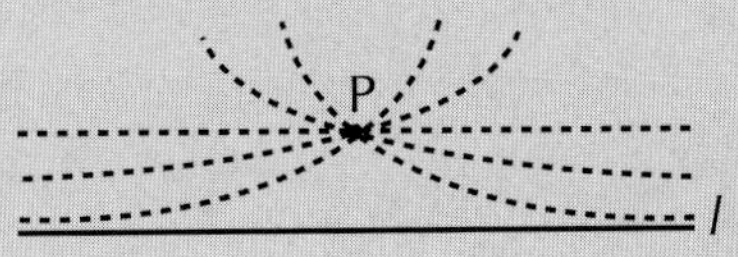 More than one line can be drawn through *P* and parallel to *l*.	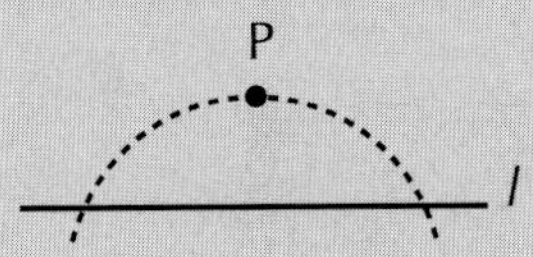 No line can be drawn through *P* and parallel to *l*.
Line	Lines have infinite lengths.		Lines have finite lengths.
Triangles	 Triangles can have the same angles but different length sides (Similarity as well as congruency).	 Triangles with the same angles must have the same length sides (Congruency only).	
	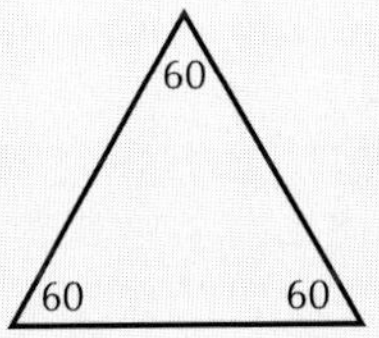 The sum of angles of a triangle is 180°.	 The sum of the angles of a triangle is less than 180°.	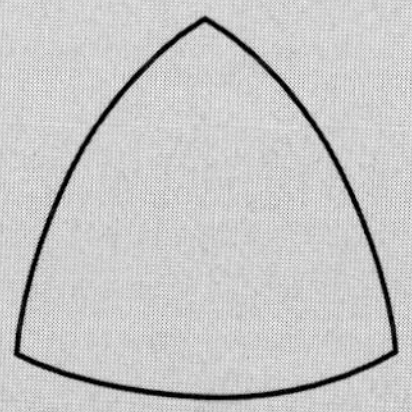 The sum of the angles of a triangle is more than 180°.

Procedure for Constructing an Illusion Sphere on the Plane

Given a square grid:

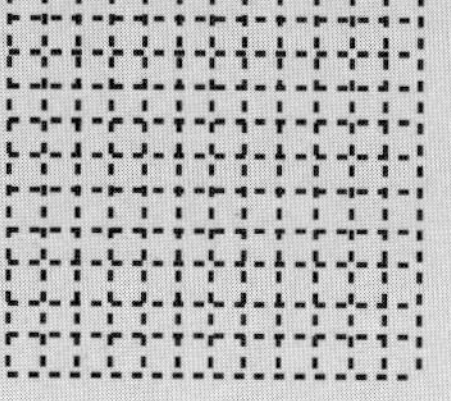

1. On the grid, construct a square whose sides have an odd number of units, in this case 13. The sides of the square will serve as the boundaries of the illusion sphere.

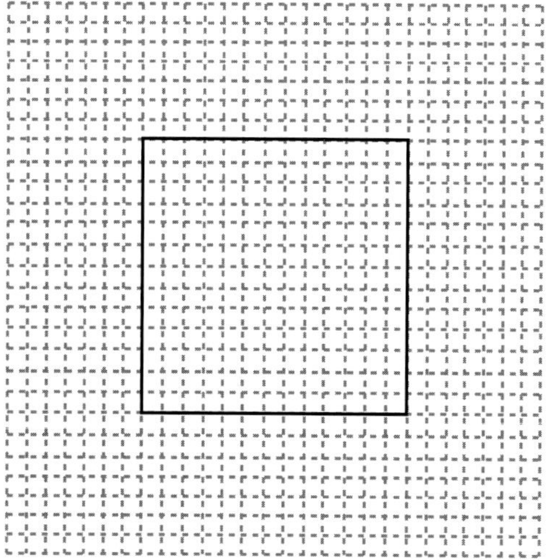

2. With O as the center and OA as the radius, construct a circle inside the square. Outside the circle, lightly shade the columns and rows that intersect at O. Label the four pairs of points where these intersect the circle A, B, C, D, E, F, G and H, consecutively. Eliminate the grid lines from the interior of the circle.

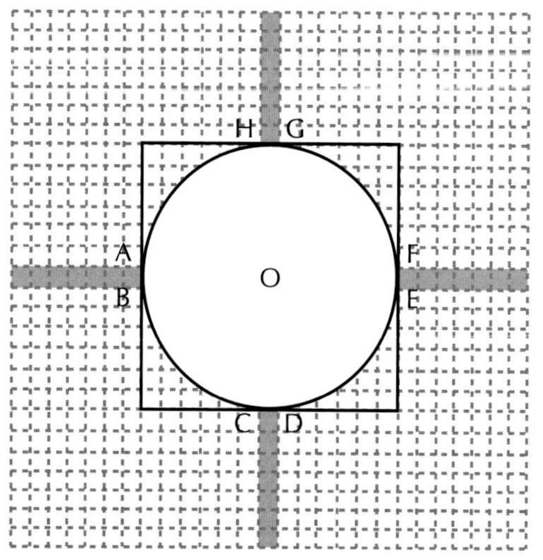

3. The required arcs within the sphere will be constructed from the center of the circle out, using the lines on the grid as endpoints, and reducing the length of the radius each time, until it approaches that of the original circle. The following steps give the specific as in our example. All line segments mentioned will be those in the shaded rows and columns.

4. Place the metal tip in the exterior of the circle on the midpoint of the line segment 7 units away from line segment CD and with the pencil tip on A, draw arc AF. Without changing the setting, and staying 7 units away from the square, move around in a counterclockwise direction, and draw arcs CH, EB, and GD.

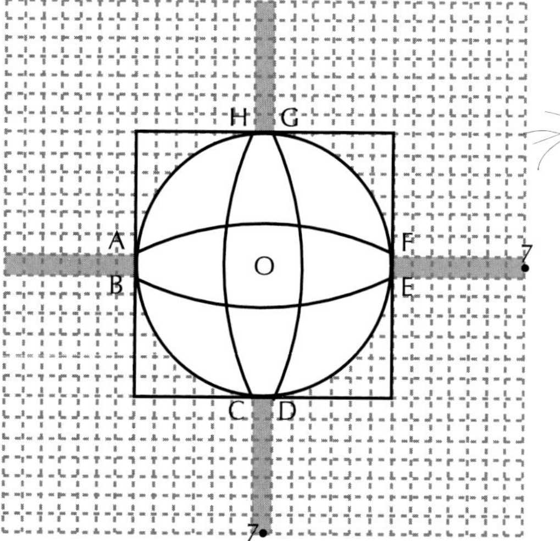

5. To construct the next set of arcs place the metal tip on the midpoint of the segment 5 units away from line segment CD and draw an arc whose endpoints lie on the next line out from the center after line AF. Continue around the circle as before.

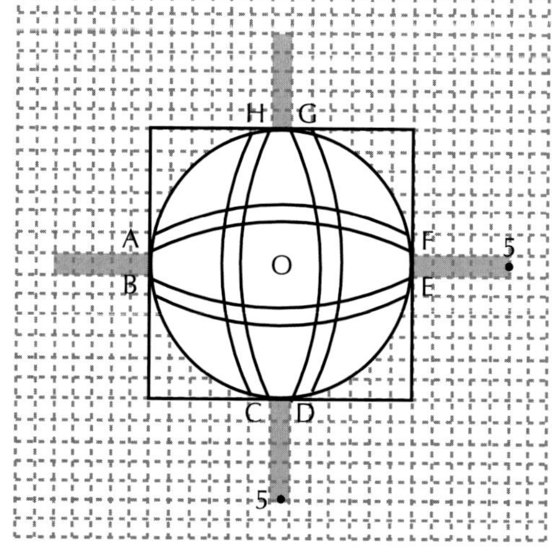

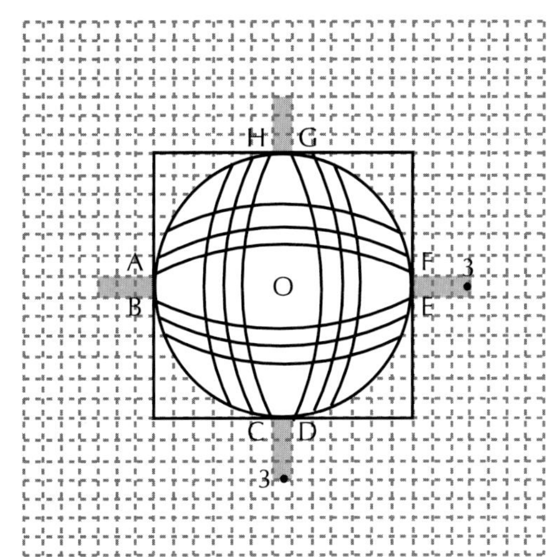

6. The remaining sets of arcs will be centered on the midpoints of the segments 3 units, 2 units and 1 unit from line segment CD, respectively, and drawn as before.

Now the figure is an illusion sphere.

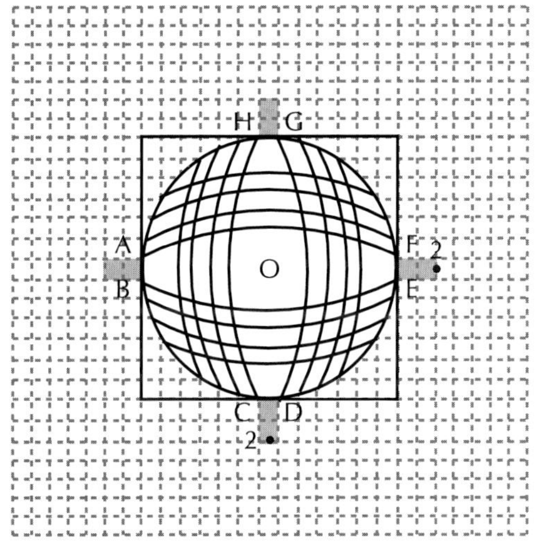

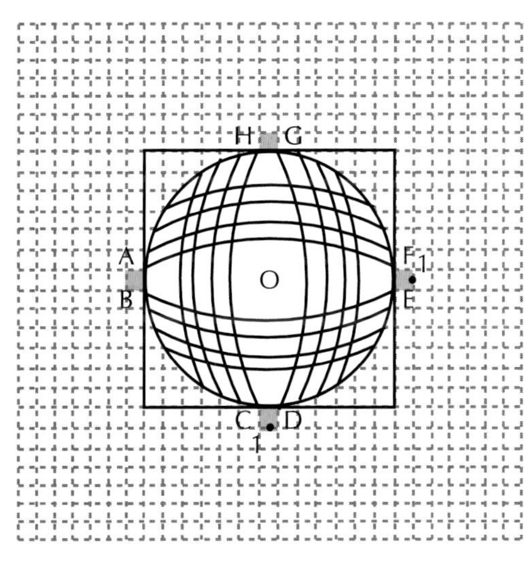

This page:
Photograph of a convex ceil-
ing mirror in an antiques
shop. Baltimore, Maryland.

The above photograph is of a convex ceiling mirror hanging in an antique shop in Baltimore, Maryland. The photographer got caught up in his own reflection. Notice that the straight lines in the architecture and objects become curved in the image. Can you identify any of the objects found within this space? The bounded circle of the mirror hints at the way the hyperbolic plane works. The objects at the center of the image appear larger than those at the boundary edge of the circle. No straight lines exist. These become arcs of circles. Try doing a drawing based on what you see here. Use a circular format and pencil tone only. Eliminate elements that you think would make your drawing too cluttered to be visually legible. Research the self-portrait in a mirror ball created by M.C. Escher. Compare his drawing to this photograph.

You could also set up a group of still life objects made of metal such as coffee pots, pitchers, creamers, etc. and see what kinds of reflections occur in objects that have both convex and concave surfaces.

Hyperbolic Geometry

Let us look at just one geometry that is related to the sphere but which is different from spherical geometry. It is called hyperbolic from the word hyperbole, meaning excess. In this situation, the excell refers to the number of parallel lines. Through any point *not* on a straight line, infinitely many straight lines can be drawn having no point in common with the given line. F. 3.6

Hyperbolic space is an abstract spatial concept developed by mathematicians for solving particular types of problems. It does not exist in the physical sense (at least on this planet at this moment in time). Yet we know from other discoveries in the history of mathematics that what starts out as solely a theoretical problem eventually relates to an actual physical experience.

Hyperbolic reality has a relationship to Euclidean reality in many properties. The first four postulates of Euclid work in this geometry too. The basic terms of point, line, surface, and plane are the same. We shall look at a schematic model of hyperbolic geometry developed by the mathematician Henri Poincaré that uses the circle instead of rectangular portions of the plane. Circles on a sphere map to circles on a plane so a circle can function as a descriptive visual of what happens on the three-dimensional surface of a sphere.

In the Poincaré model, the Euclidean plane is replaced by the interior of a circle. Lines are represented by arcs of circles that cut the big circle at right angles. The twist in this game is how distance is measured

inside this circle. You can think of his model as a world in which your measuring rod shrinks as you move away from the center of the circle. The bounding circle is "infinitely far away". You can never reach it, no matter how long your stride or how fast your run. The edge of the world is ever-elusive. It is a world in which Lewis Carroll's Alice would feel very much at home.

The problem for the visual artist when looking at this model is that it is a contradition between what is known and what is seen. You have to disregard the visual and operate at the conceptual level. Figures in this model appear curved but are considered straight. Figures appear larger in the center of the circle, gradually get smaller, and are smallest at the boudary edge but these figures are all to be considered congruent.

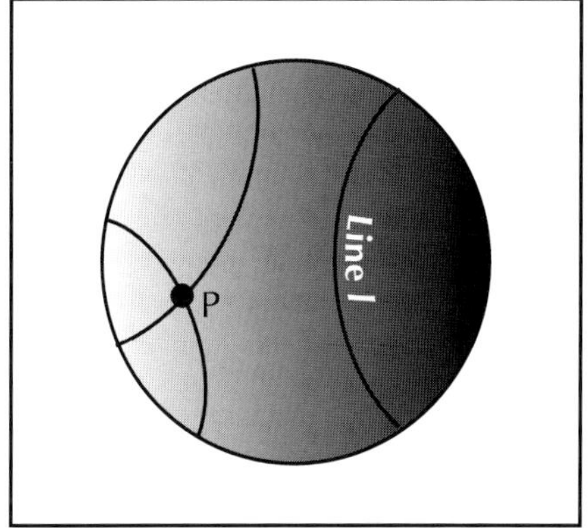

This page;
Above: Fig. 3.6
On the sphere, lines through
P which do not meet line l.

Below:
Photo of a sphere made from
woven elements. Notice that
the overlapping reeds suggest
a tiling of equilater triangles.
Geometry of a sphere is called
elliptical geometry.

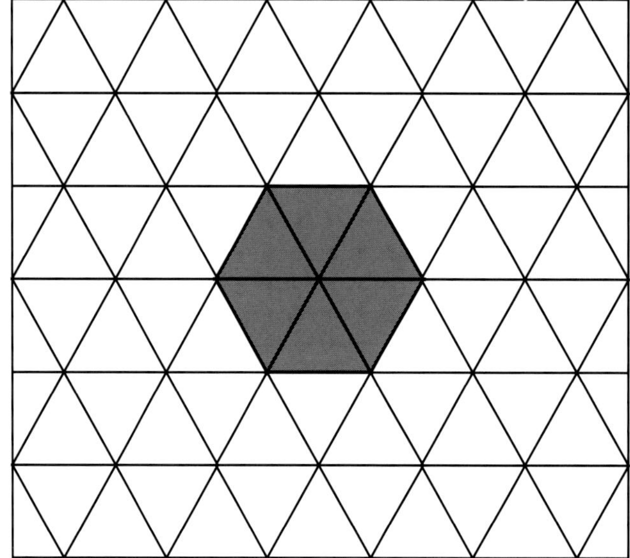

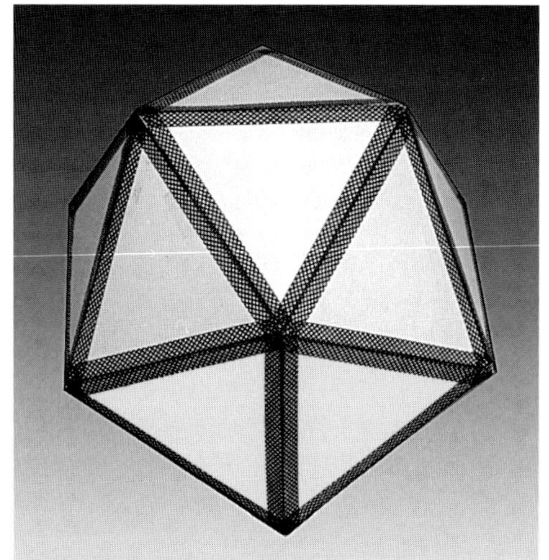

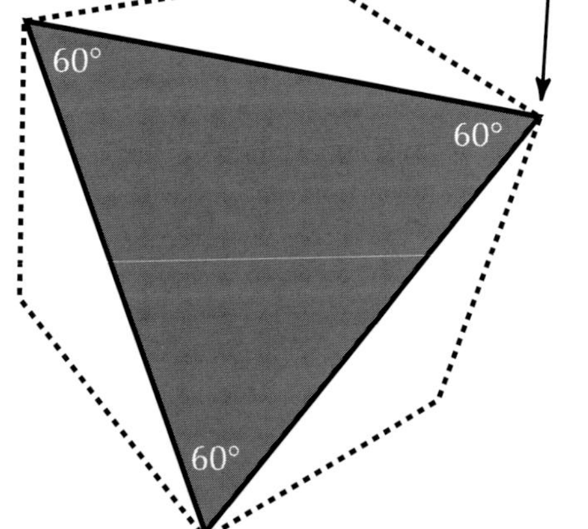

In tiling the Euclidean plane with equilateral triangles, six is the maximum number required at each vertex. The plane lays flat. Fig. 3.7

If five equilateral triangles are used at every vertex, the group will soon close up to give the three-dimensional figures called the icosahedron as seen in the photograph to the left. It is a member of the large family of polyhedra that will be discussed in the next chapter.

To construct a model of a hyperbolic surface with equilateral trianggles, you need a minimum of 7 at each vertex as in the photograph on the facing page. Where elliptic geometry takes place on the surface of a sphere and Euclidean geometry takes place on a plane, hyperbolic takes place on a surface of negative survature. To see what this looks like, try the following based on tiling equilateral triangles together.

To do this for yourself, use a copy of Fig. 3.9, leave a slit along the edge at the dotted line. Score all the other lines and insert an additional triangle as indicated. You could try adding two, three, or more at this point. Do this at all the vertices. Continue to add groups of 7 until you get tired of the game. To complicate the visual aspect, not the structural, add drawing, pattern, texture, or color to the surfaces.

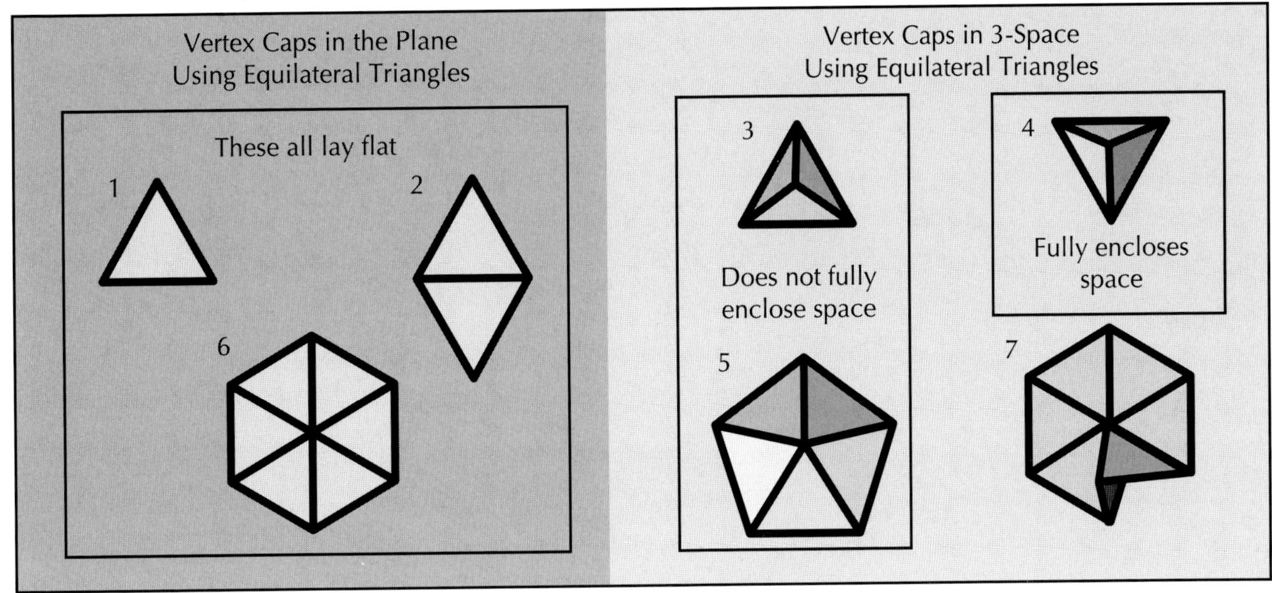

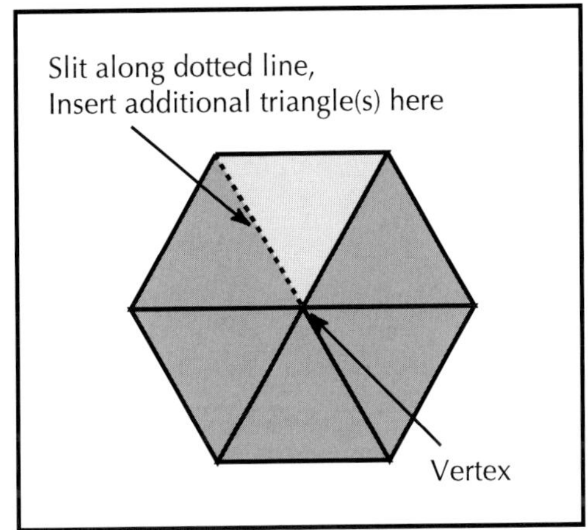

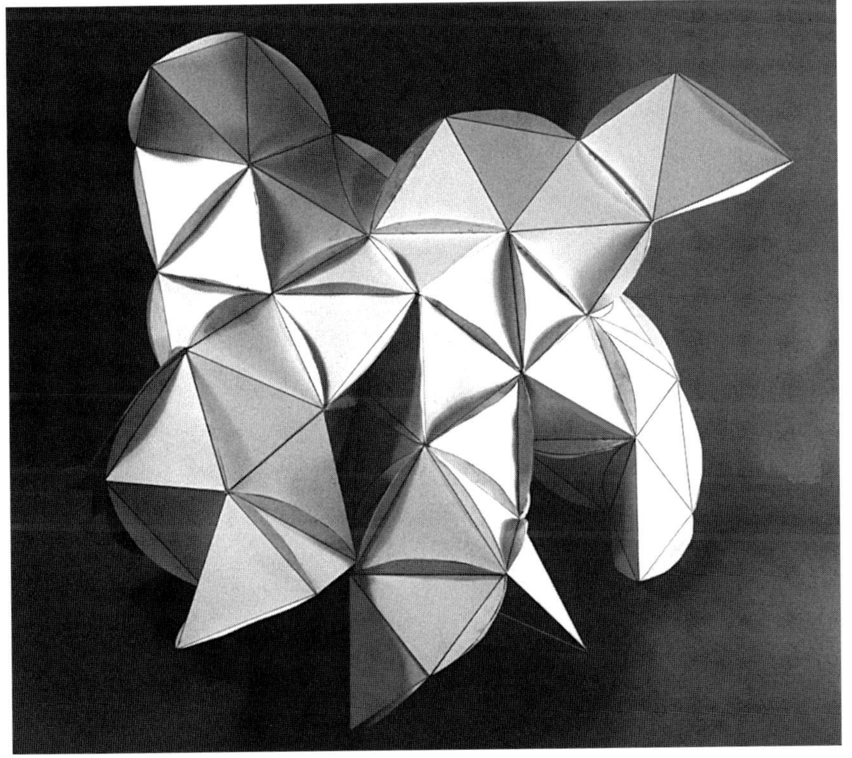

This page:
Top: Chart showing the various vertex caps, from 1 to 7, using equilateral triangles. Notice that you need a minimum of three triangles to enclose space. Four triangles enclose space entirely.

Center: Fig. 3.9 Template for six joined equilateral triangles with edge marked for adding more triangles.

Bottom left:
Student work. Doug Hooten. A portion of the plane in which 7 equilateral triangles are used at each vertex. Notice that there are concave and convex areas.

Comparison Between Euclidean & Hyperbolic Space
(The terms point, line, straight line, surface, and plane surface are the same in both geometries)

This page: Comparison chart between Euclidean and Hyperbolic space.
Hyperbolic geometry is a consistent axiomatic system. According to mathematician Richard Trudeau, "A system is consistent if no contradiction can be deduced from its foundation of primitive terms, defined terms, and axioms." A model is then developed to be the simplest interpretation of the terms and the axioms.

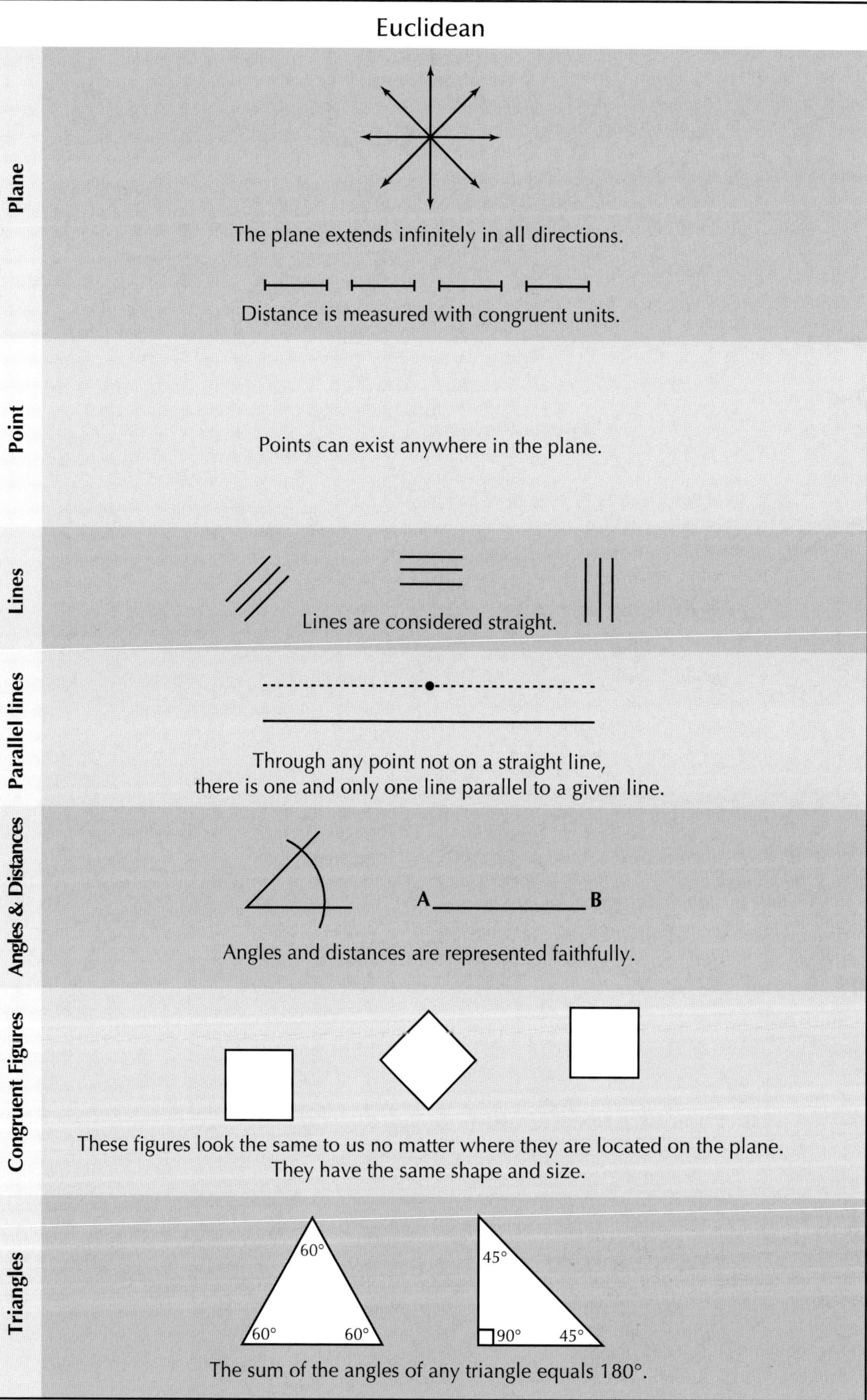

Comparison Between Euclidean & Hyperbolic Space

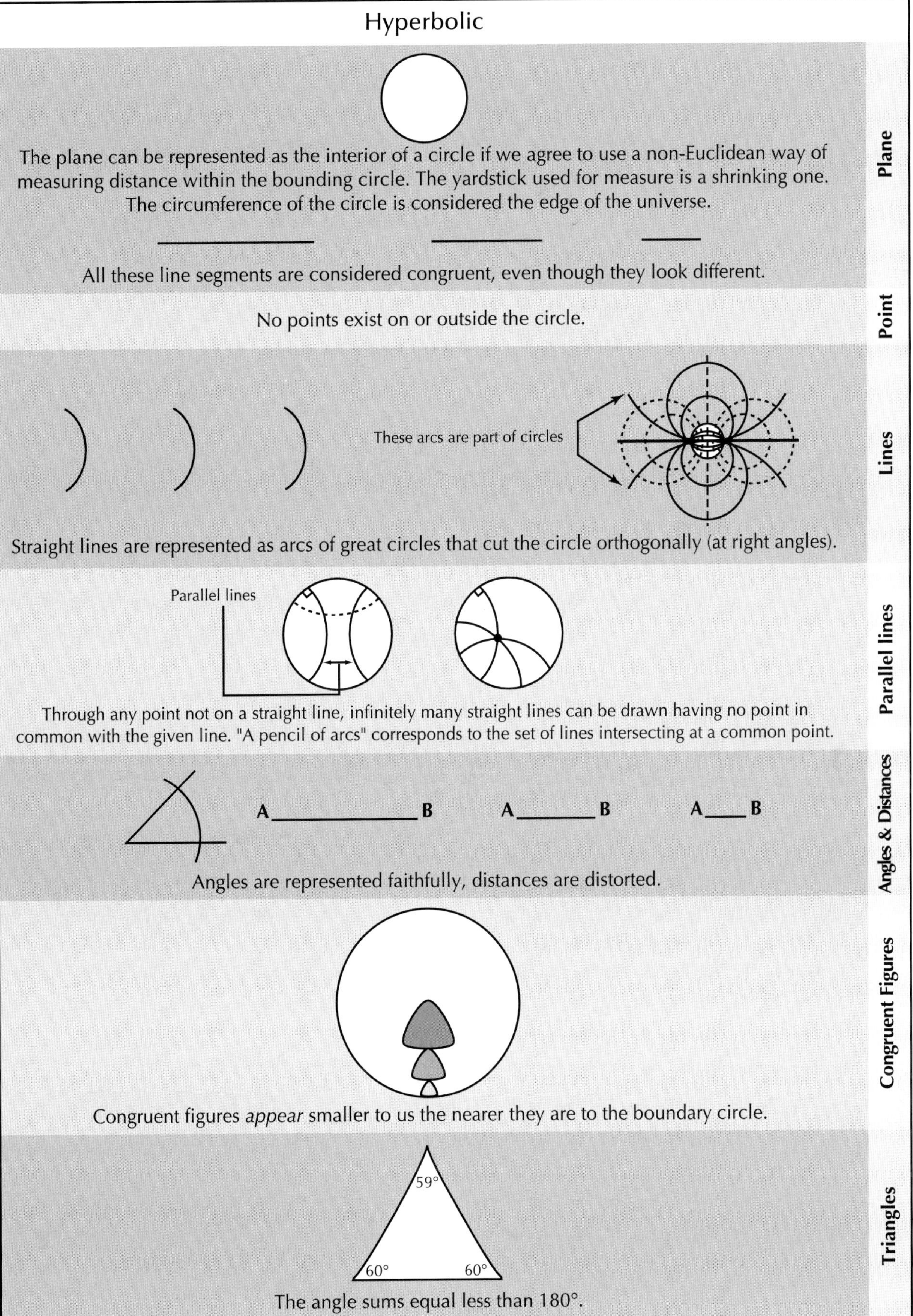

The Euclidean plane can be tiled in so many wasys despite the fact that there are only the three regular tilings of the equilateral triangle, the square, and the regular hexagon. A regular tiling is defined as a covering of the plane by regular (congruent lengths and congruent angles) polygons so that the same number and kind of polygons meet at each vertex. As we have said before, six equilateral triangles come together at each vertex; four squares come together at a vertex; three regular hexagons come together at a vertex. Each of these polygons can be manipulated in a number of ways and still tile the plane in their altered states.

On this and the facing page are three black and white tilings of the Euclidean plane. The Twentieth Century artist, M.C. Escher, used these to great artistic advantage by joining pure geometry with representational imagery.

He also explored hyperbolic til-ings because he found that this finite schematic structure could be used to represent his sense of the infinite. He did a number of artworks enti-tled "Circle Limits" that explore this type of tiling. We suggest you re-search the very many good books that are available on this artist.

The hyperbolic plane is shaped differently from the Euclidean plane, therefore, it can be tiled by any reg-ular polygon, including the penta-gon, and in many more ways than the Euclidean plane. Like the Archimedean tilings in Euclidean space, these hyperbolic tilings use two kinds of polygons around any given vertex.

Despite the fact that it is a much easier task to develop hyperbolic tilings using a computer program, we always like to offer a humble hands-on compass and straightedge approach, if at all possible. On the following papers we offer two dif-ferent artful constructions, one in a square, and the other in a circle.

This page:
Two tilings of the Euclidean plane in black and white only. Artworks by Richard and Rochelle Newman. Marker or acrylic paint on watercolor paper.

Top: Richard Newman

Bottom: Rochelle Newman

All quadrilaterals, four-sided fig-ures, tile the plane. A square is a member of the quadrilateral family. The square can be tilted to give a rhombus and that rhombus will tile the plane also. Sections can be add-ed and removed from the quadrilat-eral and through the symmetry oper-ations of translation or rotation, a new iregular unit, having the same area as the square, will also be able to tile the plane. See the Design Ap-pendix for information on symme-try.

Triangle and hexagons, as well, can be manipulated through the use of the symmetry operations. New and interesting figures will emerge. Research books on tilings and then try some works of your own.

Using the Square
to Suggest the Poincare Plane

Given square ABCD

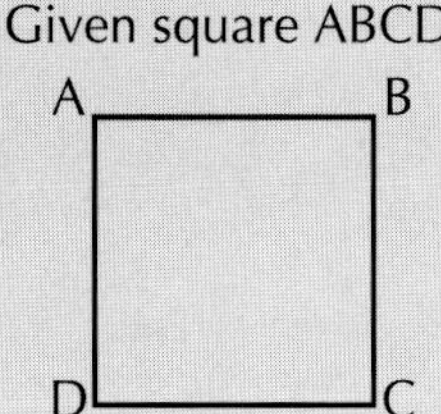

1. Bisect square ABCD. Label the points of intersection E,F,G,H,I respectively.

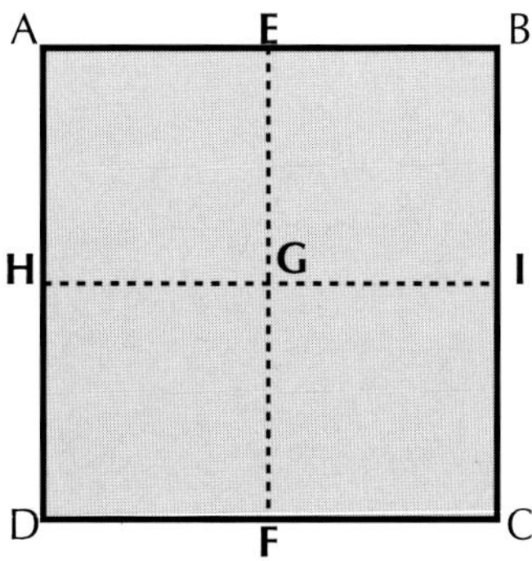

2. Choose one of the quadrants just generated, in this case AEGH, and bisect it. Label the points of intersection J,K,L,M,N respectively.

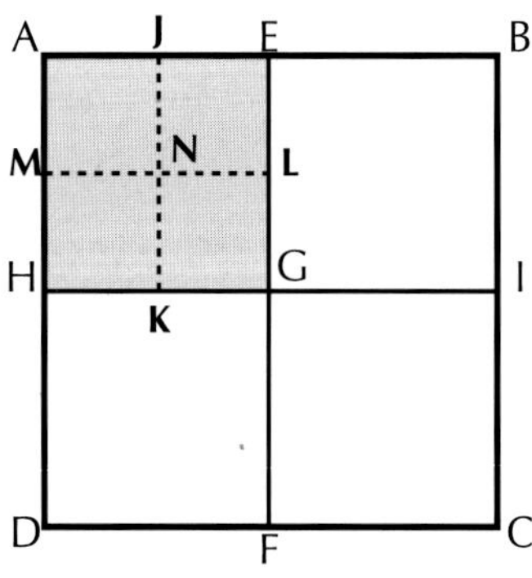

3. Choose one of the quadrants just generated, in this case AJNM, and bisect it. Label the points of intersection O,P,Q,R,S respectively.

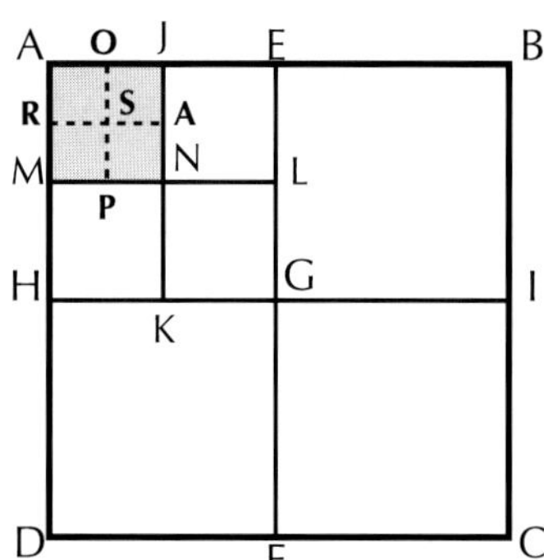

As you can see this process can go on indefinitely. It relates to the property of self similarity found in fractal geometry.

4. This process can be duplicated in the three remaining quadrants. Now you have an armature on which to construct a design using purely abstract shapes or representative images as did the artist M.C. Escher in his works such as *Square Limit*, 1964.

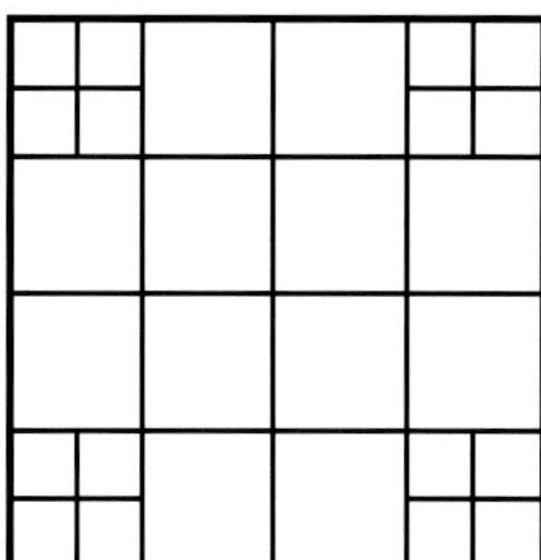

On the following page, we show you an image which has been based on this particular Escher work and uses the armature below.

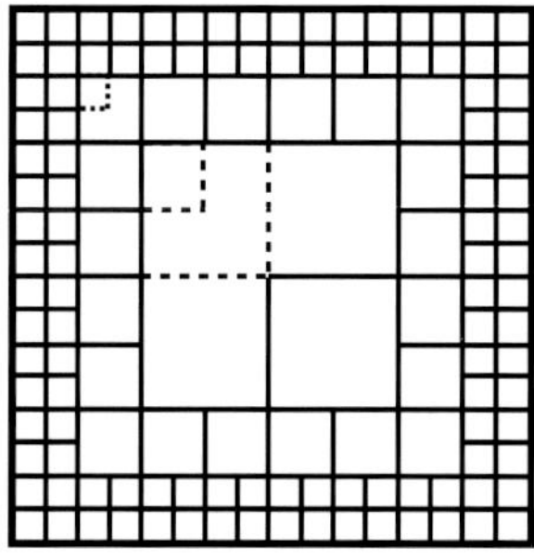

The above diagram, the last step of the construction given on the previous page, can be used as the basis for an artwork as you will see on the following two pages. When creating your own artwork, photocopy the diagram, lay tracing paper over the grid, and develop your image, abstract or representational.

The entire grid could be skewed to produced rhombuses and the grid, while looking different, would still function in the same way. Or the entire grid could be stretched either vertically or horizontally. What other ways can you develop that would change the grid and still keep the essential information the same? Think of internal and well as external manipulations. Consider the grid and its image as a mapping as was the face way back in Chapter 1.

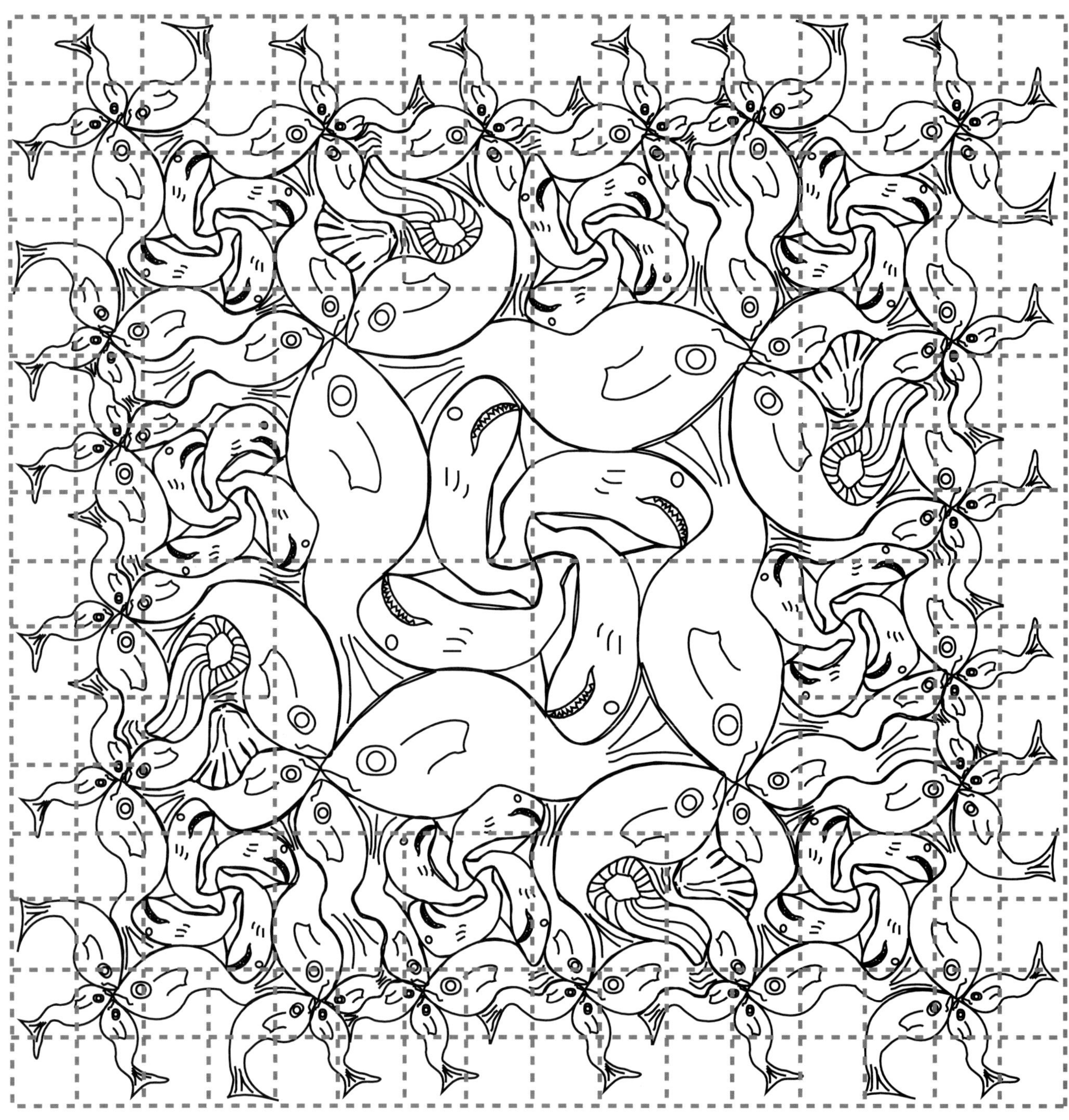

Above:
Computer line drawing constructed within the square grid outline.

Using the graphic work of M.C. Escher entitled, "Square Limit, 1964", as a point of departure, computer artist, Claire Belanger, used a fish motif to fit within the square unit. The fish is a simple form to recognize and adaptable to manipulation. This may be why Escher probably chose it in the first place.

The fish unit is given a four-turn symmetry operation in order to pro-duce the large central square. The initial unit is reduced one-fourth of its original size. This unit is then placed along the edges of the large central square. Subtle adjustments have been made for artistic effect so that the positive and negative shapes fit together well. This process is repeated again and units are placed along the edges. Notice the subtle alterations so that the fish interlock.

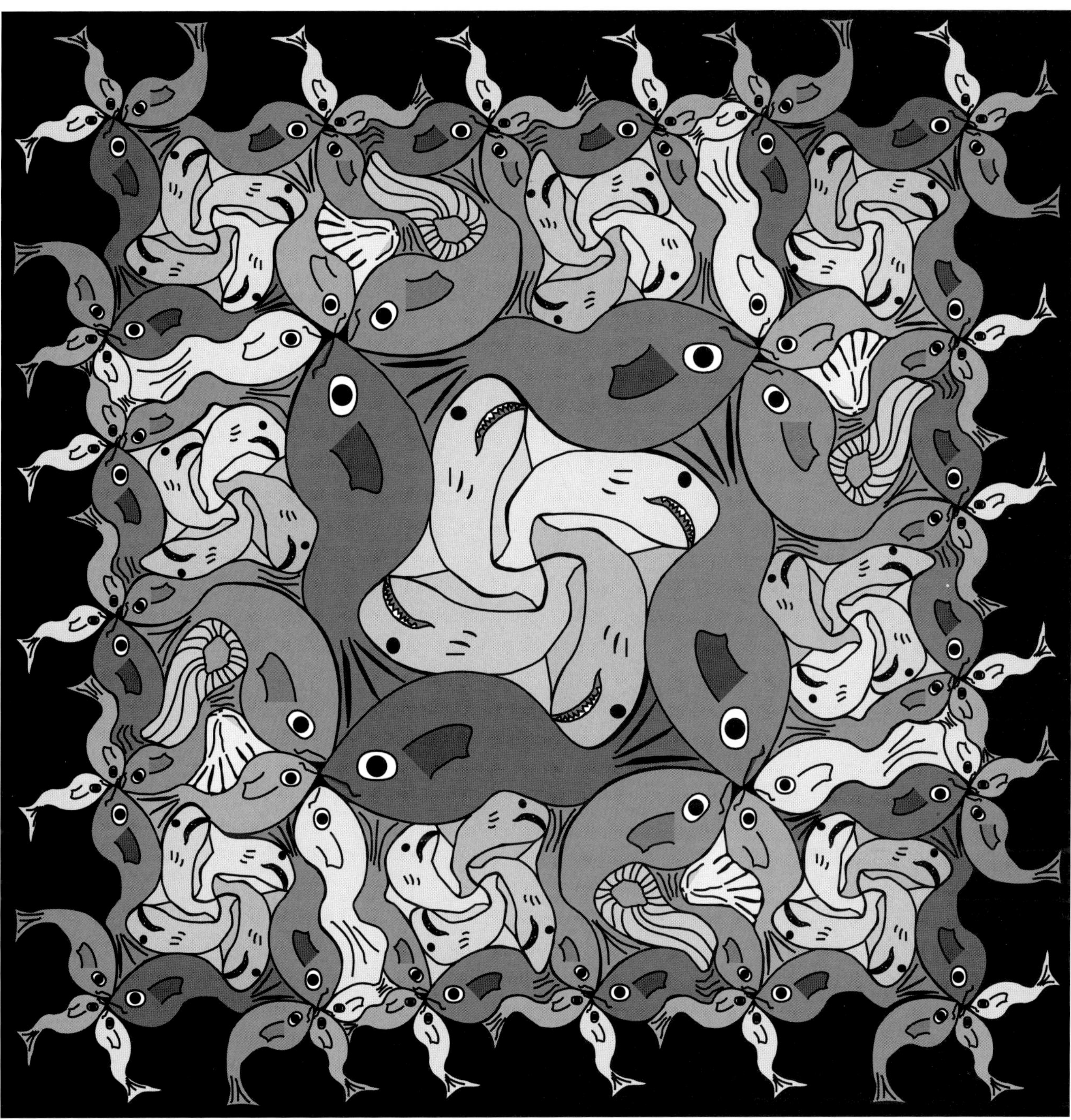

The completed undersea image, "Spirit Guardians of the Seas" by Claire Belanger. Notice how the underlying grid is sensed but is not easily comprehensible to analysis without a great deal of work using tracing paper overlays.

The various gray tones help to unify a complex image. What would happen if colors were substituted for the various tones? Would the image get too chaotic looking or would the colors add another artistic dimension? Lay tracing paper over this image and with colored pencils or markers, convert the tones to colors. A monochromatic color approach would be closest in mood to the tones while a complementary color structure would set a strong contrast. What if you used an acetate material to obtain a stained glass effect.

Above:
Finished computer generated artwork by Claire Belanger entitled "Spirit Guardians of the Sea".
The artist used various tones of gray only to create her finished work.
Lay tracing paper over the above and try a color variation of your own.

133

Since the circle is used for developing the Poincare Schematic for the hyperbolic planes let us give some properties of the circle in Euclidean space.

Circle: The set of all points in a plane at a fixed distance from a given point.

Arc: The continuous unbroken portion of the circle which lies between two points on a circle.

Chord: A line segment which has its endpoints on the circle.

Circumference: The distance around the circle. This denotes a real number rather than a geometric figure. To find the circumference use the formula:

$C = \pi d$ or $C = 2\pi r$
(C= circumference, d= diameter, r= radius, $\pi \approx 3.14$)

To find the area of the circle use the formula:

$A = \pi r^2$ (A= area, r= radius, $\pi \approx 3.14$)

Degrees: A circle has 360°.

Diameter: A chord that passes through the center of the circle.

Quadrant: A sector that contains one-quarter of the circle.

Radius: A line segment from the center to a point on the circle.

Secant: A line that contains a chord of the circle.

Sector: A region in the interior of the circle bounded by two radii and the circle itself. A sector is called a quadrant if it contains one-fourth of the interior.

Tangent: A line outside the circle that intersects the circle in exactly one point and lies in the same plane.

A circle subdivides the plane into three distinct regions.

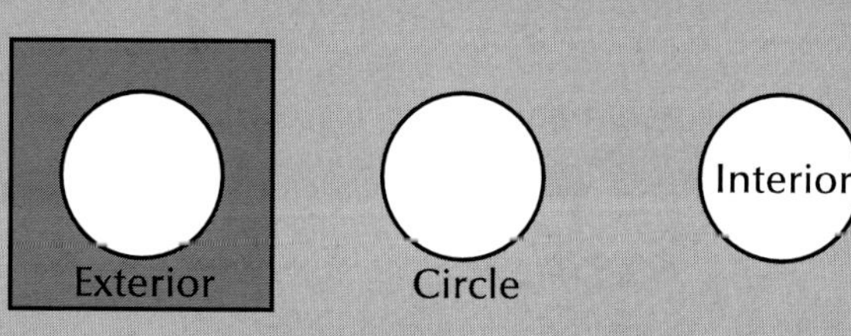

Every circle contains 360°, we need only to divide 360° by the number of congruent arcs required to find the measure of the central angles. For example, to divide a circle into 13 congruent arcs, we divide 360° by 13 to get 27.8°. We draw a central angle of 27.8° and then cut the circle consecutively with the corresponding arc length using a compass to obtain the 13 congruent arcs.

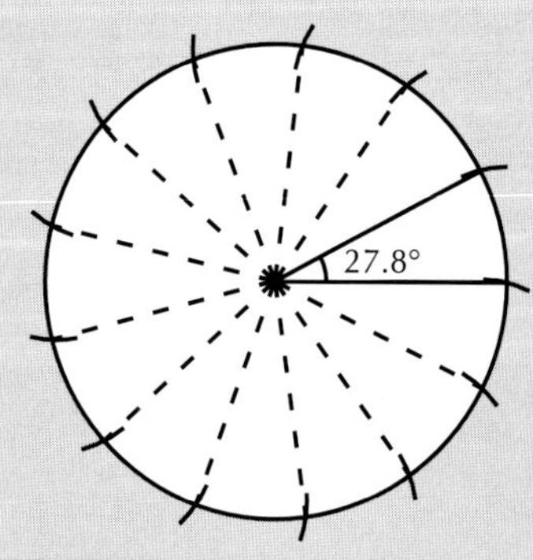

This page:
Photographs of artful circles.

Top: Crocheted doilly of circles surrounding circles.

Center: Plan view of a woven basket using natural materials.

Bottom: Celtic interlace of stonework.

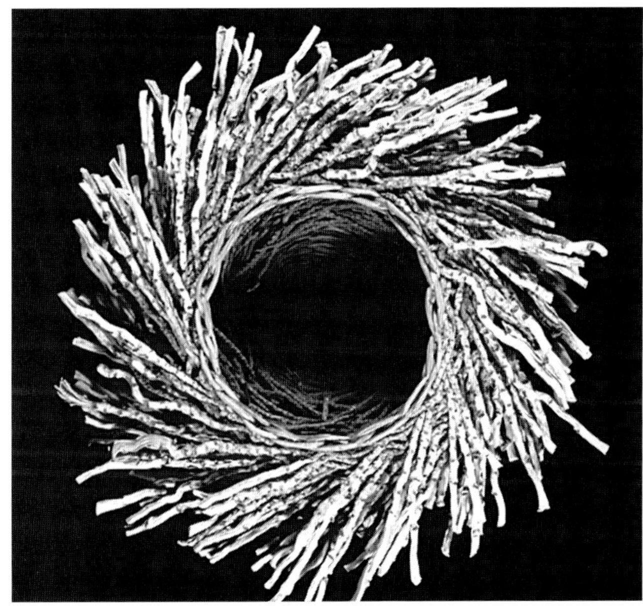

The Poincare Schematic for the Hyperbolic Plane as Represented on the Euclidean Plane

Given a circle with center O

1. The given circle is divided n-times, the choice is arbitrary (Try using a division of 8). In this example we have chosen to use a division of 6, (360° ÷ 6 = 60°). This circle is called the bounding circle which is a Euclidean circle somewhere in the Euclidean plane. The radius "R" is considered large enough to allow for a good sized population of two-dimensional figures to live inside the circle generated by R.

 The population is not aware it is living within a circle. The interior of the circle stretches infinitely in all directions just like the Euclidean plane.

 The viewer, who is very much oversized, is standing on a plane outside the circle and observing what is happening within the circle.

Measuring devices such as a one meter stick when placed at the center of the bounding circle is considered to be truly one meter. It shrinks as it moves away from the center in a particular ratio. We, the observers, notice that the figures are shrinking but the figures within the circle are not aware of any differences of the length of their meter sticks as they approach the boundary of the circle. The figures will never approach the edge as it is infinitely far from them.

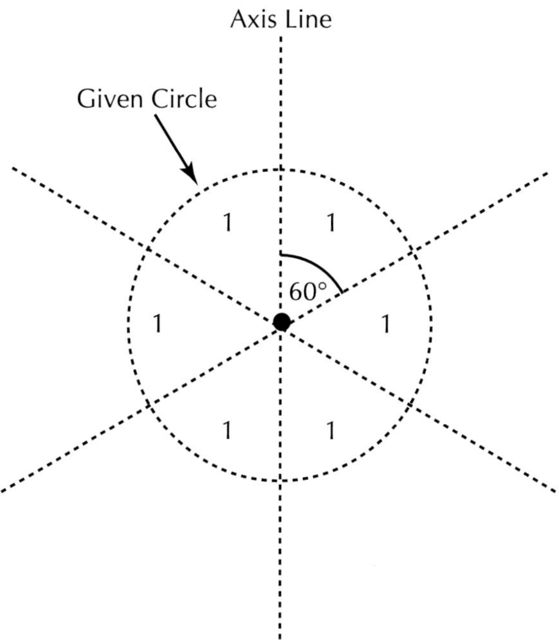

2. Bisect each sector of the circle so that there are a total of twelve axes. These will be used to set radii of circles.

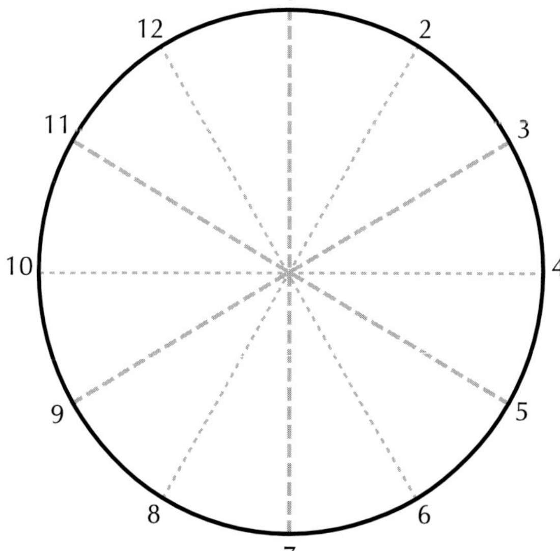

3. The given circle is used to establish the first ring of six overlapping circles. The first circle in that ring needs to be set orthogonally, which means at right angles, to the original bounding circle.

    Any line that the outside viewer perceives as curved will be perceived by the inside inhabitants as straight.

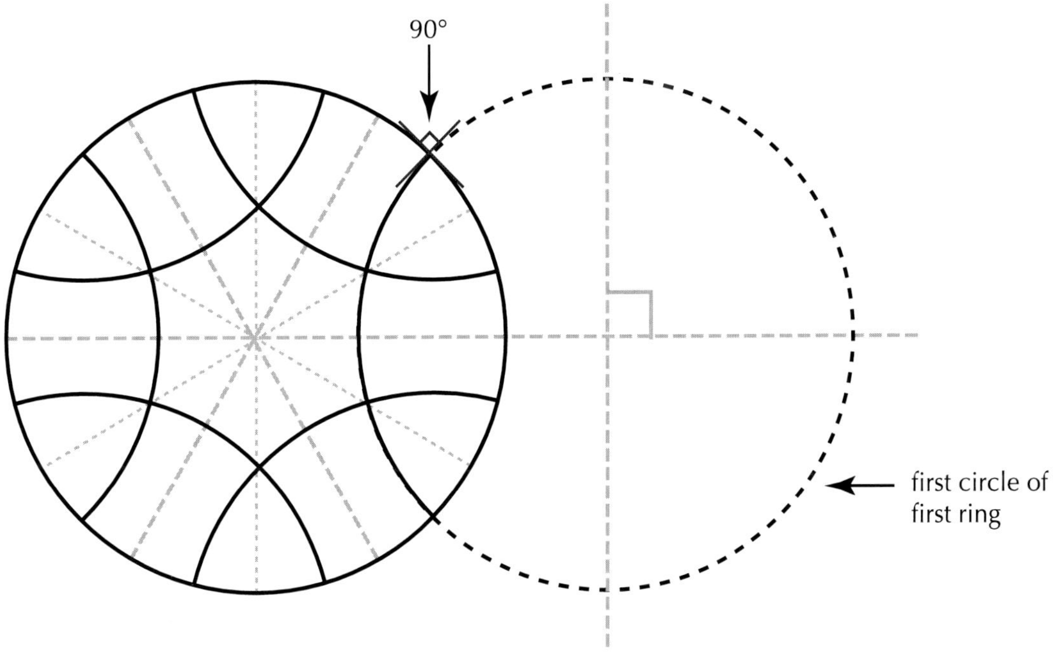

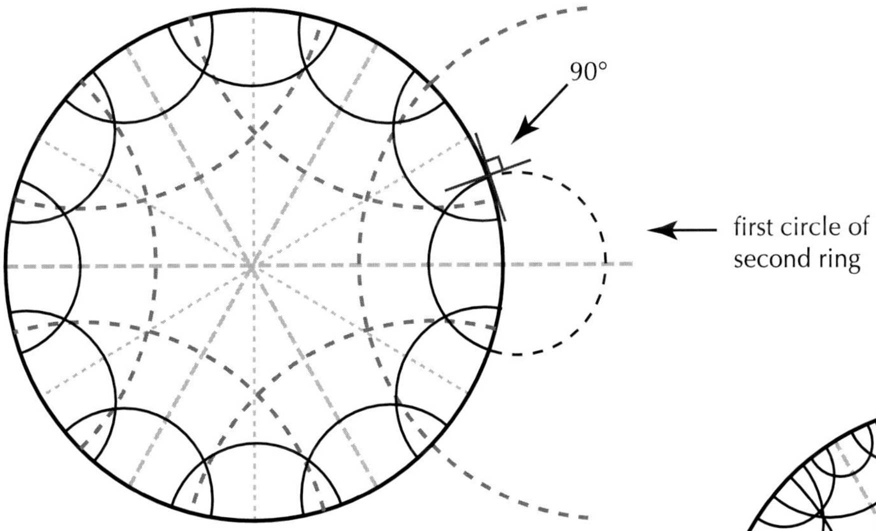

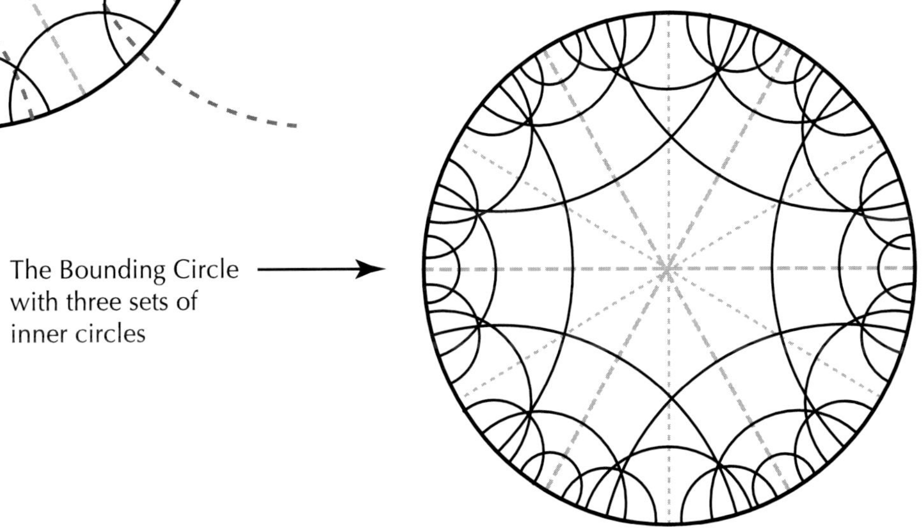

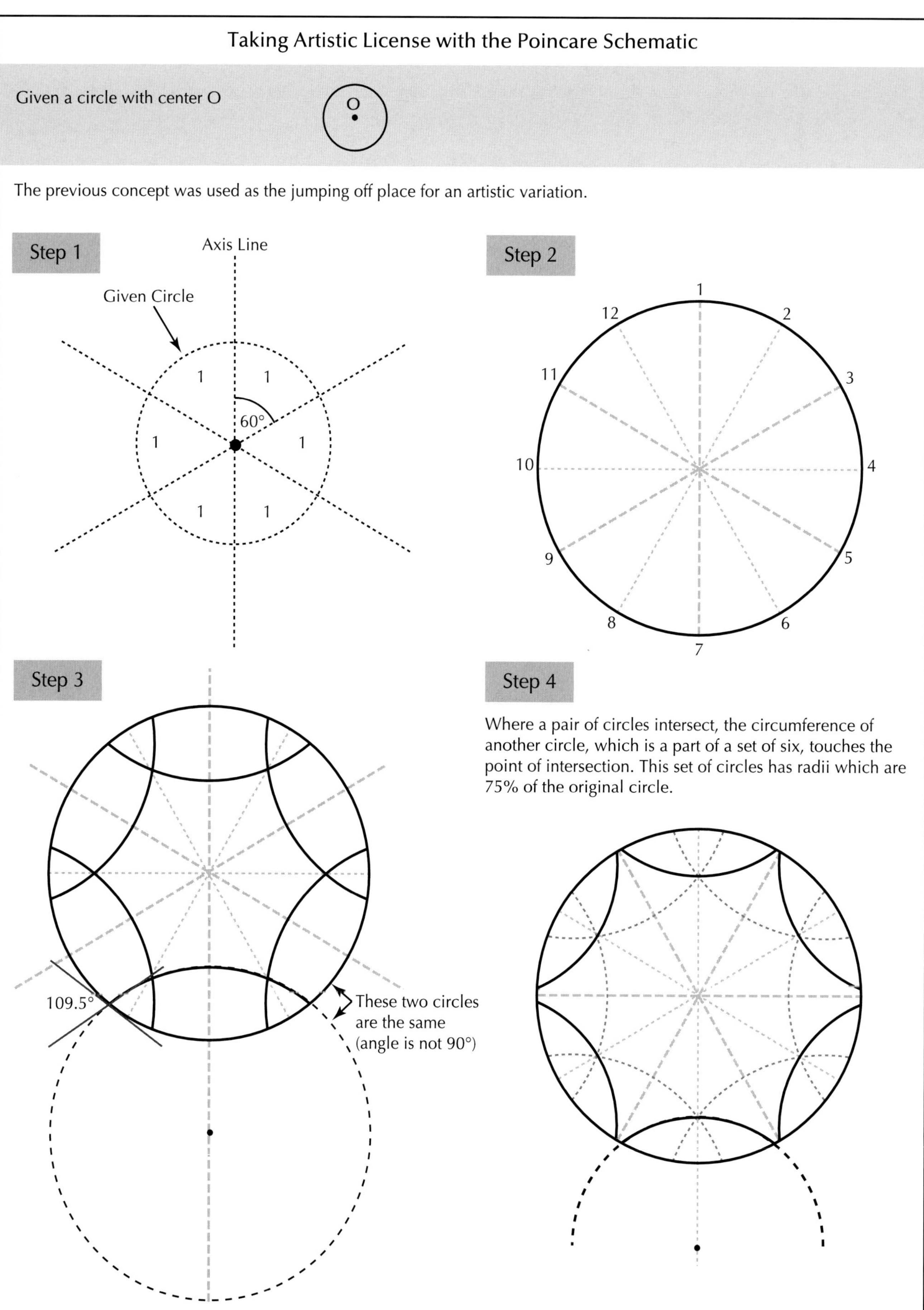
Taking Artistic License with the Poincare Schematic
Given a circle with center O
O
The previous concept was used as the jumping off place for an artistic variation.
Step 1
Axis Line
Given Circle
1
1
1
1
1
1
60°
Step 2
1
12
2
11
3
10
4
9
5
8
6
7
Step 3
109.5°
These two circles
are the same
(angle is not 90°)
Step 4
Where a pair of circles intersect, the circumference of another circle, which is a part of a set of six, touches the point of intersection. This set of circles has radii which are 75% of the original circle.

Step 5

The distance between circle C and circle B is half of the
distance between the circumference of circle B and the
center of circle A.

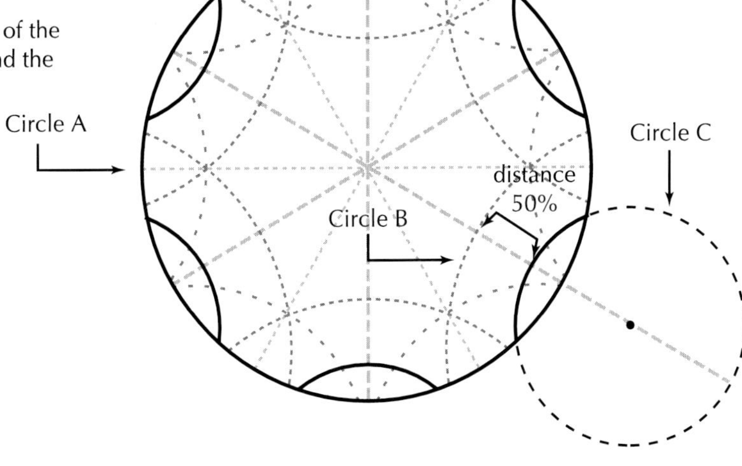

Step 6

Where a pair of circles intersect, the circumference of another
circle, which is a part of a set of six, touches the point of
intersection. This set of circles has radii which are 50% of the
original circle.

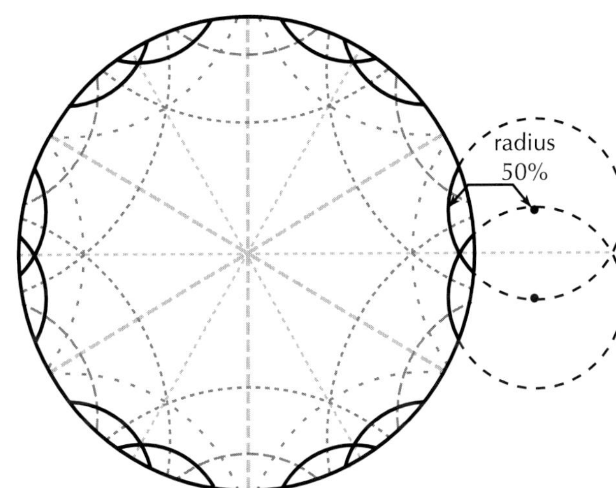

Step 7

The distance between circle D and circle C is half of
the distance between the circumference of circle C
and the circumference of circle B.

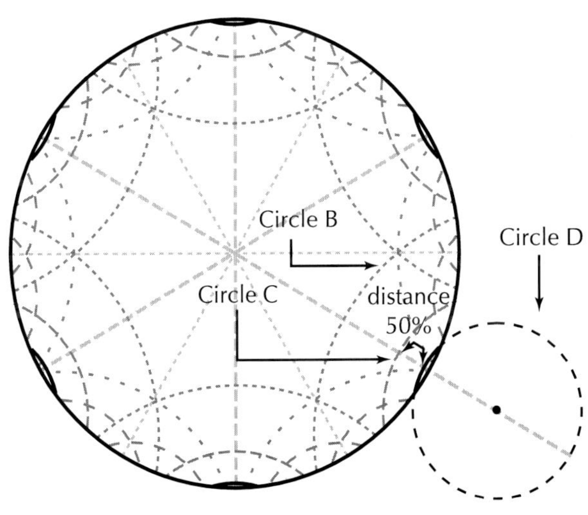

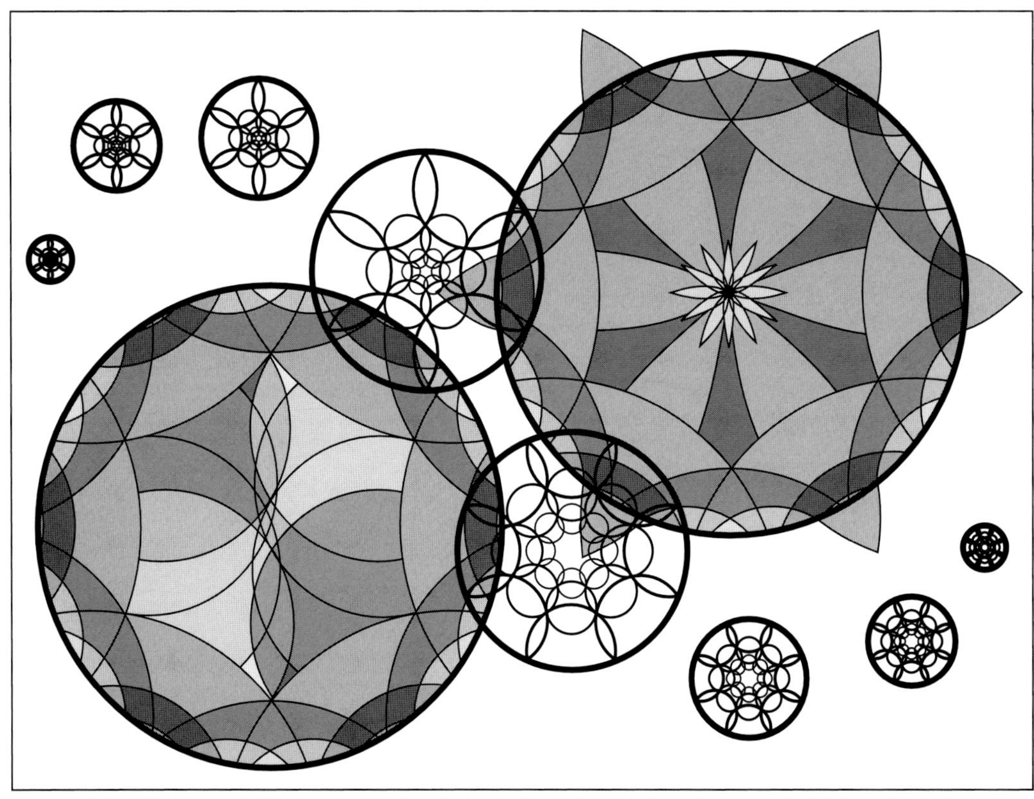

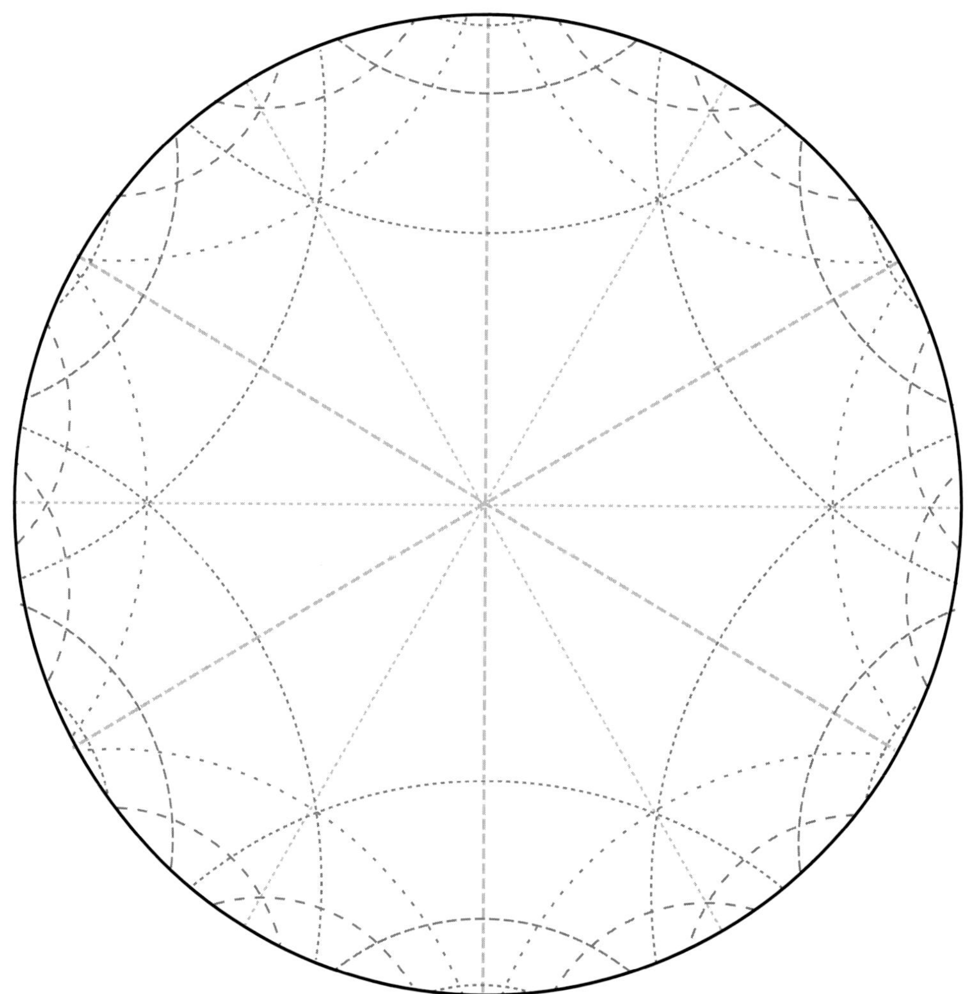

This page:
Top:
 Several artful manipulations of the computerconstruction given here.

Right:
The completed construction from the previous two pages with extraneous information removed. Enlarge to use as a template for your own artwork.

140

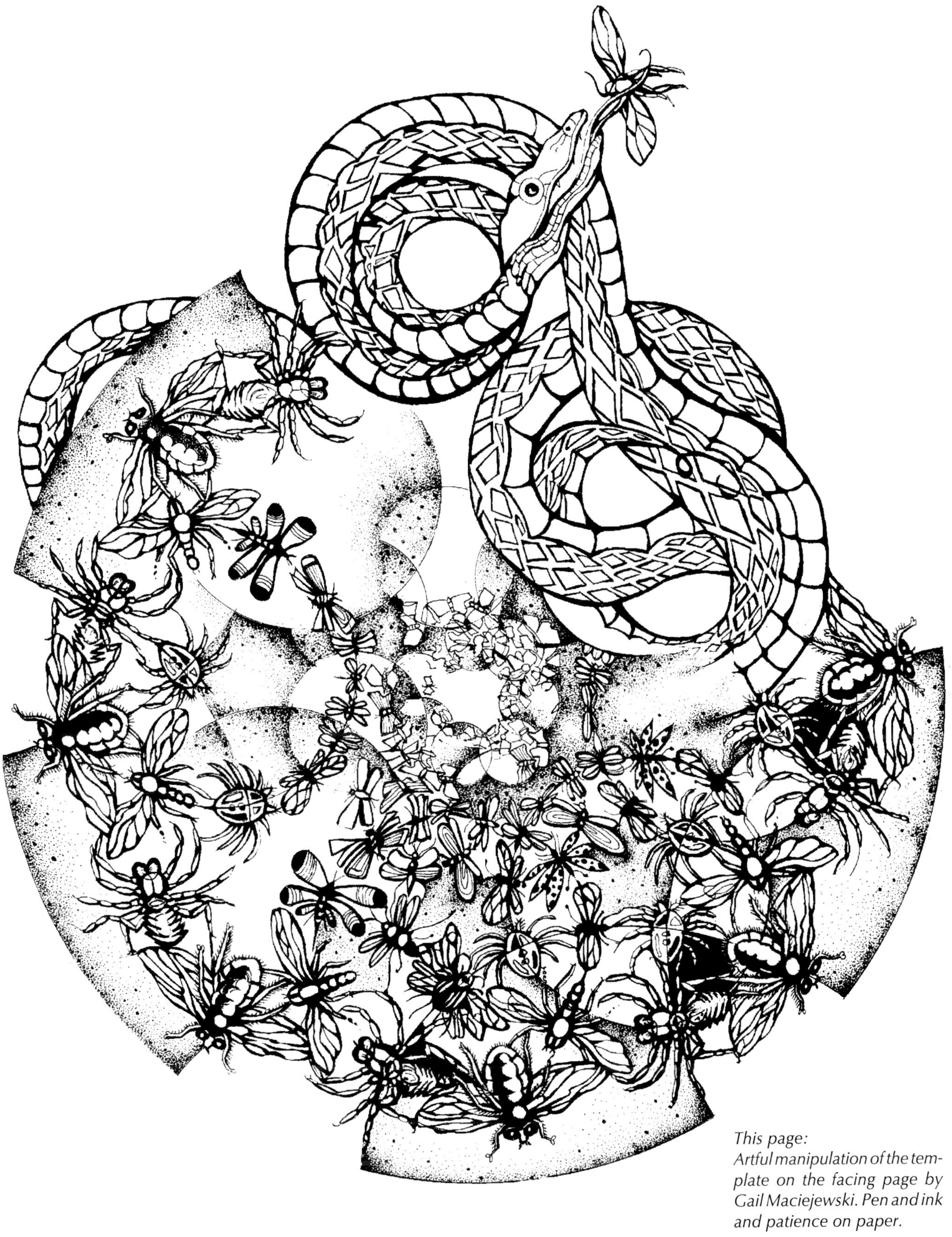

This page:
Artful manipulation of the template on the facing page by Gail Maciejewski. Pen and ink and patience on paper.

Problems, Projects, and Play

1. Research the cultural period at the time of Gauss, Lobachevsky, Bolyai. In a written essay, discuss what other ideas were coming to the fore that would change the thinking patterns of culture?

2. Develop a Non-Competitive Game for n-players.

This is not an easy project either in the thinking stage nor in the execution phase. What you are creating is a working model of a game based on cooperation. This game must combine the concepts of a grid with the rules of game strategy. It must contain the following elements:

a. Decisions about the age level of the players of the game.

b. The number of players involved in the game.

c. The size, shape, and structure of the gameboard.

d. The rules of the game.

e. A mock-up version of the gameboard and its rules.

3. Research the subject of "decision trees" and develop one of your onw. Then research how nature develops a tree. In writing and images, compare the two.

4. Locate a copy of a "Circle Limit" artwork by the artist M.C. Escher. With a written and visual analysis, explain its structure, symmetry, color, technique, content. etc.

5. Develop a card game that is cooperative and not competitive.

6. Research, and then present ;in an appropriate form, the them of games and contests found in literature. Suggestions: Poseidon and lAthena; Achilles; Gilgamesh, Ariadne, King Arthur. lWhat do these all have in common and ho do they differ?

7. Find the rules for a favorite game of yours. Change some of them and develop a game that is a variation on the original.

8. Write out the rules and information for an artist to follow in developing an artwork. Research the art of a person like Sol ;Lewitt of the Twentieth Century.

9. Develop a comparison chart in which you analyze three games such as basketball, baseball, and football in order to see what the commonalities and the differences are. From this comparison, develop a new game that incorporates elements from all three games.?

10. Analyze the ever popular game of "monopoly" and discuss in writing what you think it tells about our particular culture and its particular values? Compare it to a comparable game of another civilization. Suggest and develop an alternative game to monopoly..

11. Research the function and form of mandalas as the basis for a gameboard. Design and execute a prototype, include the rules.

12. Locate a computer program that generates hyperbolic tilings. Check the Internet for same. Print out an image that you like and do a variation on it using colored pencils, markers, or paints.

13. Develop a protoptype for a chess set in which the pieces are variations on polyhedra. See the next chapter for information on them.

14. Develop a gameboard, and rules, for a hyperboic plane. Base it on information you research on hyperbolic geometry.

15. Design and execute four face cards out of a full deck. Use a nature theme for the imagery. Can you find out why a deck of cards has 52?

16. Research the Tarot cards and do a contemporary version of them.

17. Take a popular game and convert it from a two-dimensional board to a three-dimensional space game. Write out the rules and have some friends play it with you. Find out where the glitches occur and then adjust the rules. There is already a three-dimensional version of tic-tac-toe, so do not do that one.

18. Research the subject of regular tilings of the Euclidean plane. Develop one in black and white. Then do a color variation of it. Use the same tiling and see if you can devlop a hyperboic variation of it.

19. Develop a prototype for a board game on the subject of the environment. Call it "Earthwatch". Have some rules which involve competition and some which involve cooperation.

20. Do a prototype for a children's book that introduces them to the notion of the hyperbolic plane.

21. Do a prototype for a variation on the board game "Monopoly" which reverses all the rules, images, choices, etc.

22. Using the symmetry operations of rotation, reflection, and translation, develop a full set of playing cards that uses a circular format.

23. Do a set of drawings of the human face, or figure, in black and white that begins ;with the plane and moves onto other surfaces.

24. Set up a still life of a group of objects that have both a curved and a reflective surface. Do a detailed pencil drawing of the objects paying special attention to how the curved surfaces affect the reflections.

25. Using a material called "self-hardening clay", develop a sculptural form that uses changes in surface that begins with the plane and moves on to more and more curved areas.

26. Using the information in this chapter, develop a drawing for a surface composed of 8 equilateral triangles at every vertex. Execute the drawing in black and white, tone, or color.

27. Read "Alice in Wonderland or "Through the Looking-Glass" by Lewis Carroll. Then do a drawing or collage of lan imaginary landscape that is alos a gameboard. Place Alice and other characters in the image.

28. Do a drawing, collage, painting, or sculpture that inclues at least four of the geometric constructions introduced in this chapter.

29. Do a visually interesting page layout that introduces the first four postulates of Euclid's geometry to the reader. Make it so exciting that the reader wants to know more and more and more....

30 Obtain a copy of Euclid's "Elements". With computer skills or hand drawing or calligraphy, do a contemporary artful verion of several pages. Base them on aspects of the book that you find interesting.

This page:
Suggestion of human form built from geometric constructions. Drawing in pen and ink by Gail Maciejewski.
Try your own figure variations using geometric elements and the information on proportion found in Chapter 2.

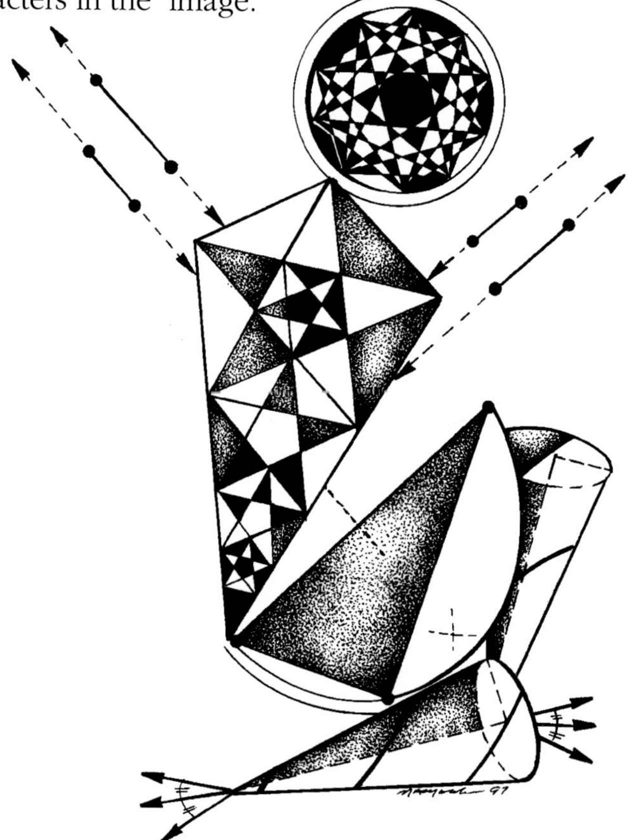

Further Reading

Burton, David M., *History of Mathematics: An Introduction,* Third Edition. Bston: Wm. C. Brown Publishers. 1995

Courant, Richard and Herbert Robbins. *What is Mathematics?*. New York: Oxford Unversity Press. 1941.

Drescher, Melvin. *The Mathematics of Games of Strategy.* New York: Dover Publications, Inc. 1991.

Eigen, Manfred and Ruthild Winkler.Laws of the Game. Princeton: Princeton Unversity Press. 1993.

Greenberg, Marvin Jay. *Euclidean and Non-Euclidean Geometries,* 2nd Edition . New York: W. H. Freeman and Co. 1982.

Grunbaum, Branko and G.C. Shephard. *Tilings and Patterns.* New York:W. H. Freeman and Co. 1986.

Kline, Morris. *Mathematics for the Non-mathematician.* New York: Dover Publication, Inc. 1985.

Kline, Morris. *Mathematics in Western Culture.* New York: Oxford University Press. 1953.

Trudeau, Richard J.*TheNon-Euclidean Revolution.* Boston:Birkhauser. 1987.

Velleman, Daniel J. How to Prove It. Cambridge: Cambridge University Press. 1994.

4 *Perplexing Polyplexes Plus*

The generic term "polytope, introduced into English in the late 19th century by a woman mathematician, Alicia Boole Stott, refers to the large category of geometric figures bounded by portions of lines, planes, or hyperplanes. The families found under this category include the polygons (2-polytopes), the polyhedra (3-polytopes), the polyplexes (4-polytopes), and the *n*-polytopes (*n* standing for any dimension).

Here we are still looking at Euclidean figures even when examining dimensions beyond three. The Euclidean plane is considered to be flat and polyhedra are built from planar polygonal faces. There is an essential difference between a polyhedron and a polyplex. The former is composed of polygonal faces while a polyplex is composed of polyhedral cells.

Since we humans actually live in a space of three-dimensions, we cannot experience dimensions beyond three. But we can try to deduce properties of four-dimensional space using analogy, and we can try to develop our spatial intuition using computer graphics, but we are still locked into our three-dimensional world.

In this century, the computer has come to be the assistant par excellence in our ability to visualize figures from higher dimensions. What are the advantages to these? For the scientist and the mathematician, with each new dimension, the space becomes more open and forms can be moved about more freely. What is impossible in a lower dimension becomes possible in a higher one. In the plane of 2-space, figures can only rotate about a point. In 3-space, figures can rotate about a line. And in 4-space, figures rotate about a

There are antelopes
    and cantaloups,
And bethrothed couples who can't elope.
There are popes with copes,
And strings and ropes,
And downhill skiers on downhill slopes.
There are those who mope
    without a hope.
But here, we have to finally grope
with the ever-elusive polytope.

plane. Scientists, who look at the origins of the Universe, believe that it might have existed in ten dimensions before the "Big Bang" and then fragmented into what we have now.

In order to understand spatial dimensions, let us begin with the concept of a point which has no length or depth or height. Move that point to generate a line which has only length. Move that line in a direction perpendicular to itself to generate a square which has both length and width. Move the square at right angles to itself through a distance equal to a side in order to obtain a cube in three dimensions which has length, width, and depth. Now, imagine moving this cube a distance equal to its edge length in a direction perpendicular to each of the three directions defined by one of the corners of the cube. This will trace out a 4-dimensional hypercube. This process can be continued theoretically, even if we cannot physically visualize the result. For example, moving a 4-dimensional cube a distance equal to its edge length in a direction perpendicular to each of the four directions defined by one of the 4-dimensional corners of the hypercube, will generate a 5-dimensional cube.

In the physical world of things in space, how do we count objects or describe the location of any particular object? How do we measure its length, its width, its mass, its distance from us? We could take a length of rope and knot it at intervals or notch a stick so as to give us a useful measuring device. Somewhere before 5,000 BC, humans came to grips with the need for measurement which relates to size comparison. The distance between each point on our measuring stick is called a unit length. That unit, unless tied to a particular measuring tool, is an arbitrary length which, when once established, becomes the module for a particular project. If standardized, it becomes the measurement system for a community, a culture, a continent, a planet. The "cubit", which is the length of the distance from the human elbow to the end of the middle finger of the outstretched arm, became a standard device for the ancient world. In the construction of the three Great Pyramids of Egypt, the Royal Cubit, based on the body of the Pharaoh, was used for measurement. The idea of the cubit was handed down to the Greeks and Romans. But it was the Greeks who first thought of subdividing the cubit into two spans, six palms, 24 digits. Human body parts became useful measuring units, but since bodies differ considerably from one person to another, which body would become the selection of choice? In Ancient Egypt, there was no question about the answer.

Think of the metric system. What is it derived from? What is its advantage or disadvantage? Do you use it? Do you prefer the system of inches, feet, yards, etc? Why? Can you suggest other systems of measurement?

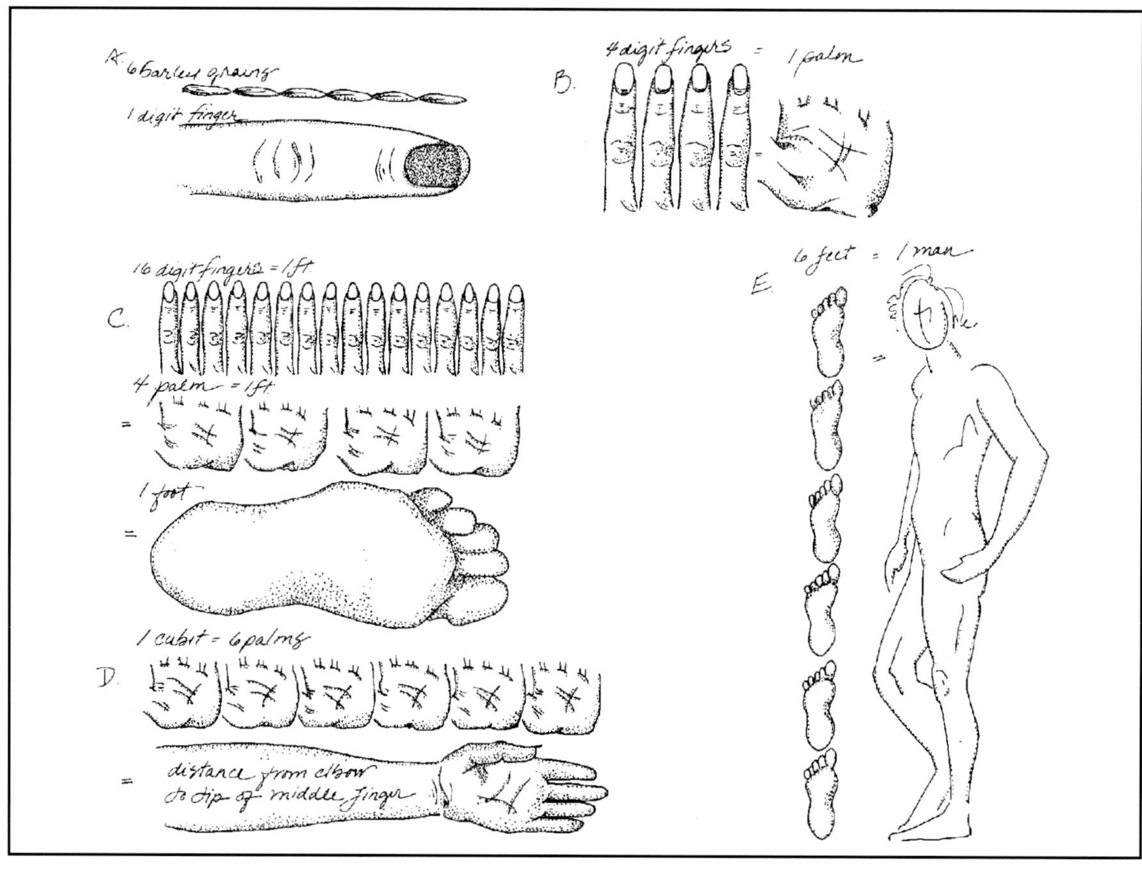

In order to mark points on a line segment, points in the plane, or points in three-space, we have need of a system. That system has to be consistent to be useful. In Fig. 4.1A a line is divided into lengths defined by the unit measure. We have marked point *P* on it.

What happens in the plane? Once points are marked, we need a way to coordinate them. Points are located and then marked in respect to a beginning point called the origin, O. For that system, we have to thank philosopher/mathematician Réne Déscartes. It was his genius that wedded algebra to geometry. The two direction axes, called x and y, divide the plane into four regions that are named quadrants. Every point in the plane is located either on an axis or within a quadrant. Each located point is indicated by a pair of numbers which are called an *ordered pair*. These pairs are called the coordinates. The first number of the pair refers to the horizontal x-coordinate and the second number refers to the vertical y-coordinate. Fig. 4.1B.

We can add a third axis, called the z-axis, which is perpendicular to the other two axes and allows for the indication of the third dimension. The three axes divide space into eight regions called octants. Every point in 3-space is either on an axis or within an octant. The co-ordinates of each point are an *ordered triple* of numbers. The first number indicates the x-coordinate, the second number indicates the y-coordinate and the third number indicates the z-coordinate. By carrying this into 4-space, how would you write the coordinates?

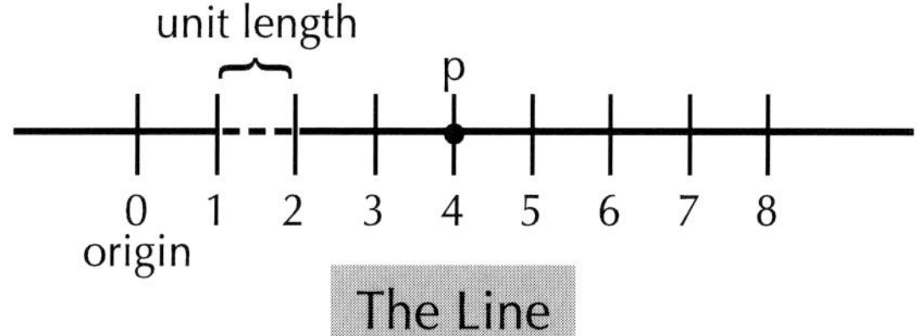

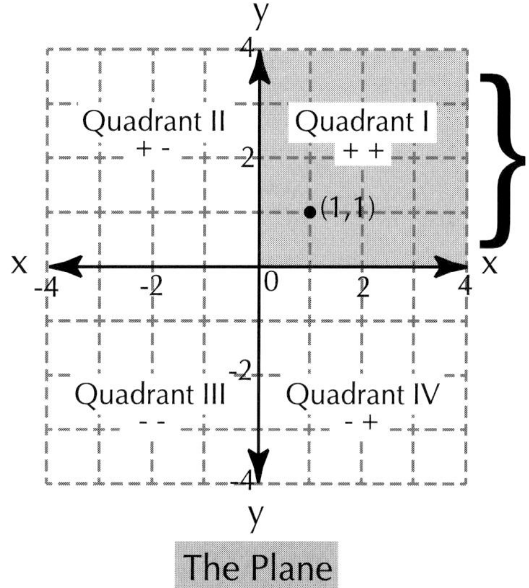

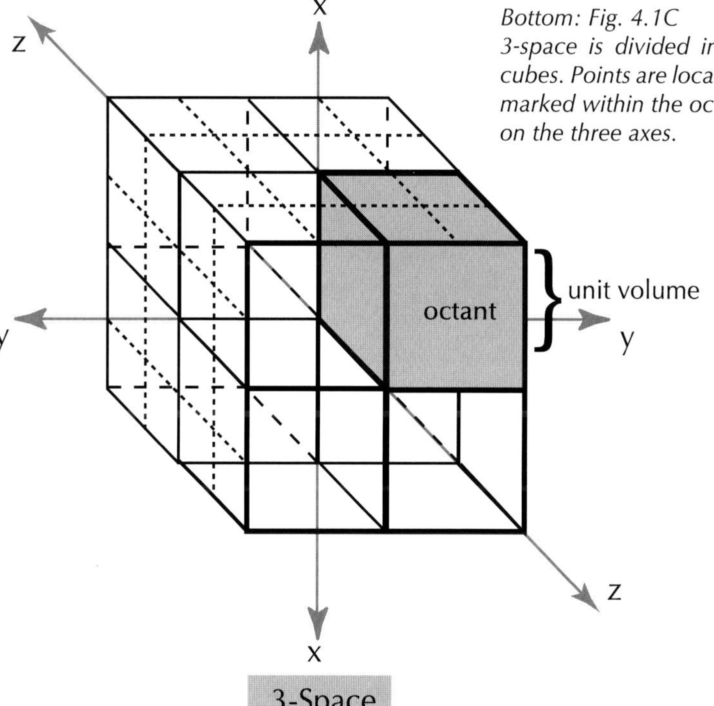

This page:

Above: Fig. 4.1A
A line divided into lengths determined by the unit measure. A point, P, is marked on it. The unit measure is an arbitrary choice.

Center: Fig. 4.1B
The plane is divided into unit squares. Points are located and marked within the quadrants or on the two axes.
The 'point' is defined to be the pair (x,y). The pair is defined to be the set.

Bottom: Fig. 4.1C
3-space is divided into unit cubes. Points are located and marked within the octants or on the three axes.

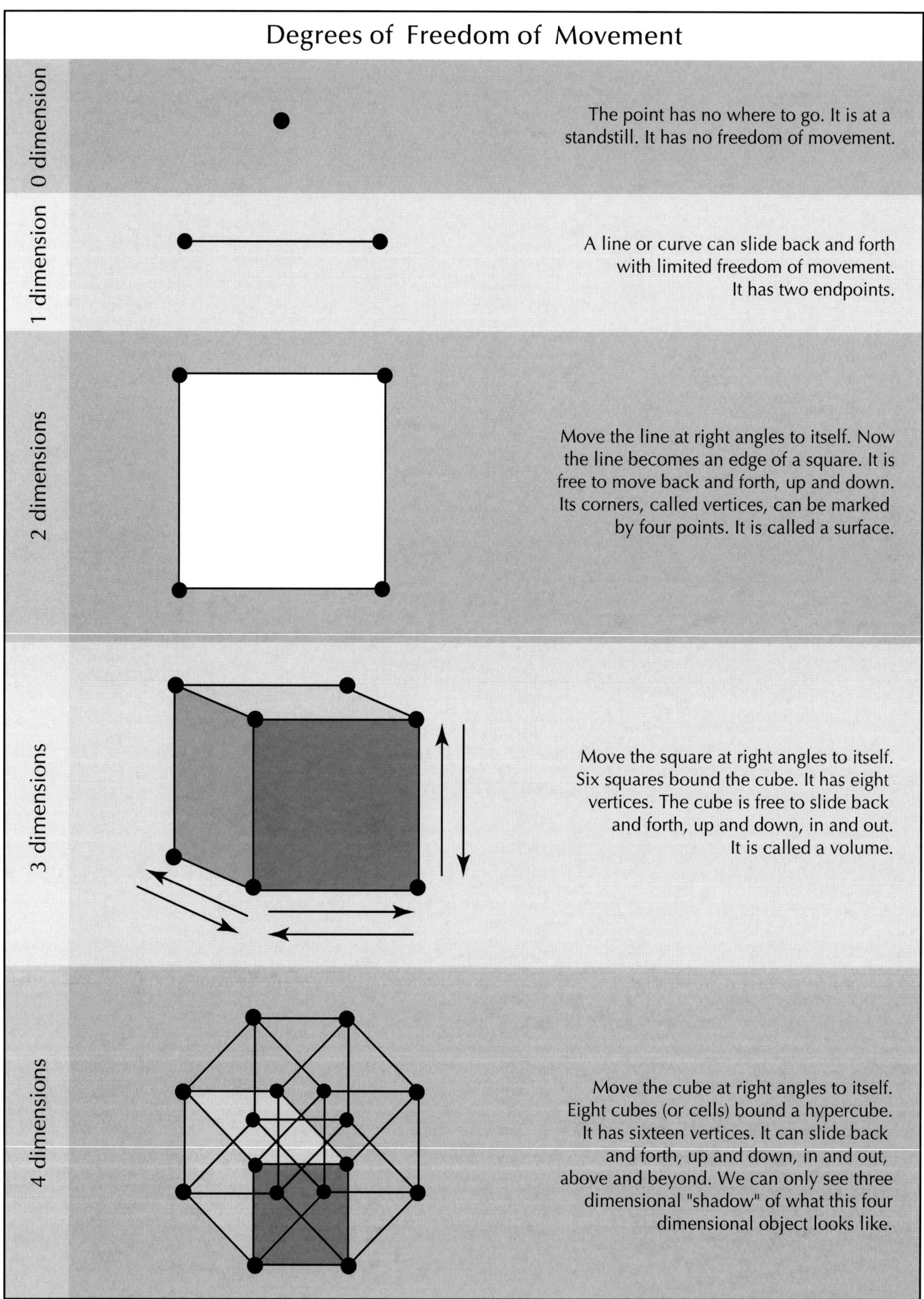

Degrees of Freedom of Movement

0 dimension

The point has no where to go. It is at a standstill. It has no freedom of movement.

1 dimension

A line or curve can slide back and forth with limited freedom of movement. It has two endpoints.

2 dimensions

Move the line at right angles to itself. Now the line becomes an edge of a square. It is free to move back and forth, up and down. Its corners, called vertices, can be marked by four points. It is called a surface.

3 dimensions

Move the square at right angles to itself. Six squares bound the cube. It has eight vertices. The cube is free to slide back and forth, up and down, in and out. It is called a volume.

4 dimensions

Move the cube at right angles to itself. Eight cubes (or cells) bound a hypercube. It has sixteen vertices. It can slide back and forth, up and down, in and out, above and beyond. We can only see three dimensional "shadow" of what this four dimensional object looks like.

Coordinates Required for the Different Dimensions

No coordinates are required.

One coordinate, x, is sufficient for specifying the location of a point along a one-dimensional line or curve.

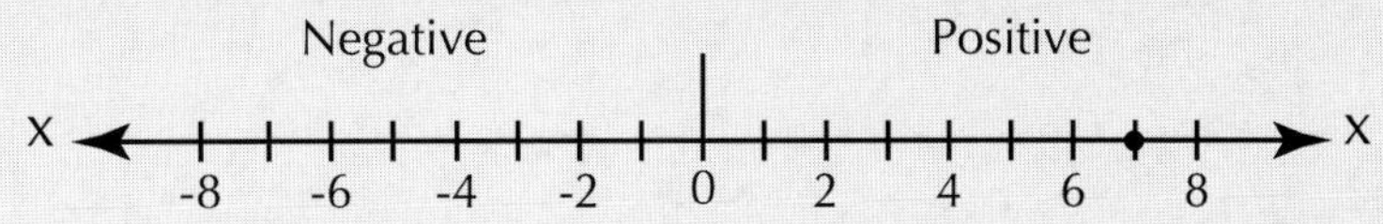

Two cordinates, x and y, are sufficient for specifying a location of a point on a two-dimensional surface. A point in two dimensions is specified by assigning a numerical value to the two coordinates.

A circle of radius 1 around the origin (0,0) is the collection of all points (x,y) that satisfy the equation $x^2 + y^2 = 1$.

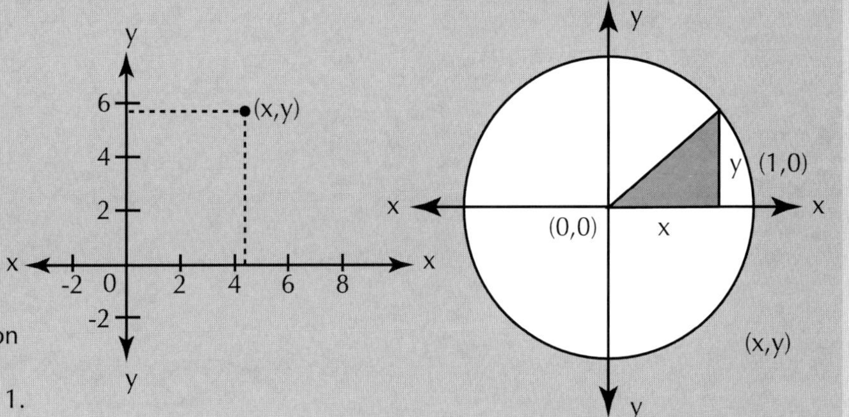

Three coordinates, x,y,z are sufficient for specifying a point's location in three dimensions.

The surface of a sphere of radius 1 around the origin (0,0,0) is the collection of all points that satisfy the equation $x^2 + y^2 + z^2 = 1$.

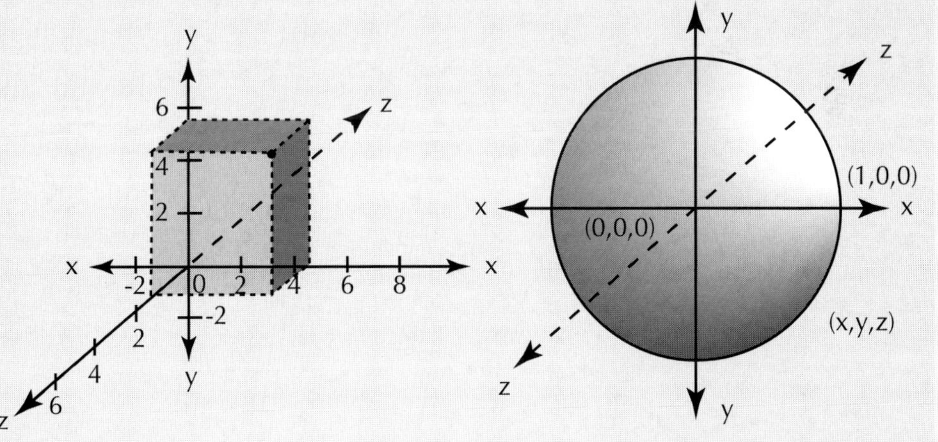

This procedure can be continued. Four coordinates are sufficient for specifying a point's location in four dimensions.

Can you write the equation for a 4-D sphere?

The invariant of dimension refers to the number of coordinates that are needed to specify a point in any given space. How many coordinates would you need to describe the position of a person in an apartment on the 13th floor of a high rise building which is located at 34th St. and Main St.?

Procedure for Constructing a Regular Polygon Given the Edge Length

Given line segment AB as the base of the *n*-sided polygon.

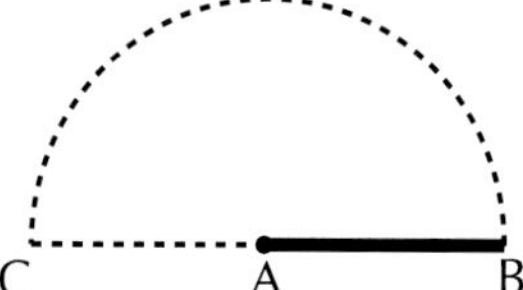

1. Use line segment AB as the radius for semicircle CB.

2. Choose the polygon you wish to construct. In the example, we have chosen to use a 5-sided polygon called a pentagon. Mark off *n*-divisions of equal intervals on the circumference of the semi-circle. To do this you must use a protractor. You must also know that a line segment is considered to have 180°. Since in this example 5 intervals are needed, 180° ÷ 5 = 36°. Lay the protractor along line segment CB with the 90° set at point A. Reading the protractor from left to right, or right to left, locate 36°. Use this measure to mark off all the intervals.
Label the points of intersection 1,2,3,4,5 respectively.

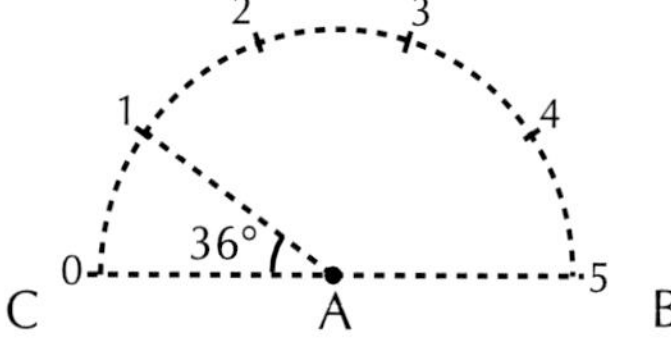

3. From A, draw line segments that extend through the arc divisions of 2,3, and 4.

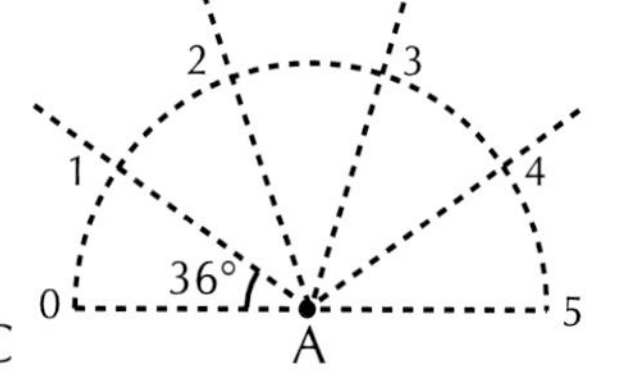

4. Beginning at point 2, and moving clockwise, cut off a side length equal to the measure of AB from point 2 to where it intersects line segment 3. Continue this process until all the necessary segments are joined.

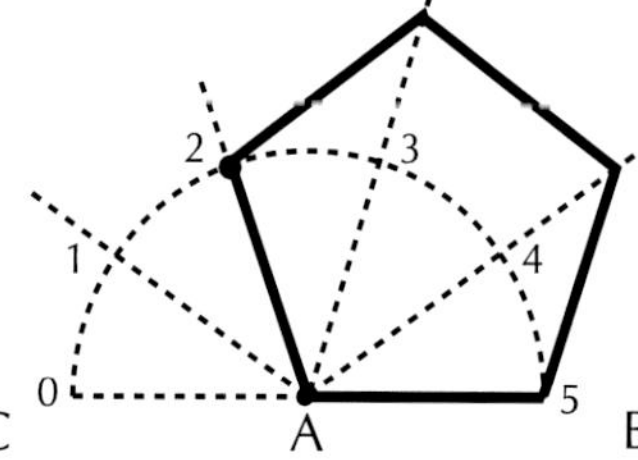

Now you have a polygon with five sides called a pentagon.

Can you prove mathematically why this construction works?

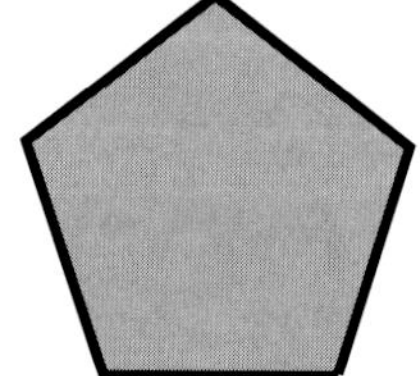

The polyplexes of four space can be visualized by looking at cross sections or projections of these figures as they occur on the plane. They can also be "unfolded" in the third dimension.

Before moving on to examine some of the polyplexes, the following pages and charts give some polygons (2-d), and some polyhedra (3-d) that relate to their higher-dimensional counterparts. If you would like to build some of these two-dimensional tilings and three-dimensional figures, we give templates for some polygons and directions for the joining of these polygons to construct polyhedra. On this page we show a procedure to construct your own regular polygon with whatever edge length you choose. Using this method, try drawing all of the ones we have given templates for on subsequent pages.

In this chapter we offer you four important families of forms in 3-space. They are the 8 convex Deltahedra, the 5 convex regular Platonic Solids, and the 13 semi-regular Archimedean and Catalan Solids. These families are built by following particular game rules as to what type and how many polygons come together at a vertex. The rules define the families. If, however, you just begin polyhedra playing by joining polygons all having the same edge length, you will find that there are many combinations that work well for architecture design and sculpture. Many architects are exploring variations on polyhedra to develop architectural complexes that are efficient in their use of materials and land space, as well as being aesthetically satisfying.

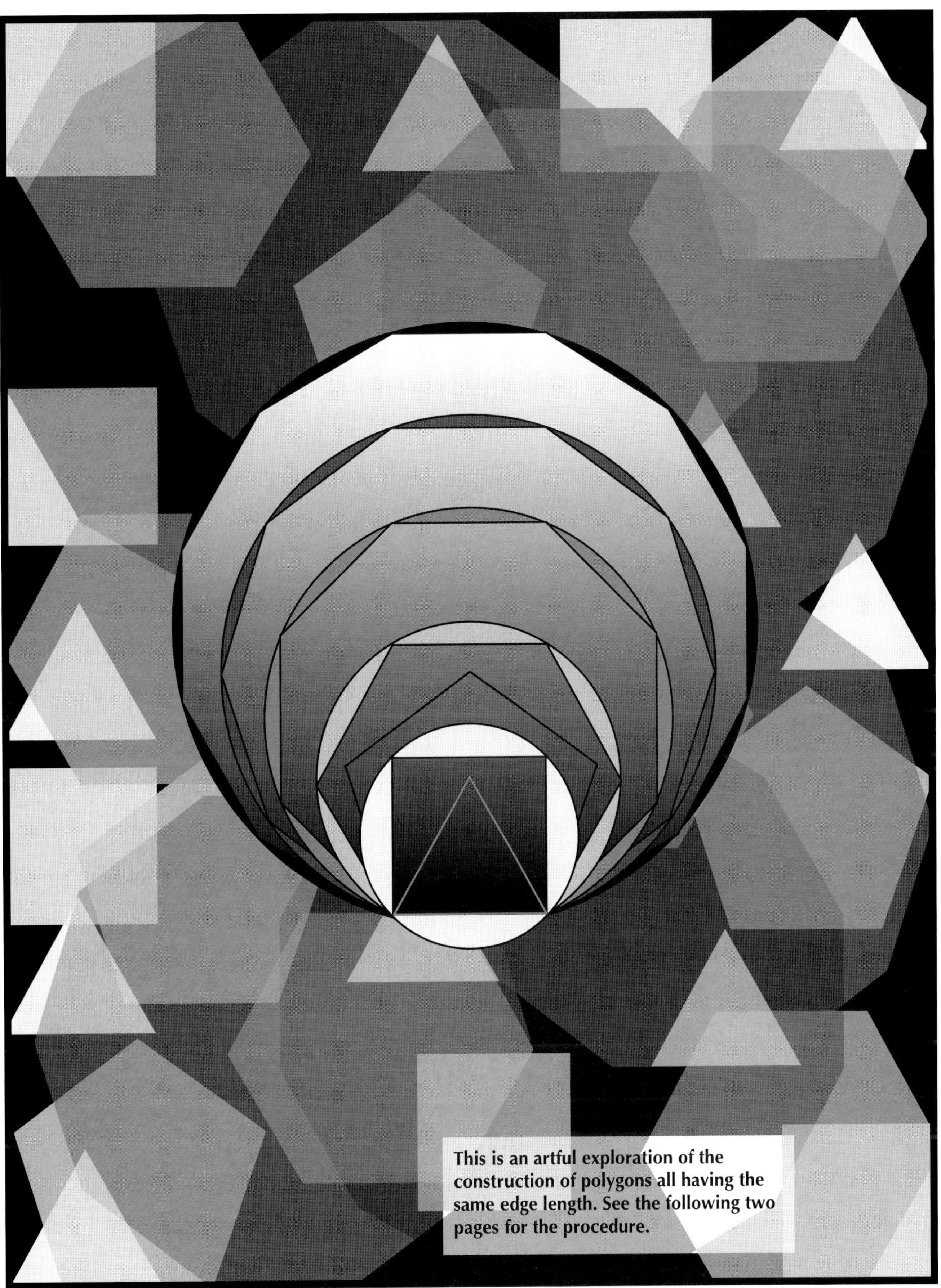

This is an artful exploration of the construction of polygons all having the same edge length. See the following two pages for the procedure.

Procedure for Constructing a Family of Regular Polygons Having the Same Edge Length

Given line segment AB as the base of the *n*-sided polygon.

A———B

1. Construct two congruent intersecting circles, A and B, using line segment AB as the radius of each. (Extend line AB in both directions for use in subsequent steps).

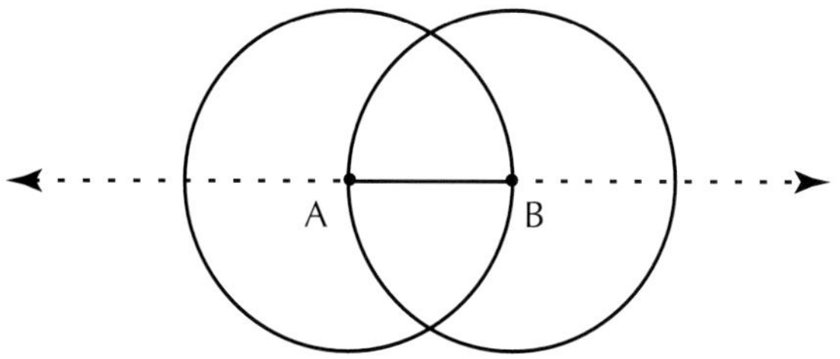

2. To construct the equilateral triangle:
Complete the figure by joining A and B, respectively, to C, one of the points of intersection of the circles.

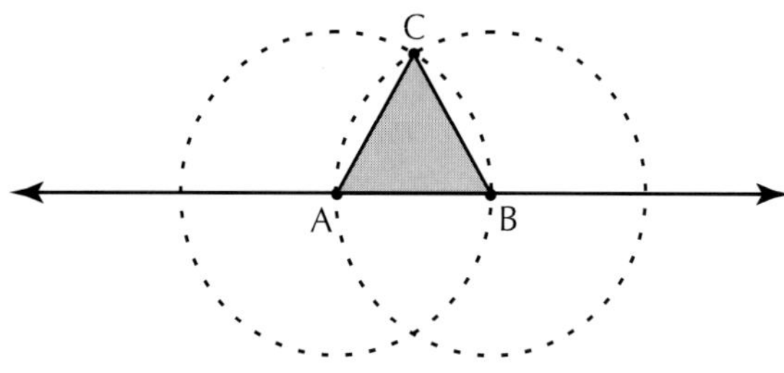

3. Draw the perpendicular bisector of line AB by joining C and the other point of intersection C^1, of the circles. Extend in both directions well beyond the two circles.

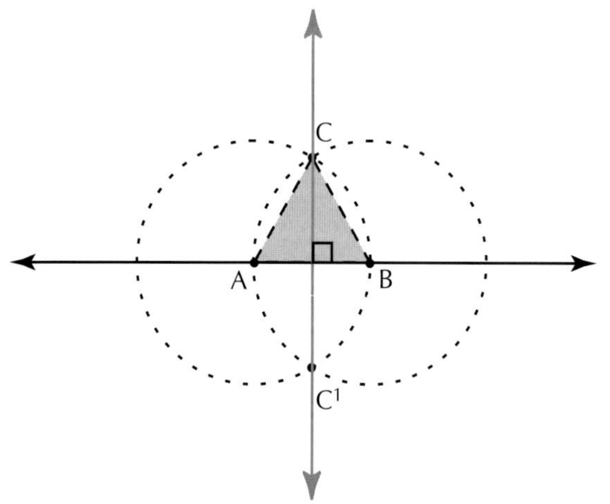

4. To construct the square:
Erect perpendiculars to line AB at points A and B and extend them parallel to the line drawn in Step 3. Complete the square by joining D and E, the points of intersection of the circles, and the perpendiculars just constructed.

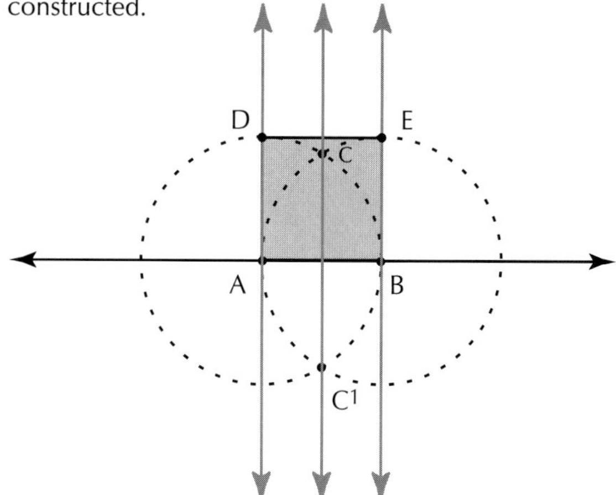

5. Construct diagonals AE and DB. With the point of intersection F as the center, construct a circle with radius FE. The circle circumscribes the square.

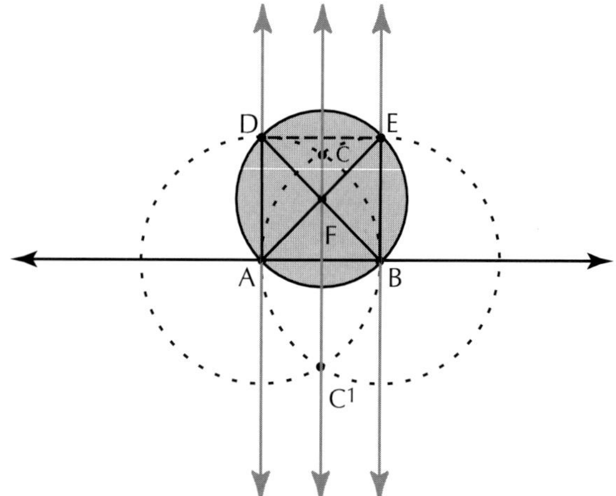

6. To construct the pentagon:
a. Place the metal tip on X, the midpoint of line AB. Place the pencil tip on D, and cut a semicircle with endpoints Y and Z on line AB.

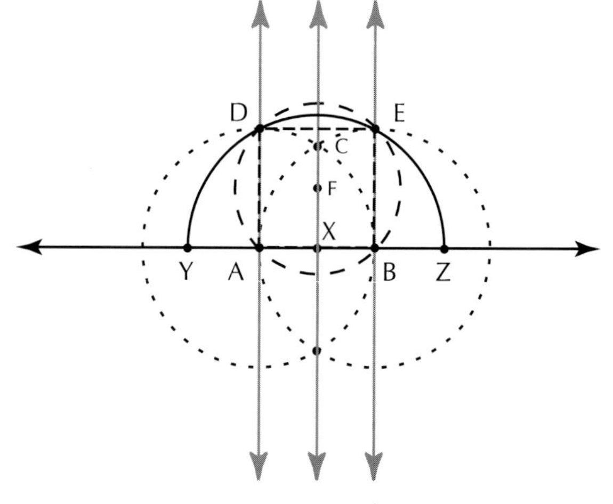

b. Place the metal tip on A, the pencil tip on Z, and cut an arc that intersects ray FC at W. Repeat with pencil tip on Y and metal tip on B. W should lie on both arcs just constructed and on ray FC.

c. Using length AB, place the metal tip on W and cut an arc that intersects circle A, and then intersects circle B. Label the points of intersection R and S respectively.

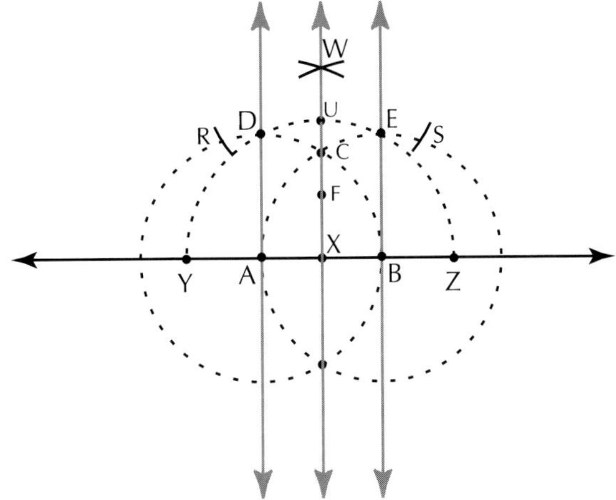

d. Connect points W, R, A, B, and S to form a pentagon.

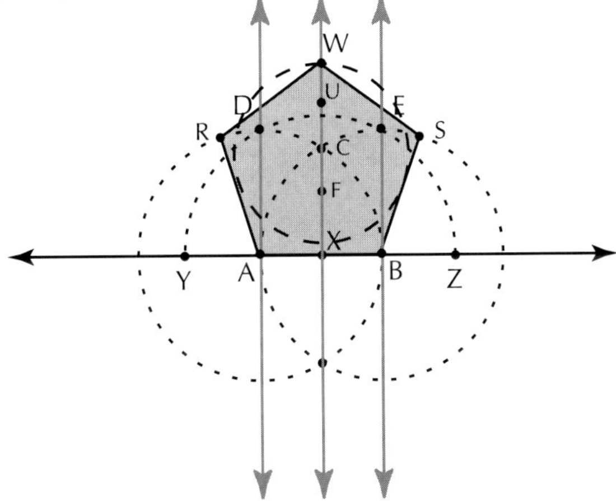

7. To construct the hexagon, octagon, decagon, and dodecagon:

The circle for each different polygon is contructed in the same way. The metal tip is placed on the center of each circle and the pencil tip is placed on either points A or B. Once the circle is drawn, the sides of the particular polygon (of length AB) can be marked off in succession on it. Each of the above mentioned polygons will have a side parallel to line AB with its endpoints on ray AD and ray BE.

a. C is the center of the *hexagon* and CA is the radius of the circle.

b. U is the center of the *octagon* and UA is the radius of the circle.

c. W is the center of the *decagon* and WA is the radius of the circle.

d. V is the center of the *dodecagon* and UA is the radius of the circle. (This polygon is not constructed on this page).

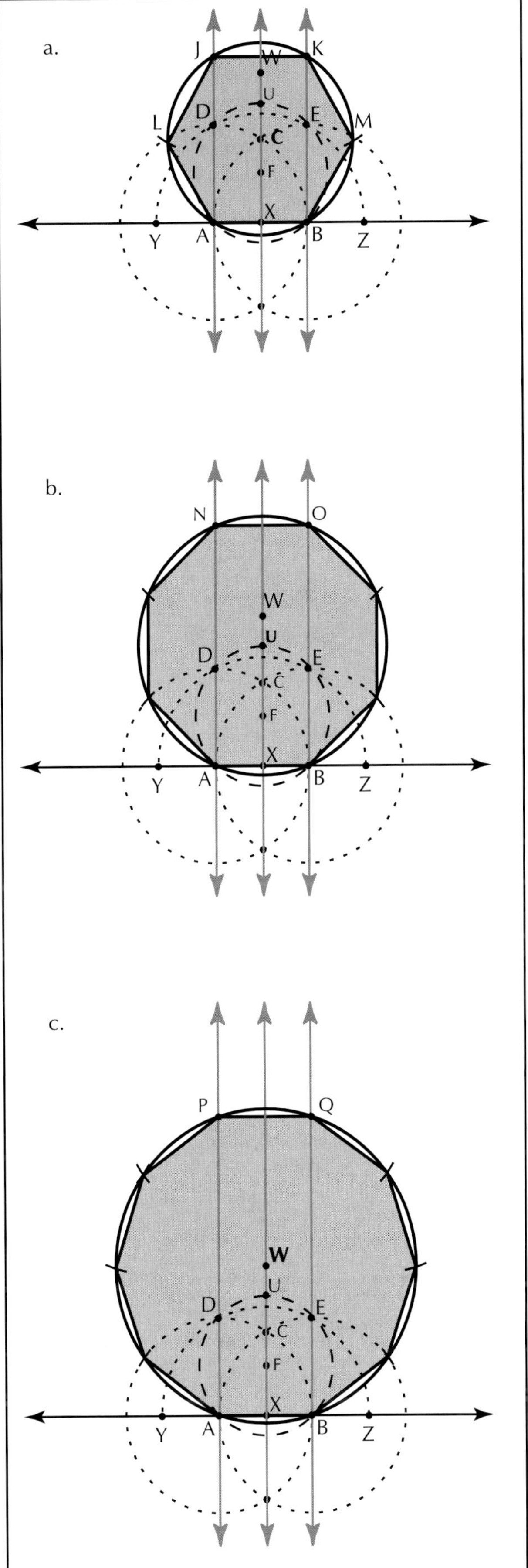

Polygons for Building Polyhedra

*These facing pages:
Chart of Polygons.
The basic elements of a polygon are: the face, the edge, the vertex.
The polyhedra and tilings found on subsequent pages can be built by using combinations of the polygons provided here. Enlarge these polygons until they are a comfortable working size. Too small and they are difficult to put together. A 3" or 4" edge length works well. Make sure that all the edge lengths are the same. If not, the different polygons will not have edge lengths that match up. If you are using only one type, this is not a problem. The Deltahedra require only groups of equilateral triangles. The Platonic Solids require one type for each of the five members of the family. It is with the Archimedean group that you will need to be careful.
Photocopy the required type and number of polygons for your chosen project. Acetate paper made for copiers, bristol, construction, manilla, and watercolor papers can be used in the photocopier with good results. Experiment with different kinds of paper. Cut out the requisite number of polygons needed for each figure. With a blunt instrument, press down along the edges of the polygons so that there is a good crease. This is called scoring.
Apply glue to the glue tabs. Use a bobby pin, paper clip or a binder clip to hold the edges of the glue tabs of a pair of polygons together until they dry. Work only a pair of polygons at a time and work with only one vertex at a time. Patience is definitely a virtue in this game.*

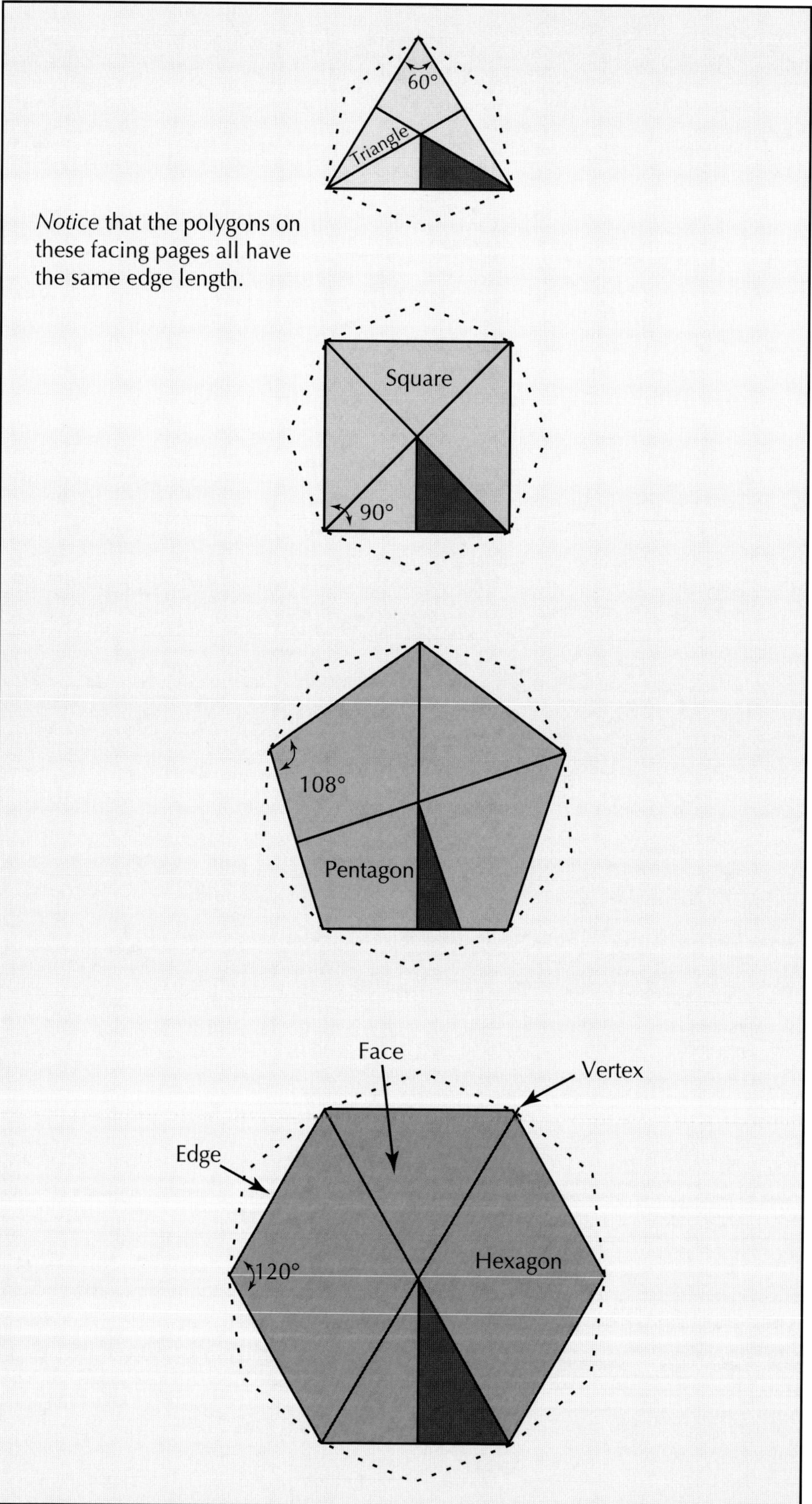

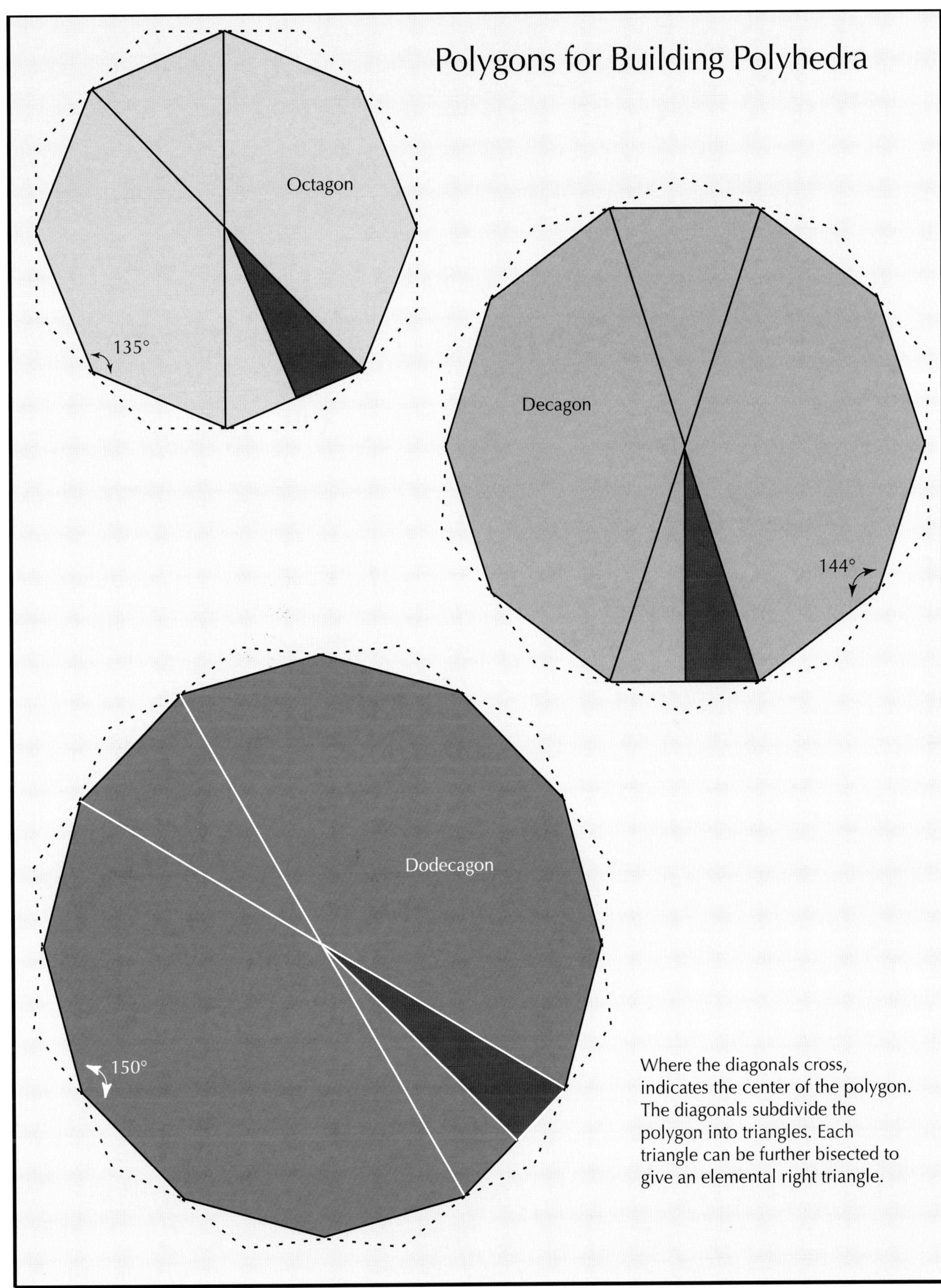

Where the diagonals cross, indicates the center of the polygon. The diagonals subdivide the polygon into triangles. Each triangle can be further bisected to give an elemental right triangle.

This page:
Student works. Polyhedra built from combining two or more polygons. Surface manipulation was left to the choice of the student.

Above:
Rebecca Whitehill. Decagons and triangles. The glue tabs were left on the outside of the form as part of the surface manipulation as opposed to being hidden within the figure.

Below:
Eugene Bourassa. Squares and triangles. The surface was treated with a crackle white texture paint.

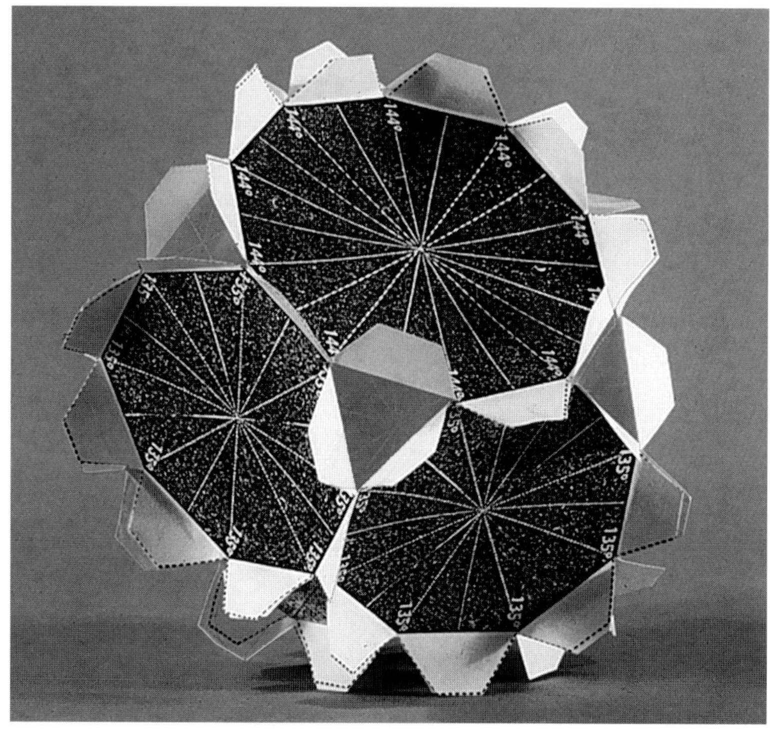

These facing pages show polyhedra exploration by students using polygons made of paper. The student was required to use two or more different polygons, with the same edge length, to construct a sculpture that would enclose space. There were no preliminary drawings but a direct hands-on approach to enclosing space with interesting form. The surfaces were to be finished in any number of ways according to the preference and skill of the student.

Try building one for yourself. Could you develop a set of "rules" for your construction procedure? What would happen if you replaced regular polygons with non-regular ones? Or you chose to add caps to some faces and not others? Or you opened up areas on some faces and not others? The more variables you add, the more complicated the game becomes. Could you get complexity using only very simple rules? Perahps reading up on fractal geometry would give you some clues.

This page:
Student works using combinations of regular polygons. What is the advantage of using regular versus non-regular figures?

Above: Paul Melchin. Hexagons, squares, and triangles are used to build a very complex figure. The surface is given a texture.

Below: Doug Hooten. Triangles and squares used to construct a highly symmetric form. Surface is colored and sanded.

Right Triangles and the Pythagorean Theorem

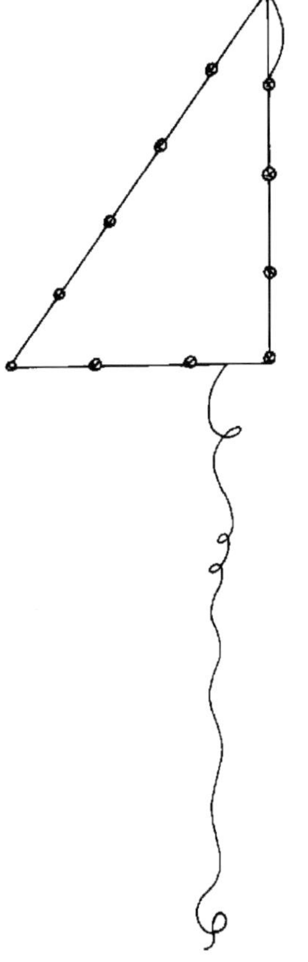

This page:
The 3-4-5 Right triangle is sometimes referred to as the "Rope Knotters Triangle". If a rope with 13 equidistant knots is wrapped tautly around three pegs, a right triangle is formed in the arithmetic progression of 3-4-5. Obtaining a right angle was important information for the Ancient Egyptians when constructing the Great Pyramids. If the sides of a triangle are related so that $a^2 + b^2 = c^2$, then the angle that is opposite side c is a right angle.

Polygons can be decomposed into congruent right triangles which can be considered a most basic polygonal building block. A triangle is a very stable figure especially for building in 3-space. The Pythagorean Theorem describes the relationship among the three sides of a right triangle in relationship to area.

Geometry investigates all possible spaces so that Euclidean geometry is just one of many possibilities. It has worked exceedingly well for a great many situations but, as you have seen in Chapter 3, there are other spatial conditions that require other geometries. Mathematics involves itself with both number and space. They are a complementary pair. These two arenas of investigation do not grow at the same pace. Throughout the history of mathematics at times they were separated and at other times they were together.

The scholars in Classical Greece (600 BC-300BC) gave geometry its initial form, structure, and substance. Theirs was a desire to understand the workings of the universe--a worthy task for any of us to undertake. The Greek thinkers gravitated toward problems in geometry, in part, because of the difficulty with the idea of an irrational number, i.e. a number that is neither a whole number nor a ratio, like a fraction, of whole numbers. Pythagoras (died circa 497BC) and his followers believed that whole numbers were the basic components of reality. Irrational numbers could not be reconciled with this worldview. So the Greeks concentrated on the study of geometry and space using proofs that do not require the use of numbers.

The Pythagorean Theorem, also known as the 47th Proposition of Book I of Euclid's "Elements" and the Carpenter's Theorem, gives the relationship between the lengths of the sides of any right-angled triangle. It states the relationship as thus, $a^2 + b^2 = c^2$, a and b are the two legs of the triangle and c is the hypotenuse. Now, if a right triangle has two sides of one (1) unit each, the hypotenuse, the longest side, will be an irrational number. The Pythagoreans also knew that a triangle with length of sides in the arithmetic relationship of 3 units to 4 units to 5 units is the smallest right triangle which has whole numbers and the only right triangle whose integers are consecutive.

But a more general question arose: Where there other right triangles that had sides in that same relationship? This theorem has been examined by the Chinese, the Islamics, the Hindus, but it is Pythagoras' name that has been attached to it for these many centuries because it was he who generalized and abstracted it, around 540 BC.

In order to prove this theorem, there are four kinds of demonstrations, one of which is geometric. It is the geometric proof that we are interested in here since spatial relationships are the concern of this book. The proofs are based on the comparison of areas. There are ten types of geometric figures from which a proof can be deduced. All of them relate to the unit square. But even with this tight limit, the number of proofs is limitless. See the following pages for examples.

Two Procedures for Constructing the 3-4-5 Right Triangle

Given the unit, AB, construct a 3-4-5 Triangle.

A———B

1. Extend ray AB through point B.

2. Open the compass so that it measures the length of AB.
Place the metal tip of the compass on B and cut an arc that intersects line segment AB on the side of B and opposite A. Label the point of intersection C.

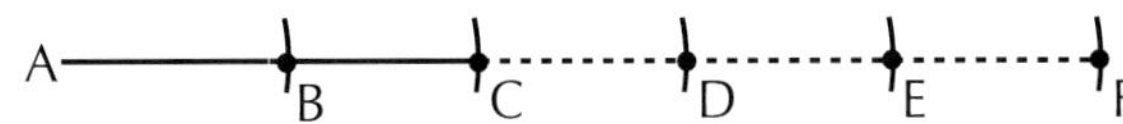

3. Mark off three more congruent line segments. Label the points of intersection D,E,F respectively.

4. Place the metal tip of the compass on A and the pencil tip on E. Cut an arc on one side of ray AB.

5. Place the metal tip of the compass on F and the pencil tip on C. Cut an arc that intersects the one just drawn. Label the point of intersection G.

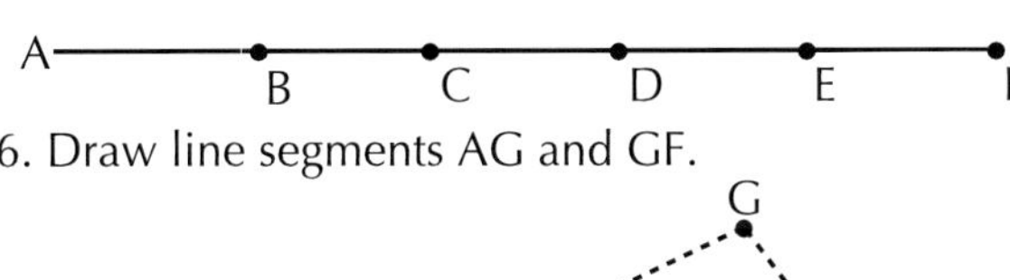

6. Draw line segments AG and GF.

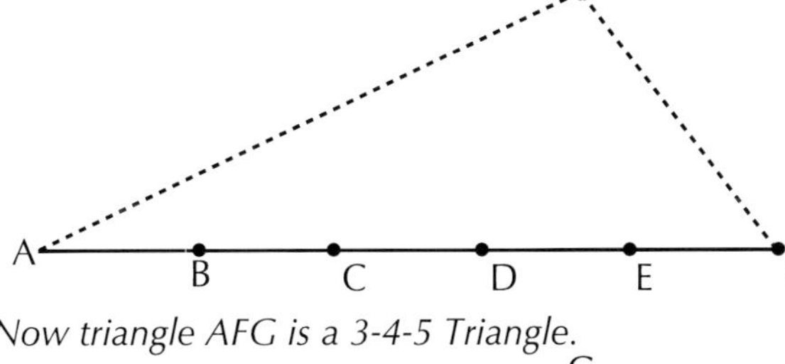

Now triangle AFG is a 3-4-5 Triangle.

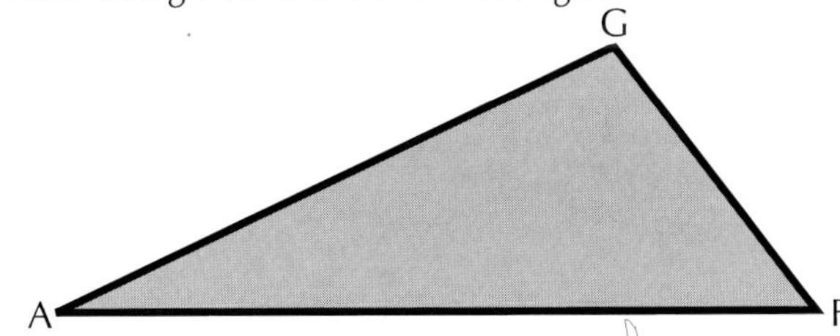

Given the longer leg, AB, construct a 3-4-5 Triangle.

1. Bisect line segment AB and label its midpoint C.

2. Bisect line segment AC and label its midpoint D.

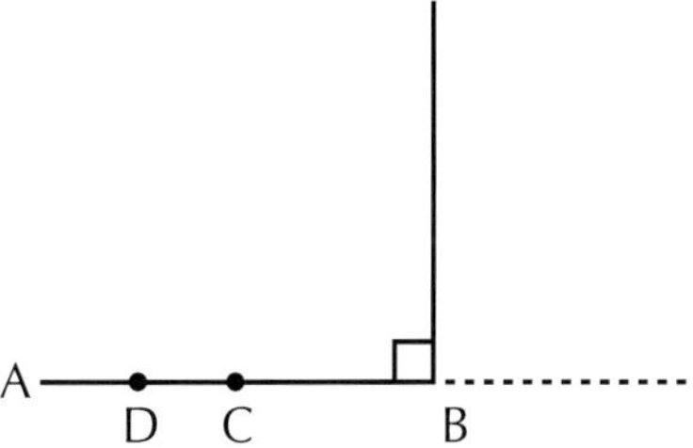

3. Extend ray AB through B. Construct a perpendicular to line AB at B.

4. Place the metal tip of the compass on B and the pencil tip on D. Cut an arc that intersects the perpendicular drawn in step 3. Label the point of intersection F.

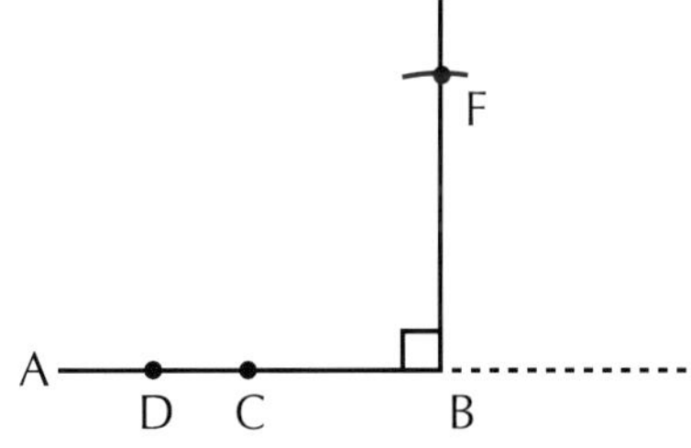

5. Draw line segment AF.

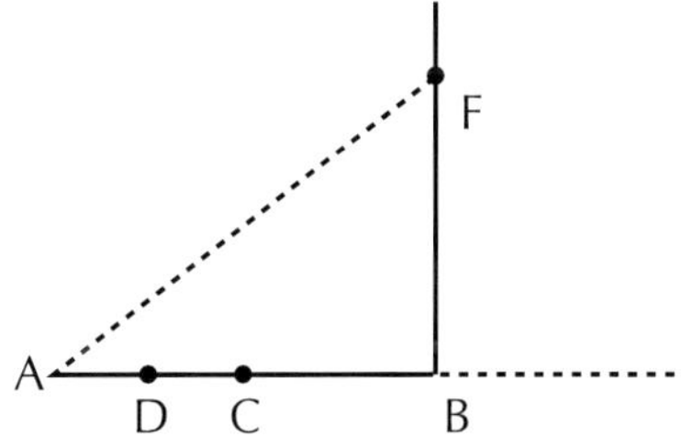

Now triangle AFG is a 3-4-5 Triangle.

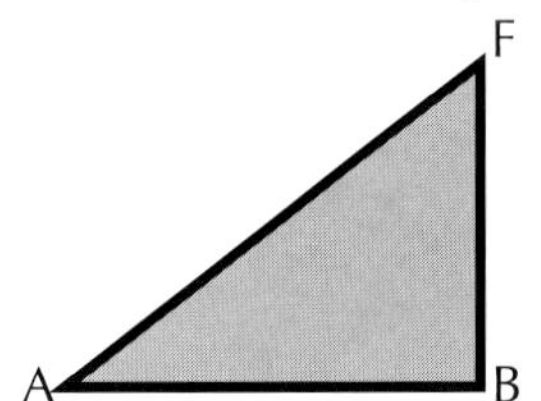

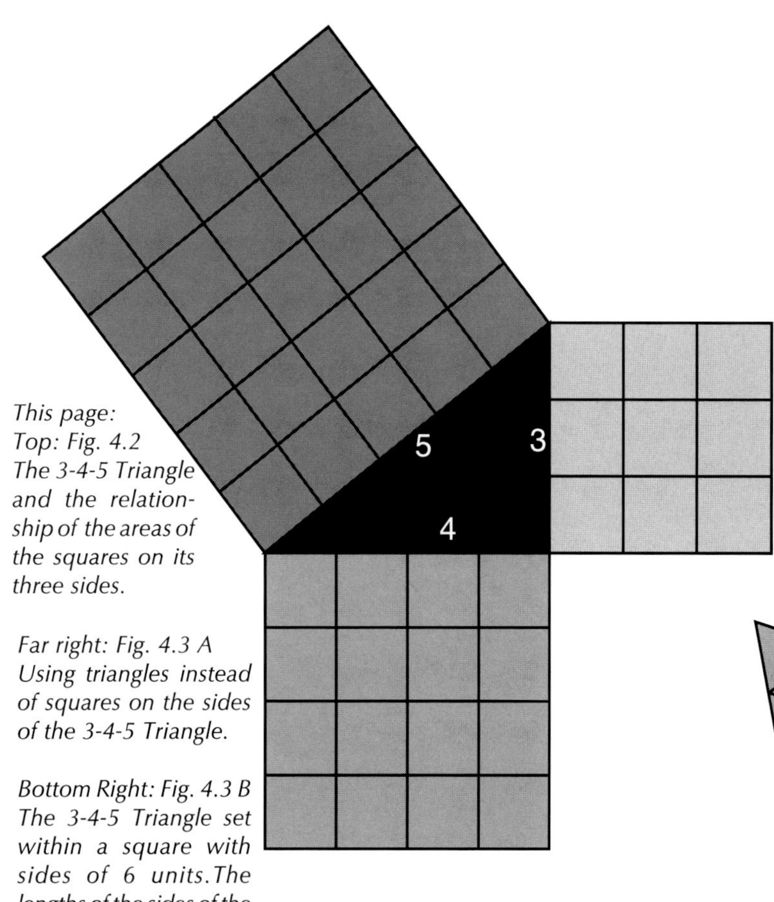

In any right triangle, the area of the square on the hypotenuse equals the sum of the areas of the squares on the two other sides. $c^2 = a^2 + b^2$. Fig. 4.2 shows this relationship in the 3-4-5- Right trangle. This is the simplest right triangle whose sides are in an arithmetic progression and whose solution yields a whole number. In Fig. 4.3 we show this relationship within a square and using triangles instead of squares. The 3-4-5- Triangle can also be set within a square with sides equal to six units. The sides of the triangle are then used as the diameters of three circles as seen in Fig. 4.4.

This page:
Top: Fig. 4.2
The 3-4-5 Triangle and the relationship of the areas of the squares on its three sides.

Far right: Fig. 4.3 A Using triangles instead of squares on the sides of the 3-4-5 Triangle.

Bottom Right: Fig. 4.3 B The 3-4-5 Triangle set within a square with sides of 6 units. The lengths of the sides of the triangle are used as the diameters of the circles.

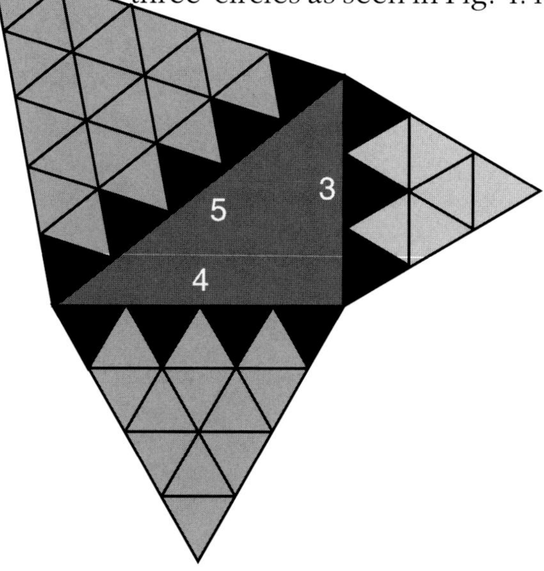

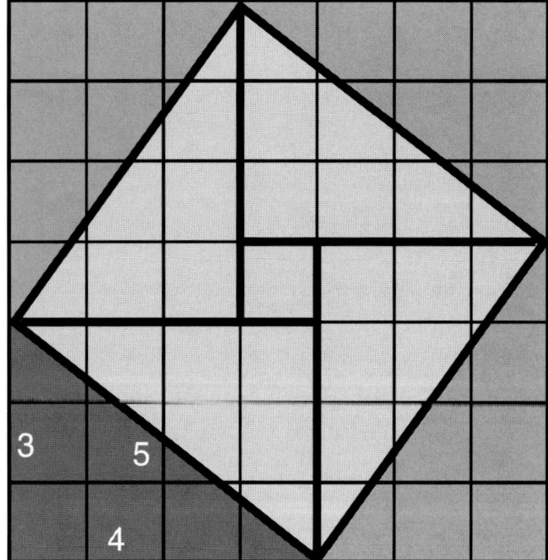

Fig. 4.4
Bottom left: Ancient Chinese proof of the 3-4-5 Triangle called the hsuan-thu.

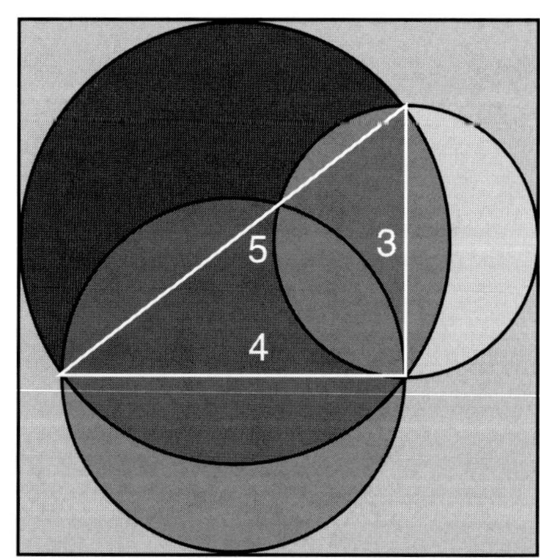

The shortest geometric proof was given by the Indian mathematician, Bhaskara, in the 12th century. He gave a simple diagram with the word "Behold", Fig. 4.5 Some of us, however, we need a few more visuals in order to see the proof.

Given a 3-4-5 Triangle ABC.

A. Construct a square on the hypotenuse of the 3-4-5 Triangle.

B. Use four copies of the 3-4-5- triangle and place them within the square constructed in the previous step.

C. The center square that remains after drawing the four triangles, has sides that equal the difference between two sides of the right triangle. This square and the four triangles can then be rearranged to make up the areas of the two squares, the lengths of whose sides correspond to the legs of the right triangle. Now we can behold!

Another geometric proof was given in the 20th century by the humanist scholar and mathematician, Dr. Jacob Bronowski. By moving two triangles, a and b, the square on the hypotenuse is transformed into two adjacent squares on the sides enclosing the right angle. Fig. 4.6

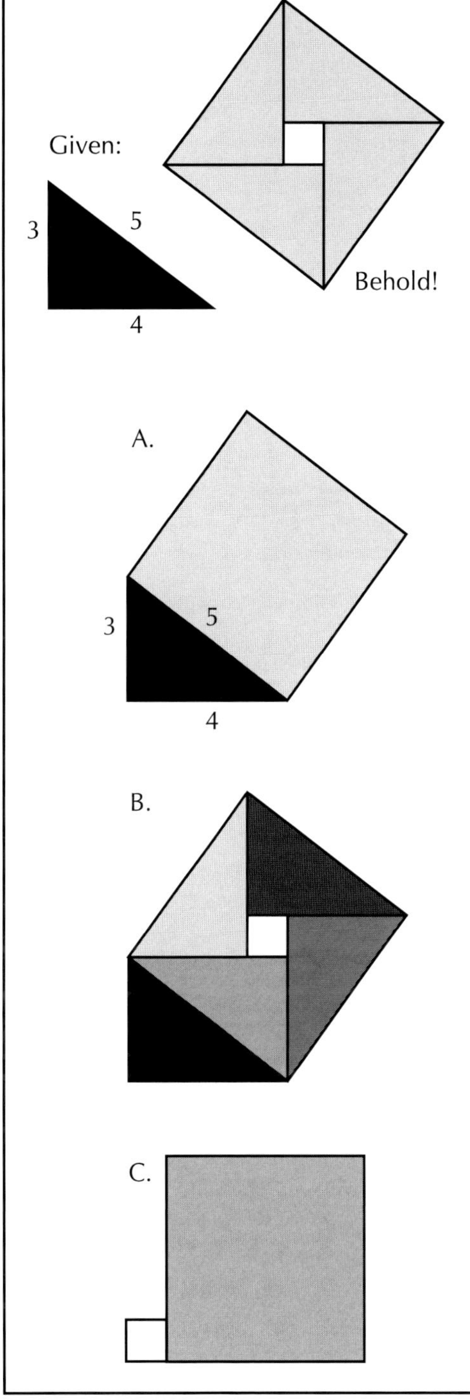

This page:
Left: Fig. 4.5
Bhaskara's Proof of the
Pythagorean Theorem

Bottom: Fig. 4.6
Bronowski Proof of the
Pythagorean Theorem.
The area of a triangle is one-half that of its containing rectangle and is a property of a right triangle.

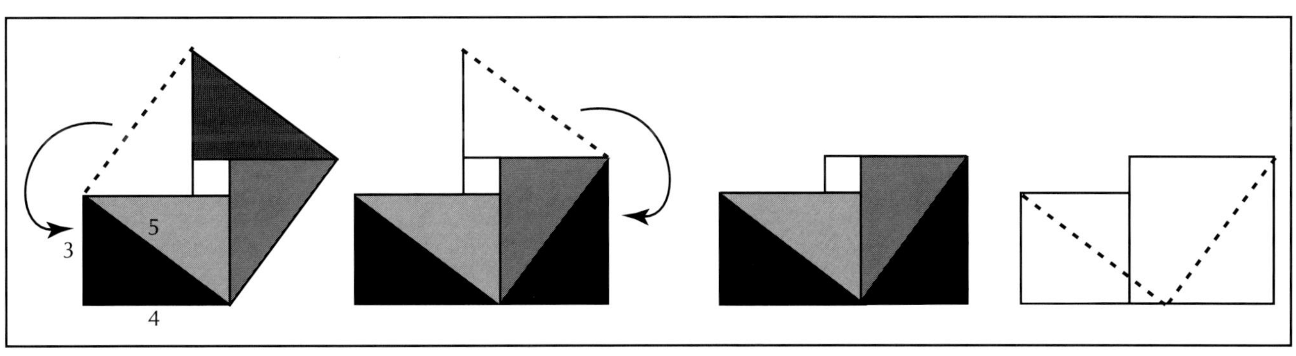

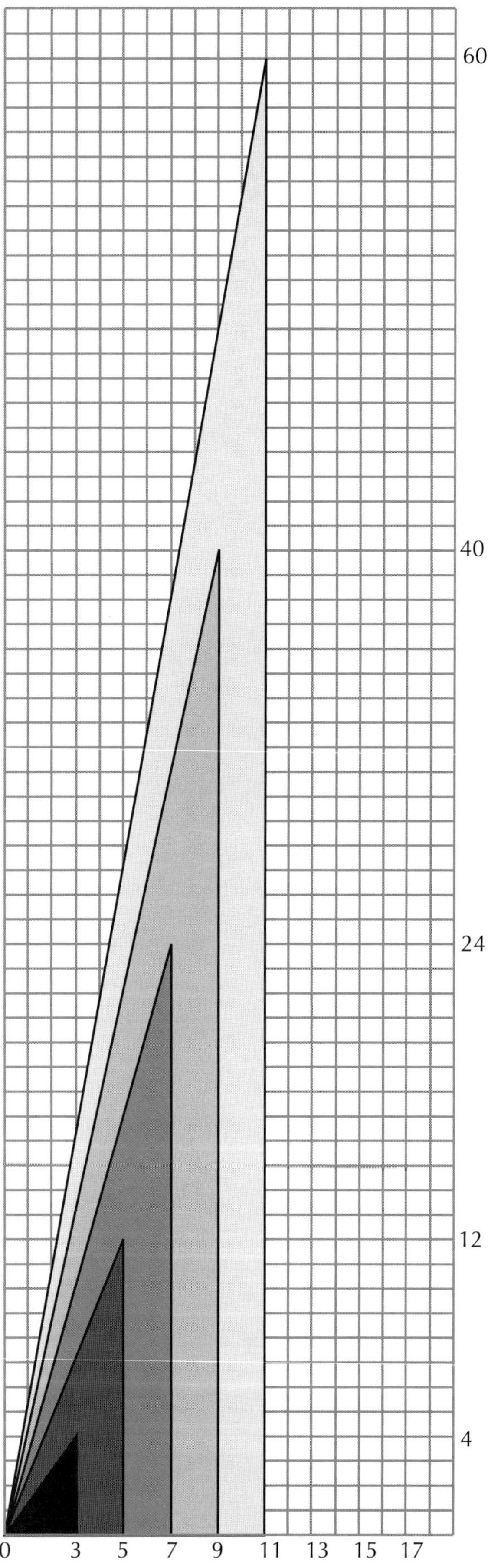

The 3-4-5 Triangle is the beginning of an infinite series of right triangles which have whole numbers for their sides. These can be extremely useful for design purposes. But how can these others, beyond the ones shown here, be determined? Thanks to mathematicians there is a game rule which can be applied so that you can generate others in the series. Fig. 4.7

Here is how:
If n is any odd number, such as 5, then the other two numbers in the equation $a^2 + b^2 = c^2$ can be obtained.

For the second number use:

$$\frac{n^2 - 1}{2} = \frac{5^2 - 1}{2} = \frac{24}{2} = 12$$

For the 3rd number use:

$$\frac{n^2 + 1}{2} = \frac{5^2 + 1}{2} = \frac{26}{2} = 13$$

Thus the triangle has sides of 5, 12, 13.

On the facing page is a chart with some design possibilities for the right triangles.

For the mathematically inclined, research the connection between these "Pythagorean Triples" and Fermat's Last Theorem.

side a	3	5	7	9	11	13	15	17
side b	4	12	24	40	60	84		
side c	5	13	25	41	61	85		

Using the formula above, fill in the blanks.

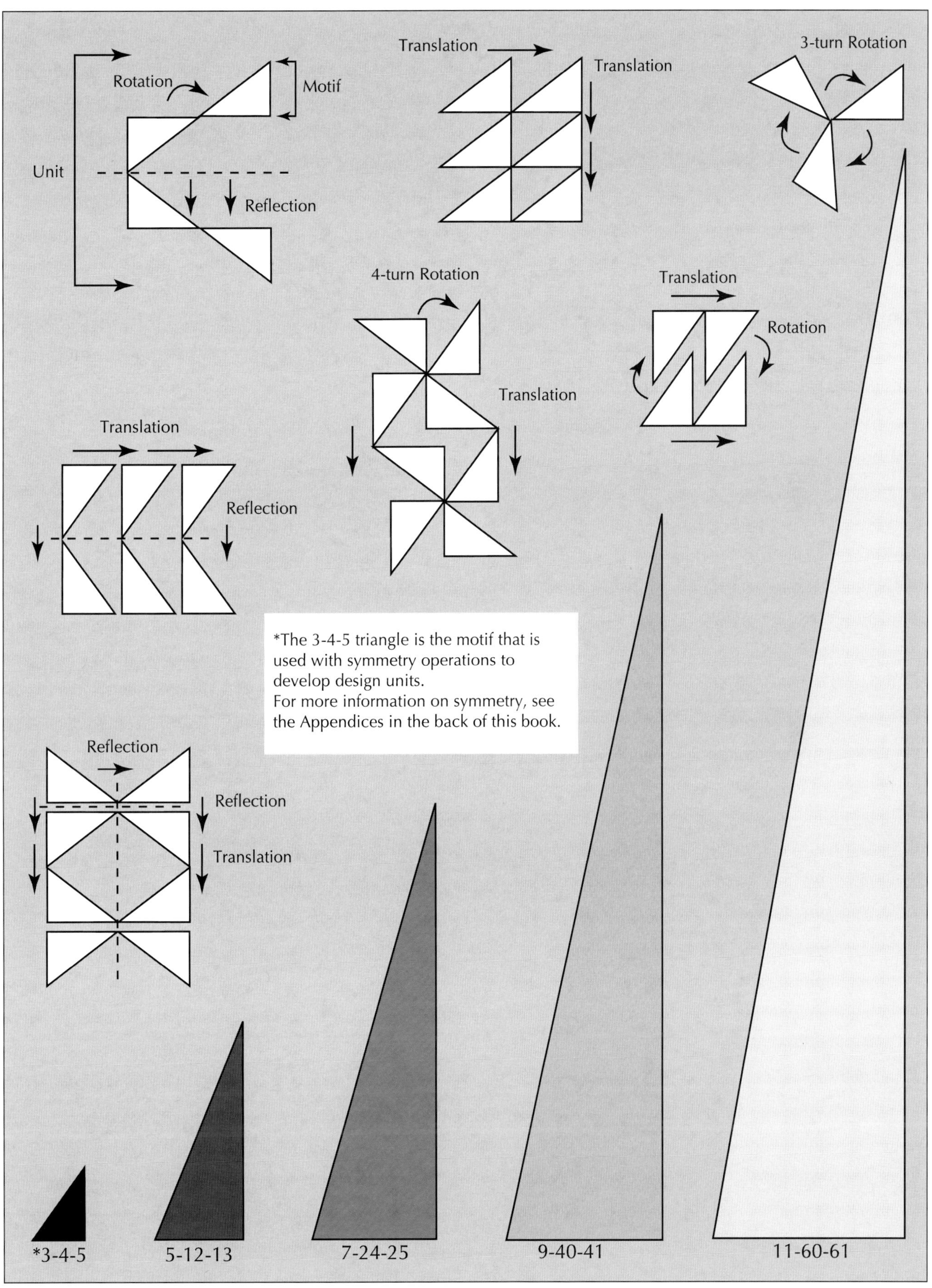
Rotation
Motif
Unit
Reflection
Translation
Translation
3-turn Rotation
4-turn Rotation
Translation
Translation
Rotation
Translation
Reflection
*The 3-4-5 triangle is the motif that is
used with symmetry operations to
develop design units.
For more information on symmetry, see
the Appendices in the back of this book.
Reflection
Reflection
Translation
*3-4-5
5-12-13
7-24-25
9-40-41
11-60-61

*This page: Fig. 4.7
A chart of some triangles for
your artistic explorations.*

*A. The short Golden Triangle
(Isosceles) with angles of 36°,
36°, and 108°.*

*B. The tall Golden Triangle
(Isosceles) with angles of 72°,
72°, and 36°.*

*C. the Isosceles Right Triangle
with angles of 45°, 45°, and
90°.*

*D. The 30-60-90 Triangle with
angles of 30°, 60°, 90°.*

*E. The equilateral triangle with
angles of 60°, 60°, and 60°.*

Triangles are extremely useful for design in any dimension. Since in Euclidean geometry, all types of triangles will cover the plane without leaving gaps between the tiles, each of the triangles given below can be used to develop a plane tiling. Given particular rules, triangles can be altered so that irregular shapes can also tile the plane. The sum of the angles in any given triangle in Euclidean geometry is 180°. Also the sum of the internal angles is also two right angles. Try playing with any triangles on any plane surface.

Design a pattern structure within any triangle. Use black and white or color. Copy the tile unit as many times as required for your design. Can you combine more than one type of triangle as long as you keep the edge lengths the same? Can you use three types of triangles? Would mathematics help you to solve these questions before putting artistic pen to paper? What questions would you put to a mathematician to help you with your problem?

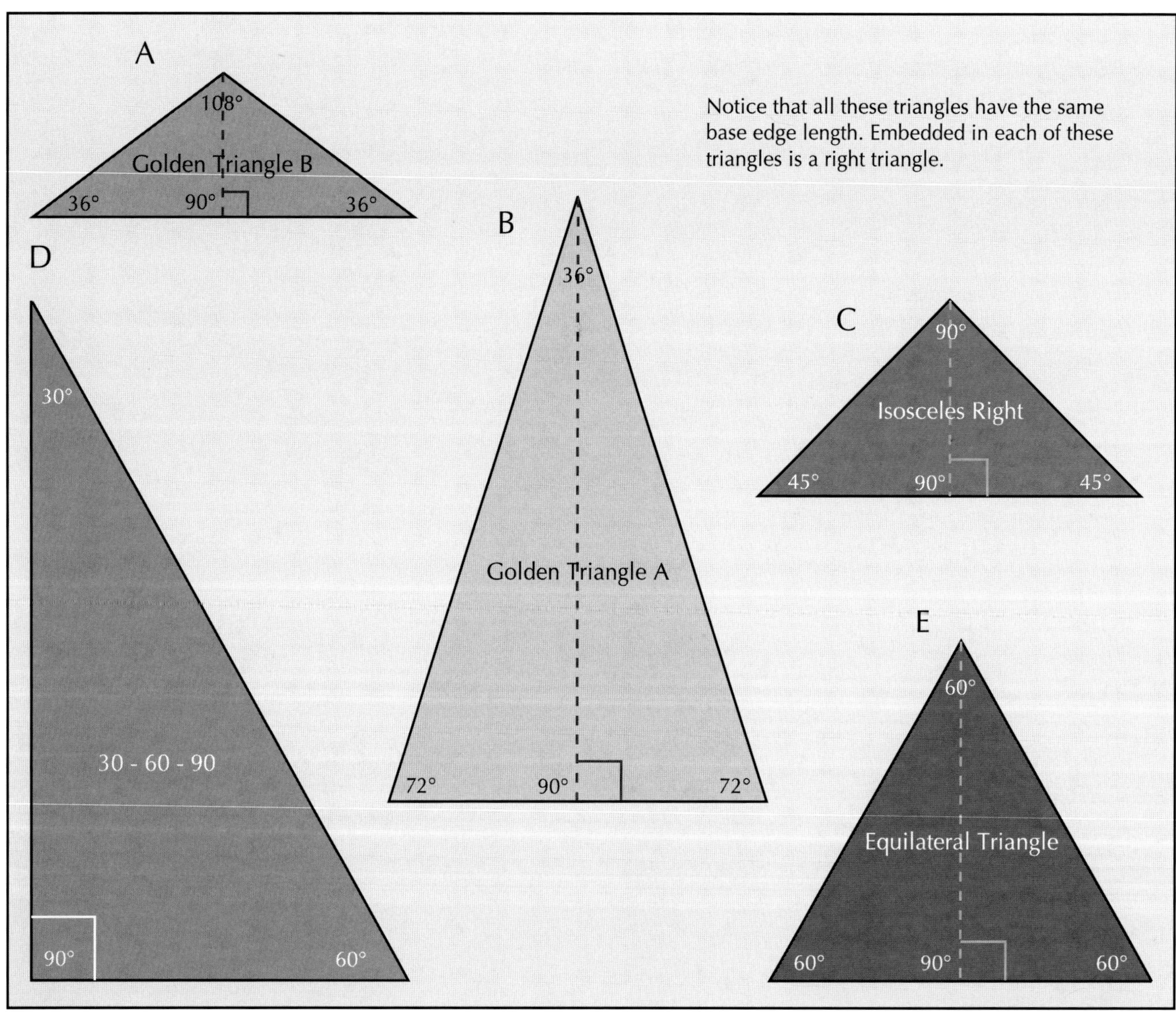

The right triangle and the Pythagorean Theorem can help in understanding 3-space and beyond. In 1854, the mathematician George Berhard Riemann gave a lecture "On the Hypotheses Which Lie at the Foundation of Geometry" that challenged the primacy of Euclidean geometry. It was a most important breakthrough for mathematics. Euclidean geometry could no longer be the only game in town. Riemann found connections between higher dimensional spaces and the Pythagorean theorem by generalizing it.

In two space, the equation states $a^2 + b^2 = c^2$, giving the length of the diagonal of a right triangle in the plane, Fig. 4.8 top. The theorem can be carried into the third dimension by using it to obtain the length of the longest diagonal within a rectangular solid. In Fig. 4.8 bottom, we have used a cube. This diagonal is the hypotenuse of a right triangle having one side along the edge of the cube while the other hypotenuse is found along a diagonal of the rectangular face of the cube. Apply the Pythagorean Theorem, $a^2 + b^2 + c^2 = d^2$. Generalizing the theorem, this states that the sum of the squares of three adjacent sides of a cube is equal to the square of the diagonal of the cube. If, a, b, and c are the sides of a cube, d is the length of the diagonal.

The Pythagorean Theorem can then be extended to dimensions beyond three by the addition of more variables. You add a variable for each new dimension. $a^2 + b^2 + c^2 + d^2 + ... = z^2$. Notice that the geometric object cannot be visualized in the higher dimensions but requires the ability to rely on abstract reasoning and algebra.

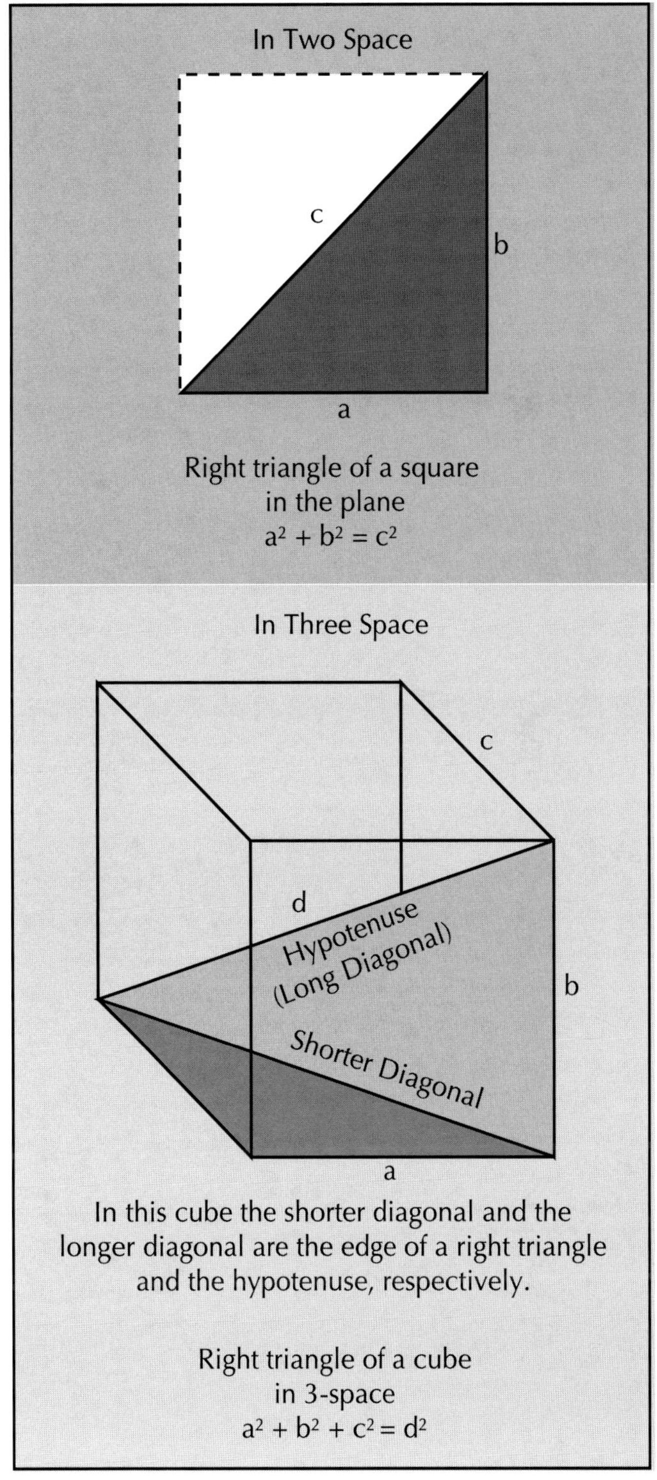

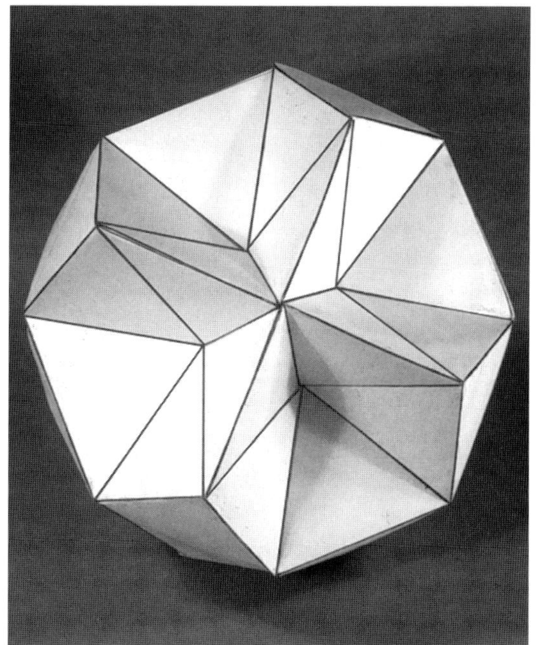

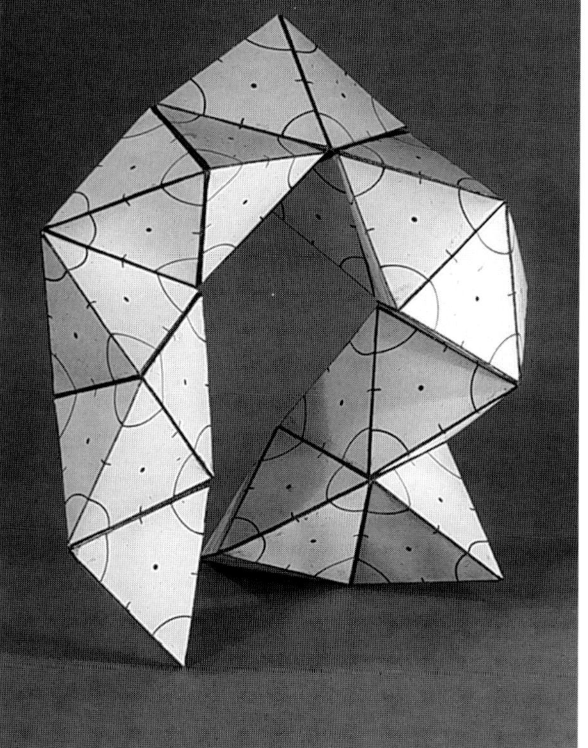

This page:
The triangle in 3-space
Student works exploring the use
of only equilateral triangles us-
ing paper as the medium.
Above:
Paul Melchin. Two view of the
same work.
Center:
Paul Melchin. Bridge form.
Bottom right:
Michelle Briggs. Paper with sur-
face treatment of paint and sand.

*This page:
Richard Newman.
"Time Capsule". 5"x2"x2".
Mixed-Media Sculpture of
wood paint, aluminum, photograph using triangles as the
basic compositional structure.*

The Eight Convex Deltahedra Family

This page:
Photograph of paper models of the 8 convex Deltahedra. The edges have been covered in graphic tape so that you can see how they join at the various vertices. The octahedron and icosahedron are highly symmetric and have the same number of triangles coming together at each vertex. Other figures are not so regular.

The facing page:
A chart giving the nets for the convex Deltahedra. The name of each is indicated under the particular net.
The "net" of a polyhedron is obtained when you unfold the faces of the polyhedron and lay them flat. The nets serve as a blueprint of the polyhedra. When glue tabs are added, they are called templates.

In chemistry, the borane compounds, composed of boron and hydrogen, have polyhedral structures, some of which are deltahedra.

The equilateral triangle is the most versatile and the most stable, of the triangle family because of its congruent edge lengths and congruent angles. It can be used to build an infinite family of concave structures but it also gives a finite family of convex solids as you can see in the above photograph. This is the convex Deltahedra group that has only 8 members. Each is composed of a given number of equilateral triangles joined in a particular way. On the facing page you will find nets for all 8 members. In order to build the forms, glue tabs need to be added to the edges of the triangles. They also need to be enlarged to provide a comfortable building size.

Score along the lines, add glue and join. Or you can construct each net by using multiples of a single triangle and arrange as shown in the chart. Consider putting patterns on the faces of the polygons before joining, or add extensions to the surfaces, or cut out portions of the faces. Think of other surface treatments. Magically, the three-dimensional solid will appear, that is, if you have been careful all along with each part of the process.

Different arrangements in the plane will effect the way the 3-d form comes together. It is important, therefore, to lay the triangles according to the plan. Notice that there is no 18-Delta. A mathematician would be able to prove to you why this is so. Can you try to figure it out for yourself?

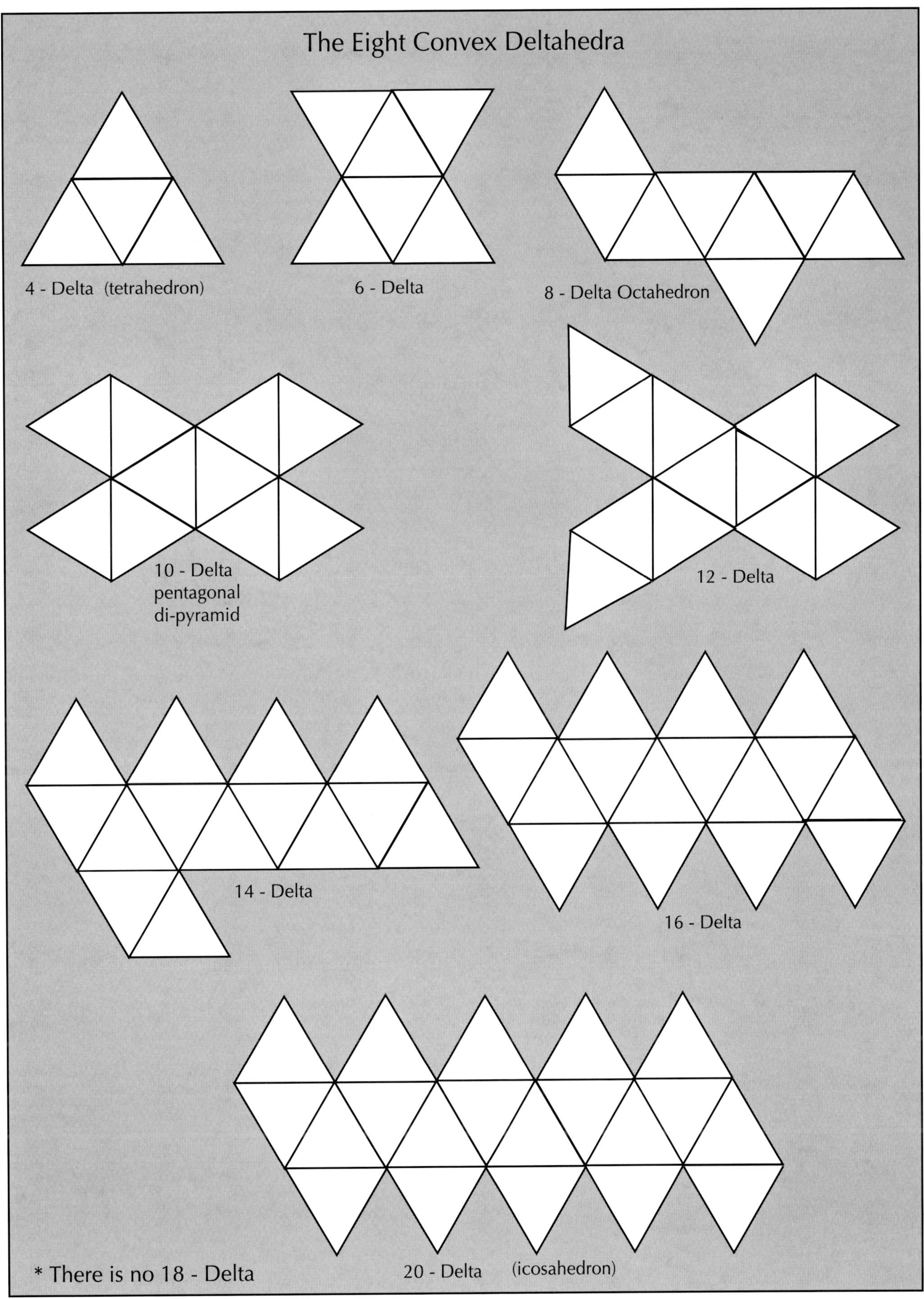

The Eight Convex Deltahedra
4 - Delta (tetrahedron)
6 - Delta
8 - Delta Octahedron
10 - Delta
pentagonal
di-pyramid
12 - Delta
14 - Delta
16 - Delta
* There is no 18 - Delta
20 - Delta (icosahedron)

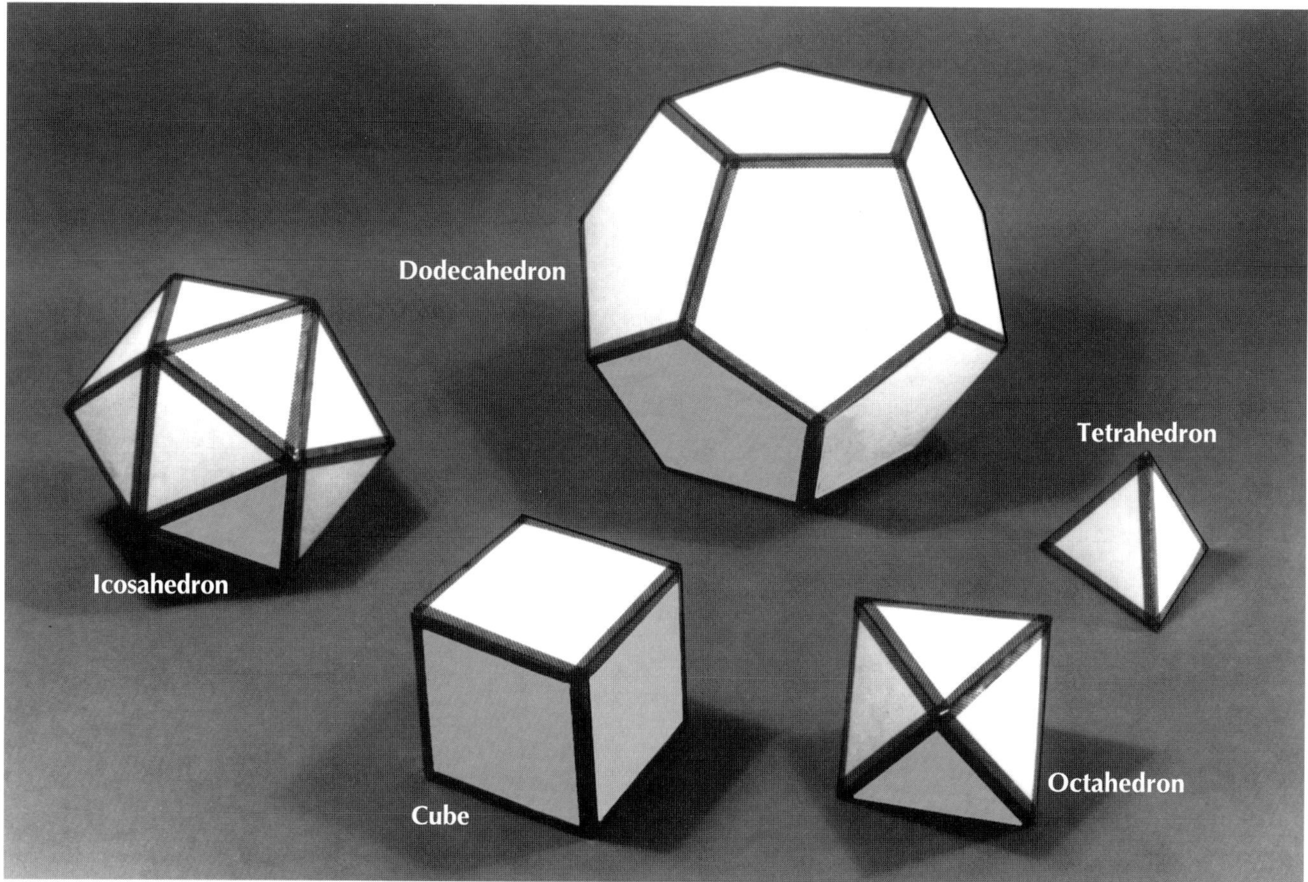

Tetrahedron
(3-3-3) Total faces = 4.
Hexahedron (Cube)
(4-4-4) Total faces = 6.
Octahedron
(3-3-3-3) Total faces = 8.
Dodecahedron
(5-5-5) Total faces = 12
Icosahedron
(3-3-3-3-3) total faces = 20.

The Platonic Solids

There are nine regular polyhedra in Euclidean 3-space that use only a single type of regular polygon for their construction. They are the five Platonic solids, as seen in the above photograph, and the four Kepler-Poinsot stellations, pages 172 and 173. The Platonic Solids, named after the philosopher Plato, have been known since Classical Greece, 500 BC. Euclid wrote about them in book XII of the "Elements". He also proved that there were only five regular members to the family.

In order to build a set of the Platonic Solids, you need to know which type of polygon is required for each particular figure and how many are found at a vertex. The caption on this page gives you this necessary information.

A toothpick and gumdrop model of the cube, seen on the facing page, shows how a solid can be represented by only edges and nodes with the faces only implied. Notice the shadows that this model casts. Build some polyhedra for yourself. Set up some lights and with a camera, take some photographs from different views. Include one shot directly above the center of a face. These shadows can be seen to be the basis for diagrams of these solids. Schematic maps can be obtained when a light is shown down upon the figure from a point directly above the center of one of its faces. One of the planar faces is used to frame the others and must be included when counting the total number of faces. Fig. 4.9

In the chart below, each of the Platonic Solids is seen by the use of a schematic diagram using only nodes and edges to indicate the particular polyhedron. The center of projection is at a point a short distance from the center of a face. It projects into the plane of that face. The number of faces is counted by including the outer enclosing face and the inner subdivisions.

Notice that three of the five solids use equilateral triangles. Only the cube is composed of square faces and only the dodecahedron uses regular pentagonal faces.

The mathematician Leonhard Euler made the connection between the number of faces (F), Vertices (V), and Edges (E) of all convex polyhedra. F + V - E = 2.

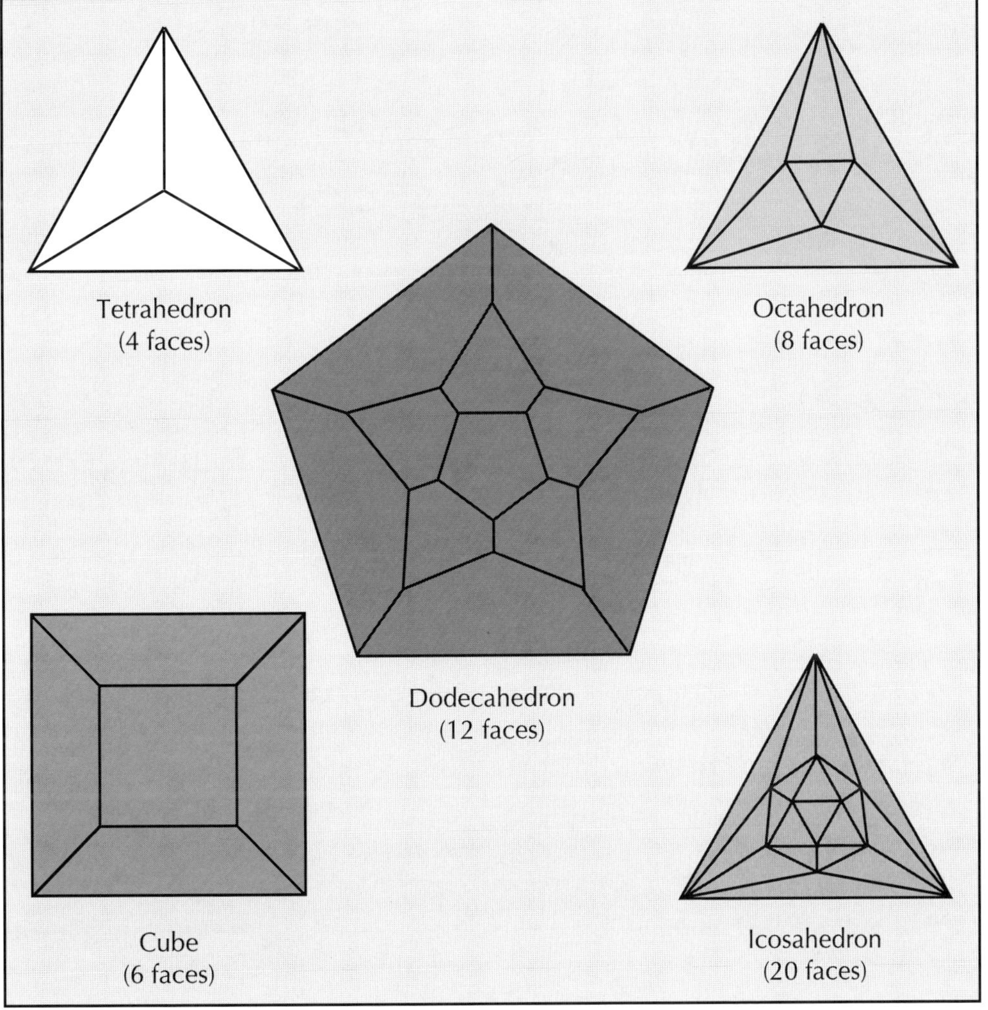

This page:
Above:
Photograph of a toothpick and gum drop model of a cube. The gumdrops function as nodes while the toothpicks function as edges. Try constructing a full set of the Platonic Solids using these humble materials. Try to eat as few as possible throughout the process. Weather will effect the stickiness of the gumdrops.

Left: Fig. 4.9
Schematic diagrams of the 5 Platonic Solids using lines and nodes. Each can be considered a graph of a polyhedron. Topology uses the concepts of graph extensively.

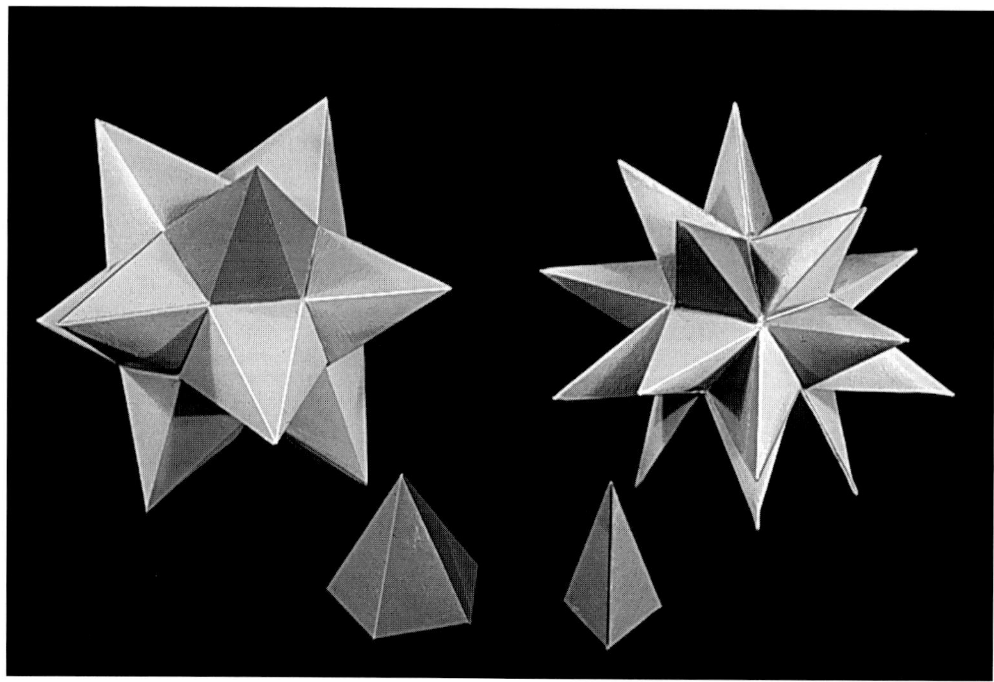

KEPLER PAIR
Small Stellated
Dodecahedron
(12 units required)

Great Stellated
Dodecahedron
(20 units required)

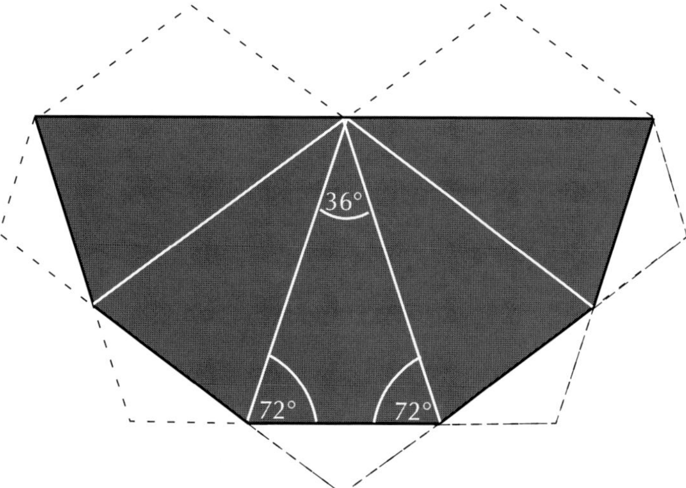

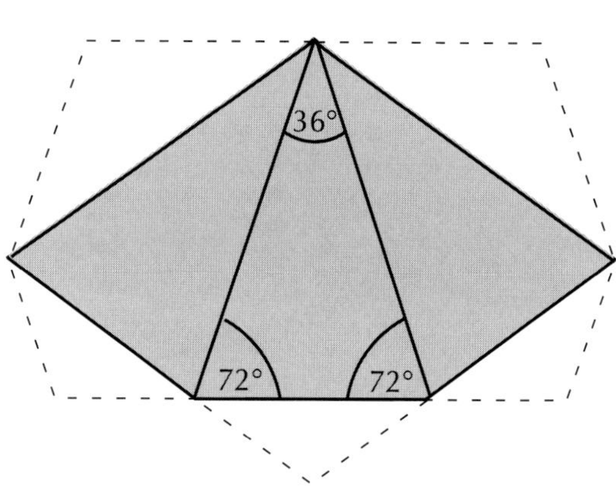

This page:
Above:
Photograph showing models of the Kepler pair of stellations. The solid on the left is the Small Stellated Dodecahedron, three caps come together at each vertex. The one on the right is the Great Stellated Dodecahedron, five caps come together at each vertex.

Center: Fig. 4.10
Templates for each cap. Score fold lines (solid lines are mountain folds and dotted lines are valley folds.) Cut out the requisite numbers of caps. Add glue to the tabs. Glue all caps first. Then, join the right number at each vertex.

Described in his book, Harmonies Mundi, it was Johannes Kepler in the 17th Century who discovered two of the stellations named after him and Louis Poinsot, two hundred years later, who discovered the other two that completed the set of regular polyhedra. These are related to the dual pair of Platonic Solids, the dodecahedron and the icosahedron. Think of these forms as having caps or cups added to their faces. Actually stellation is a particular kind of polyhedral growth. Refer to the Further Reading list. There are three stellations of the dodecahedron and fifty-nine of the icosahedron. According to mathematicians, there are higher space analogues to the Kepler-Poinsot polyhedra as well as the Platonics. In fact, there are ten of them, all in four-dimensional space. These are explained later on.

On these facing pages we show you photographs of each pair and the templates needed to construct the required caps or cups. In the above photograph of the Kepler pair, the individual caps are shown too. Notice that the finished figures appear more complex than the individual caps suggest.

174

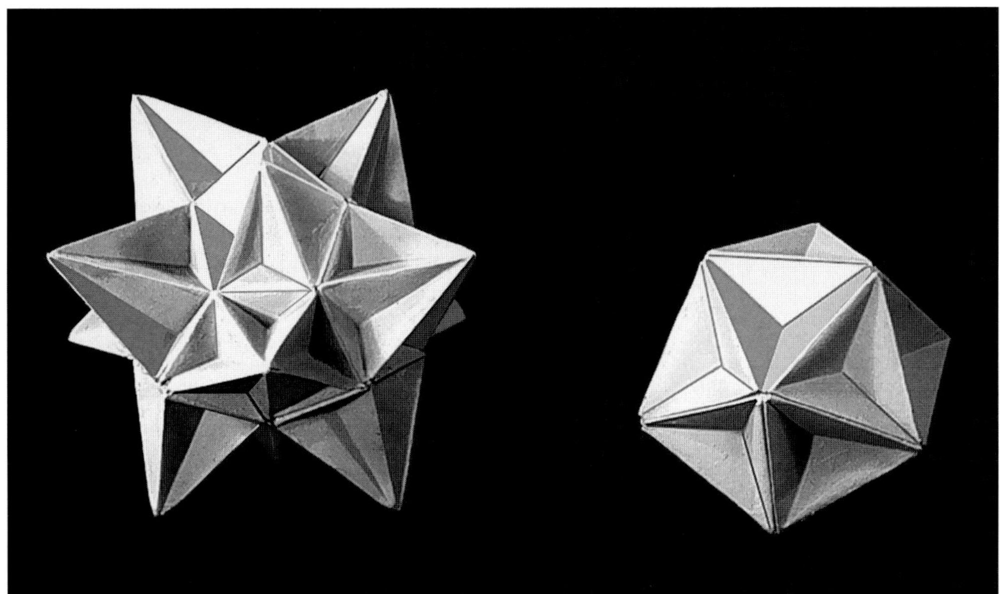

POINSOT PAIR
Great
Dodecahedron
(20 units required)

Great Icosahedron
(12 units required)

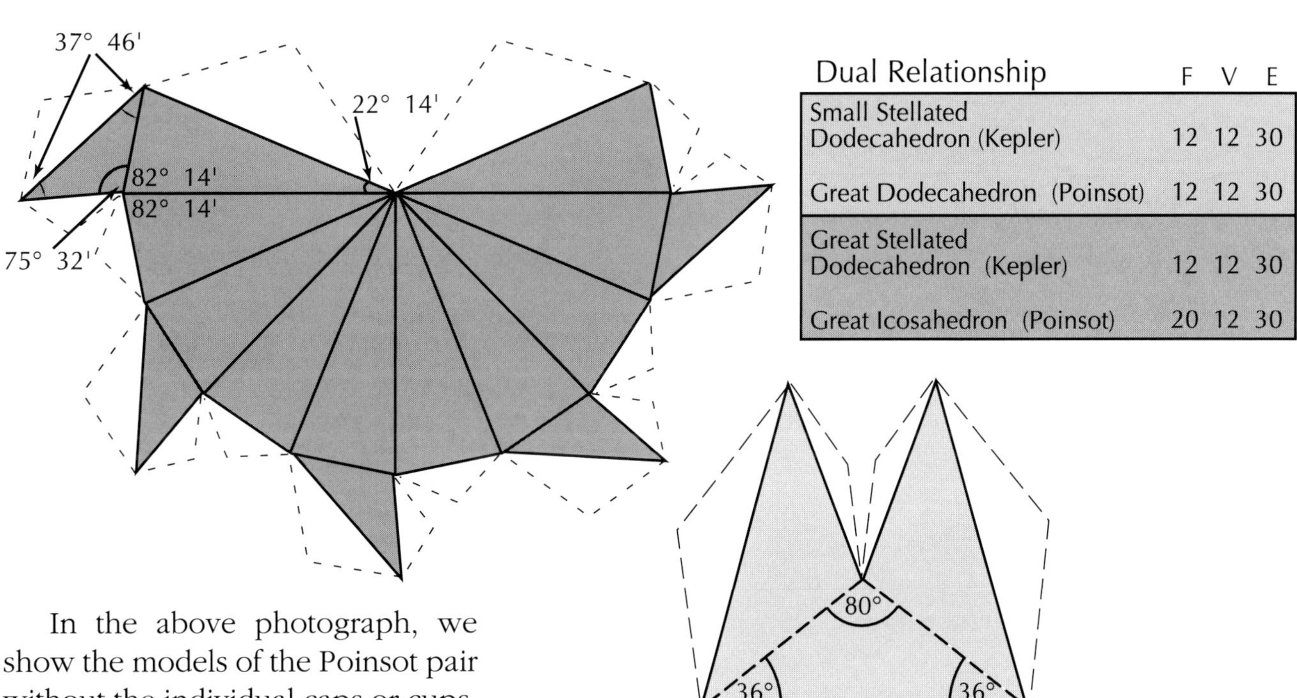

Dual Relationship	F	V	E
Small Stellated Dodecahedron (Kepler)	12	12	30
Great Dodecahedron (Poinsot)	12	12	30
Great Stellated Dodecahedron (Kepler)	12	12	30
Great Icosahedron (Poinsot)	20	12	30

In the above photograph, we show the models of the Poinsot pair without the individual caps or cups. The finished forms look more complicated than the individual parts. If you build some, make sure you enlarge the templates to a good working size or you will have difficulty in gluing the components together. Be patient and work slowly. The results will be worth the effort. Use a heavy type paper. Consider painting the model after it is put together. Acrylic paint works well. You could also spray it with metallic paints. Devel-

op a pattern for each face that enhances the star-like projections. Consider using the new polymer clays to cover the surface. Follow the manufacturer's directions for best results. Make multiples for a hanging mobile. Think of other variations and materials that would show off the forms to their best advantage. Research and make models of other stellations.

This page:
Above:
Photograph showing models of the Poinsot pair. The solid on the left is the Great Dodecahedron, three caps come together at each vertex. The solid on the right is the Great Icosahedron, five cups come together at each vertex.

Center: Fig. 4.11
Templates for each. Complete the necessary steps for joining.

The 13 Archimedean and 13 Catalan Solids

This page:
Photograph of models of all thirteen Archimedean Solids, sometimes called the semi-regular polyhedra because they are composed of more than one type of regular polygon. The above are made with patterned faces on bristol paper.

1. *Truncated Tetrahedron*
2. *Truncated Cube*
3. *Truncated Octahedron*
4. *Truncated Dodecahedron*
5. *Truncated Icosahedron*
6. *Truncated Cuboctahedron*
7. *Truncated Icosidodecahedron*
8. *Cuboctahedron*
9. *Rhombicuboctahedron*
10. *Rhombicosidodecahedron*
11. *Icosidodecahedron*
12. *Snub Cuboctahedron*
13. *Snub Icosidodecahedron*

Facing page: Chart of the 13 Archimedean Solids and their 13 dual partners called the Catalan Solids, named after Eugene Charles Catalan (1814-1894).

A family of closely related solids to the five Platonics is the Archimedean group of 13 members. Each one of the family requires a combination of polygons around a vertex. But, since these polyhedra are all symmetric, having information for one vertex gives the information for all vertices. In the chart on the facing page, we have listed each polyhedron with a shorthand notation that gives the vertex polygons. For instance, Archi 3-6-6, which we will talk about in greater detail later on, requires one equilateral triangle plus two regular hexagons around a vertex. In the chart we have also indicated their dual partners from the Catalan family.

In the above photograph, the models of the Archimedean Solids are all made of paper and they are built using a constant edge length so that you may see the relative sizes of family members. Each polygon was given a different face pattern so that the various polygons would be more evident. If you want to construct any of these, use the polygon templates given at the beginning of the chapter. Enlarge the template to at least a three inch edge length. Work slowly with two polygons at a time, and one vertex at a time.

On the pages following the chart, we have given templates for all thirteen Catalan Solids as well as photographs of some of the family members. Make a set for yourself.

*Dual Partnership Relationship
Archimedeans / Catalans

F=Faces
V=Vertices

*If the vertices of one polyhedron can be in a one-on-one correspondence with the centers of the other polyhedron, they are duals.

The Archimedean Solids	F	V	Edges	F	V	The Catalan Solids
Truncated Tetrahedron (Archimedean 3-6-6)	8	12	18	12	8	Triakis Tetrahedron
Truncated Cube (Archimedean 3-8-8)	14	24	36	24	14	Triakis Octahedron
Truncated Octahedron (Archimedean 4-6-6)	14	24	36	24	14	Tetrakis Hexahedron
Truncated Dodecahedron (Archimedean 3-10-10)	32	60	90	60	32	Triakis Icosahedron
Truncated Icosahedron (Archimedean 5-6-6)	32	60	90	60	32	Pentakis Dodecahedron
Truncated Cuboctahedron (Archimedean 4-6-8)	26	48	72	48	26	Hexakis Octahedron
Truncated Icosidodecahedron (Archimedean 4-6-10)	62	120	180	120	62	Hexakis Icosahedron
Cuboctahedron (Archimedean 3-4-3-4)	14	12	24	12	14	Rhombic Dodecahedron
Rhombicuboctahedron (Archimedean 3-4-4-4)	26	24	48	24	26	Trapezoidal Icositetrahedron
Rhombicosidodecahedron (Archimedean 3-4-5-4)	62	60	120	60	62	Trapezoidal Hexacontahedron
Icosidodecahedron (Archimedean 3-5-3-5)	32	30	60	30	32	Rhombic Triacontahedron
Snub Cuboctahedron (Archimedean 3-3-3-3-4)	38	24	60	24	38	Pentagonal Icositetrahedron
Snub Icosidodecahedron (Archimedean 3-3-3-3-5)	92	60	150	60	92	Pentagonal Hexacontahedron

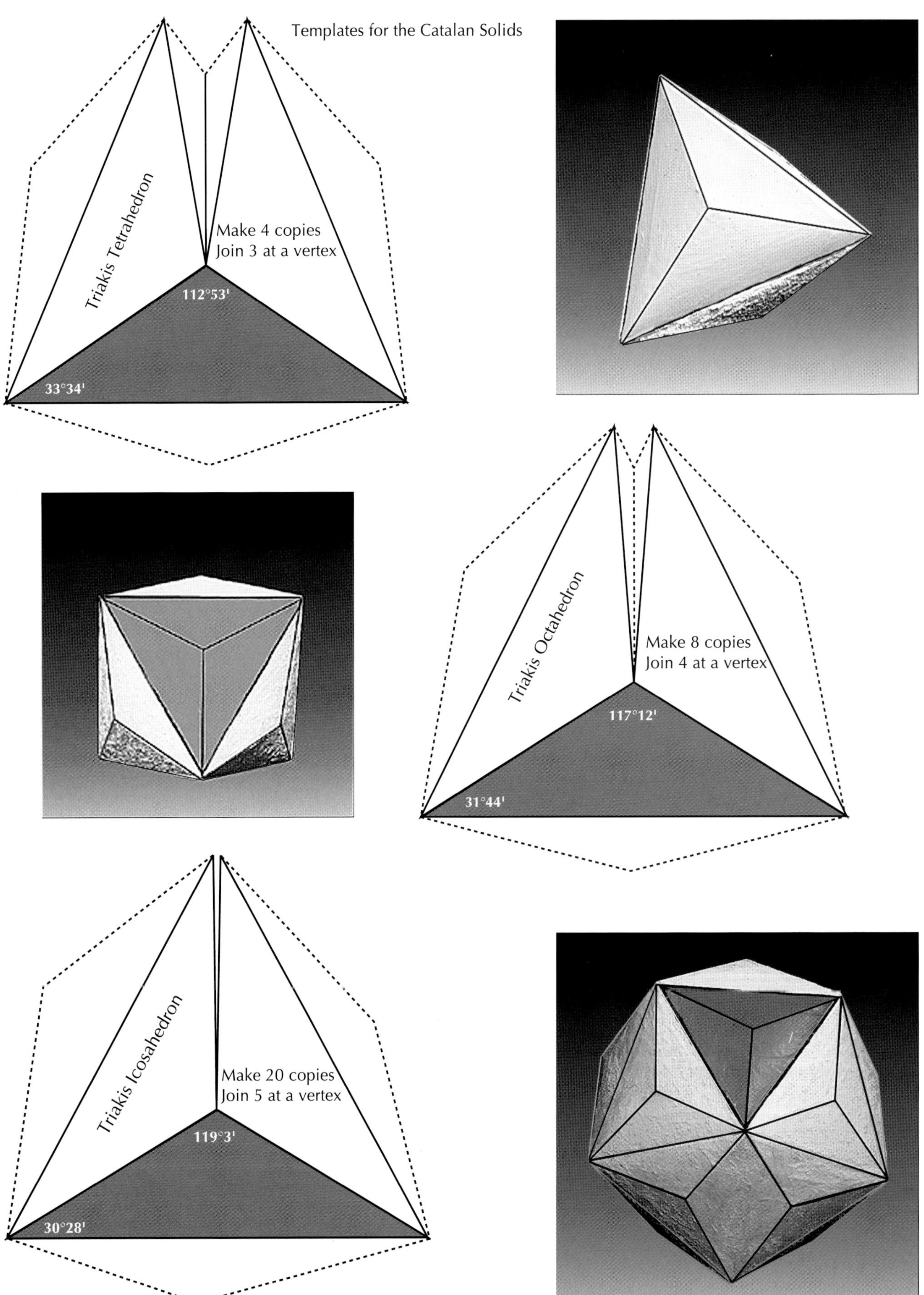
Templates for the Catalan Solids
Triakis Tetrahedron
Make 4 copies
Join 3 at a vertex
112°53'
33°34'
Triakis Octahedron
Make 8 copies
Join 4 at a vertex
117°12'
31°44'
Triakis Icosahedron
Make 20 copies
Join 5 at a vertex
119°3'
30°28'

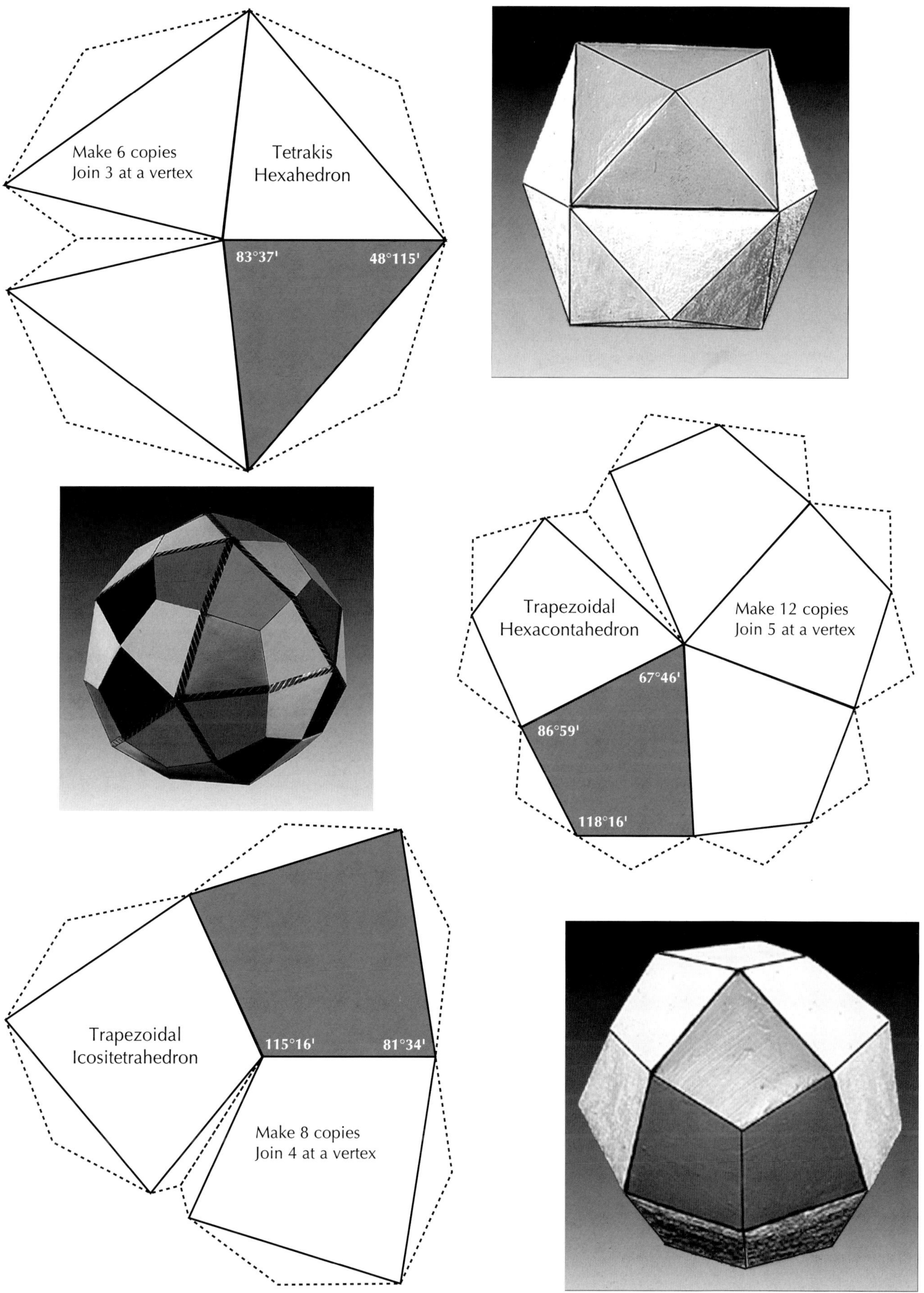

Make 6 copies
Join 3 at a vertex
Tetrakis Hexahedron
83°37'
48°115'
Trapezoidal Hexacontahedron
Make 12 copies Join 5 at a vertex
67°46'
86°59'
118°16'
Trapezoidal Icositetrahedron
115°16'
81°34'
Make 8 copies
Join 4 at a vertex

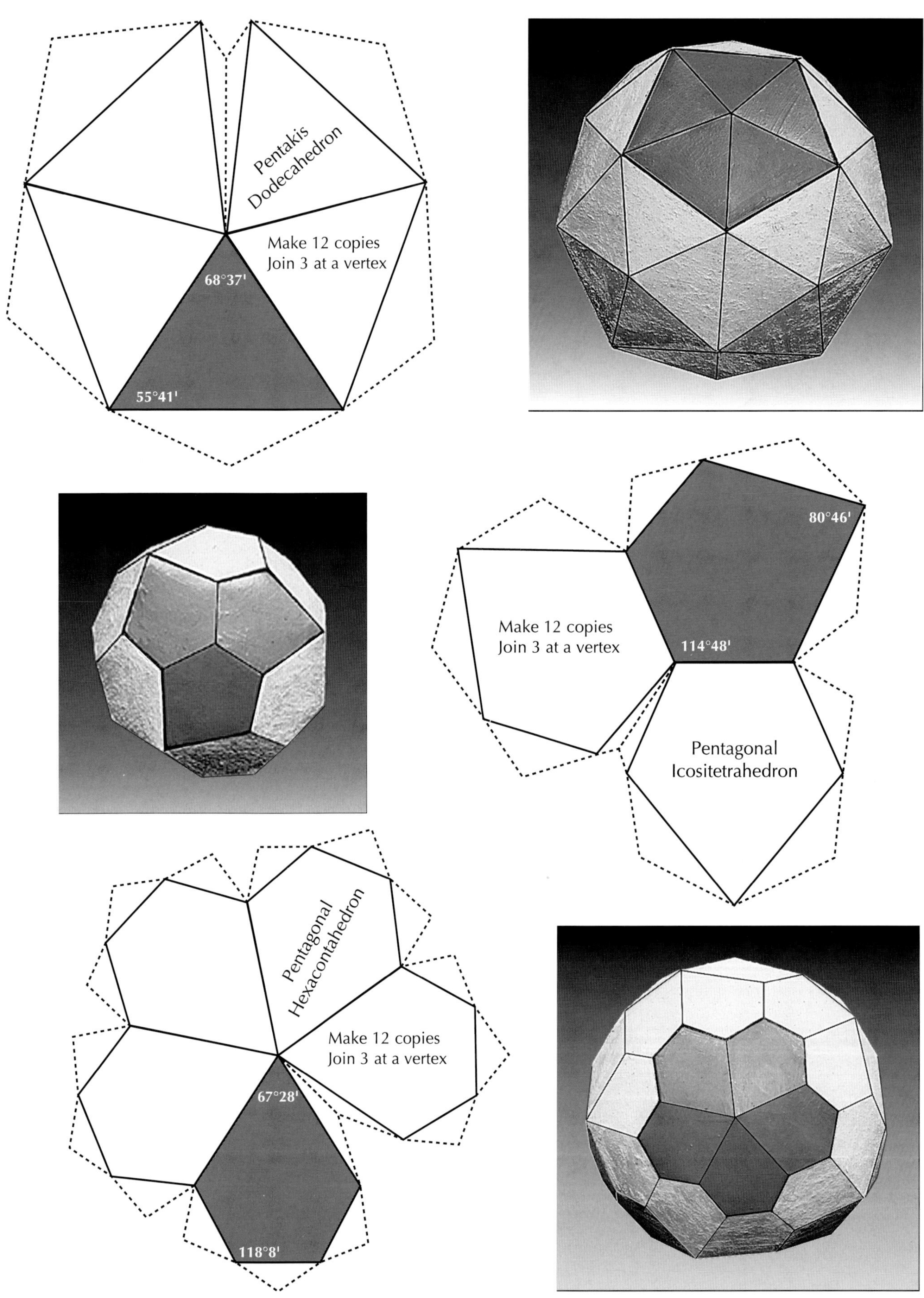

Pentakis Dodecahedron
Make 12 copies
Join 3 at a vertex
68°37'
55°41'
Make 12 copies
Join 3 at a vertex
80°46'
114°48'
Pentagonal Icositetrahedron
Pentagonal Hexacontahedron
Make 12 copies
Join 3 at a vertex
67°28'
118°8'

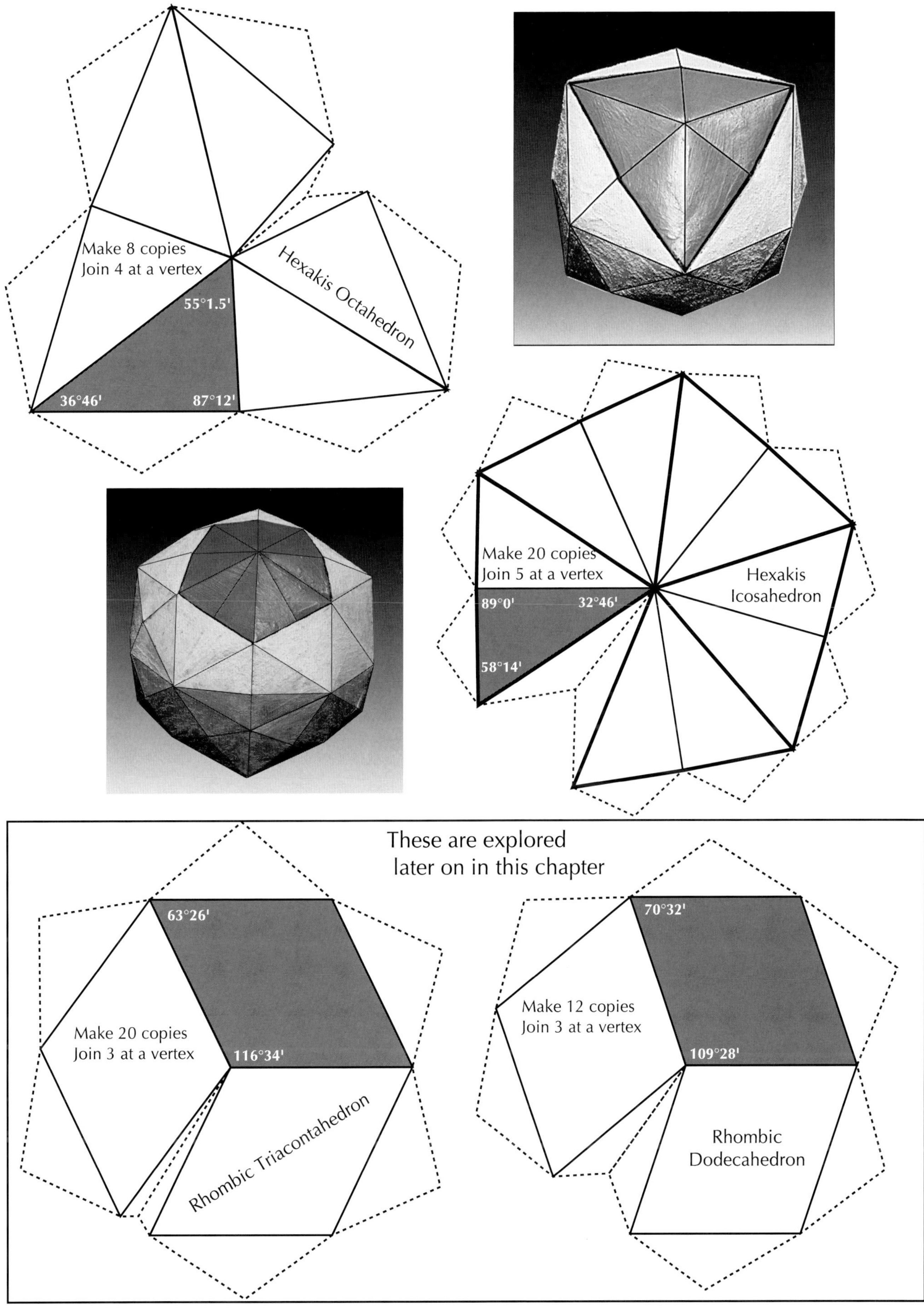
Make 8 copies
Join 4 at a vertex
Hexakis Octahedron
55°1.5'
36°46'
87°12'
Make 20 copies
Join 5 at a vertex
Hexakis Icosahedron
89°0'
32°46'
58°14'
These are explored
later on in this chapter
63°26'
Make 20 copies
Join 3 at a vertex
116°34'
Rhombic Triacontahedron
70°32'
Make 12 copies
Join 3 at a vertex
109°28'
Rhombic Dodecahedron

Polyhedra can be considered tilings in 3-space. The Catalan Solids are composed of non-regular polygons having three, four, or 5-sided units while the 13 Archimedean Solids relate to eight tilings of the plane, 2-space. These are called semi-regular tilings because they require more than one type of polygon at each vertex, just like their 3-d counterparts. The edge length of each polygon is kept constant. Every vertex is surrounded by the same number and type of polygons. These tilings can be constructed by using the polygon templates, without the glue tabs, given at the beginning of this chapter.

If you were to make a set of the Archimedean Solids in a transparent material and then shine a light on any one of them, each would cast a shadow of polygons onto the plane which would be a network of lines and nodes, sometimes called a graph. Which Archimedean Solid relates to which Archimedean tiling? Why are there thirteen solids and only eight tilings? How could you mathematically prove the relationship between the dimensions? Is there a comparable set of Archimedean Solids and tilings in 4-space, 5-space, etc?

What happens when you have more questions than answers? You spend a lot of time thinking.....and thinking some more.....

In each Archimedean tiling on the facing page, if you use the centerpoint of each polygon as a vertex and then connect the vertices you develop a dual partner tiling. Lay a piece of tracing paper over each of the Archimedean tilings and see what new tilings occur. Use these duals as the basis for an artwork, or many artworks.

The Eight Archimedean Plane Tilings

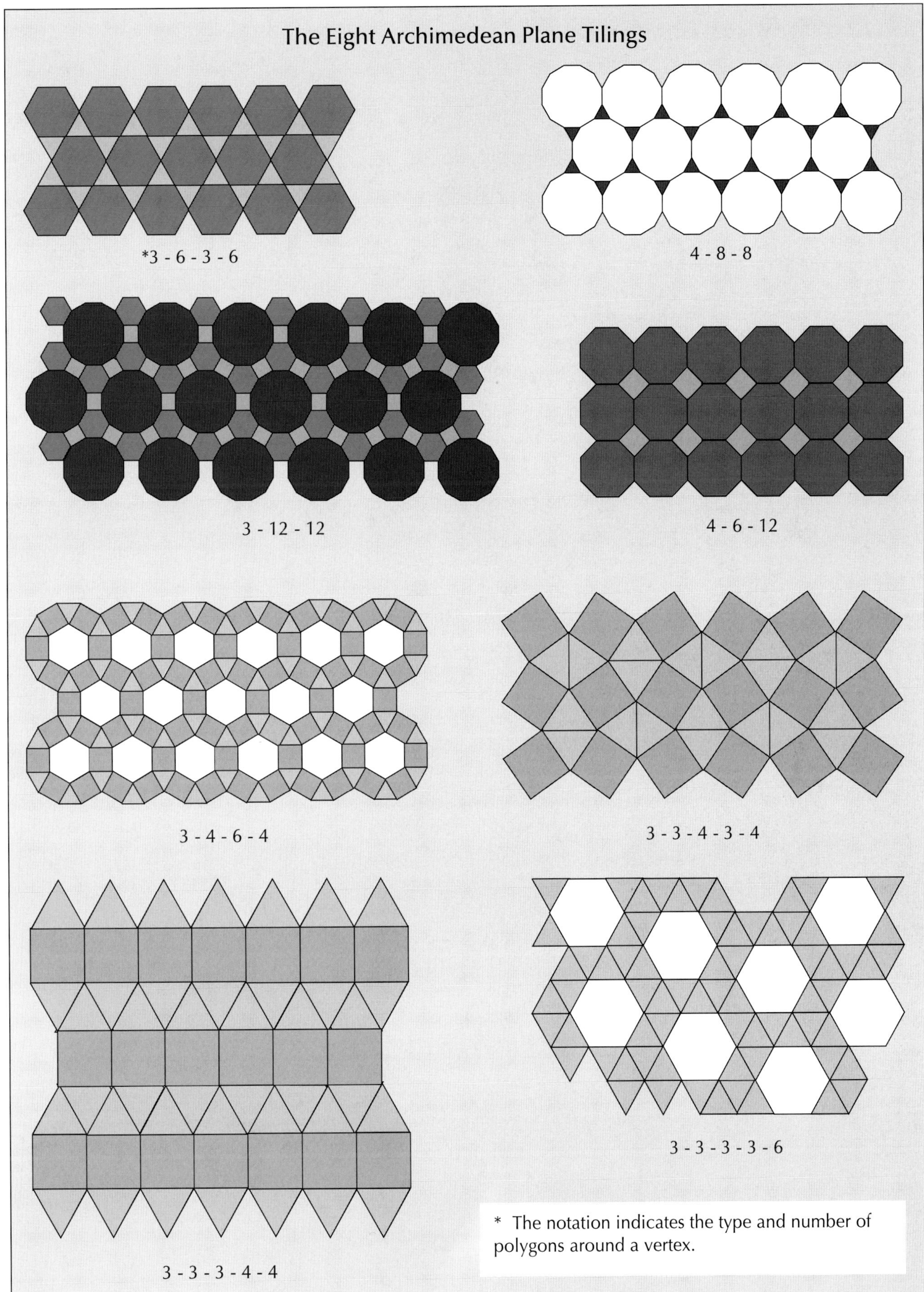

* The notation indicates the type and number of polygons around a vertex.

The Fabulous Fullerenes

Factors of 60	Factors of 10
60/30 = 2	10/10 = 1
60/20 = 3	10/5 = 2
60/15 = 4	10/2 = 5
60/12 = 5	10/1 = 10
60/10 = 6	
60/6 = 10	
60/5 = 12	
60/4 = 15	
60/3 = 20	
60/2 = 30	
60/1 = 60	

Archi 5-6-6, the truncated icosahedron, which is a polyhedron composed solely of pentagons and hexagons, each pentagon being surrounded by five hexagons, has recently been found to be a special member of the carbon molecule family. It represents a new aspect of molecules architecture. It has been named the C_{60} and affectionately called the "Buckyball". Students know it as the familiar soccer ball. See the above photograph.

Carbon is an abundant chemical element in nature. It has peculiar chemical properties which make it crucial to all organic life. It is manufactured from the helium which resides inside the core of large stars in the heavens of our universe.

Each carbon atom takes up a position at the vertex of the solid. There are 60 vertices, hence the name C_{60}. The base 60 system is extremely useful in mathematics. think of the 60 seconds and the sixty minutes in the hour. Sixty can be divided by many more numbers than 10. See chart.

Members of this carbon group are composed of even numbers. A rule for carbon atoms is that they must have three bonds per atom radiating from them. Each bond is chared between two atoms. Thus, the number of edges must be one and one-half times the number of vertices. $E = 3/2\ V$. There is the C_{44}, C_{70}, C_{76}, C_{82}, C_{84}, C_{240}, C_{540}. The last two molecules have the same symmetry as the C_{60} but two and three times the diameter, respectively. At this time, scientists are busy with the process of trying to catalog all the molecules in this group.

Scientists explore the structures of these molecules through the use of models. Model-making is essentially an art process and these carbon cages are beautiful. In order to build them, however, both scientists and artists need to follow the rules of the nature game. Only pentagons and hexagons can be used in construction. You must have 12 pentagons while there can be many more hexagons, but only even numbers of them. No pentagons may be adjacent to one another. Interestingly, while groups of hexagons must always lay flat, the introduction of the pentagon, allows the structure to fold up.

How would one go about exploring the possibilities of building these various models, if one only had this rudimentary information?

The photographs of paper and paint models on this page show some of the ways in which hexagons and pentagons joint at a single vertex. The photo to the right shows vertex nets using combinations of pentagons and hexagons. In the center two photographs are models exploring different kinds of "caps". On the left, a central hexagon is surrounded by a combination of hexagons and pentagons with no pentagons being adjacent to each other. On the right, a central pentagonis surrounded by a combination of hexagons and pentagons. The larger fullerenes are not perfectly spherical.

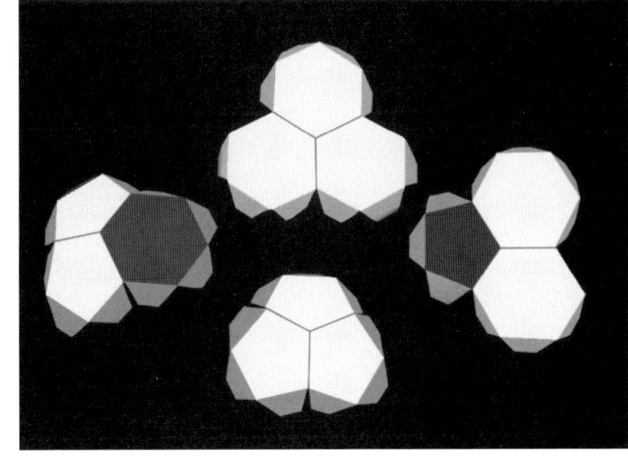

All life is based on carbon chemistry. The chain-like molecules of all living organisms are systems that carry out metabolism, reproduction, and organize into 3-d structures.

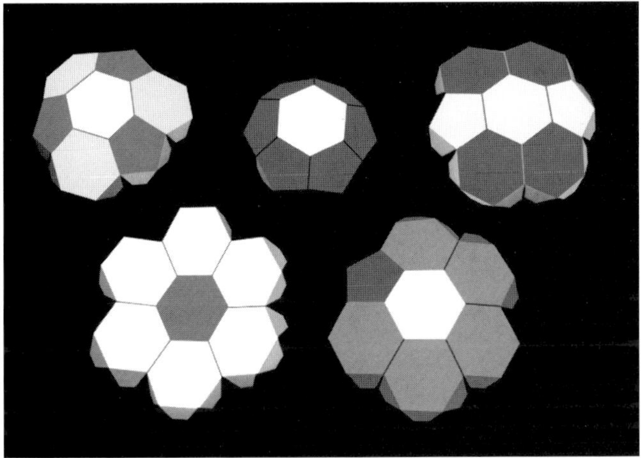

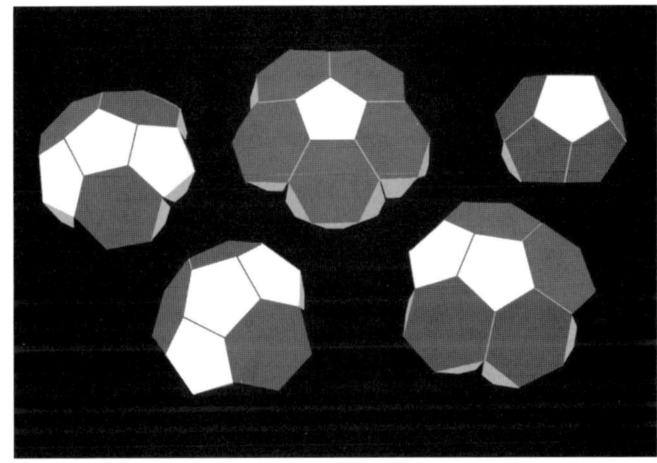

Below left shows models exploring different combinations of pentagons and hexagons used to enclose space. The model on the right is a work by student Immer Cook using magazine elements to construct his form.

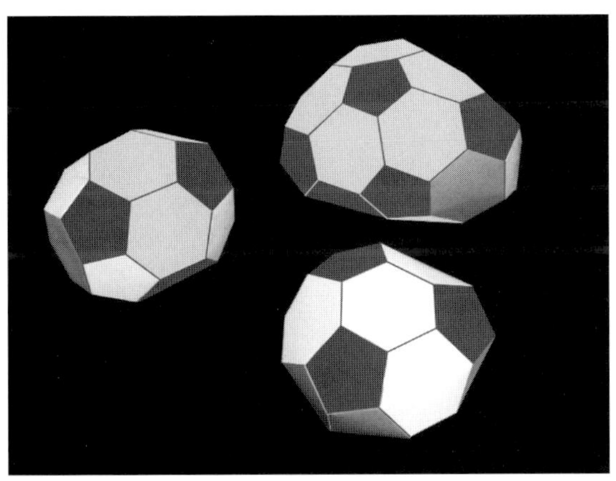

The Hypercube

The Cube
(6 squares)

The Unfolded Cube

The Hypercube
(8 Cubes)

The Unfolded Hypercube

We have just looked at some families of polyhedra in our familiar 3-space. Scientists and mathematicians believe, however, that four spatial dimensions physically exist despite the fact that we cannot see that dimension. But, by analogy, we can explore this space and others beyond four.

In four dimensions there are also Euclidean solids which we are calling polyplexes. It is obvious that we cannot actually see these figures but aspects of them can be visualized in both three ands two dimensions. The computer is a most powerful tool for this exploration but hands-on work is still possible in the other more familiar dimensions.

The square is the basic building unit in 2-space. Six square faces are the units needed for building a single 3-space cube. The cube can be unfolded so that you can see the square faces. See Fig. 4.11. This cube is the basic building block in 3-space.

Moving the cube perpendicular to itself creates the hypercube of 4-space. See Fig. 4.12 This is a conceptual move since we cannot physically do it. Eight cubes (cells) are needed in order to build this hypersolid. It can be unfolded in 3-space just like the cube was unfolded in 2-space. To see what this form looks like, make eight copies of the cube template from chapter 1 and join them together in the manner seen in the diagram at the left.

Would you think, by analogy, that the same process would hold true for polytopes of 5, 6, 8, 10, etc. dimensions? If there are six unfolded squares in 2-space, 8 unfolded cubes in 3-space, would there be 10 unfolded hypercubes in 4-space? Could you prove this mathematically?

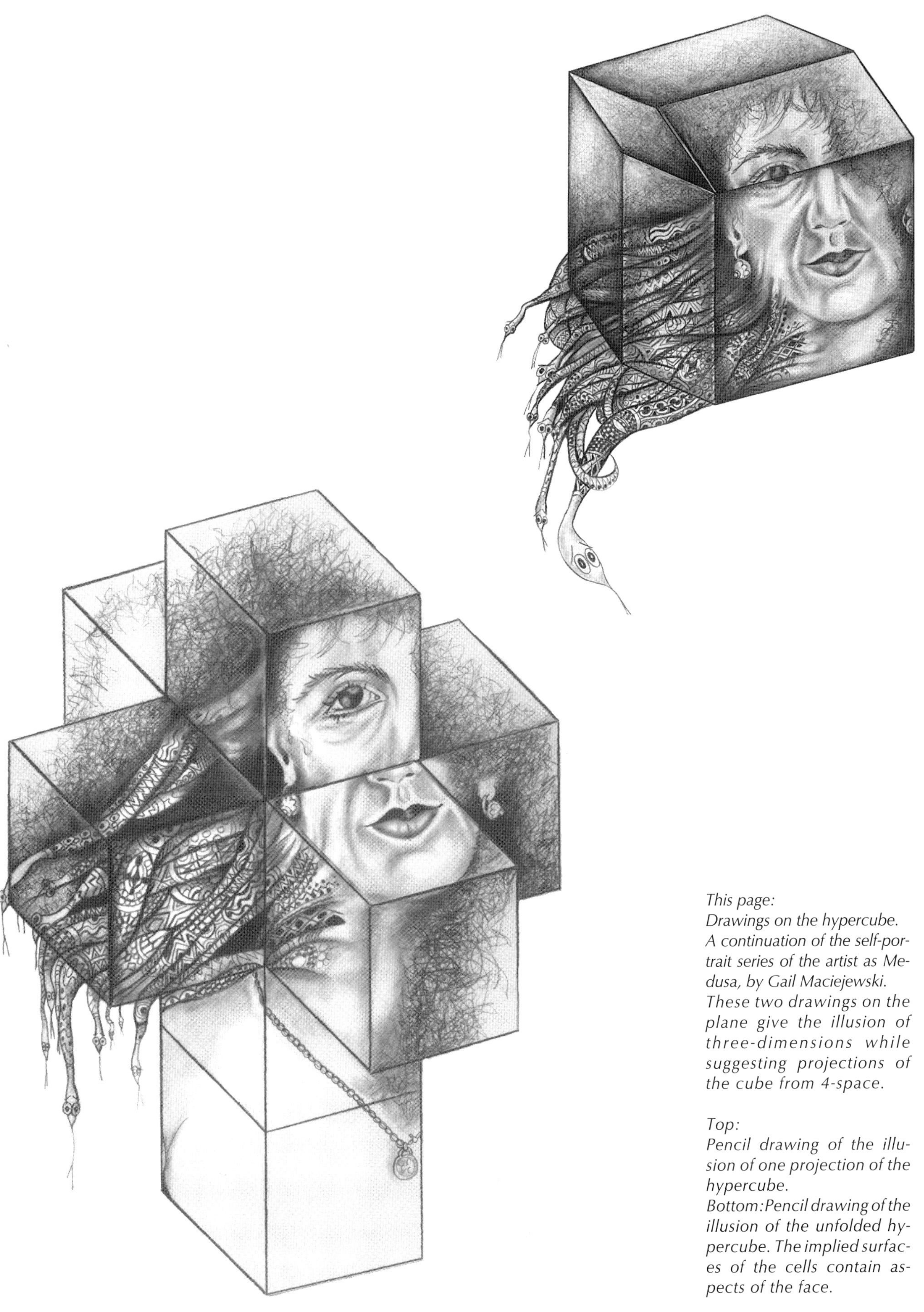

This page:
Drawings on the hypercube.
A continuation of the self-portrait series of the artist as Medusa, by Gail Maciejewski.
These two drawings on the plane give the illusion of three-dimensions while suggesting projections of the cube from 4-space.

Top:
Pencil drawing of the illusion of one projection of the hypercube.
Bottom: Pencil drawing of the illusion of the unfolded hypercube. The implied surfaces of the cells contain aspects of the face.

187

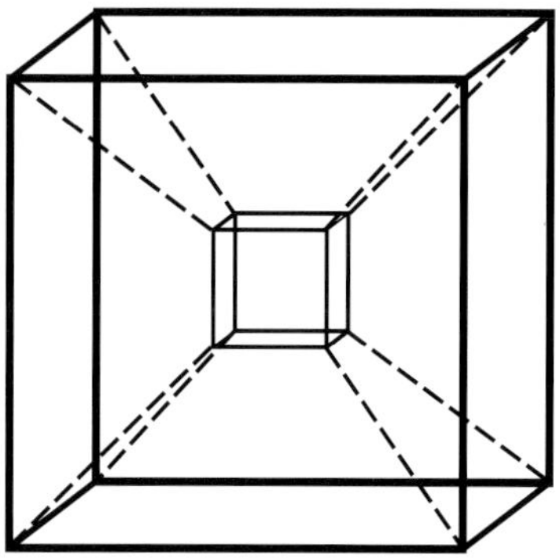

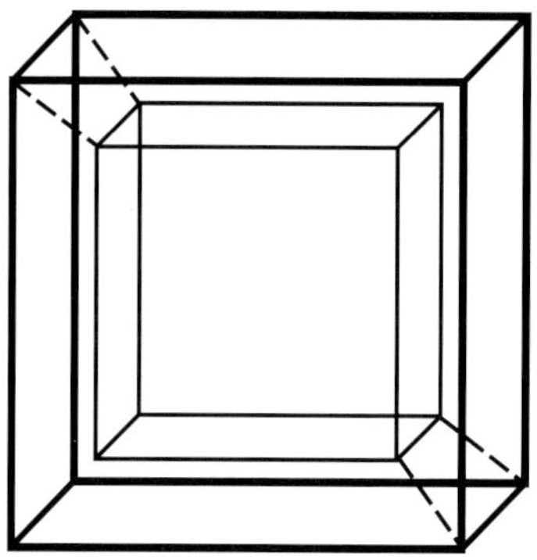

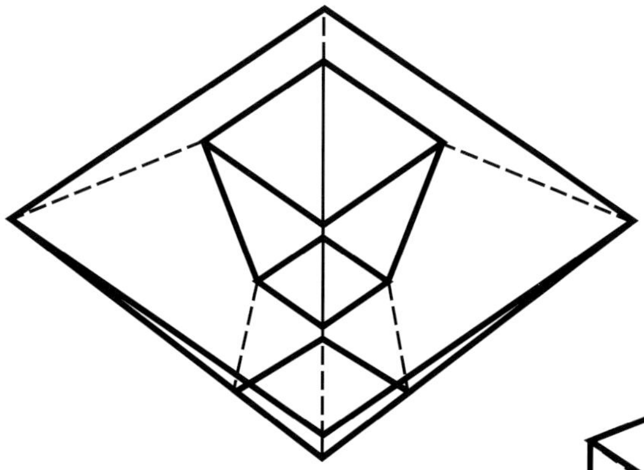

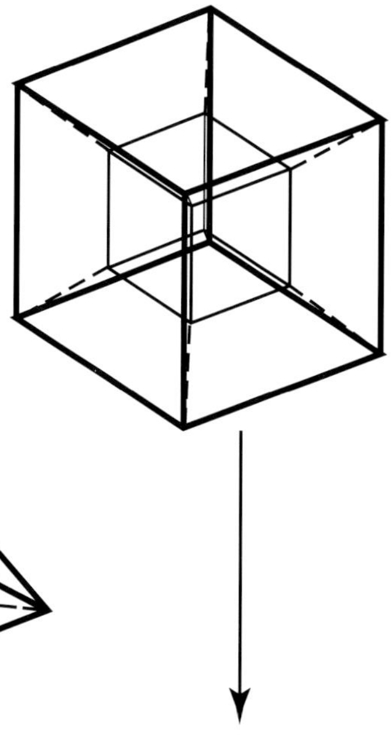

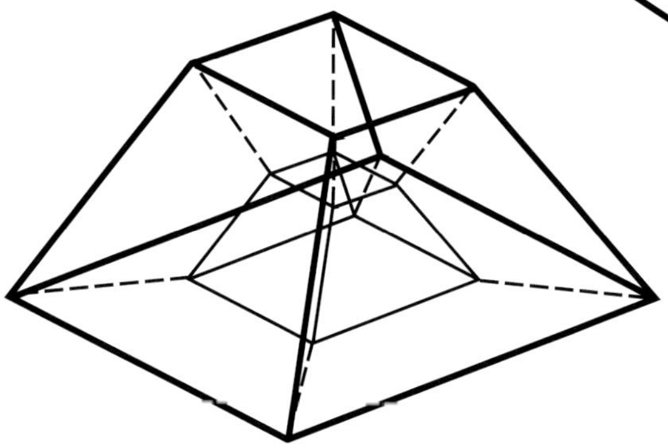

The hypercube also casts shadows in both three-and two-dimensions. Since it is a more complex geometric object because of the added dimension, the shadows are more complex. The computer may be the best tool for investigating these projections since the mathematics is programmed into the machine and the computer can generate images on the two-dimensional screen in a matter of minutes. With imagination, however, a visual artist with a variety of materials and tools can also generate visual variations of the hypercube.

On this page, we show you several projections of the hypercube as its rotates in space. Fig. 4.13. Mathematician Thomas Banchoff has done much of the definitive work in this field so some research might be in order here.

You will notice that in some rotations the framing shape of the figure is a quadrilateral. In other rotations it is a hexagon, and in others it is an octagon. You can, therefore, treat each figure as a two-dimensional potential tile unit to cover the plane. There are rules that govern the game of tiling which you may want to research. On the next three pages, we show you some artful tilings that have been generated from the basic hypercube.

This page:
Fig. 4.13. Several projections
and a tiling of the hypercube.

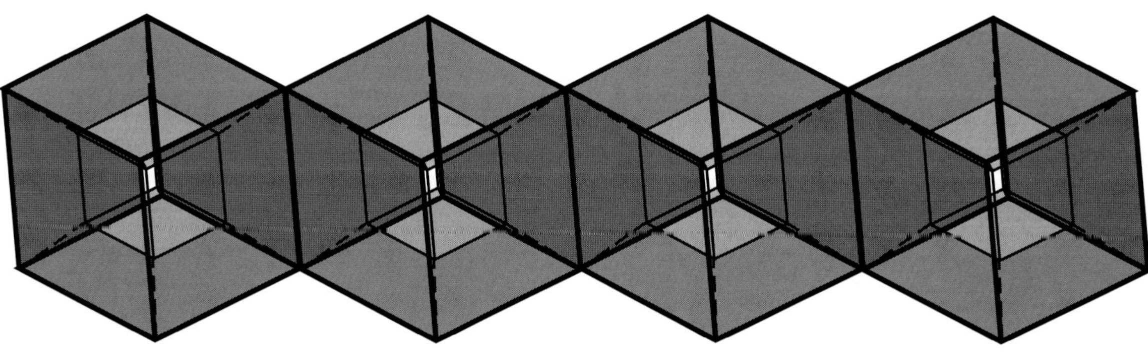

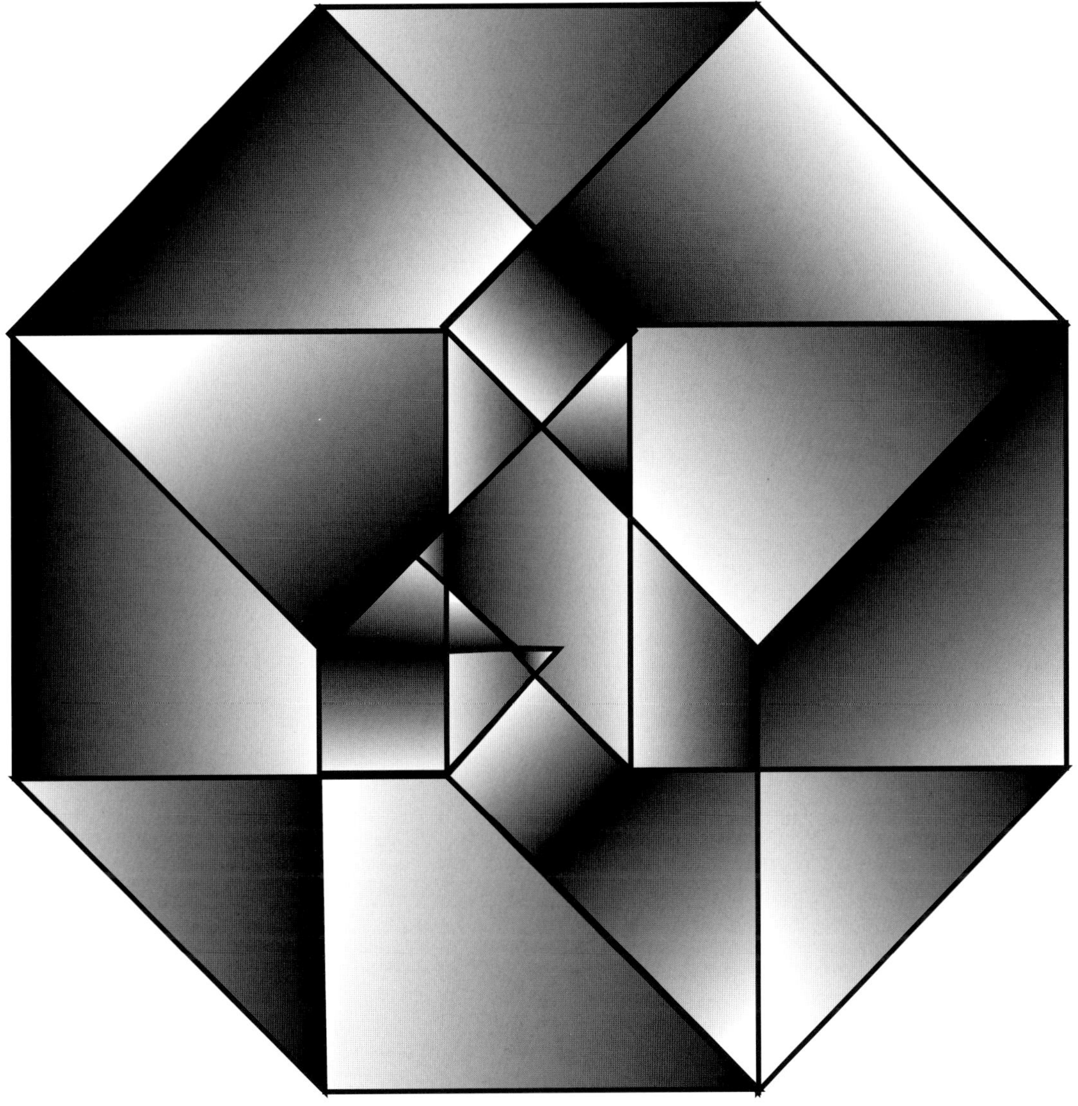

This page:
Artistic transformation of a single hypercube using variations of tone.

The hypercube unit on this page, and the tilings on the following two pages, are the combined efforts of graphic design artists Claire Belanger and Mary Calnan. The latter developed the geometric structure while the former chose and added the gray tones. Naturally, these tones could also be in color.

Consider doing a variation yourself. The above single unit is based on a most symmetric projection view of the hypercube. It was computer generated but could also be done through handwork. Markers, colored pencils, inks, acrylic paints, and colored papers would be appropriate media. Or these media could be worked together.

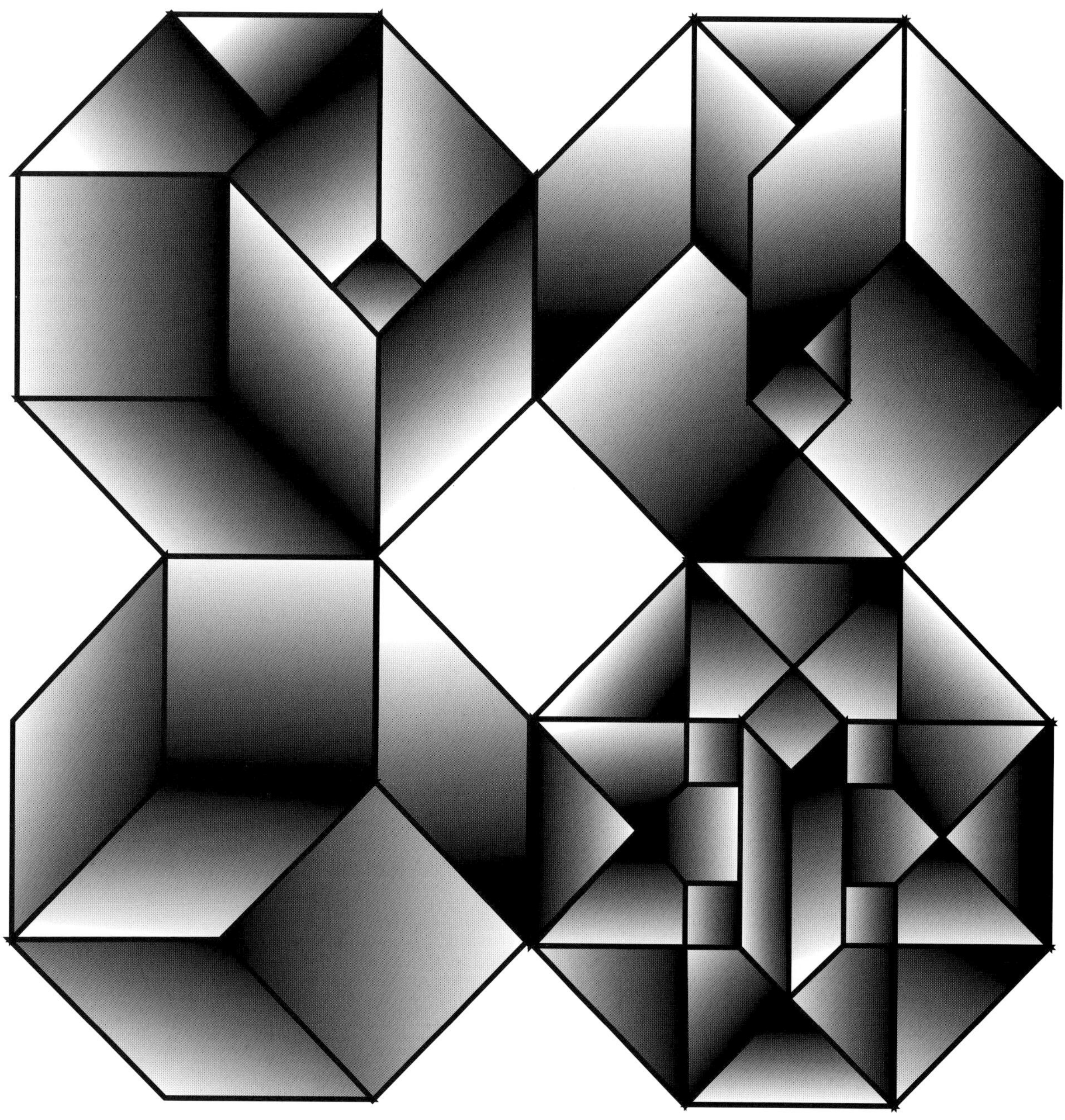

This page:
Hypercube tiling using a vari-
ation of each of the four units.

Notice that this tiling relates to one of the semi-regular Archimedean tilings using a combination of octagons and squares. Here each octagonal unit has been transformed in both structure and tone. Consider other ways in which the above units could be changed. Could you curve some of the elements? Could you skew the entire tiling?

Consider using any of the others of the Archimedean group to make an artistic tiling. Compare these tilings to ones created by Islamic artists.

This page:
A hypercube tiling having 16 hypercubes within a square format.

Notice that there is both complexity and simplicity withing this tiling. It employs both the symmetry operations of reflection and rotation. Can you find elements of translation within? What is the essential tiling unit that is used to develop the design? You could translate this entire unit to cover the plane as many times as you wished. What would happen if each tone was changed into a color? How would you keep the design from becoming too chaotic? Would you have to take away elements? Would you have to add elements? Would you have to change size or scale?

Cubes in N-Dimensions

	Number of Vertices	Number of Edges	Number of Faces	Number of Polyhedra	Number of Polytopes
POINT	1	0	0	0	0
LINE	2	1	0	0	0
SQUARE	4	4	1	0	0
CUBE	8	12	6	1	0
HYPERCUBE	16	32	24	8	1
5-D CUBE	32	80	80	40	10
6-D CUBE					

This page:
Chart for finding the faces, edges, and vertices of the cube in n-dimensions.

When examining polyhedra in 3-space, one must look at vertices, edges, and faces. The same principle holds true for other dimensions. Here we have conceptualized, through the use of a chart, what happens with vertices, edges, and faces in 3,4 and 5-space.

The numbers in the chart are obtained from the logical extensions of thinking about moving a figure in space. Moving a line at right angles to itself generates a square, moving a square at right angles to itself generates a cube; moving a cube at right angles to itself generates a hypercube, etc. Moving a hypercube....

For a cube you have the four edges of the square in its initial position, the four edges of the square in its final position and the four edges traced out by the moving vertices. For a hypercube, you have the 12 edges of the cube in its initial position, the 12 edges of the cube in its final position and the 8 edges generated by moving the 8 vertices of the cube for a total of 32 edges.

For each entry on the chart in the edge column, you obtain twice the number of edges as in the previous figure, plus the number of vertices in the previous figure. For example, for a five-cube, the number of edges is 32 + 32 + 16 = 80.

Notice, also, in the chart we have that the number of faces of an *n*-dimensional cube = 2x, the number of faces of an (n-1) -D cube plus the number of edges in an (n-1) D cube.

The number of cells = 2x (number of cells in figure of dimension 1) less number of faces in figure of dimension 1 less. The number of 3-d cells in a 4-cube = 1 + 1 + = 6 = 8.

By looking at the chart, can you find a pattern of numbers that will help you determine the vertices, edges, and faces for 6-space?

Some Questions to Ponder

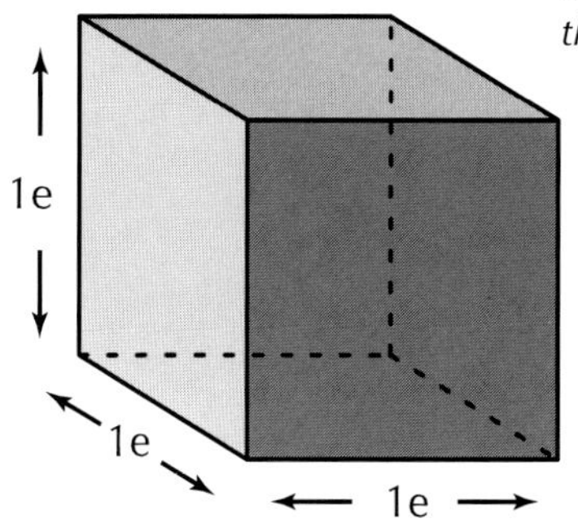

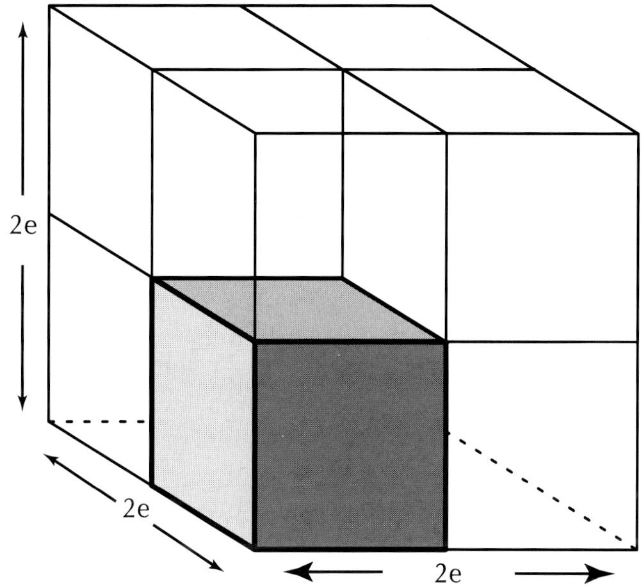

What happens to scale in the fourth dimension? In 3-space, constructing something that is larger cannot be done by merely inflating the original. What is required is a change in the basic design of the original structure. Look at an elephant and an ant. Compare the relationships of the parts to the whole. What happens to the proportions? What about the overall shape? What about the internal structure? What about the organs? What about the brain?

In all matter, the essential materials are the same basic building blocks of atoms, molecules, and cells. It is the combinations that differ to create structures, surfaces, skeletons, shapes, etc. If one were to scale up from an ant to an elephant it would require a factor of several thousand. Thus, making a bigger ant would not do. Everything about the ant would have to change.

Look at the simple form of a unit cube that has a one foot edge length. The pressure on the bottom face of the cube is the weight of the cube divided by the area of the bottom face. Fig. 4.13A. Now double the edge length of the cube. Notice the increase in the number of internal cubes. Each time the edge length of the cube is doubled, the pressure exerted on each square foot of the bottom face is also doubled. Fig. 4.13 B. Soon this process reaches its limit. The structure cannot support its own weight. The volume of material, which is related to its weight, increases faster than the area of the base that must support that weight.

One way of making a structure larger is to change the proportions as the size grows. The more material to be included, the squatter and broader the structure will have to be. The strength of any material is related to how well it can withstand external forces and this relates ultimately to the atomic structure of each material. Consider the human form. What would happen if you quadruple the size of a 6 foot tall male?

What happens to materials in the fourth dimension? What happens to gravity or surface tension, or atomic structure? We do not have the answers to these questions here. How does one go about finding them? What information do you need? What skills do you need to have in order to understand and use the information?

The bottom face of the cube is responsible for supporting the entire weight of the cube.

1e
1e
1e

2e
2e
2e

This page:
Fig. 4.14A,
Fig 4. 14B

Above: Fig. 4.14A
The area of the face of the cube = e^2. Volume = e^3. The pressure on the bottom face of the cube = $e^3/e^2 = e$.

Center: Fig. 4.14B
Now, double the dimensions of the cube. Each time the dimension of the cube is doubled, the pressure exerted on each square foot of the bottom face is doubled.
The area of one face of the cube = $4e^2$. The volume of the cube is $8e^3$. The pressure on the bottom face of the cube = $8e^3/4e^3 = 2e$.

Six Regular Polyplexes in 4-Space

In 4-space there are six regular polyplexes bounded by regular three-dimensional cells. Each of them has the same number of regular polyhedral cells around each edge. The 4-dimensional Platonic solids were discovered in the second half of the 19th century by the mathematician Schlafli. Five of them can be related to Platonic polyhedra in 3-space. Regular polyplexes must have boundary cells of regular polyhedra. A polyplex is called an n-cell if it is bounded by n-polyhedra.

Euler's formula about faces, edges, and vertices works in four dimensions.
$V - E + F - C = 0$ (Vertices minus edges + faces minus cells $= 0$).
Let C_0 = the number of 0-dimensional elements in the figure (i.e. vertices);
Let C_1 = the number of 1-dimensional elements in the figure (i.e. edges);
Let C_2 = the number of 2-dimensional elements in the figure (i.e. faces).
This can be continued. So Euler's formula becomes. $C_0 - C_1 + C_2 - C_3 + C_4 .. C_n = 1$.
The generalized formula is written: $C_0 - C_1 + C_2 - C_3 + C_4 - ... \pm C_n = 1$

1. The 5-Cell
This polyplex is related to the regular tetrahedron in 3-space. It has five regular tetrahedral cells, three of which fit around an edge. This is analogous to the three equilateral triangles which fit around a vertex in 3-space to give the regular tetrahedron.

2. The 8-Cell
This polyplex is related to the cube in 3-space. It has eight cubes, three of which fit around an edge. This is analogous to the three squares which fit around a vertex in 3-space to give the cube.

3. The 16 Cell
This polyplex is related to the regular octahedron in 3-space. It has 16 regular tetrahedral cells, four of which fit around an edge. This is analogous to four equilateral triangles which fit around a vertex in 3-space to give the regular octahedron.

4. The 24 Cell
This polyplex has *no* corresponding figure in 3-space. It has twenty-four regular octahedral cells, three of which fit around an edge.

5. The 120 Cell
This polyplex is related to the regular dodecahedron in 3-space. It has 120 regular pentahedral cells, three of which fit around an edge. This is analogous to the three regular pentagons which fit around a vertex in 3-space to give the regular dodecahedron.

6. The 600 Cell, C_{600}
This polyplex is related to the regular icosahedron in 3-space. It has 600 regular tetrahedral cells, five of which fit around an edge. This is analogous to five equilateral triangles which fit around a vertex in 3-space to give the regular icosahedron.

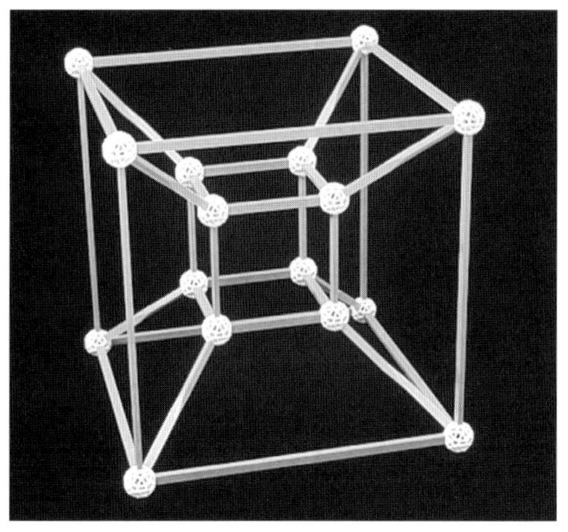

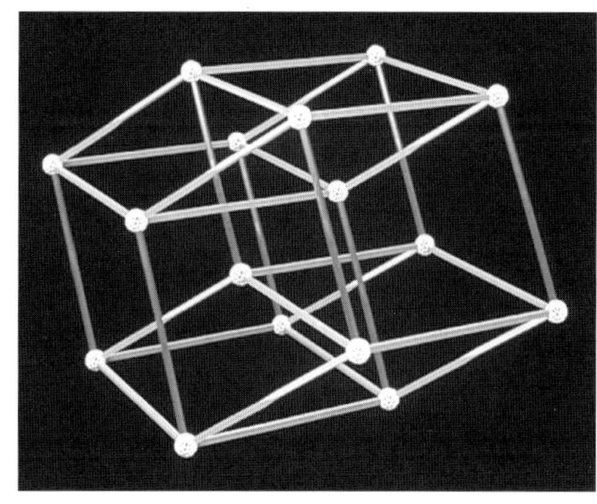

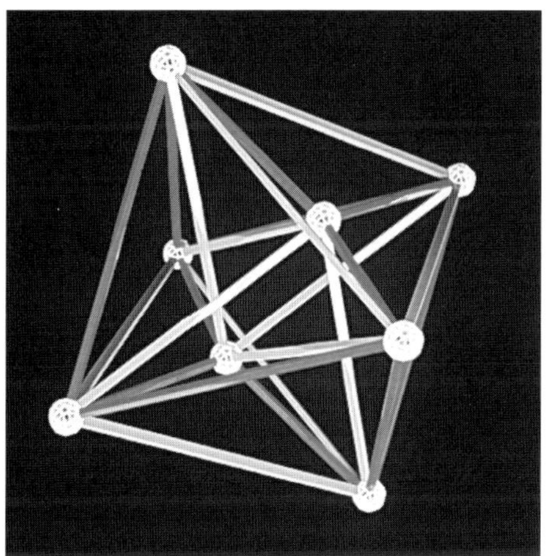

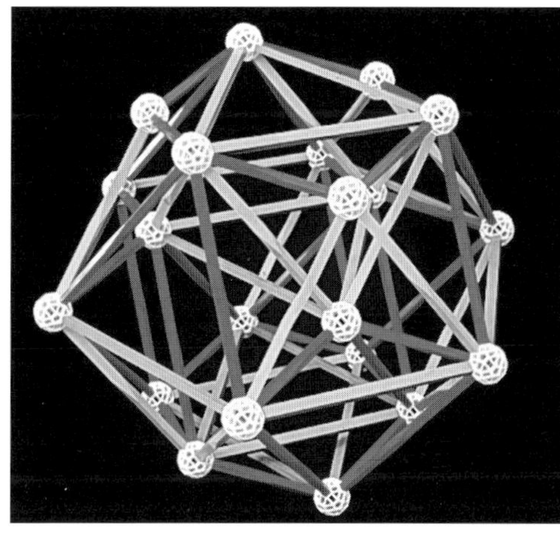

This page:
Three-dimensional models of some projections of 4-space polyplexes. These were built by the people at Zometool in Boulder, Colorado, using a modular system of special nodes and rods that they developed. Kits are available for others to also play with model making. Names by Zometool.

All images on this page are courtesy of Zometool, P.O. Box 7053, Boulder, CO., 80306, 1-303-786-9888, zometool @aol.com, U.S. Patent RE 33785.
Above:
Figure relates to the cube of 3-space.
Left: Projection of a hypercube.
Right: Projection of a hypercube.

Center:
Left: 16 Cell polyplex. This relates to the regular octahedron of 3-space.
Right: 24 Cell polyplex. There is no corresponding figure in 3-space.

Bottom:
Left: Projection of a 120 Cell polyplex. This relates to the regular dodecahedron in 3-space.
Right: Projection of a 600 Cell polyplex. This relates to the regular icosahedron in 3-space.

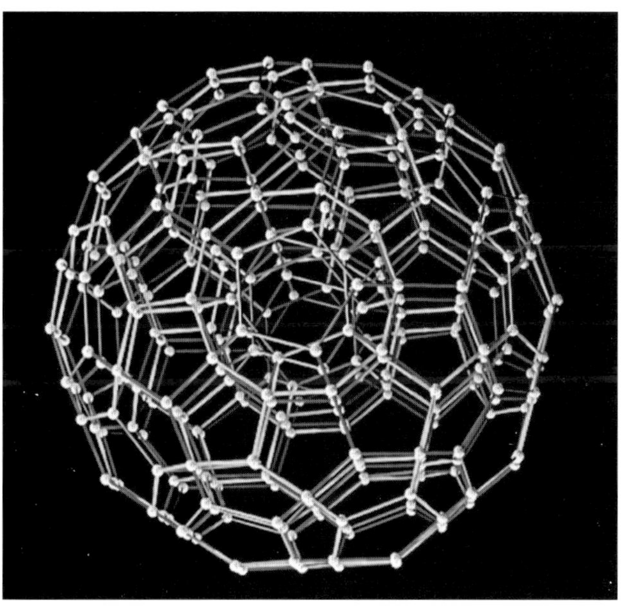

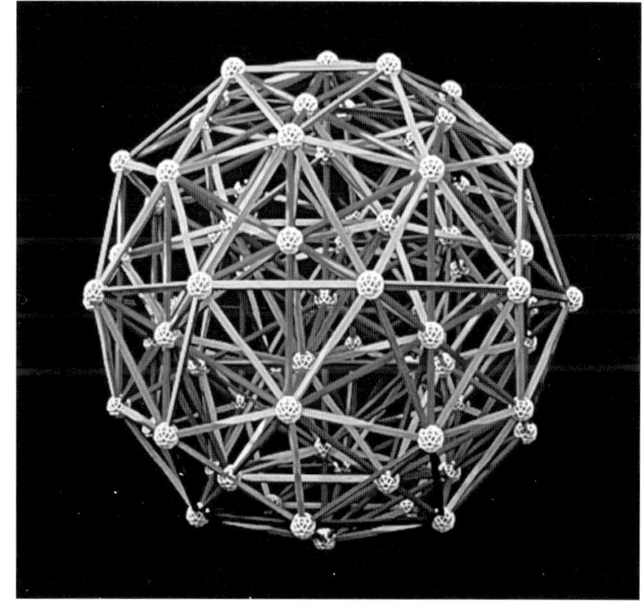

The 6 Regular 4-D Polyplexes

	Number of Vertices	Number of Edges	Number of Faces	Number of Polyhedra
Regular Simplex relates to Tetrahedron	5	10	10	5
Hypercube relates to cube	16	32	24	8
16-Cell relates to Octahedron	8	24	32	16
24-Cell (No comparable one in 3 Space)	24	96	96	24
120-Cell relates to Dodecahedron	600	1200	720	120
600-Cell relates to Icosahedron	120	720	1200	600

(partners: Hypercube/16-Cell; partners: 120-Cell/600-Cell)

This page:
Chart of the six regular 4-D polyplexes.

Above is a chart that outlines the component of the six regular polyplexes in 4-space. It compares the number of vertices, edges, faces, and polyhedra in these solids. Compare this information to the number of vertices, edges, and faces found in the five Platonic solids of 3-space.

Mathematicians have studied polytopes for dimensions higher than four and they have proved that there are only three regular polytopes found in any n-dimensions. They relate to the 3-space Platonic Solids of the tetrahedron, the cube, and the octahedron. They also relate to the polyplexes of 4-space of the 5-Cell, 8-Cell, and the 16-Cell. There are no correspond-ing figures that relate to the dodecahedron or icosahedron.

Can the shadows of these higher dimensional figures be shown by using the computer to generate images in 2-space?

The chart on the facing page summarizes the space-filling properties of polygons, polyhedra, polyplexes, and polytopes in n-dimensions. You might like to build some models either with paper and glue or on the computer using a 3-d program to explore the properties of packing space.

What kinds of polygons or polyhedra are needed to either cover the plane, fill 3-space, pack 4-space with no gaps between the geometric objects?

Uniform Space Filling

(To tile the plane with polygons or to fill space with polhedra, or cells with no gaps between units)

Dimensions

0	1	2 (plane)	3-space	4-space	n-space
point	line	Polygons	Polyhedra	Polyplexes	Polytopes
	——	3 regular tilings	1 regular space filling (only one type)		
* The number indicates the type of polygon in the particular tiling while the group of numbers indicates the vertex arrangement.		*3-3-3-3-3-3	Triangular prism (1) 3-3-3-3-3-3		There are only 3-regular polytopes that fill space
▪	——	4-4-4-4	Cube (1) 4-4-4-4		
▪	——	6-6-6	Hexagonal prism (1) 6-6-6		
▪	——		rhombic dodecahedron (1)		
Archimedean tilings		8 semi-regular tilings	N-regular (more than one type) The unit is composed of polyhedra Units fill space.		
▪		3-3-3-4-4	tetrahedron (1) truncated tetrahedron (1) }unit		
▪	——	3-4-6-4	tetrahedron (1) octahedron (1) }unit		
▪	——	3-3-3-3-6	octahedron (1) cuboctahedron (1) }unit		
▪	——	3-3-4-3-4	truncated cuboctahedron (1) octagonal prism (3) }unit		
▪	——	4-8-8	truncated octahedron (3) truncated cuboctahedron (1) cube (1) }unit		
▪	——	3-12-12	rhombicuboctahedron (1) cuboctahedron (1) cube (3) }unit		
▪	——	4-6-12	rhombicuboctahedron (1) cube (1) tetrahedron (2) }unit		
▪	——	3-6-3-6	truncated cuboctahedron (1) truncated cube (1) truncated tetrahedron (2) }unit		
▪	——		truncated octahedron (1) cuboctahedron (1) truncated tetrahedron (2) }unit		
▪	——		rhombicuboctahedron (1) truncated cube (1) octagonal prism (3) cube (3) }unit		

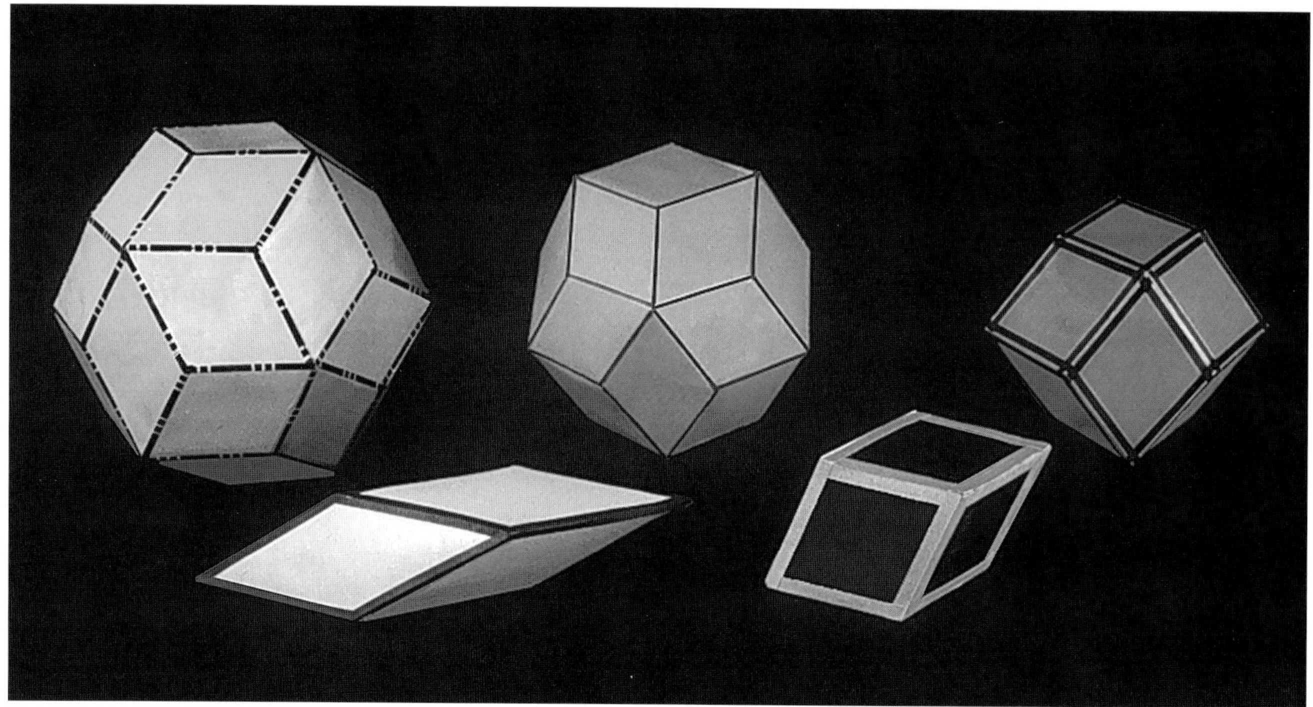

The Golden Zonohedra

This page:
Above:
Photograph of paper and paint models of the five members of the Golden Zonohedra family. The edge lengths have been kept constant so that you can see their relative size. These are convex polyhedra bounded by parallelograms and they possess central symmetry.
In the top row are the rhombic triacontahedron, the rhombic icosahedron, and the rhombic dodecahecron.
In the bottom row are the two basic rhombohedra. The "fat" rhombohedron on the right is composed of six "fat" Penrose tiles. The "skinny" rhombohedron on the left is composed of six "skinny" Penrose tiles. These can be found in both left and right-hand versions.

Center: Fig. 4.15
The pentagon with its interior pentagram. Golden Cuts abound within this figure. The two Penrose tiles are derived from triangles within the pentagon and pentagram. These are the basis for the Penrose rhombohedra.

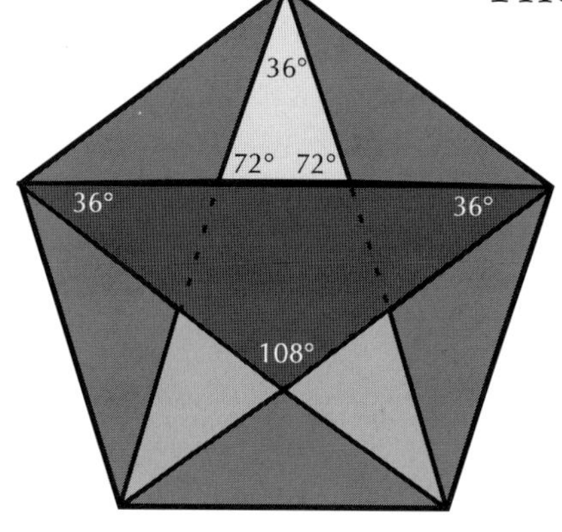

There is a family of five polyhedra that is related to the cube of higher dimensions. The members of this family can be thought of as projections from higher space as well as three-dimensional figures in their own right. It is named the Golden Zonohedra group which has connections to the Golden Ratio. See the Design Appendix in the back matter for detailed information on this special ratio. In two-dimensions, the Golden Ratio is found within the regular pentagon and its interior pentagram. See Fig. 4.15. The two triangles in the diagram generate the two rhombic Penrose tiles that continue

to carry the harmony of this ratio. These tiles further generate the two 3-space Penrose rhombohedra that carry the Golden Ratio.

In 2-space these rhombohedra relate to zonogons, a 2-n sided polygon where pairs of opposite sides are parallel and congruent. These zonogons generate three-dimensional solids called zonohedra. A zonohedron is a convex polyhedron in which all the faces are parallelograms. The Golden Zonohedra are constructed from combining the two types of Penrose rhombohedra.

For art purposes, zonohedra can be expanded, contracted, and distorted with ease. Like geodesic domes, based on the triangulation of the sphere, they have important connections to the architecture of this century and are worthy of your research time.

198

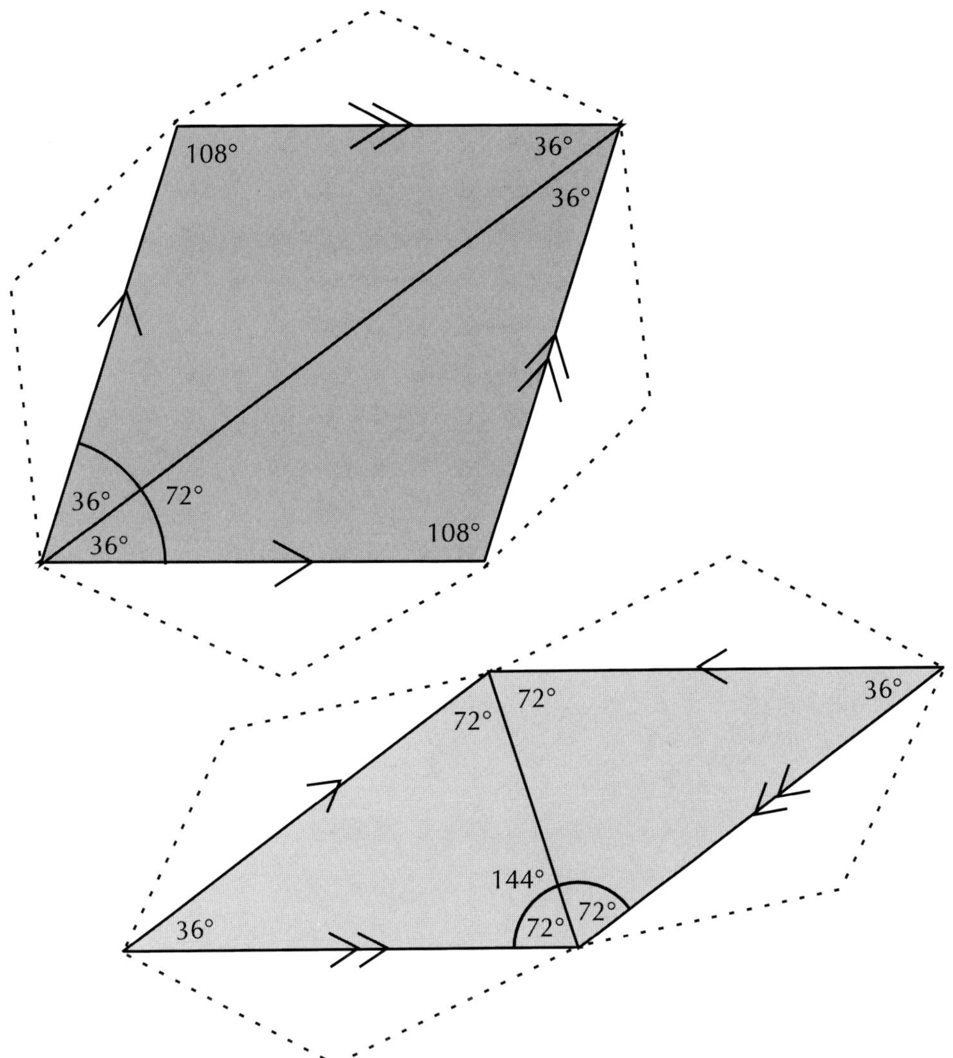

This page:
Fig. 4.16
The two Penrose rhombuses are derived from the triangles within the interior of the pentagon and its pentagram.

Upper left: The "fat" Penrose rhombus tile template. Two triangles form the rhombus with angles measuring 72° and 108°.

Lower right: The "skinny" Penrose rhombus tile unit. Two triangles form the rhombus with angles measuring 36° and 144°.

Each of the tiles, alone, will cover the plane periodically. When used together, however; they will cover the plane aperiodically.
Together as tiling units in the plane, the arrows at the edges indicate matching. Single arrows must match with single ones; double arrows must match with doubles. Remember that the edge length of the tiles must be the same.

The Five Members of the Golden Zonohedra Family

1. The "Fat" rhombohedron. It can be found in a right or left-handed version.

2. The "Skinny" rhombohedron. It can be found in a right or left-handed version.

3. The Rhombic Dodecahedron. It was discovered by Johannes Kepler in 1611. This solid is composed of two each of the "fat" and "skinny" rhombohedra. It can also be built of twelve single type rhombuses as shown further on.

4. The Rhombic Icosahedron. This solid is composed of five each of the "fat" and "skinny" rhombohedra. It can also be built of twenty units composed of squares and rhombuses as shown further on.

5. The Rhombic Triacontahedron. It was discovered by Johannes Kepler in 1611. This solid is composed of 10 each of the "fat" and "skinny" rhombohedra. It can also be built of 30 single rhombuses as shown further on.

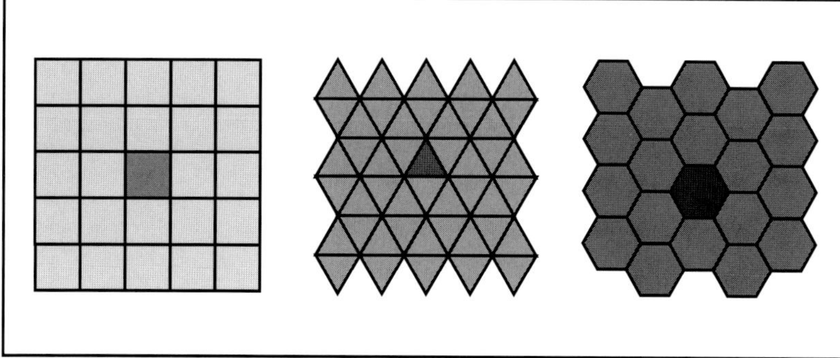

In a regular plane tiling (all angles and all lengths are congruent), using squares, triangles, or hexagons, everywhere on the plane the tiling is the same. In order for the tiles to lay flat with no gaps between the tiles, the number of degrees around each vertex needs to add up to exactly 360°. See Fig. 4.17.

This page: Fig. 4.17
Above: The three regular periodic tilings of the Euclidean plane which use squares, equilateral triangles, and regular hexagons.

Center:
The two Penrose rhombuses, the "skinny" and the "fat", used for plane tiling both periodically and aperiodically.

Bottom:
Chart showing the eight ways of joining the two Penrose tiles to produce a Penrose tiling unity that will fill the plane aperiodically. Notice that two look alike but the arrow placement is different.

Make copies of each unit and try tiling these. What happens?

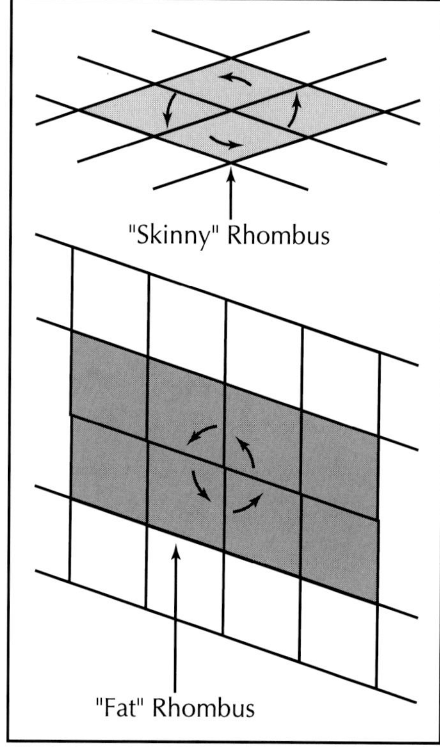

The two Penrose rhombuses can be related to the square. You can think of them as stretched squares, but they are not regular polygons. The lengths are congruent, but the angles are not. Since all quadrilaterals can tile the Euclidean plane, multiples of the same type Penrose rhombus can be used to tile the plane so that everywhere looks the same. Four come together at a vertex.

If the two different rhombuses, however, are used together, they must be arranged according to particular matching rules which specify how they must join. Overall, the tiling exhibits both areas of repetition and areas of difference. The chart at the bottom of this page shows the eight possible ways in which the two Penrose tiles can meet to produce a Penrose unit. New tiles are added at the edges according to the matching rules. On the facing page, there is one example of a aperiodic Penrose tiling.

The ratio of "fat" to "skinny" rhombuses happens to be a Golden Ratio. You need approximately ten of one kind and 16 of the other. Interesting!! Why???? Perhaps, it is because the Penrose tiling relates to tilings with regular pentagons. These, however, will have gaps between tiles as you can see on the facing page. If, however, you use non-regular pentagons, the tiling has no gaps. In the example on the facing page, the pentagon has angles of 66, 147°, 112°, 79°, and 136°.

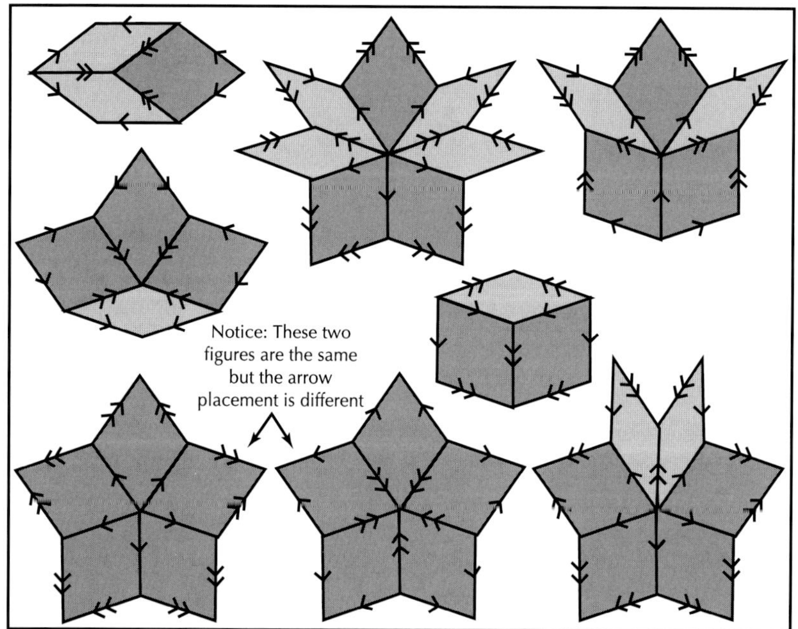

This tiling, Fig. 4.18, was discovered by the amateur mathematician Marjorie Fox. Can you find other non-regular pentagons that can also fill the plane? (Hint: Any pentagon having a pair of parallel sides will tile.) Find the tile and generate the tiling. Then enhance it artfully by using tones or colors or textures or patterns. The tones within the images on this page were computer manipulated. Scan in your tiling and create some variations on a theme. Are there two different non-regular pentagons that would also tile the plane?

Mathematicians know that there are 14 types of convex pentagons that can tile the plane. They have yet to definitely prove that these are the only types. Mathematicians are also interested in all kinds of other surfaces that can be covered by tiles such as the hyperbolic surface looked at in Chapter 3. In the next chapter, we will look at the torus and Mobius band that can also be tiled.

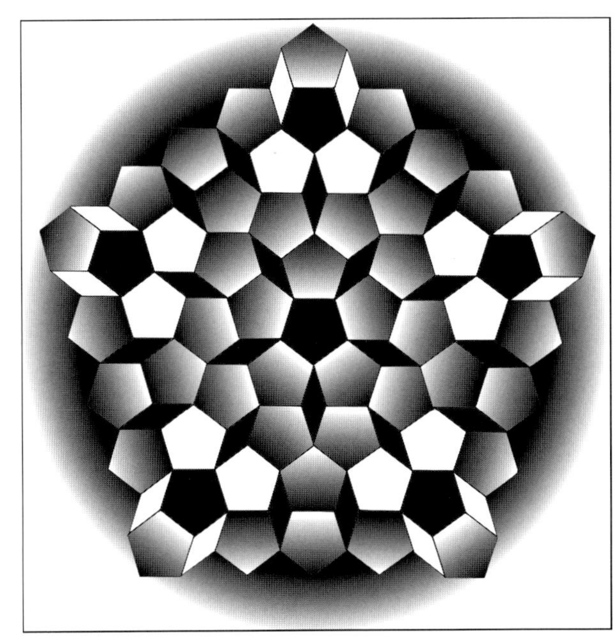

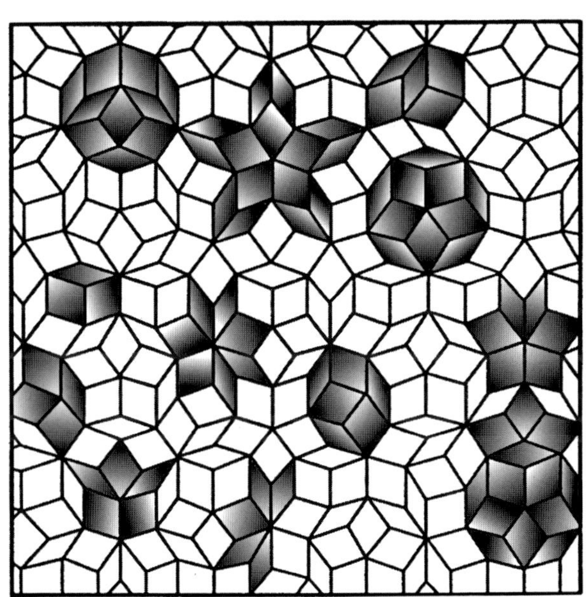

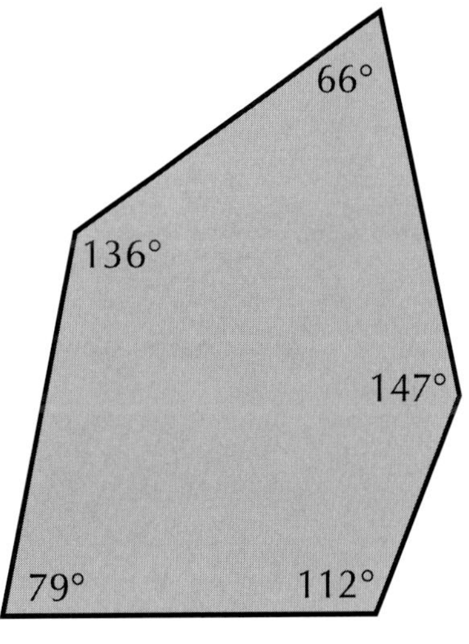

This page:
Above:
A tiling of the plane using regular pentagons. Notice the gaps between the tiles.
Try a different arrangement of pentagons and see what you can come up with.
Center: A Penrose tiling using the two rhombuses. Notice the various vertex clusters.

Bottom: Fig. 4.18
A tiling of convex non-regular pentagons, along with the single unit used for the tiling. this unit was discovered by amateur mathematician Marjorie Fox.

Fig. 4.19
The three regular tilings of the plane (the square, the equilateral triangle and the hexagon) move into the stacking of 3-space with the cube, the triangular prism, and the hexagonal prism. Notice that each makes use of a square face but the triangular and hexagonal prisms require another polygon to complete the solid.
This information is useful for architecture and package design.

Stacked cubes

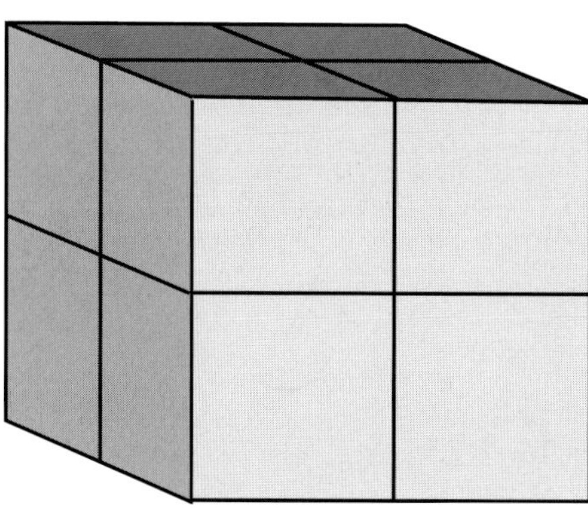

Stacked triangular prisms

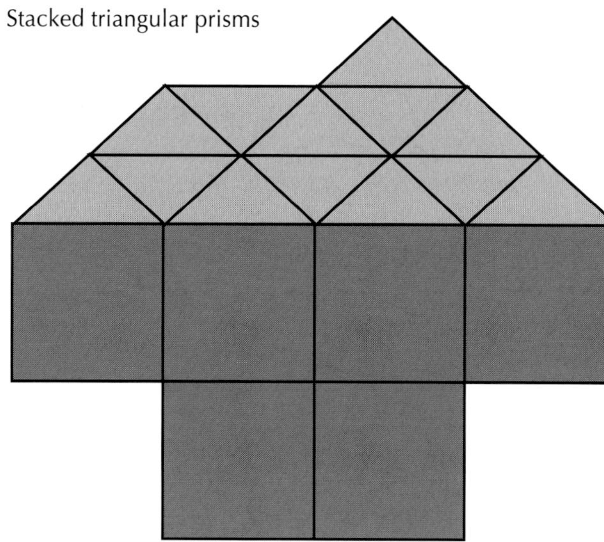

Stacked hexagonal prisms

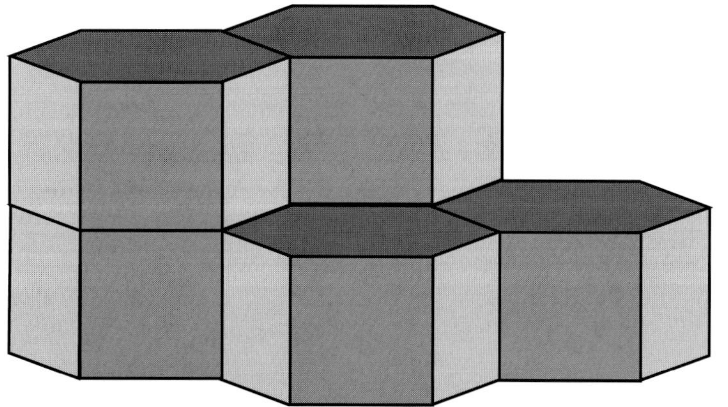

In 3-space the regular tilings of the plane become the three regular space-filling solids. Between each 3-d unit, there are no gaps just like their 2-d counterparts. This provides a most efficent stacking system. The square becomes the cube, the equilateral triangle becomes the triangular prism, and the regular hexagon becomes the hexagonal prism. The latter two prisms require the use of two types of polygons. The triangular prism uses the equilateral triangle and the square while the hexagonal prism uses the regular hexagon and the square. Fig. 4.19

The Penrose rhombuses also have three-dimensional equivalents. They generate the two Penrose rhombohedra as seen on the facing page. Fig. 4.20. Once constructed, these can be stacked in three-dimensional space analogous to the tiling of two dimensions but everywhere in space is not the same. Like their 2-d tiling counterparts, they relate to the Golden Ratio in the number of rhombohedra required of each type. They also have to come together with matching rules. At this point in time, mathematical research suggests that to build three-dimensional stackings with these pairs, the "fat" rhombohedron must carry 14 markings for edge matching while the "skinny" one must carry 8 marks.

In order for either rhombohedron to be built correctly, the rhombuses have to be arranged in a particular way. So, on the facing page, we have given templates for both. Fig. 4.20. In order to obtain both left and right hand versions, turn the templates over.

Build some for yourself. Try stacking them together and see what kinds of sculpture and architectural

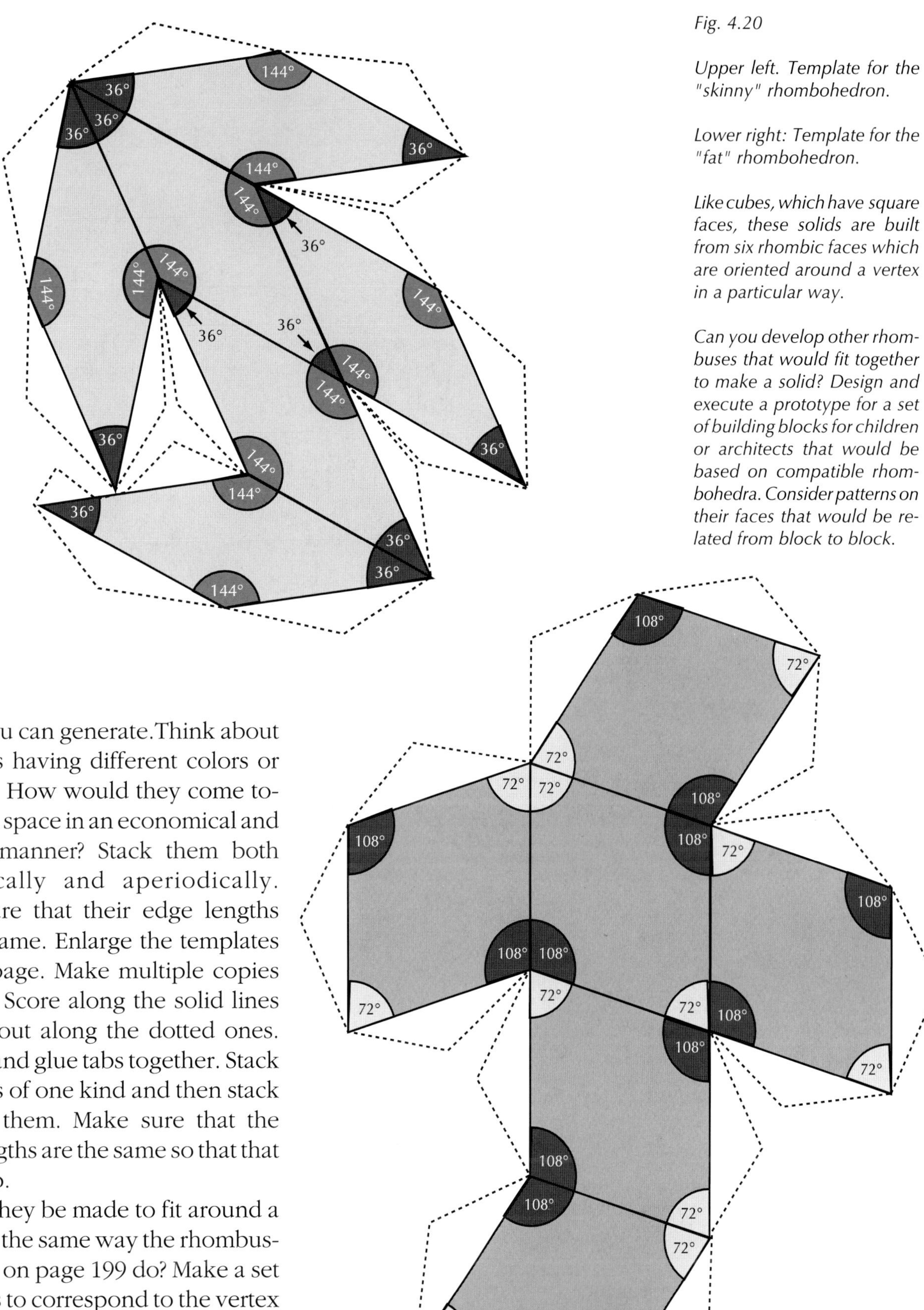

Fig. 4.20

Upper left. Template for the "skinny" rhombohedron.

Lower right: Template for the "fat" rhombohedron.

Like cubes, which have square faces, these solids are built from six rhombic faces which are oriented around a vertex in a particular way.

Can you develop other rhombuses that would fit together to make a solid? Design and execute a prototype for a set of building blocks for children or architects that would be based on compatible rhombohedra. Consider patterns on their faces that would be related from block to block.

forms you can generate. Think about the faces having different colors or patterns. How would they come together in space in an economical and elegant manner? Stack them both periodically and aperiodically. Make sure that their edge lengths are the same. Enlarge the templates on this page. Make multiple copies of them. Score along the solid lines and cut out along the dotted ones. Fold up and glue tabs together. Stack multiples of one kind and then stack pairs of them. Make sure that the edge lengths are the same so that that match up.

Can they be made to fit around a vertex in the same way the rhombuses found on page 199 do? Make a set of blocks to correspond to the vertex groups in the plane tiling. Work with the groups to see what happens in the stacking process. Have fun and good luck! Patience....patience.....

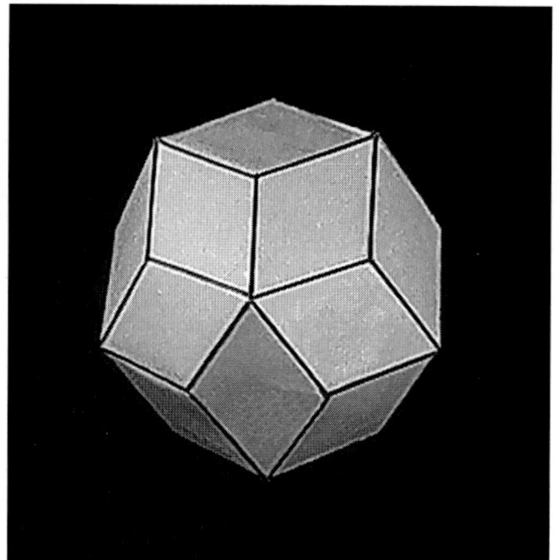

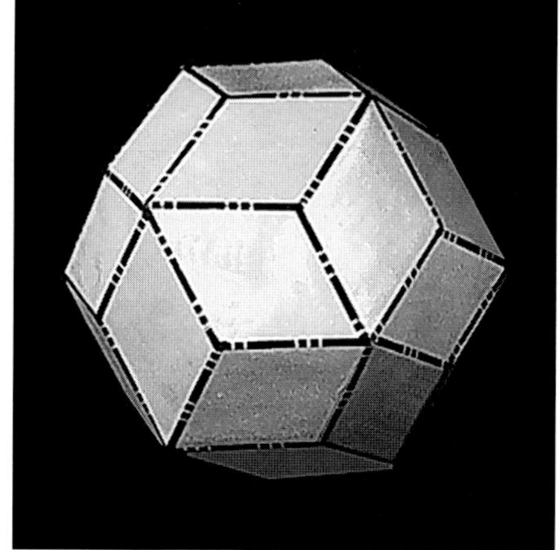

The Rhombic Dodecahedron 12 faces (one type rhombus, given here)

The Rhombic Icosahedron 20 faces (10 squares & 10 rhombuses of the type given here)

The Rhombic Triacontahedron 30 faces (one type rhombus, given here)

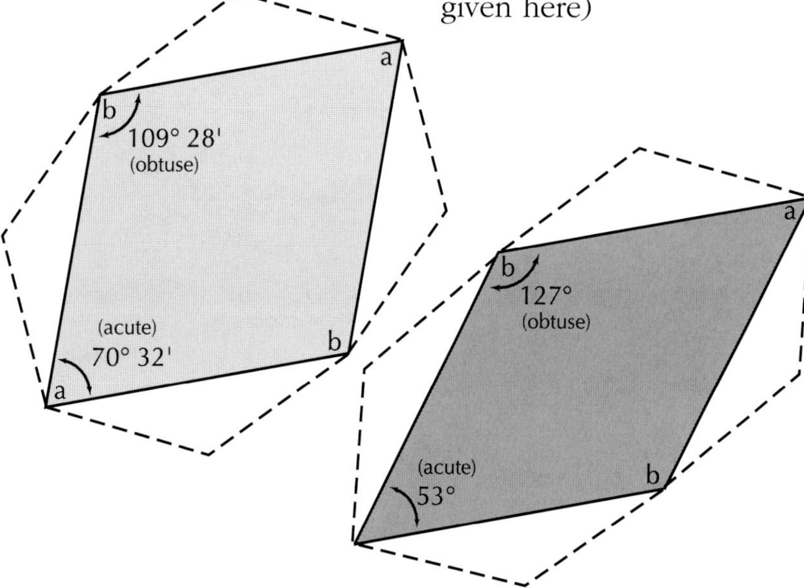

This page: Fig. 4.21
Templates for three solids
Left: *The Rhombic Dodecahedron.*
(Requires two "fat" and two "skinny" rhombohedra to complete) or
This figure requires 12 rhombuses as given.
Make a cap of 4 where the acute angles meet at the vertex. Add one rhombus at a time at the obtuse angles to ring the solid. Add another cap of 4 rhombuses.

Center: *The Rhombic Icosahedron.*
(Requires five "fat" and two "skinny" rhombohedra to complete) or
This figure requires 10 rhombuses as given and 12 squares. Keep the edge lengths constant.
First make a cap of 5 rhombuses, then a ring of 5 squares, another ring of 5 squares, another cap of 5 rhombuses.

Right: *The Rhombic Triacontahedron.*
(Requires 10 "fat and" 10 "skinny" rhombohedra to complete) orThis figure requires 30 rhombuses as given. Make a cap of 5 with acute angles meeting at the vertex. Add one rhombus at a time at the obtuse angles.

On this page we give you the template units in order to build the Rhombic Dodecahedron, the Rhombic Icosahedron, and the Rhombic Triacontahedron. These can be generated by using only a square and three different rhombuses instead of clusters of "fat" and "skinny" rhombohedra. Enlarge the templates to a comfortable working size.

The problem building with the solids is to find matching rules that guarantee that the rhombohedral units can be fitted together properly with the symmetry of an icosahedron, six axes of rotation each having five turns.

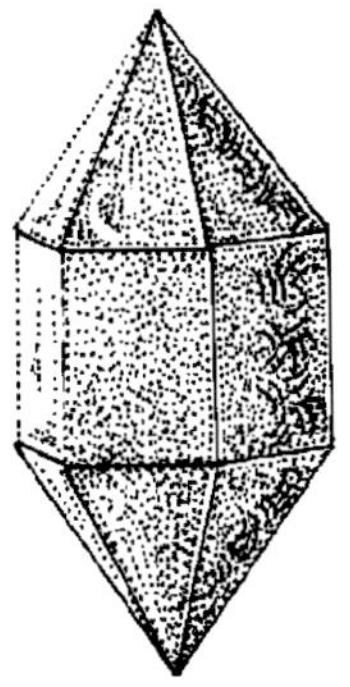 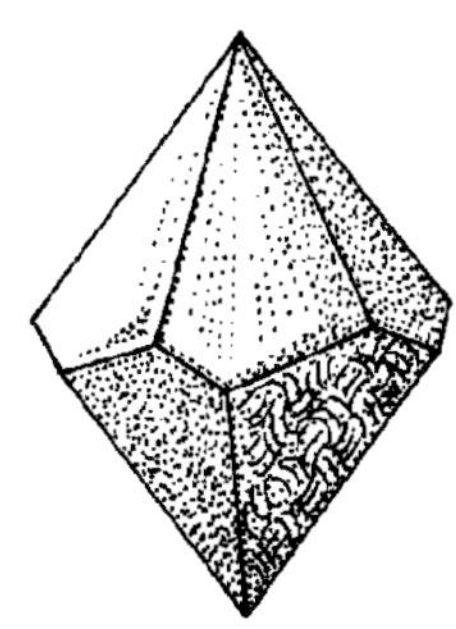 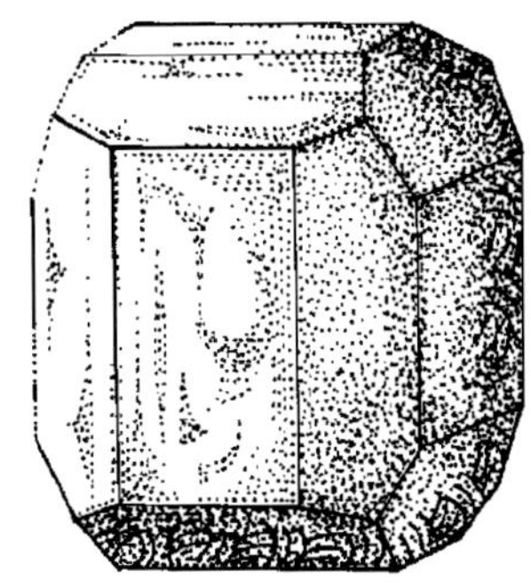

The zonohedra relate to a group of natural substances called the "quasicrystals" because their symmetry is related to that of a pentagon rather than the more "normal" crystals that have only 2,3,4, or 6 turn symmetry. These materials seem to defy the logic of crystal structure that scientists have worked so hard to establish. Crystal lattices were considered always to have symmetries that were related to the three regular tilings of the square, equilateral triangle and the hexagon . There are no gaps between the units. But five-turn, 10-turn, eight-turn, and twelve-turn were strictly not allowed. Tilings of pentagons, and its relative the decagon, have spaces between units.

Quasicrystals can be generated on a computer by using an algoritm. The mathematician John Conway, in 1993, discovered a single convex polyhedron that can fill space aperiodically and completely but only when built layer by layer. Each layer is periodic but in order for the layers to fit together, each layer requires a rotation by a particular angle. Consider doing some research on the work of this mathematician and see if you can build a working model out of paper.

Plane tilings can be considered the cross-sections of 3-d stackings in space. The position of atoms in quasicrystals cannot be orderly over long distances. At places, these crystals exhibit the symmetry of an icosahedron, which is the dual of the dodecahedron. This means that there are six axes of symmetry about which can be seen five-turn rotational symmetry. Therefore, dodecahedra and icosahedra could not be stacked, it was thought. If the two Penrose rhombohedra, however, are stacked properly, their cross- section is a Penrose tiling which has 5-turn symmetry.

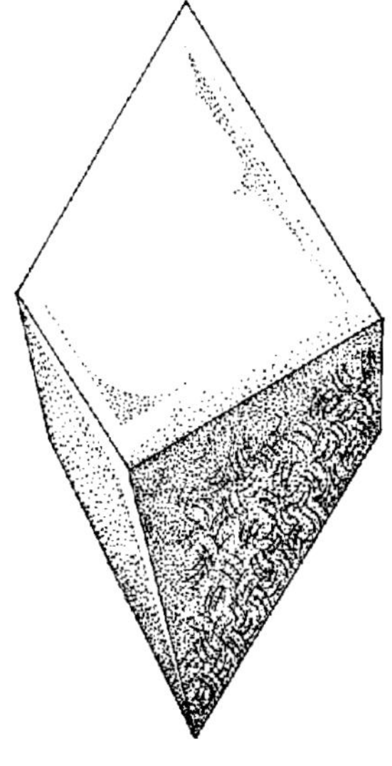

Physicists were playing around with supercooling liquids and new things were happening. In 1984 scientists were working with alloys of aluminum and manganese. The two molten metals were cooled very rapidly so that the structure of the mixture was dramatically different from a mixture that was gradually cooled. These were called quasicrystals because these substances do not truly possess a unit cell which repeats throughout the material, like a stacking of cubes. They are somewhere between a liquid and a solid in their makeup. Inside the crystal reveals that from place to place there are differences like what happens in a very large group of the two-dimensional Penrose tiles or the three-dimensional Penrose rhombohedra.

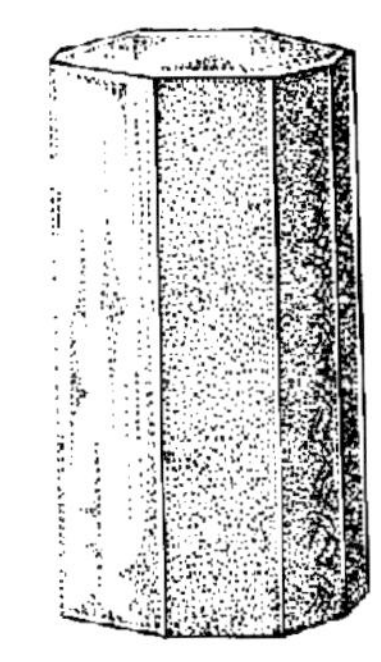

There is a belief among scientists that there is a relationship between the stacking of hypercubes and quasicrystal structures. The investigation is wide open.

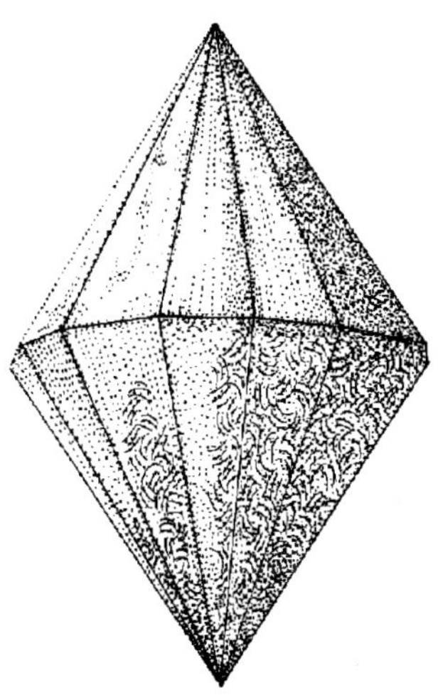

Termespheres

The artist, Dick Termes, has used the faces of all five Platonic solids (tetrahedron, octahedron, cube, dodecahedron, and icosahedron), as well as the rhombic dodecahedron, to create his three-dimensional artworks.

Each of the Platonics can be set inside a sphere so that the vertices of the polyhedron touch the circumscribing sphere at equidistant points. The two solids that concern him for the development of six-point perspective are the cube and the octahedron. These two are dual partners. The number of vertices in one figure becomes the number of faces in the other. The cube has six faces while the octahedron has six vertices. Both have twelve edges. The vertices of the octahedron are used to mark the six perspective points that are used on the sphere.

In this last section of this chapter we introduce you to an artist who uses the surfaces of both polyhedra and spheres as the basis for his representational artworks. For many centuries now artists have used the two-dimensional surface to represent a selected portions of three-dimensional space. Artists have painted on two-dimensional rectangular surfaces, and on circular disks called tondos, but it wasn't until the 20th century that an artist, Dick Termes, began explorations of painting all 360° of space on the entire surface of an actual sphere. Allen Branum, a psychologist, has likened these works to a three-dimensional mandala, a sacred sphere for the contemplation and meditation of the spirit.

As an elementary and high school art teacher, back in 1964-68, Termes was involved with teaching traditional central vanishing point perspective and was trying to find interesting ways in which to present the material. A year later, 1969, at graduate school for his Master's Degree, Termes became totally involved with exploring this topic. His first work involved curved lines in two-dimensional perspective. A fellow student suggested that his plane perspective exploration looked as if it truly belonged on the surface of an actual sphere. That was the seed idea that began an ongoing journey into interior and exterior space. For over a quarter century, Termes has created complex hand-painted works on about 150 spheres.These are meant to hang in space and rotate slowly so that the viewer watches as a full 360 degrees move past his eyes. In this visual experience, the participant finds that the object has ambiguities which make for visual intrigue. Concave and convex exchange places. What is rotating to the left suddenly rotates to the right. What was outer becomes inner.

As the reader has seen in Chapter 2, with traditional one-point perspective the viewer is able to see only a partial section of 3-space depicted on the surface of the 2-space artwork. The objects being painted usually have one face parallel to the viewer. With two-point perspective, the vista is increased but it still includes only 90° of the full volume of space at eye level. Objects are seen with an edge facing front, instead of a face, parallel to the viewer. Three-point perspective is reserved for either looking up at or looking down upon objects from a fairly extreme distance. Notice the differences between the diagrams in Fig. 4.22.

But, what if one wants to include the entire horizon within a single artwork? According to Dick Termes, the artist would need to paint on a cylinder. This would be easy enough to execute by first drawing or painting the image on a gridded rectangle and then rolling it up until the edges joined to form the cylinder. One would then be able to include four points for the perspective. Termes began to see the potential for exploring perspective on polyhedra and on spheres since they have a close connection to each other.

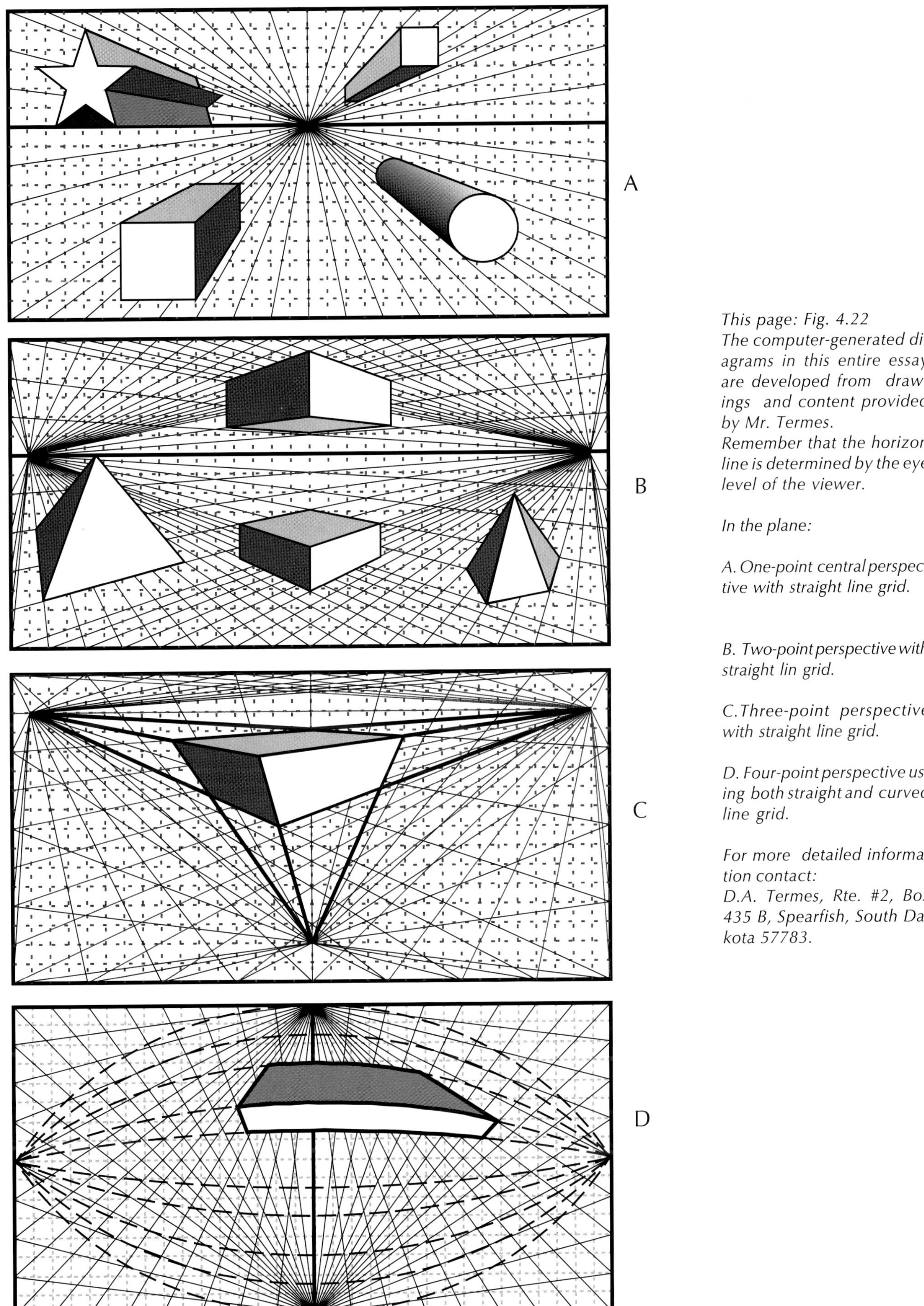

This page: Fig. 4.22
The computer-generated diagrams in this entire essay are developed from drawings and content provided by Mr. Termes.
Remember that the horizon line is determined by the eye level of the viewer.

In the plane:

A. One-point central perspective with straight line grid.

B. Two-point perspective with straight lin grid.

C. Three-point perspective with straight line grid.

D. Four-point perspective using both straight and curved line grid.

For more detailed information contact:
D.A. Termes, Rte. #2, Box 435 B, Spearfish, South Dakota 57783.

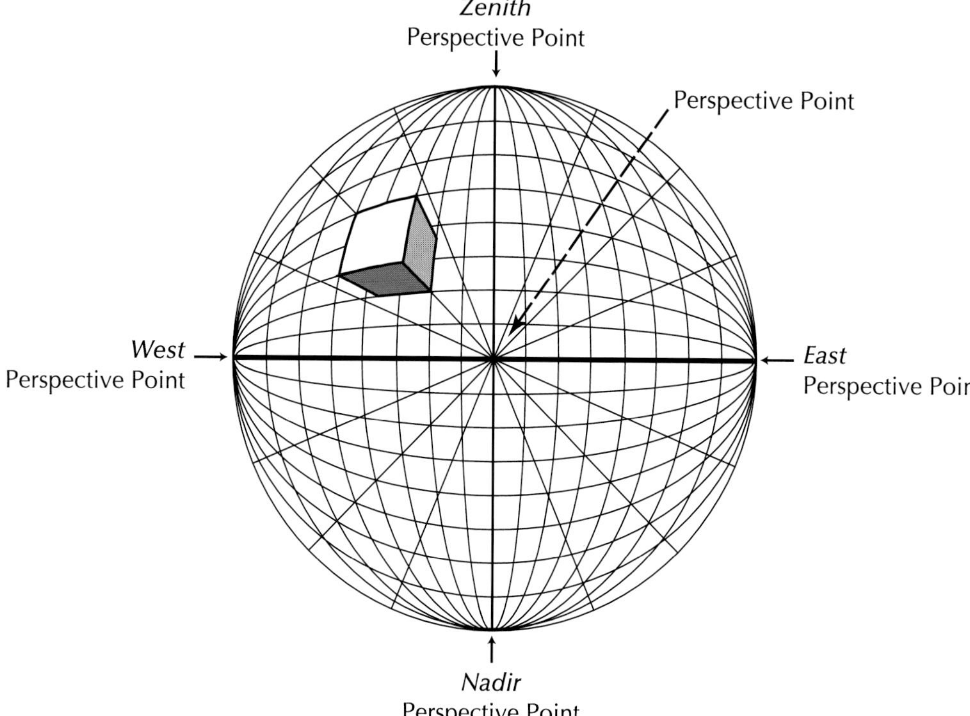

This page:

*Above: Fig. 4.23
A circle with the grid for a
five point perspective image.*

*Bottom: Fig. 4.24
A sphere set within a cube
with the six faces marked on
it (ceiling/floor; north/south;
east and west walls).*

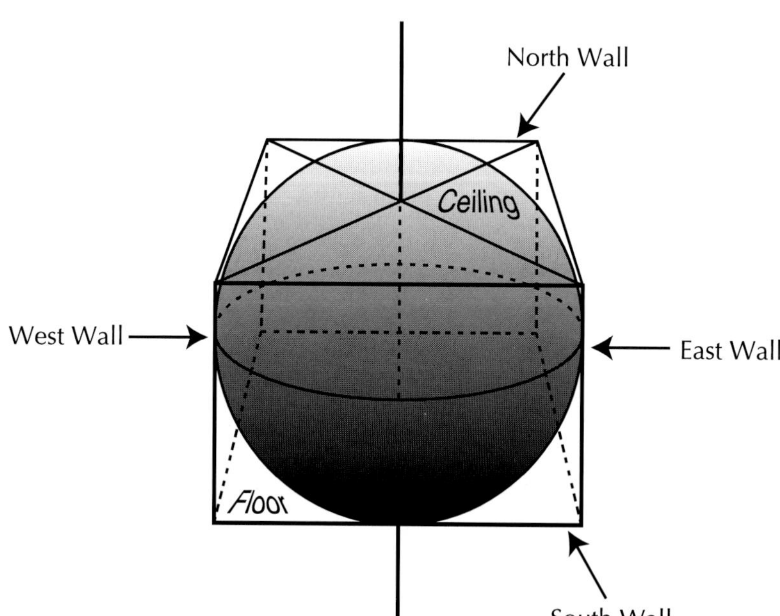

In order to suggest a curved hemispherical space on the plane, the artist needs to use a five-point perspective system on a circular format rather than a rectangular one. A diameter is used for the horizon line while one perspective point is set at the center of the circle and the other four points are set equidistant around the circumference of the circle. A grid of both straight and curved lines is developed within the circle as seen in Fig. 4.23.

The basic illusion cube is set within the area of the circle and is treated as if it were a one-point perspective image with the converging lines meeting at the vanishing point at the center of the circle. Depending upon the location of the cube, the lines will either appear to be straight or more and more curved. What happens on this surface is similar to the effect produced by the fish eye lens of a camera. The center of the image bulges out toward the viewer. For one construction method we refer the reader to an article by Michael Moose, LEONARDO, Vol. 19, No1 pp 61-64, 1986. He writes of people doing research into curvilinear grids so that you may look into the subject till your heart's content.

Imagine a sphere filling the volume of a cube. The midpoints of each face touch the sphere at 6 points. Now imagine the sphere shrinking down to become a painting surface. Fig. 4.24.

Dick Termes wanted to express the full surround of the physical world in his artwork. The concept of east to west; north to south, and up to down suggested to him that he needed to use six points in his perspective rendering of any scene. But where would these six points come from and where would they be located? First, he knew that much of human experience with space was related to the cubic volumes of rooms. A single cube has six faces which can stand for the six perspective directions he required.

When inside a room (a cube), Dick Termes imagines that he is in the center of the space. Fig. 4.25. What he sees of all six faces of the room is transferred onto the outside of his actual sphere which is a lightweight, thin, strong, plastic much like the material used for the construction of astronauts' helmets.

The artist does rough studies on available small globes such as children's toy rubber balls. In his finished works, he may work on spheres as small as 16" in diameter as well as on ones that are 7.5' across.

The artist prepares this plastic sphere by first roughing up the surface with sandpaper and then applying 6 or 7 coats of gesso primer before painting with acrylics. Each layer of gesso must be sanded smooth so that the layers of paint can be applied evenly. This is slow and time-consuming work, but well worth the effort.

This page: Fig. 4.25
Looking down through a transparent ceiling into a room. The above four diagrams show how the artist visualizes the connection between the cubic space of a room and his spherical canvas. The six plane surfaces of the room are transferred to the six perspective points on the sphere. The sphere is set in the center of the room. The artist stands in front of a particular wall and draws what is seen of that wall onto the section of the sphere directly in front of him.
Using the cube template from Chapter 1, reconstruct the above room on the faces of the cube. Add ceiling details to complete.

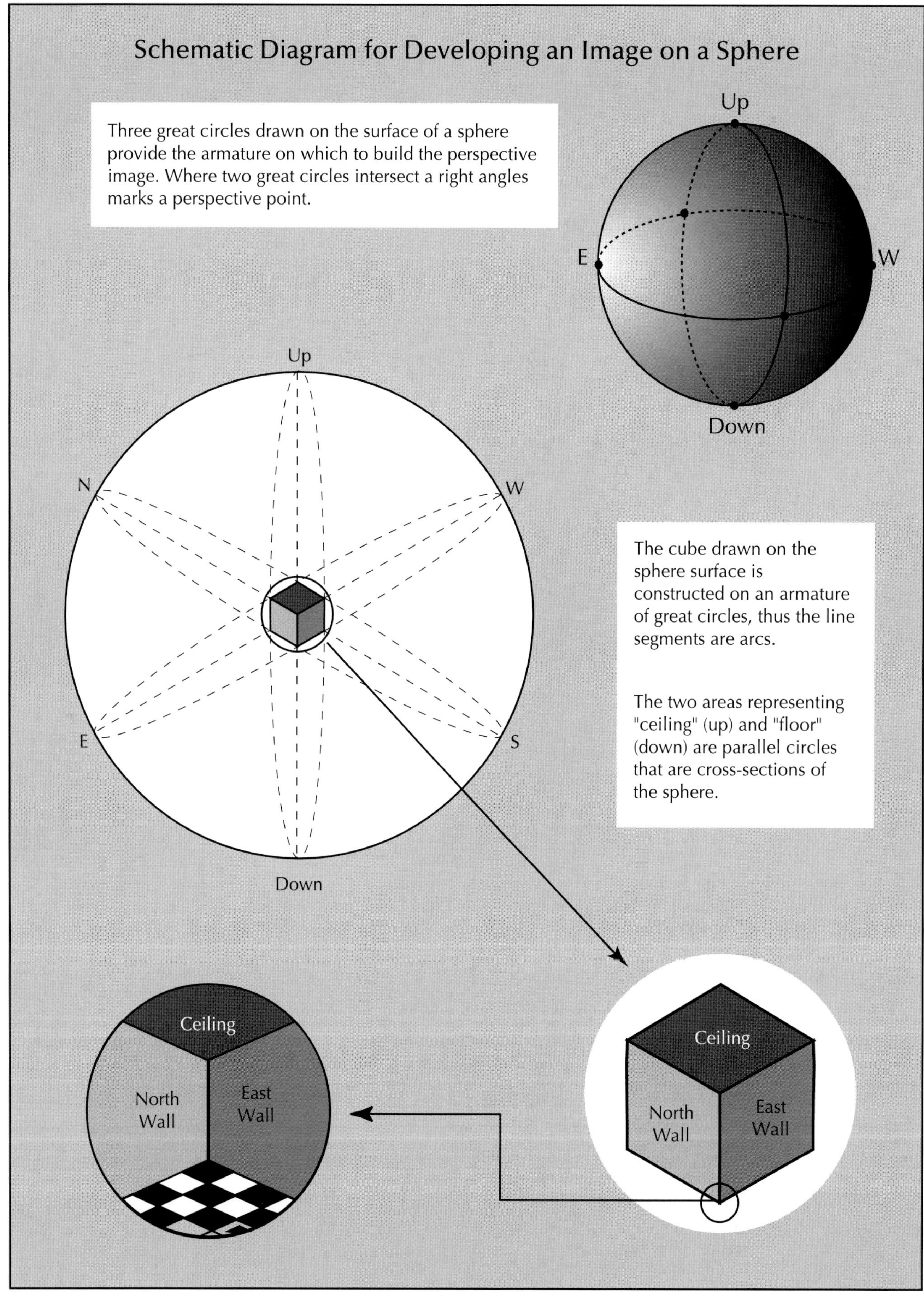

Schematic Diagram for Developing an Image on a Sphere
Three great circles drawn on the surface of a sphere provide the armature on which to build the perspective image. Where two great circles intersect a right angles marks a perspective point.
Up
E
W
Down
Up
N
W
E
S
Down
The cube drawn on the sphere surface is constructed on an armature of great circles, thus the line segments are arcs.
The two areas representing "ceiling" (up) and "floor" (down) are parallel circles that are cross-sections of the sphere.
Ceiling
North Wall
East Wall
Ceiling
North Wall
East Wall

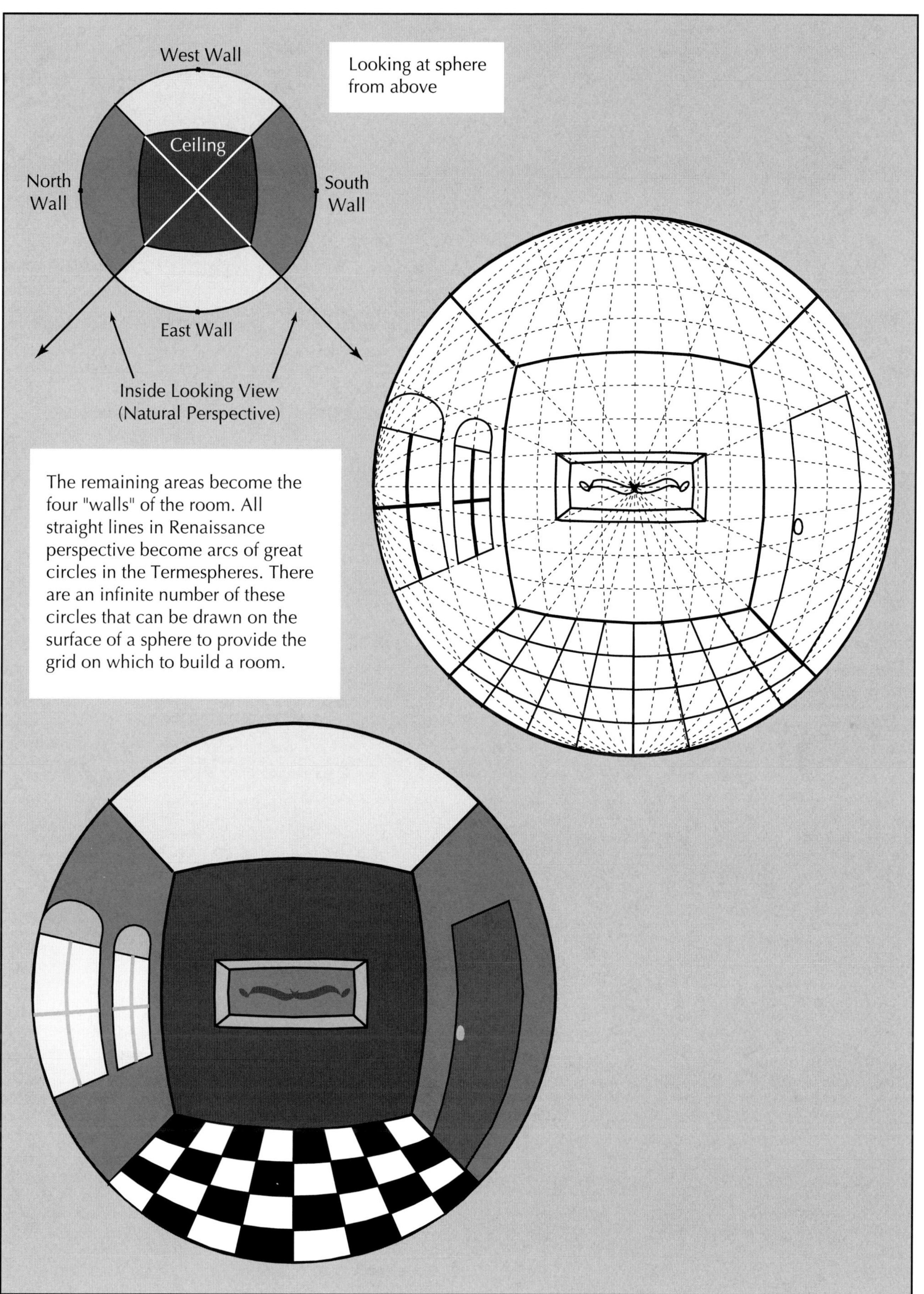

The remaining areas become the four "walls" of the room. All straight lines in Renaissance perspective become arcs of great circles in the Termespheres. There are an infinite number of these circles that can be drawn on the surface of a sphere to provide the grid on which to build a room.

This page:
Two views of the San Loretto Church in Santa Fe, New Mexico.

On this page we show you a comparison of a 35 mm photograph of the famous staircase of the San Loretto Church in Santa Fe, New Mexico and a view of the same staircase as represented by Mr. Termes in a painting on a sphere. You will notice obvious differences.

In order to obtain a sense of the entire space of the church, the photographer using the 35 mm camera would have had to take a minimum of six different views of the interior: the floor, the ceiling, the two side wall, the back wall and the front entrance way.

This page:
"Loretto in the Round" by Dick Termes.
Three views of the San Loretto church rendered on a sphere. These are only three aspects of potentially an infinite number that could be shown here. The photographs are courtesy of the artist.

Locate a garden gazing globe and take a number of photographs of it from several viewpoints. Compare your results to what Mr. Termes has done in his painting. Try doing a drawing, collage, or painting using your photographs as reference.

Or attempt to render aspects of it on the faces of a cube. Or be daring and do a variation on one of the Archimedean, Catalan, or Zonohedra solids.

Problems, Projects, and Play

1. Research the topic of 4-space. Would the 8 Convex Deltahedra exist in that dimension? Can you prove you answer? Could you write a computer program that would generate these solids in three and/or four dimensions? Print our your results. (Who said this was an easy problem?)

2. Research the four Kepler-Poinsot Solids. How do they relate to 4-space? What would happen to them in spaces beyond four? (This isn't easy either.)

3. Research the paintings of Salvador Dali, 20th Century artist and his interest in figures from 4-space. Document your findings in writing, drawing, collage, etc.

4. Research the problem of scale in the natural world by looking at the subject of trees, or giraffes, or mice, etc. Write and draw about your findings.

5. Research and write about the relationship between higher dimensions and the field of telecommunications.(Scientific American Magazine would be a good source of information.).

6. Build a working set of models of the Golden Zonohedra. Make the units out of sturdy material.

7. Find a computer program that can generate images of 4-dimensional objects. Print out a group of them and write about what you see.

8. research the history and origins of the metric system. Do a written comparison and contrast between this and the inch-foot-and -yard system.

9. Read Edwin A. Abbott's book, *Flatland*, and Dionys Burger's book, *Sphereland*. Write a short dialogue on a hypercube encounter with a 3-space cube and a 2-space square.

10. Read the books of Lewis Carroll and compare the content with the mathematics of Dodgson. Write about what you discover. (Another not-so-easy problem.)

11. Take the equation for the Pythagorean Theorem up to 6 and 8-space. assign real world numbers and work the equations for these two spaces.

12. Using the information found in this chapter on the Rhombic Dodecahdron, Icosahedron and Triacontahedron, make the necessary tempates and build a model of each.

13. Research the connection between hyperspaces and the theories on the origin of the Universe. Write about your findings.

14. Build a sculpture, that is not a member of the Archimedean family, using only two types of polygons. Consider developing a pattern for the faces or an interesting surface finish.

15. Construct a full set of the Deltahedra using both a color and pattern progression from the tetrahedron on through to the icosahedron.

16. Constrct a full set of the five Platonic or the four Kepler-Poinsot solids using only linear elements and nodes. Set up lights and a camera. Take a number of different views of these solids. Use the camera to record their changing shadows and present a visual essay. In a written essay, record your observations about the process and results. Present to an audience.

17. In drawing and/or painting, do a visual essay from the point of view of an ant who lives his entire life on a twisted leaf

18. Using the information given in this chapter about the tiling of the hypercube, develop a variation tiling. Then do a finished version in marker, colored, pencil, paint, quilting, or embroidery.

19. Build a model of an unfolded hypercube. Use it as a basis for a single sculpture or a connecting architectural lattice or a garden colonnade. Enhance it through your choice of art materials.

20. Using the information on 6-point perspective, do a drawing of your living room on one of the Platonic solids.

21. Using the information on 6-point perspective, do a drawing of an historic building in your neighborhood on a sphere.

22. build a modular sculpture using only either the "fat" or the "skinny" rhombohedron of the Golden Zonohedra family. Consider a 2-dimensional pattern for the six faces. The pattern could stay constant or change gradually from face to face. Do a black and white version, a toned variation, or a colored group.

23. Read the science fiction story, *And He Built a Crooked House* by Robert Heinlein. Do a drawing or animated film interpreting the content.

24. Research the Catalan family of solids which are dual partners to the Archimedean family. Build a set of models of them, making each artful.

25. Use one of the projections of the six regular 4-d polyplexes as the basis for a two or three-dimensional artwork.

26. Research the architectural work of 20th Century artist Steve Baer and his use of the zonohedra. Build a small model incorporating some of the ideas generated by your readings.

27. Using the eight possible Penrose vertex tiles to cover the plane, develop a four color prototype design for fabric, wallpaper or wrapping paper.

28. Find out more about 5-point perspective. Develop a drawing of your own based on your research.

29. Use the non-regular pentagonal tiling on page 177 as the basis for an artwork. Or develop your own tile.

30. Using 12 pentagons and n-hexagons, develop a paper sculpture that fully encloses space.

31. Using only equilateral triangles, develop a sculpture that is not one of the Deltahedra.

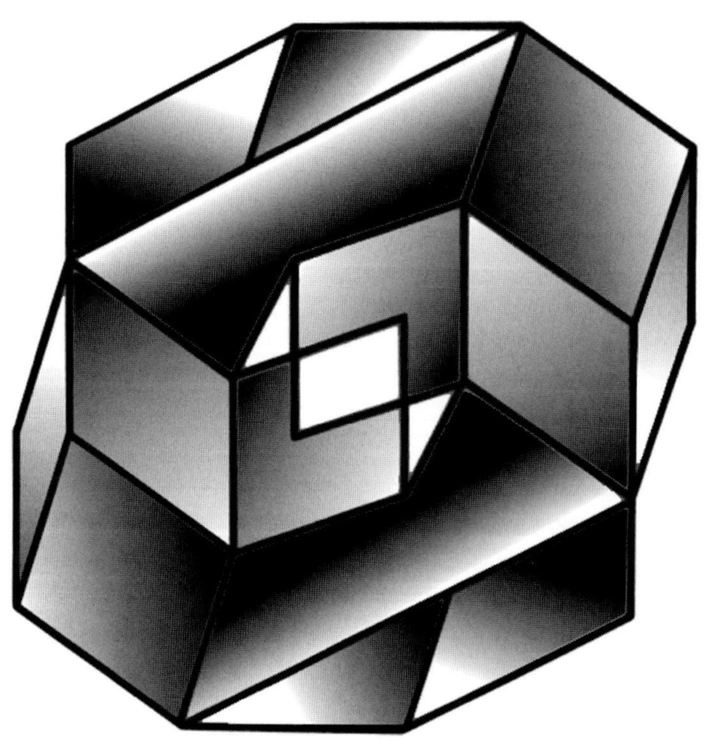

Further Reading

Abbott, Edwin A. *Flatland.* New York: Harper & Row Publishers,1983.

Ball, Phillip. *Designing the Molecular World.* Princeton: Princeton Unversity Press, 1994.

Banchoff, Thomas F. *Beyond the Third Dimension.* New York: Scientific American Library, 1990.

Burger, Dionys. *Sphereland.* New York: Barnes and Noble Books, 1983.

Boles, Martha and Rochelle Newman. *The Surface Plane.* Bradford: Pythagorean Press/Brown & Benchmark, 1996.

Coxeter, H.S.M. *Regular Polytopes.* New York: Dover Publications, Inc., 1973.

Kaku, Michio. *Hyperspace.* New York: Oxford Unversity Press, 1994.

Cromwell, Peter. *Polyhedra.* Cambridge: Cambridge Unversity Press, 1997.

Miyazaki, Koji. *An Adventure in Multidimensional Space.* New York: John Wiley & Sons, 1986.

Newman, Rochelle and Donna Fowler. *Space, Structure, and Form.* Bradford: Pythagorean Press/Brown & Benchmark, 1996.

Robbins, Tony. *fourfield: Computers, Art & the 4th Dimension.* Boston: Little, Brown and Company, 1994.

Rucker, Rudy. *Geometry, Relativity and the Fourth Dimension.* New York: Dover Publications, Inc., 1977.

5 Topological Transformations

When looking at several objects, a person can choose to look either for similarities or differences among them. For instance, we can say all human beings are the same: each has two arms, two legs, one head, one heart, etc. Men and women are human beings, thus they are the same. Or we can begin to notice the differences. Men tend to be taller, women tend to have broader hips, etc. Neither mode of observation is better than the other. Only the focus changes. With that change, different conclusions can be drawn. These conclusions then can determine how the object is to be treated, for good or ill.

Mathematicians, and scientists, look for classification systems that provide for the greatest degree of similarity among things rather than pointing out the differences. This interest has created a new field of mathematical investigation called Topology. Johann Benedict Listing (1802-1882) was the first to use the term. The word topology comes from the Greek meaning position. The subject had its beginnings in the 19th Century with the work of Gauss, Poincare, and Riemann. Ongoing is the search for a complete classification of geometric objects.

Topology like other fundamental branches of mathematics, cannot be explained simply. Its relationship to geometry lies with its examination of properties of geometric figures that are preserved despite the many transformations these figures might undergo. Topology studies the properties of objects that remain unchanged, invariant, when these objects are shrunk, stretched, bent, twisted, or distorted, but not torn or glued. It deals with questions about the basic structure of objects that do not change under these transformations. Topology does not concern itself with size relationships or the arrangement of objects. Two figures are topologically equivalent if one figure can be transformed into the other without connecting or disconnecting points. Points close together in an original figure remain close together in the transformed figure. Exact preservation of distance is not relevant nor is straightness.

The traversing of a path, or the coloring of a map, are topological problems because the topologist, unlike the artist, is not interested in the shape of the path or its boundaries. We could describe topology as being able to consider two very different things and finding that they really resemble each other in very fundamental ways.

Topologists look at surfaces. These are treated as if they were made from infinitely thin rubber sheeting. Thus the sphere, the ellipsoid, the cube, and the tetrahedron all share the same topological properties.

In this chapter we are dealing with only the aspect of topology which is called "metric or point-set topology.

A sheet of paper is considered an object that has a two-sided surface and a single edge. When it is crumpled into a ball, it still maintains the same basic properties of two sides and one edge. That is unchanging, or invariant.

Two other important invariants refer to regions that can be drawn on a surface so that each region has a border in common with every other region. If each region is given a different color, each color will border on every other color. Think of a map of the United States in which each state is assigned a color.

Networks, graphs, polyhedra, knots, and labyrinths are all aspects of the subject of topology. The theorems of topology which began with geometry now connect to other mathematical fields of investigation such as the theory of groups and the algebra of complex numbers.

Geometry involves the study of objects and their properties in space. In the physical world of 3-space, objects can exist in essentially one of three states: solid, liquid, or gas. Properties of solid objects have to do with their dimensions, their angles, their areas, and their rigidity. What happens to an object in space when it is moved? What aspects of that object stay constant and what aspects transform? Projective geometry, which relates to linear perspective drawing discussed in Chapter 2, deals with the concepts of points that lie on a line (collinearity); lines that go through a point (concurrency); section (a collection of points) which are unchanged by the still larger group of projective transformations. Euclidean and Projective geometries are very special cases of topology.

Euclidean space has geometric properties that do not vary from point to point. A Euclidean plane is a special kind of "rigid" surface in which the curvature is zero everywhere. It is flat, to put it simply. Geometric figures in that space exist independent of their orientation. If we take a cube and move it in 3-space, only its position changes. Aspects of position refer to the exploration of spatial symmetry. In symmetry, the operations of rotation, translation, and reflection are used. The cube still remains a rigid body. Euclidean geometry deals with properties of length, angle, and area that are unchanged by the rigid motions of geometric objects.

A sphere is a surface that has constant positive curvature. Surfaces of constant negative curvature relate to the non-Euclidean geometries discussed in Chapter 3. In topology a surface is called 'simply connected' if it is all one piece and has no holes in it.

Precision about meanings is most important to a topologist. Within the system of meanings and rules, does the system remain consistent in its logic?

Topology is like the Cheshire Cat in "Alice in Wonderland" by Lewis Carroll. At first you have the appearance of the full bodied cat with a grin on its face sitting within the branches of a tree. But, at the end, you are left with only the grin, the topological invariant.

In the above image, artist Claire Belanger fused the concepts of bilateral symmetry and topological transformation using the capabilities of the computer program, Adobe Illustrator. All of the forms have been developed with a group of eight transformations with a circle at one end of the process and a square at the other. Two pieces of information affect the outcome of the transformation: how close the two geometric figures are placed to each other and how many steps are used in the transformation. The artist used the blend tool of the program to develop the shape as well as the tone change. Where the cursor point of the mouse is place on the two basic figures also affects the intermediate stages of the transformation. The computer, not yet a thinking tool, is working with mathematical rules embedded within the program.

Artists do not usually take advanced courses in mathematics so they are never introduced to topics that might have some interesting connections to the arts especially through the use of the computer. Topology could provide insights into forms that the sculptor, the architect, and the designer would find meaningful for their work.

A bounded portion of the plane is an essential surface of the artist A wall, a painting, a page of a book, a billboard, are all planar pieces for the artist to manipulate . On it marks are made in relation to its shape and size and orientation. Each has unique properties on which to build an image.

On the other hand, for the mathematician, the plane has the same topology as a sphere with one pinprick hole in it. Two surfaces have the same topology if they can be transformed in such a ways as to convert one form into the other without cutting, tearing, or gluing the surface. It is the examination of the essence of what constitutes continuity by beginning with the continuity of shapes and spaces and moving further and further afield to continuity in other areas of investigation. What remains connected?

What happens when a geometric object is no longer rigid but can be considered malleable and able to take on different shapes? How can one discover aspects of groups of objects that join them into families even when they appear to be very different? These are mathematical questions but their answers could have implications for the artist as well.

There are definitions and rules to the game of topology. Topologists who study manifolds are concerned only with the surfaces of objects, not their interiors. They are extremely interested in the most fundamental of qualities which are the invariants.

The invariant of dimension refers to the number of coordinates that are needed to specify a point in any given space. A line requires one point, a plane needs two. Recall the chart found in the beginning of Chapter 4. How many coordinates would you need to describe the position of a person in apartment 13B on the 13th floor of a hirise building which is located at 34th Ave. and Main St.?

What are the invariants related to surfaces? The term manifold is used to describe the more complicated spaces. Manifolds are surfaces that logically appear flat, or Euclidean, but on a larger scale may bend and twist into very intricate forms. Manifolds may be bounded or unbounded, may be smooth or wrinkled, can transform into another without tearing. The surface of a two-manifold is the two-dimensional surface of a three-dimensional object. Three-dimensional surfaces are called 3-manifolds of a mathematical object occupying four-dimensional space.

Geometric objects are classified according to their genus. The latter is defined by the largest number of non-intersecting simple closed curves that can be drawn on a surface without separating it into pieces. Closed curves divide space into an outside region and an inside one. A sphere with no handles is called Genus O. A sphere with a single handle is called Genus 1. A sphere with two handles is referred to as Genus 2 and a sphere which begins to resemble a Swiss cheese is Genus 3 or more.

Mathematically a curve is an undefined term; however, common types of curves can be described.

A simple curve is one which can be drawn without lifting up the pencil from the surface and with-out passing through any point twice.

A closed curve is one which has its beginning and ending points the same. It is drawn without lift-ing up the pencil from the surface. Closed curves divide space into an outside region and an inside one. See the classification chart on pag-es 226 and 227 for the properties of topological objects.

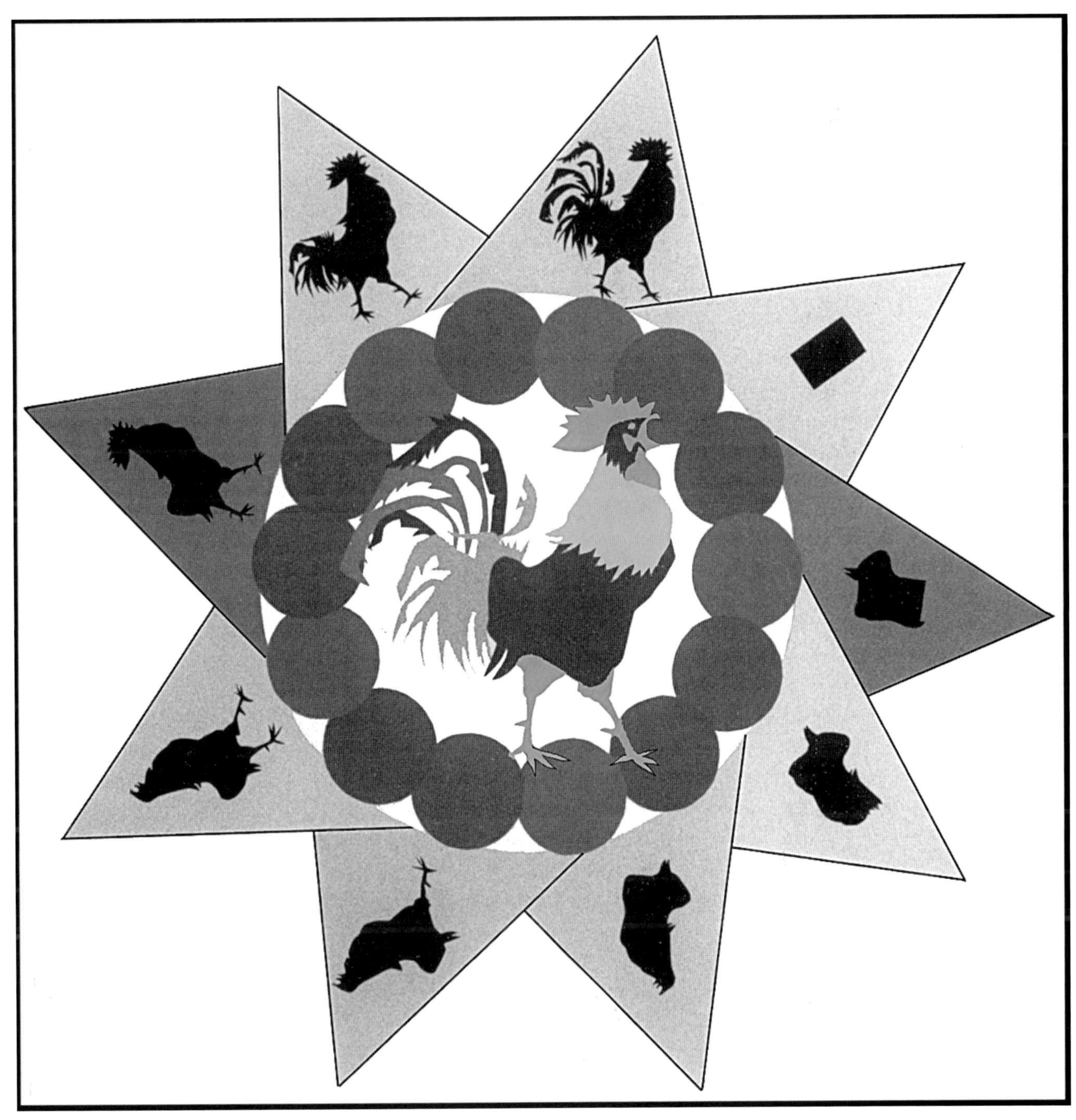

Natural Transformations

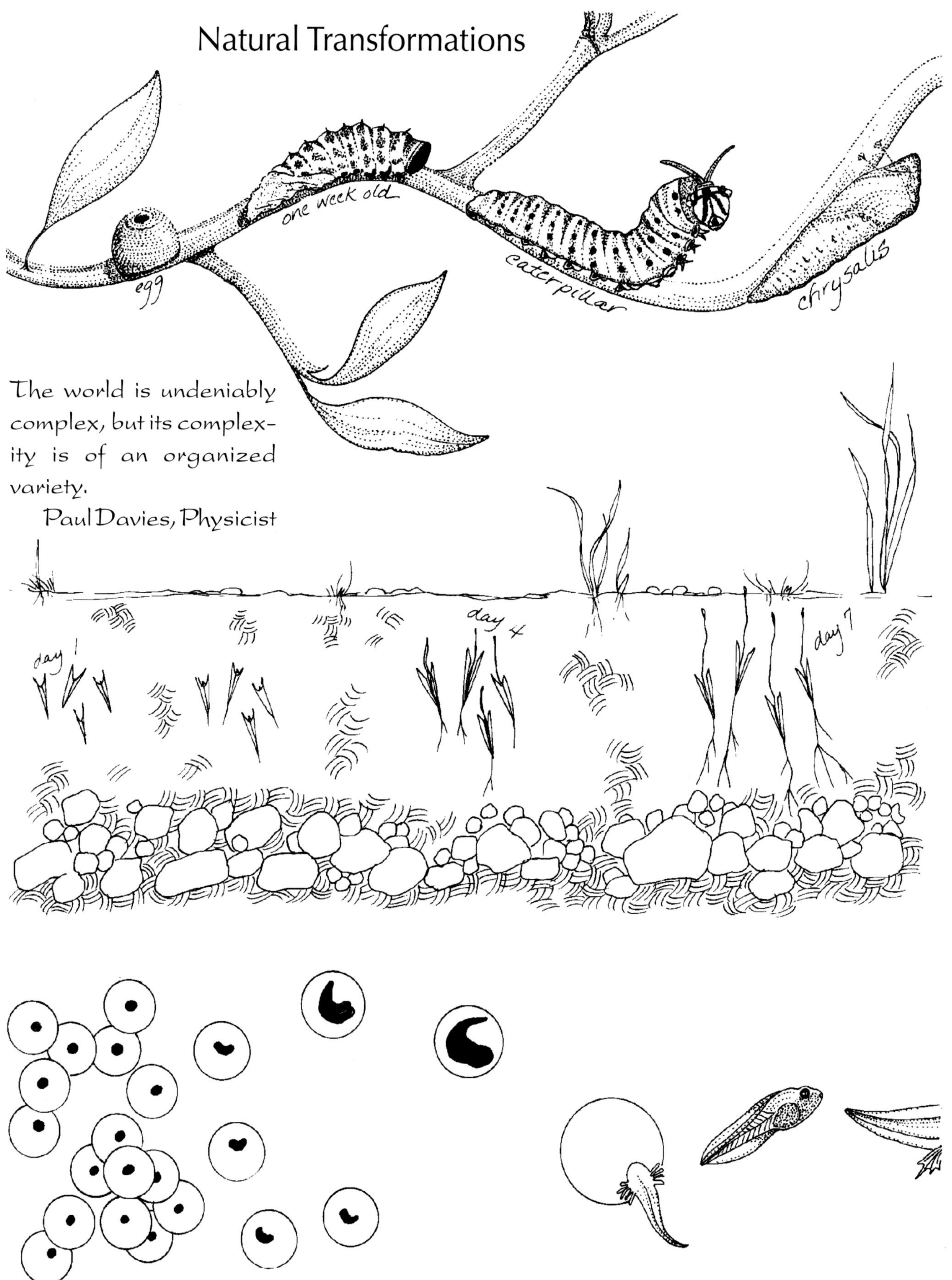

The world is undeniably complex, but its complexity is of an organized variety.

Paul Davies, Physicist

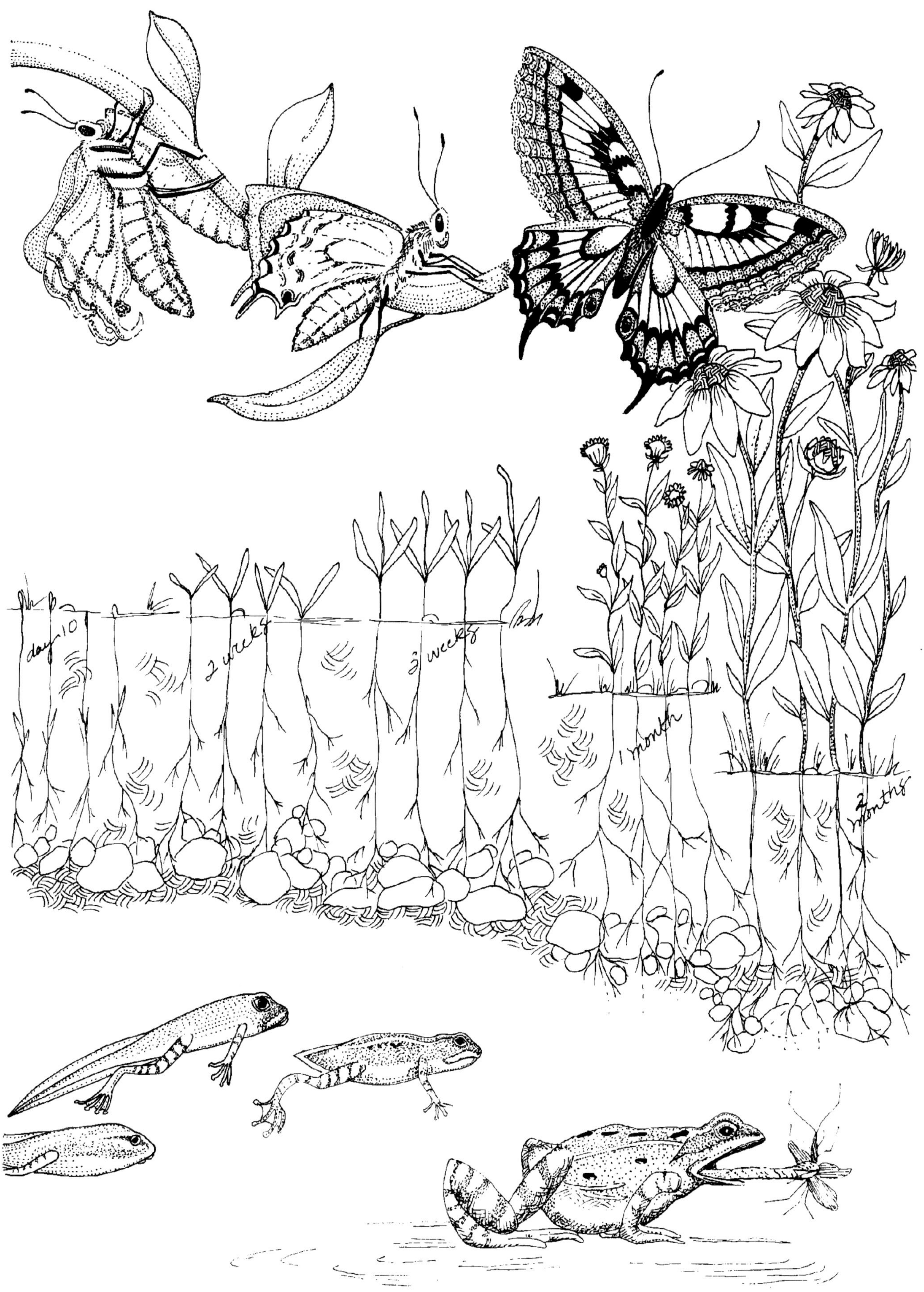

day 10
2 weeks
3 weeks
1 month
2 months

Properties of Topological Objects

The objects topologists study are called spaces and surfaces. Mathematically they relate to sets. (A set is a collection of objects). Topology is the examination of the properties of geometric figures that are left unchanged by a continuous transformation of the figure so that points that are together, (A,B,C,D...) at the beginning of the transformation remain together at the end of the transformation (A,B,C,D...). It is the order of the points that matters and not the distance between. Any two "surfaces" that share the same properties are topologically equivalent. Topologists are really interested in what happens at the beginning and the end of a transformation, not the middle.

A surface can be considered as being patched together of many small planar pieces. When zooming in on any one of the individual pieces, the surface appears as a portion of the familiar Euclidean plane, a "neighborhood". The properties found within it are the local properties. All the component pieces and their manner of assembly refer to its global properties. The term manifold is used to describe more complicated spaces. Geometer Bill Thurston is involved with developing a classification system of only-eight categories based on the particular shape of a "neighborhood".

A. *Dimension-* refers to the number of coordinates needed to specify a point in any given space.
   What spatial dimension does a particular surface exist within? A 2-manifold exits in 3-space.
   A manifold describes a complicated space. It may be bounded, unbounded, wrinkled or smooth.

B. *Boundary*
   Is the surface bounded or unbounded?
   Is the surface closed or open?
   Closed curves divide space into an outside region and an inside one.
   A circle is bounded.
   The surface of a sphere is bounded and closed but without boundary.
   The plane is not bounded.

C. *Possession of an Edge*
   Does the surface have an edge? If so, how many?
   A circle has no edge.
   A sphere has no edge.
   A line segment has an edge. It has two endpoints.
   A cylindrical band has two edges.
   A mobius band has one edge.

D. *Orientability (is a global property)*
   Is the surface orientable or non-orientable?
   This refers to properties of right and left-handedness.
   The mobius band is one-sided, thus non-orientable.
   The cylinder is orientable.
   The torus is orientable.

E. *Handles, Holes, and Crosscaps (Called Genus)*
   Genus is defined as the largest number of non-intersecting simple closed curves that can be drawn on a
                                          surface without separating into pieces.
   Does the surface have zero or more holes? An indentation is not considered a hole.
   How many holes does it have?
   The hole can be considered a handle.
   This property relates to how an object is embedded in space. Does the hole belong to the space that the figure is
                                          embedded in or is it part of the figure?
   A sphere has no holes. With handles it is a closed orientable surface. With crosscaps it is a non-orientable surface.
   A torus, or doughnut form, has one hole.
   A pretzel has several holes.
   The mobius band has 0 holes.

F. *The Euler Characteristic (Networks on a Surface) V+F-E=?*
   A surface can be related to the Euler Equation just as a polyhedron can. Faces, Vertices and Edges
                                          become lines and nodes of a network.
   The Euler characteristic of a torus is 0.
   Network on a plane (whether simple or complex)= V+F-E=1
   (Maps are networks)
   Network on the plane = V+F-E=1 when not considering the "infinite face" which surrounds the network.
   (Polyhedra are networks) Any polyhedras can be transformed into a sphere by appropriate pulling and pushing.
   Network in a hyper sphere = V+F-E=?

Modifications to Spheres (Holes, Handles, Crosscaps)

Objects are classified according to their genus. All closed, orientable surfaces can be continuously transformed into a sphere with a certain number of handles. The number of handles is called g, the genus of the surface.

All closed non-orientable surfaces require a 'cross-cap', which is constructed from the mobius band. The cross-cap is topologically equivalent to the mobius band. All closed non-orientable surfaces can be continuously transformed into a sphere with a certain number of cross-caps. It still uses g as the genus of the surface.

Orientable surfaces	Non-Orientable surfaces
(left and right handedness are different) refers to two-sidedness	*(cannot distinguish right and left handedness)* refers to one-sidedness

Handles and Holes

Genus 0

These are used to calculate the Euler characteristic. Any closed orientable surface is equivalent to a sphere with a given number of handles.

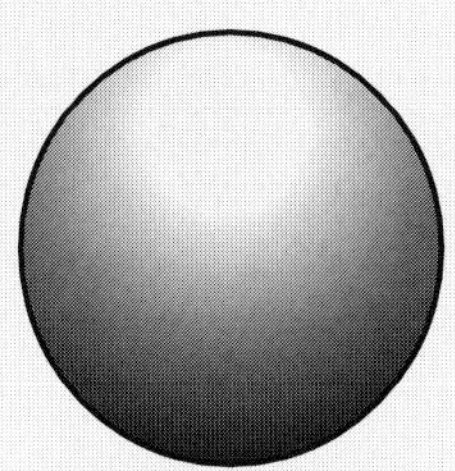

A handle is a tube which attaches to two holes on the sphere surface.

Genus 1

No closed cut can be made around the sphere without cutting it into two pieces, so its genus is 0. Only one closed cut can be made in a doughnut without dividing it, so its genus is 1. Similarly, a two-holed figure has a genus of 2.

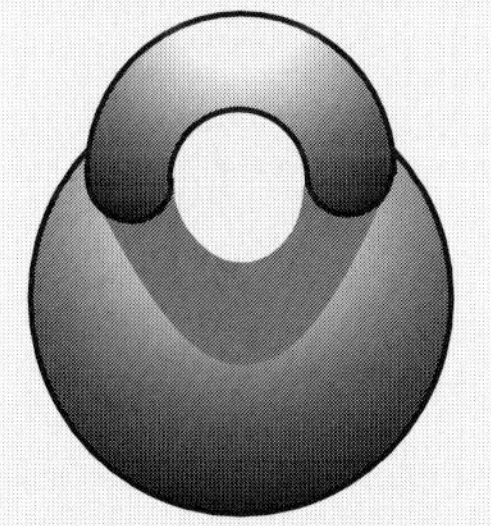

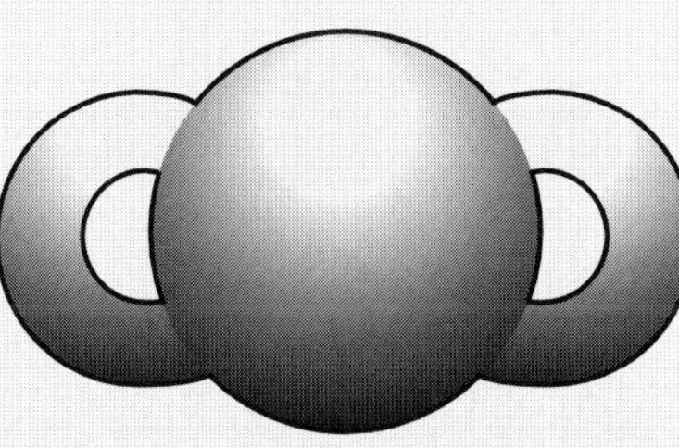

Genus 2

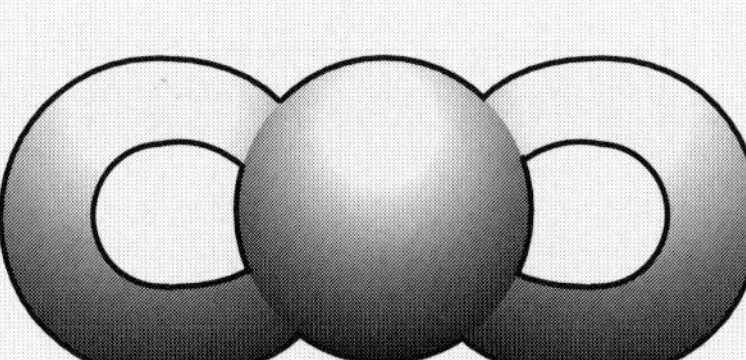

Crosscaps

Genus 0

Genus is defined as the largest number of non-intersecting simple closed curves that can be drawn on a surface without separating it into pieces. Curve is an undefined mathematical term; however, common types can be described. A simple curve is a curve which can be drawn without lifting up a pencil from a surface and without passing through any point twice. A closed curve is a curve which has its beginning and ending points the same. It is drawn without lifting the pen from the surface.

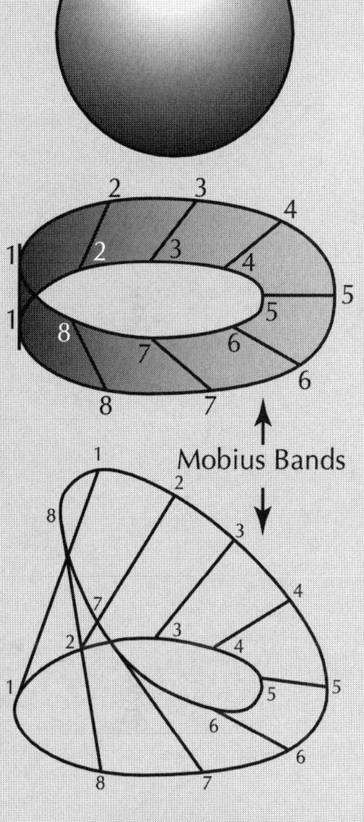

Crosscap

A crosscap is a mobius band which attaches to one hole on the sphere surface. This is a theoretical operation in 3-space since it needs the room of 4-space to be performed.

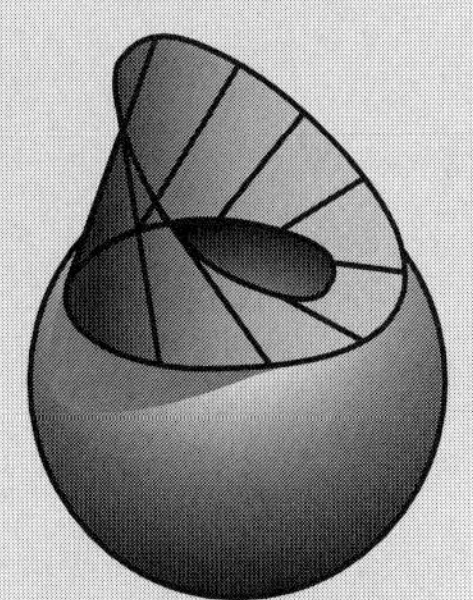

Genus 1

Euler characteristic sphere with one crosscap
V - E + F = 1
Sphere with two crosscaps
V - E + F = O

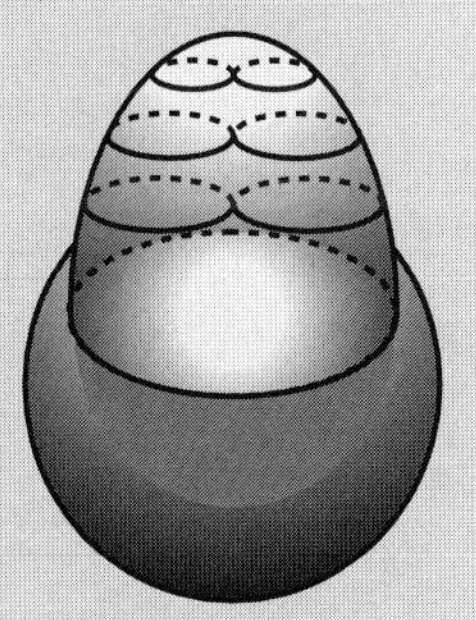

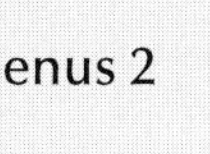

It is widely believed among scientists that beauty is a reliable guide to truth.
Paul Davies,
Physicist

*These facing pages:
Photographic essay of some familiar surfaces.
Above: Reflection of sky and clouds onto a metal ladle and bowl.
Bottom: Reflection of a tree in puddles of water. Notice that the unseen three-dimensional tree is projected onto the two-dimensional surface of the water. Could you now take this photograph and taranslate it into a pencil or pen and ink drawing? You would be mapping from one two-dimensional surface onto another two-dimensional surface. How would you transfer the drawing onto a three-dimensional surface?*

228

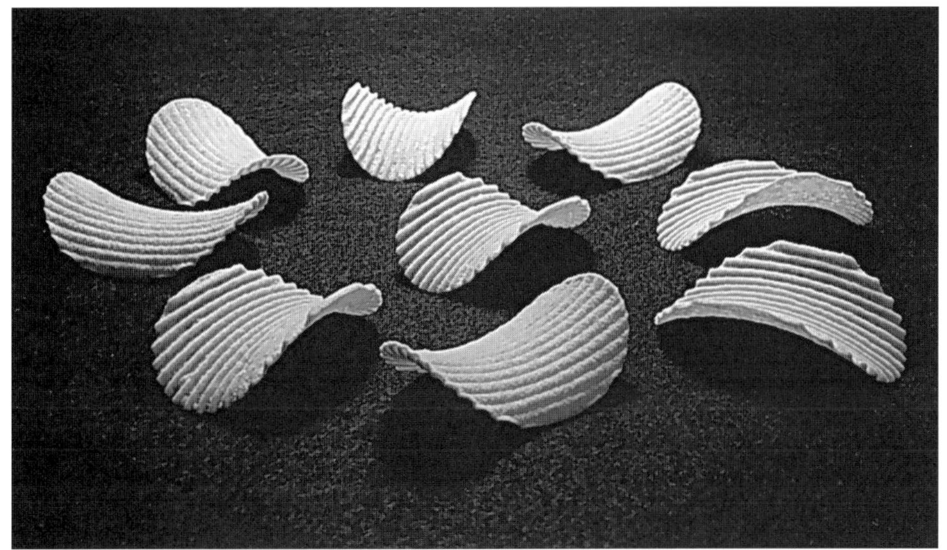

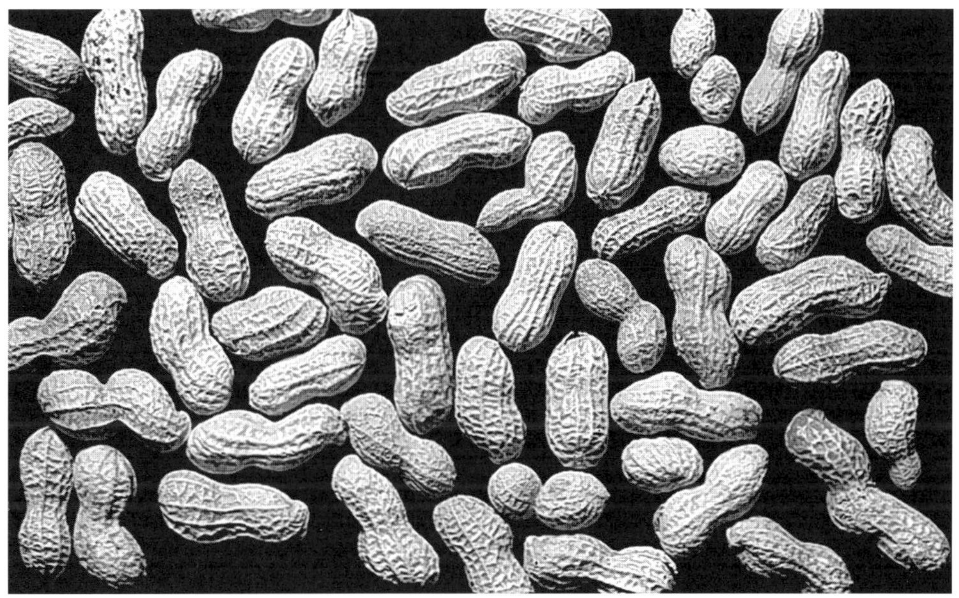

This page:
Three different types of sur-
faces in 3-space. How would
you classsify these three sur-
faces? Are they related to one
another and, if so, how?
Above:
A rondo of potato chips.

Center:
A flock of peanuts.

Bottom:
A plethora of popcorn.

Genus 0 - The Sphere
(no handles, zero holes, orientable)

In Topology one studies surfaces. A hollow sphere, or ball, is the simplest closed surface one may encounter. It has an inside and an outside. It has no edges. A rubber ball can be stretched, shrunk, pinched, pulled, and deformed, as long as it is not torn or punctured, and still maintain its topological properties. Every closed curve on a sphere divides it into two pieces. A sphere can be triangulated. That is, the surface can be built up from a finite number of triangles. Look at the geodesic spheres of geometer Buckminster Fuller. A sphere can be transformed into any number of polyhedra and still be equivalent to the original rubber ball.

What properties of the ball could not be changed by any of the above moves? These are the invariants. One such aspect is the notion of inside and outside. To turn the ball inside out would take a number of very involved moves.

Another such topological property involves the concept of a closed curve. A closed curve is defined as any line, curved or partially straight, whose ends are joined but which does not cross over itself. A closed curve is any figure topologically equivalent to a circle. Take a joined string and see how many closed curves you can produce using the above definition. A closed curve drawn onto the surface of a ball divides it into two parts.

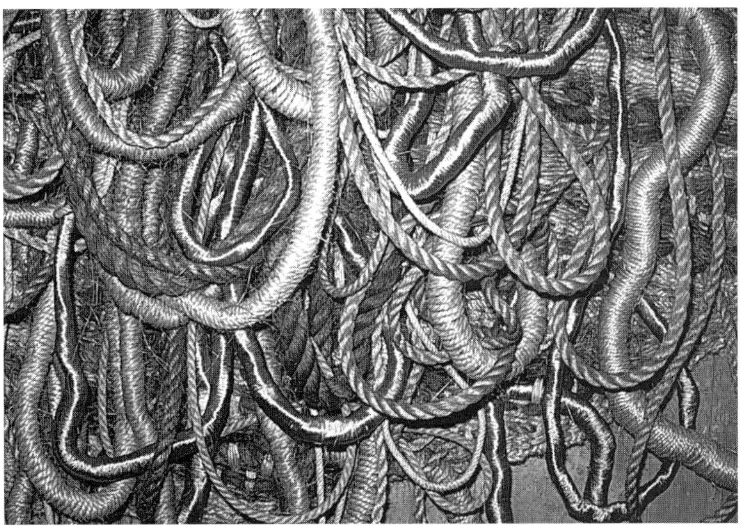

*This page:
Photographic essay of ropes which are suggestive of curves. Can you determine which rope arrangement is what kind of curve?*

Genus 1: The Torus
(one-handle, one-hole, orientable)

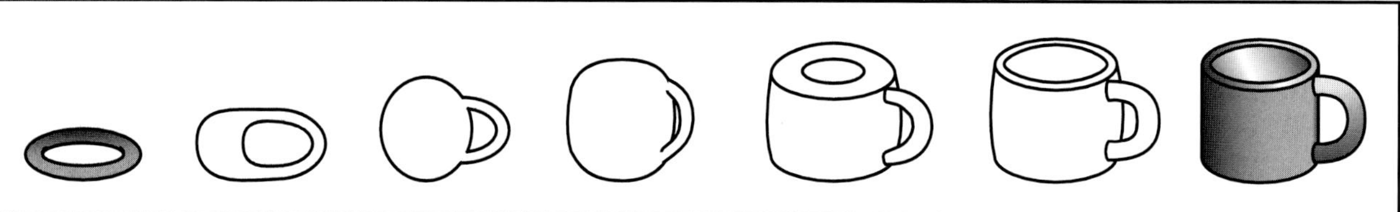

Definition of a torus:
A circle of radius a is rotated through one revolution about a line, in the plane of the circle, a distance b from the center of the circle where b is greater than a.

The torus is a geometric object. It is likened to a doughnut, an inner tube for a tire, or a bagel without butter. It is also thought of as equivalent to a cup with a single handle. Though it is a three-dimensional object, it is considered a two-dimensional surface on which two closed curves can intersect in just one point and on which a closed curve does not necessarily separate the inside from the outside. It is like the surface of a sphere because any point on the surface of the torus can be identified uniquely by giving coordinates of two numbers. The latitude shows the position of the point on the semicircle while the longitude indicates how far the semicircle has been rotated.

The torus is not, however, equivalent to a spherical surface because when two closed curves are drawn on the surface of the torus, one a horizontal cross-section and the other a vertical cross-section, neither curves separates the surface into two parts. With a piece of flexible rubber hose or a length of baking dough, we can construct a torus by joining the two ends.

Is it possible to construct a torus out of paper so that the form curves gently and smoothly? Would we have to construct it out of polygonal paper units? On the facing page is a chart with the properties of a rectangle. In the physical world that rectangle could be a piece of paper, a sheet of baking dough or a length of rubber. The latter would be the most flexible surface to use when rolling and bending the rectangle.

On pages 234 and 235, we have shown how to construct a torus using paper and a number of polygonal units that are joined together to form the torus ring. The number of units used would increase or decrease the smoothness of the curve of the figure.

There are several ways to subdivide the surface of a torus, as well as other surfaces, into regions that can be used for measurement or design or color placement. Maps on different types of surfaces require different numbers of colors for the contiguous regions. The more complicated the surface, the greater the number of colors needed. Mathematicians have proved that four colors are the most that can be used when coloring a planar map. Coloring the surface of a sphere requires the same number as does a plane. The torus is the next simplest surface. It can be triangulated and the triangles colored. It requires seven different colors so that no adjacent regions have the same color.

It has also been mathematically proved that for Genus 2, 3, 4 the required number of differents colors is 8, 9, and 10 respectively.

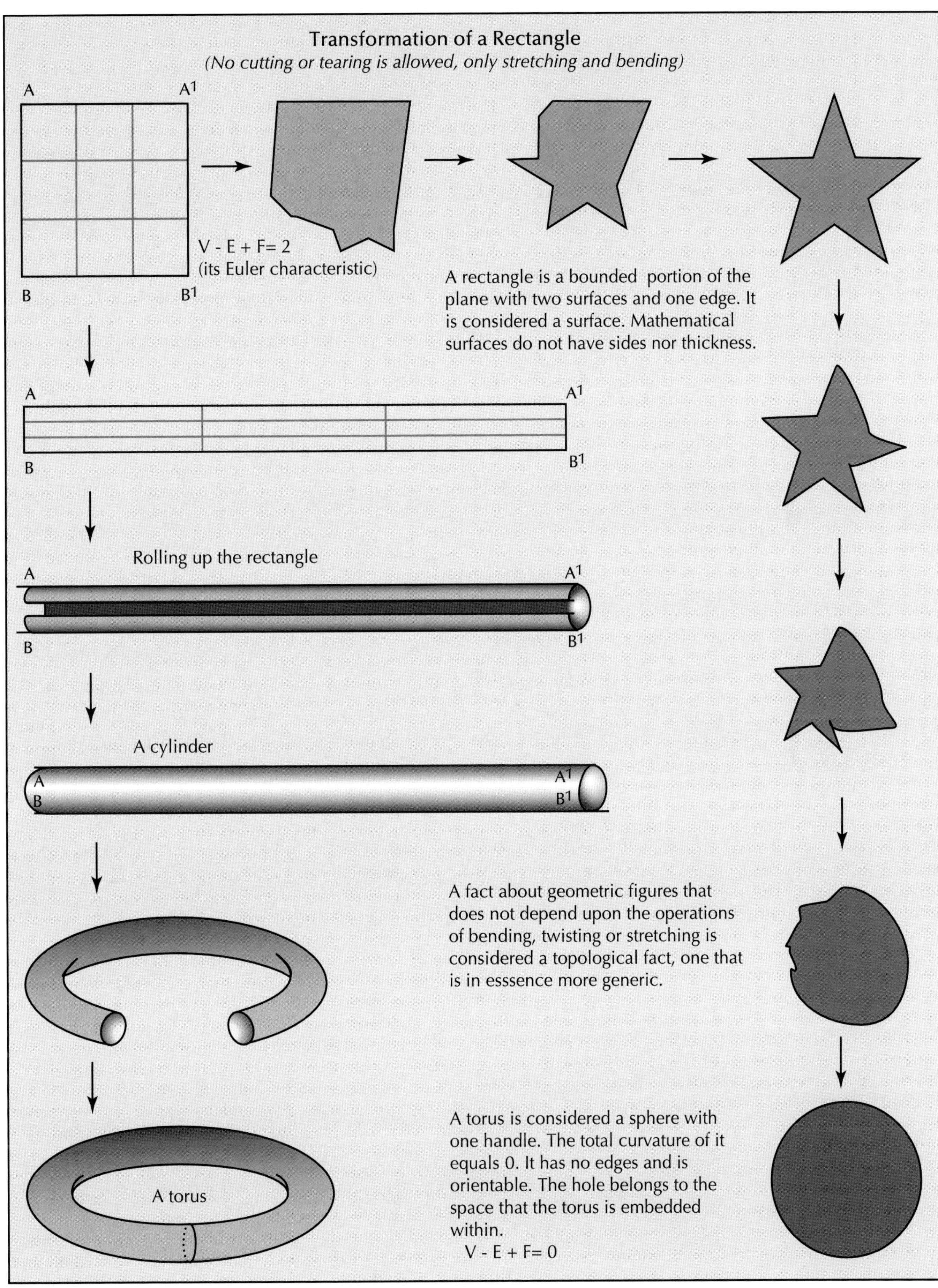
Transformation of a Rectangle
(No cutting or tearing is allowed, only stretching and bending)
A
A1
B
B1
V - E + F = 2
(its Euler characteristic)
A rectangle is a bounded portion of the plane with two surfaces and one edge. It is considered a surface. Mathematical surfaces do not have sides nor thickness.
A
A1
B
B1
Rolling up the rectangle
A
A1
B
B1
A cylinder
A
A1
B
B1
A fact about geometric figures that does not depend upon the operations of bending, twisting or stretching is considered a topological fact, one that is in esssence more generic.
A torus
A torus is considered a sphere with one handle. The total curvature of it equals 0. It has no edges and is orientable. The hole belongs to the space that the torus is embedded within.
V - E + F= 0

Constructing a Torus

A malleable material like dough or clay, can be used to shape a torus in the simplest way (see recipe in the project section). Paper and straight edges can also give the artist a chance for pattern development on a surface. On this page you will find a template for such a torus. Notice that it uses 6 tone changes. You could consider giving it a monochromatic color structure or try a textured variation. Copy it 12 times and join together to form a ring. Make several, if you choose, and stack for a sculptural form.

1. Enlarge the template (on right) for ease of manipulation and make twelve copies.

2. Cut out templates.

3. Score all lines which will then be given mountain folds.

4. Join a pair of templates together along their lengths before glueing tab A to tab B. Construct a half ring of six.

5. Make another half ring of six. Join the two halves together.

If you choose to "break the rules", try joining the units together in other ways. Some surprising sculptural forms result.

Toroids can be composed of groups of polyhedra. One possible choice, the dodecahedron of the Platonic Solid Family, can be used in multiples to form rings.

Notice that the torus below is divided into sixteen sections. The greater the number of sections, the closer the figure approximates a smooth form. The template we have provided here has six sections and requires twelve units.

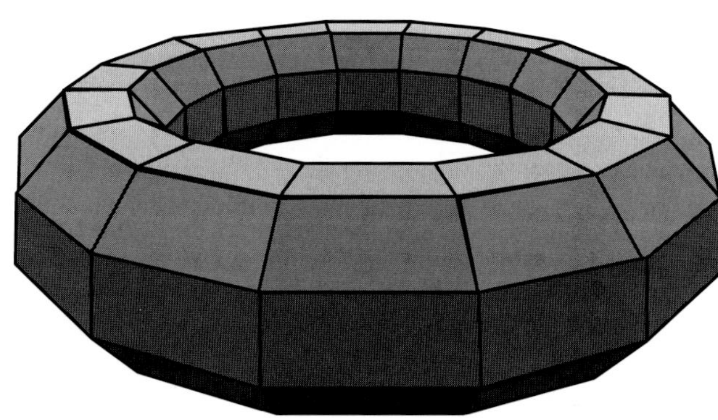

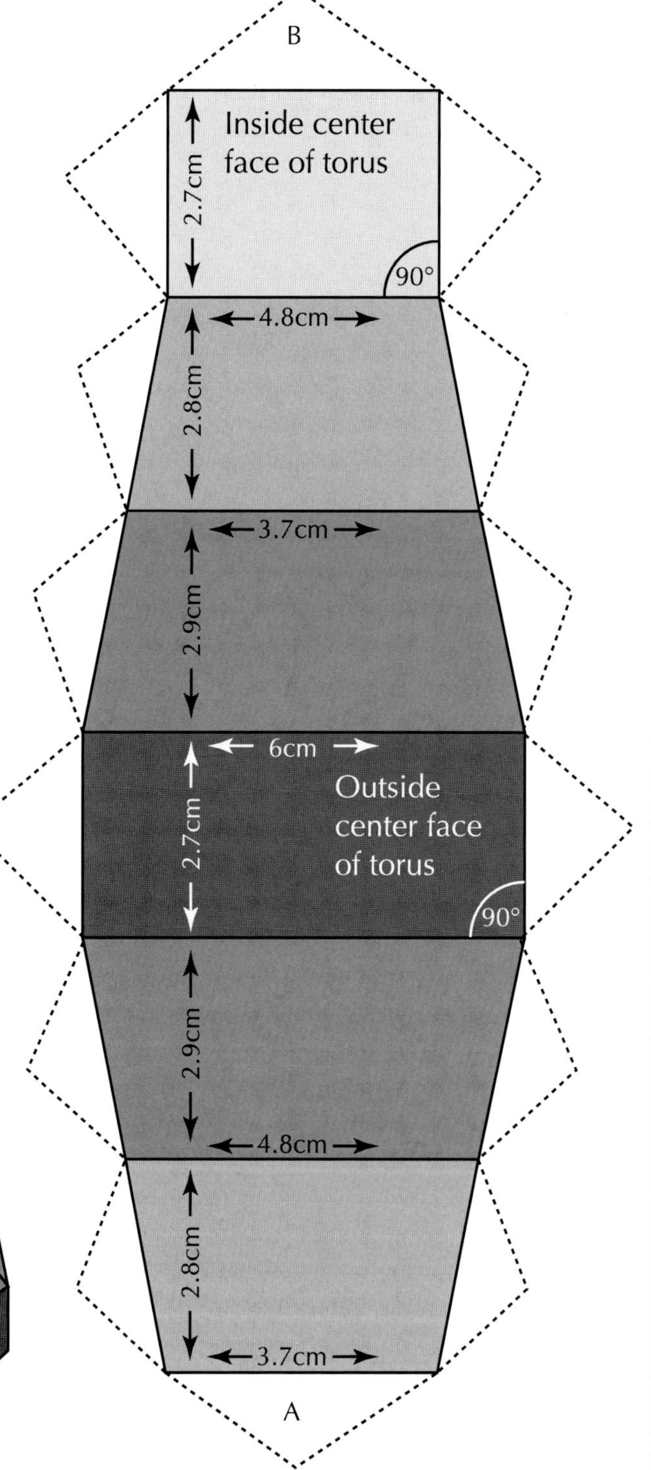

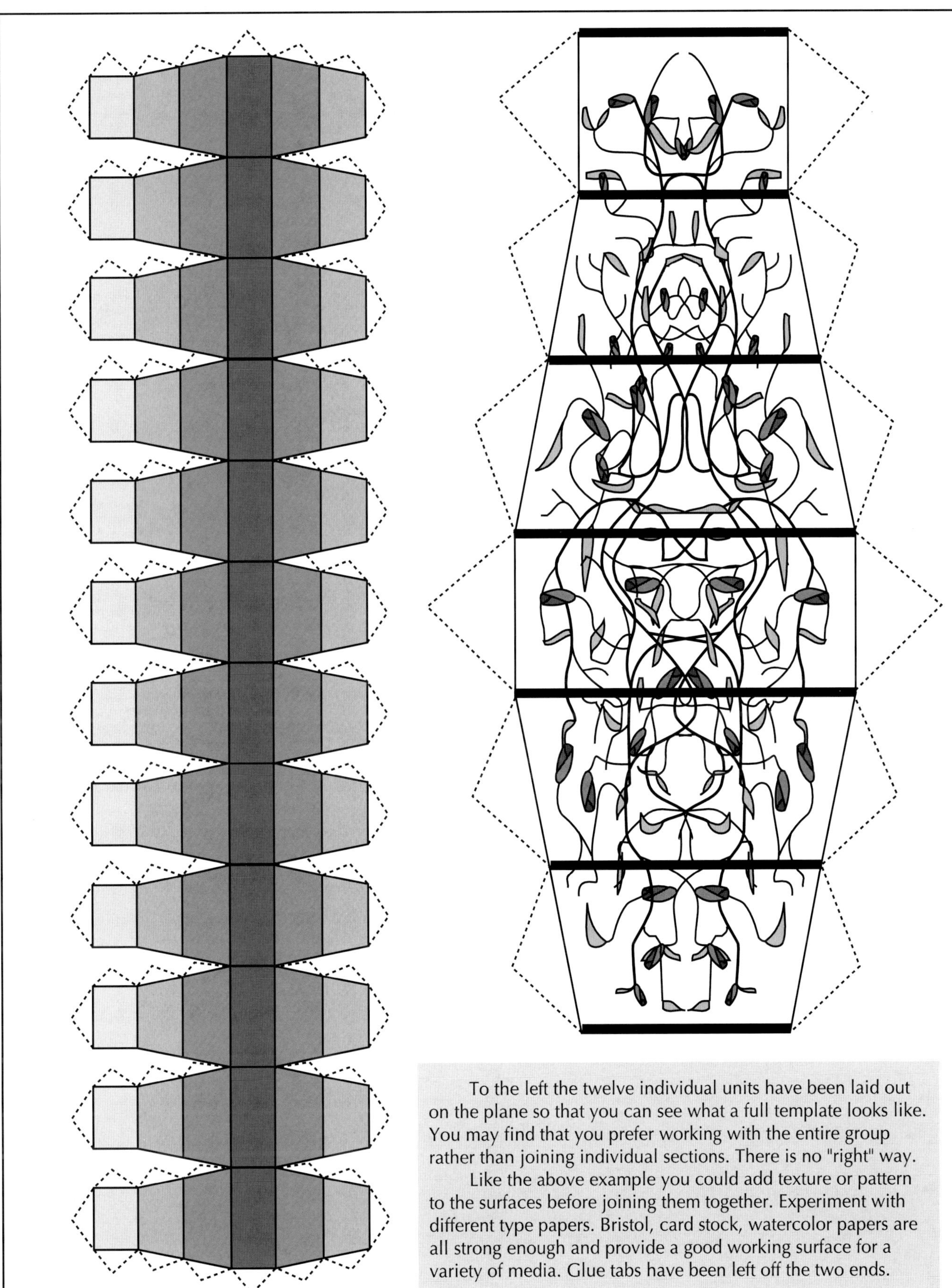

To the left the twelve individual units have been laid out on the plane so that you can see what a full template looks like. You may find that you prefer working with the entire group rather than joining individual sections. There is no "right" way.

Like the above example you could add texture or pattern to the surfaces before joining them together. Experiment with different type papers. Bristol, card stock, watercolor papers are all strong enough and provide a good working surface for a variety of media. Glue tabs have been left off the two ends.

The Mobius Band
(A one-sided surface with one edge, it is non-orientable)

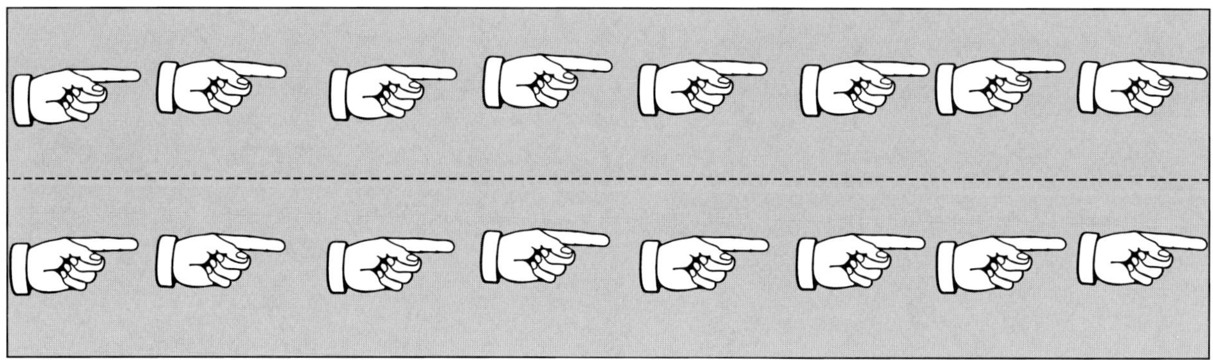

This page:
To explore the aspect of orientability, cut out copies of the above and fold along the dotted line in the diagram. First join the band into a ring and note where the finger is pointing. Then give the band a twist of 180° and join to form the Mobius band. What do you notice about the relationship of the fingers?

Between any two points on the Mobius, a continuous line can be drawn on its surface without crossing an edge. Thus, it has only one surface and one edge.

A sheet of paper is a familiar geometric object to many of us. It has two distinct surfaces with a single edge. This rectangle can be divided into two parts by any curve joining two points of its edge. When rolled up and joined edge-to-edge, this rectangle gives another familiar object, the cylinder. This object has two sides and two bounding edges with an inside and an outside.

Now, take a long strip of paper with parallel edges, and before joining it edge to edge, give the strip a single twist through 180°. This geometrical object has interesting properties and is named for the geometer who first discovered its properties in 1865. It is called a Mobius band, after August Mobius, 1790-1868. It has only one surface and one bounding edge. Between any two points on the Mobius band, a continuous line can be drawn on its surface without crossing an edge. Thus it has only one surface and one edge. Also, it has no inside nor outside. Travelling over its surface, one can go from any point on the strip to any other point on the strip in one continuous path. If one travels on the upper edge in a continuous path, one reaches a point which is on the lower edge opposite the starting point. Continue the journey and one ends up back where he or she started. Any point on this strip can be joined to any other point on the strip with a curve lying wholly on the strip and not crossing the bounding edge. A surface is said to be closed if it has no boundary and if it fits into a finite portion of space.

If the strip is now twisted 360° before joining its edges, another type of Mobius strip results. It is two-sided with two bounding curves, but now they are linked together. Try taking this long strip and give it 3, 4, 5 or n -twists before gluing the two ends together. What happens to the strip ? Cut the strips along their center-lines to see what happens now. Try cutting each along a line one-third the way down from the edge. What do you find out now about these strips?

Mobius Manipulations

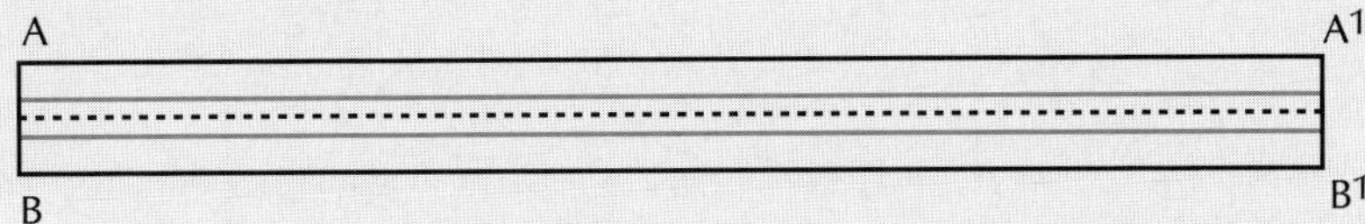

The dotted line indicates the center of the band and the solid lines indicate that the band is divided into thirds. These are guides for future cutting. Enlarge the above diagram two or three times for workable cutting and twisting length.

The band is non-orientable. It has only one surface, one edge and one kind of spin.

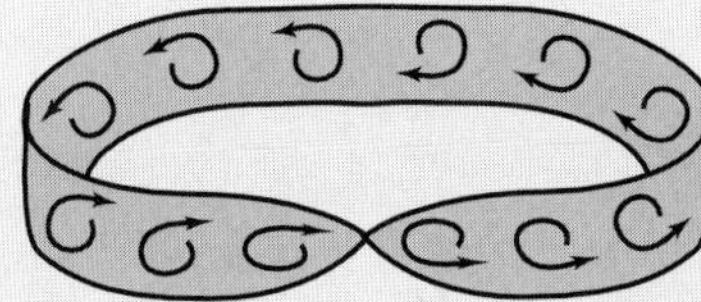

Cutting a Mobius band in half

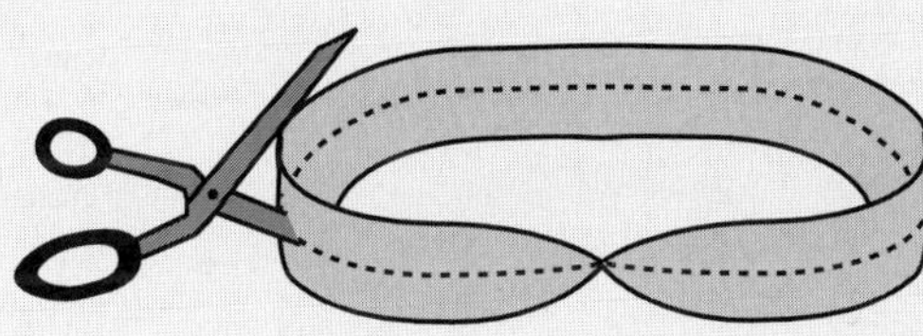

Cutting a Mobius band in thirds

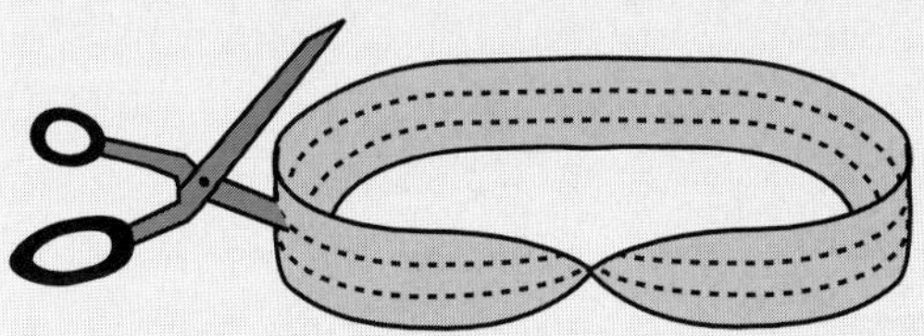

The Six-Color Map Theorem

On any Mobius band, six different colors are required to make sure that no contiguous areas will be filled with the same color. A planar map can be drawn on any rectangle or cylinder using only four different colors. If the band is twisted into a Mobius configuration it may require six colors.

Construct, or photocopy the above, several Mobius bands from long rectangles made out of paper, approximately 3cm x 25cm. Cut out one of the bands and take one end of the band marked AB, give it a 180° twist (half twist) before joining it to the other end of the band. A joins to B[1] and B joins to A[1].

To play with a Mobius band:

a. Cut the band along its centerline parallel to its edge. What happens?

Instead of obtaining two bands, a single two sided band is the result.

b. Cut the band along a line one-third of the width of the strip and parallel to the edge. What happens?

The scissors will make two full trips around the band but with only a single continuous cut. Thus, two intertwining bands occur. One of the bands has two sides while the other is another Mobius band.

c. Give the band two full twists, cut it transversely, untwist it and then join it together along the line just cut. What happens?

A continuous closed line drawn on the band will lie on both the inside and outside surfaces and will not cross an edge.

d. Take the Mobius band, give it three half-twists, join the ends together. Then, cut it along the center of the band.

The result is a single band that is twice as long as the original band and is tied into a trefoil knot (knots will be discuss later in the chapter).

Note: If n is an even number of twists, two bands similar to the original band will result. If n is an odd number of twists, only one band results. If the bands are divided into thirds, the middle band will resemble the original but the outer bands will be single or a pair of bands and they will be linked to the center band.

This page:
Mathematical surfaces do not have sides nor thickness. Thus, an object can be considered "in" a surface but not placed on a surface.
For the artist, a surface is both that and a "space".

A Mobius "rattlesnake" on an actual surface for you to manipulate.
If you wish, add color to your snake with crayons, markers, inks, or paints. Our snake is a short one due to the limits of the page. You could consider adding more length by inserting a section at the middle.

2. Cut out the two parts of the snake. Score and fold along the dotted line of the back and between the head sections.
3. Glue or tape the two parts together so that the drawings are on the outside and are back to back.
4. Give the strip a twist so that A matches up with A; B matches up with B.

Try a snake variation for yourself. Do a drawing or painting or collage on a piece of fabric or canvas. Or try an abstraction using elements from the seven symmetry line groups as seen in the Design Appendix at the back of this book.

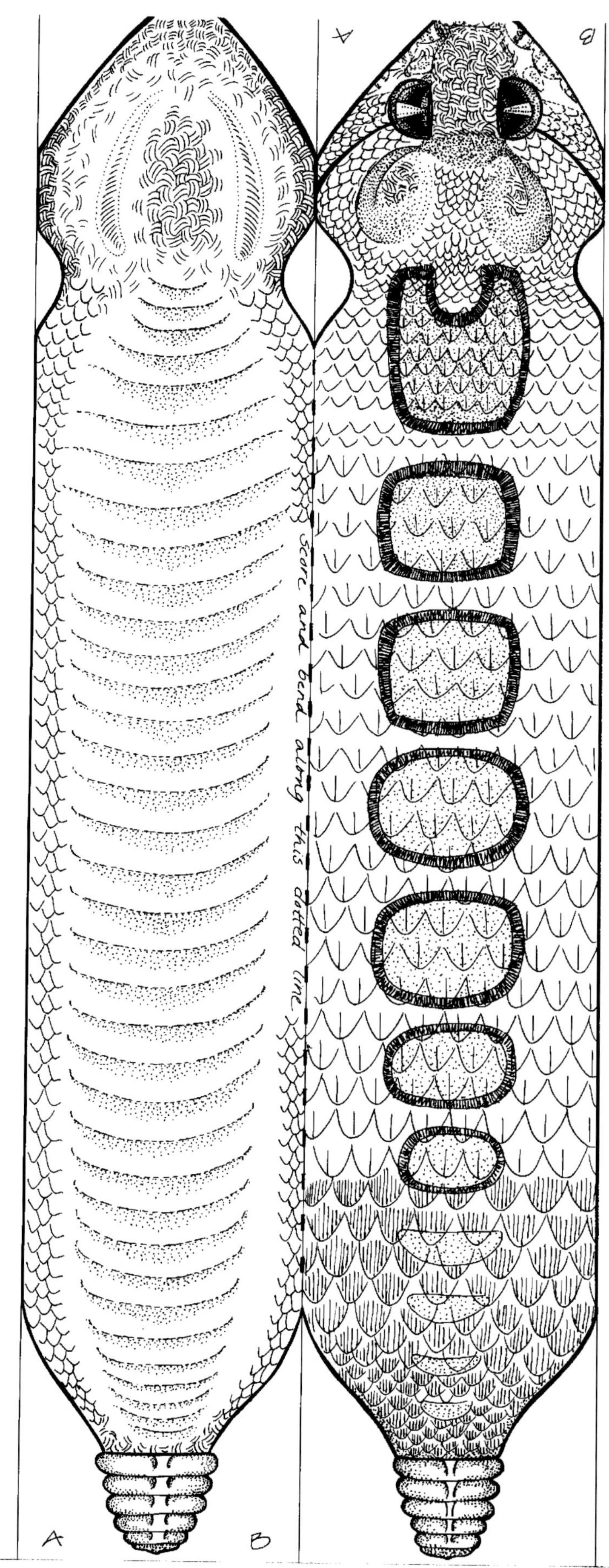

Chemists are using the field of topology to study the interconnections among the atoms within molecules. There is a new branch of chemistry called chemical topology. It is the study of the structures of the configurations of chemicals. It is this that defines the structure of the molecule. While its three-dimensional shape can be ignored as well as the angles between bonds, the number of atoms in each molecule is very important. Do the atoms form chains or rings or combinations of both? This kind of information will allow the scientist to predict the properties of substances before they are ever synthesized in the laboratory.

There are interesting connections between the Mobius band and the chemical world of molecules. In fact, there are Mobius molecules! By joining the ends of a double-stranded band of carbon and oxygen atoms, scientists have synthesized the first molecular Mobius band.

Mobius, the mathematician, discovered the band in 1858 at the same time that the chemist August Kekule discovered that carbon molecules can join together in long chains. This became the basis of the subject of organic chemistry. Then, a little over a decade later, he realized that these carbon chains could curl around and form rings. That was over one hundred years ago. It has been only within the last fifty years that chemists have observed molecules in the configuration of the Mobius band. The Mobius molecule does not occur in nature, but has been synthesized in the laboratory. Like its paper counterpart, the molecule has interesting chemical properties. Using flexible wire, develop sculpture models using the idea of rungs and twists.

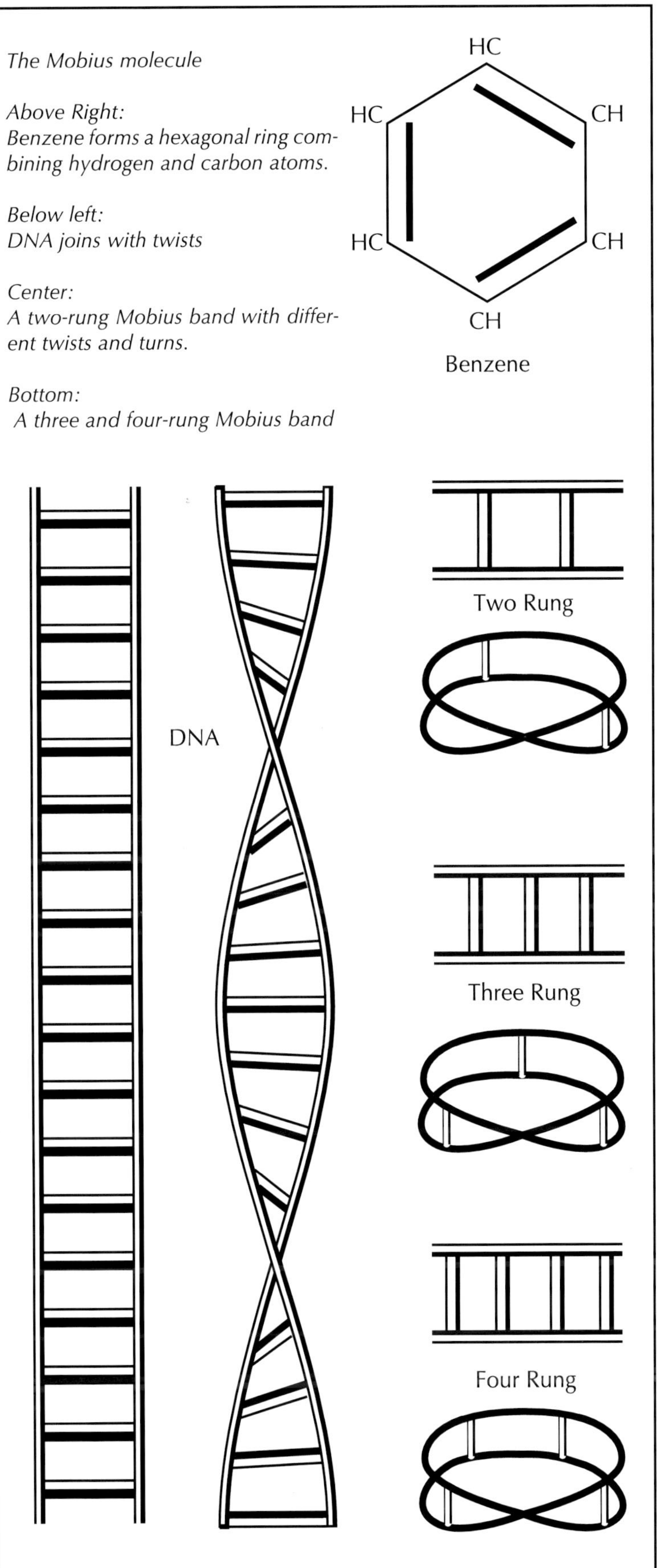

The Mobius molecule

Above Right:
Benzene forms a hexagonal ring combining hydrogen and carbon atoms.

Below left:
DNA joins with twists

Center:
A two-rung Mobius band with different twists and turns.

Bottom:
A three and four-rung Mobius band

Genus 3: The Knot

This page:
Above:
Photograph of matted grass-
es, suggestive of knotting.

In the fall when the wind blows the dried grasses up against each other, they tangle and wrap around forming groupings that gain strength and surface. They have been doing this for a very long time. So long, in fact, that perceptive humans found a way to use this visual information to gain and develop skills that have given civilization threads, ropes, knots, baskets, and textiles. The fiber arts, as well as ceramics, are major gifts that our neolithic forbears bequeathed to us. The inventiveness of skill and form, without high technology remind us that intelligence crosses time, culture, race, and religion.

Knots exist in 3-space and are related to surfaces. Knots, no matter how twisted they seem, are essentially simple circles embedded in 3-space, that is, until you have tried to untangle a group of them. A knot in which all the crossings alternate under-and-over is the simplest possible form for a knot to take. In two dimensions, there is not enough room for the one-dimensional curve to be knotted up. In 4-space, there is enough maneuverability to undo any knots. A knot cannot exist in more than three dimensional because it cannot stay knotted.

How can different knots be distinguished? In nature, knot theory relates to molecules in biology and chemistry. The helical DNA molecule, which is extremely thin and long, looks like a tangled object when it is situated outside the cell. Inside it, however, a pair of DNA strands may be very carefully connected by loops. The questions asked, and answered, by mathematicians in solving the knotty problems of knot theory may in the long run help these scientists.

240

The mathematical classification of knots goes back only to the 19th century when Lord Kelvin saw a connection that really could not be fully explored until the subjects of chemistry and biology were developed more fully. Like examining the shadows of 4-space polyplexes, mathematicians look at the simplest cast shadows of 3-space knots on the plane.

Mathematicians have just recently begun to investigate the properties of knots for the purpose of generalization and classification. Scientists are finding that "knot theory" has applications to problems in physics, chemistry, and biology. The theory also relates to networks and graphs in which the crossed strands of knots translate into nodes on graphs. Think of clover-leaf roadways on interstate highways.

Mathematicians think of knots differently from boy scouts, sailors, and artists. The knot, for them, is composed of a single loop that is twisted and has no free ends. Different types of knots simply represent the various ways that a circle can be embedded in three-dimensional space. Mathematicians think of knots as having no thickness and having a cross-section of a projection from 3-space onto the 2-space plane. Where the knot crosses itself is call a "crossing". Thus, a figure-eight knot has four crossings because the various projections of it show no fewer. A circle is called an "unknot", or trivial. It has no crossings.

The simplest knot is the overhand, or the trefoil. It has three crossings and two versions--a left and right-hand knot. There is one knot with four crossings and there are two knots with five crossings. A crossing can be over, positive, or under, negative. After that things get a bit tangled up. According to author Ivars Peterson, there are 12,965 distinct knots with 13 or fewer crossings that have been identified by mathematicians. Theories are being developed that find topological invariants for knots.

Knot theory joins Mobius bands which allow molecular biologists to investigate the properties of DNA. A single strand of this substance is very, very long. In the cell there is what seems to be an orderly looping of the folding of the strand. Outside the cell, the DNA looks disordered. So, what kinds of loops, folds, knots can the strand take?

What are you doing when you tie your shoe laces or knot your tie or wrap a Christmas present? Or do you only use velcro fastenings?

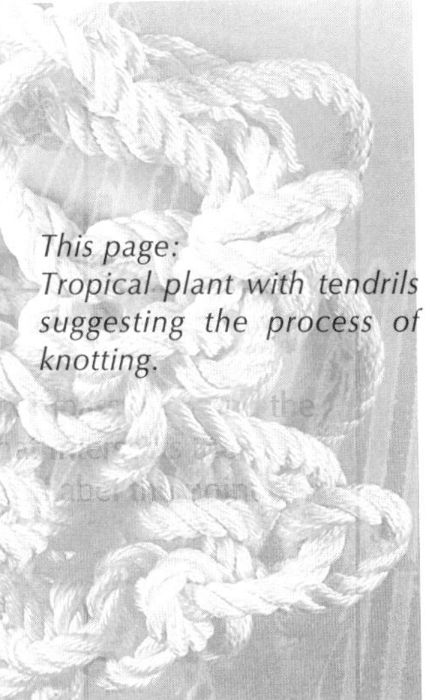

*This page:
Tropical plant with tendrils suggesting the process of knotting.*

Knot Classifications

| Projections (Shadows)
There are many different projections of the same knot. | Crossings
(Topological invarient) | A mathematical knot has no free ends. Each knot can be described by an algebraic equation. Mathematicians define a knot as a simple closed curve in three dimensional space. In two space and four space all such closed curves can be transformed into simple circles without intersecting themselves. In three space however, lies the possibility of the curve forming a knot which cannot be transformed into simple circles. *You can only tie your shoelaces in three dimensions.* |

0 — Has no crossings

Unknot
Null Knot
Trivial Knot

1 — Has one crossings

A single crossing. Can be transformed into an unknot.

Left-handed **3** Right-handed
Has three crossings.

Simplest and most basic knot. Also called clover or overhand knot. These cannot be transformed into each other. Just like you cannot switch your left shoe for the right.

4 — Has four crossings.

This distinct knot with its four crossings can be transformed into a left and right handed version, just like a pair of socks.

5 — Has five crossings.

There are two distinct kinds of knots with five crossings which alternate over and under.

6

Three different colors may be all the colors needed to distinguish knots from each other. At each crossing either three different colors come together or all the same color comes together.

A circle is considered an unknot. It does not intersect itself anywhere. It is essentially one-dimensional. It has no thickness, and a cross section of it is a point.

An apparent knot can be transformed into an unknot. By the drawing convention of breaking the lines of the shadow of a knot and indicating a right or left direction, mathematicians can keep track of the spatial placement of a knot and its crossings. Knots are categorized according to the minimum number of times in which a given loop crosses over itself.

A projection is the caste shadow from the three-dimensional object on to the two-dimensional plane.

Thirteen, so far, is the highest number of crossings which have been catalogued.

By classifying knots according to the minimum number of times in which a given loop crosses itself in a plane image of the knot, as seen in this chart, mathematicians look for patterns to help them distinguish between knots. They must find some invariant, any property of a knot that does not change when you subject the knot to any allowable manipulation of the knot.

One knot cannot be transformed into another knot.

Notice: There are no two crossing non-trivial knots.

This is a ten crossing knot. Can you locate the crossings? (Remember alternate over and under) *Hint: use white out.*

This page:
Knotting requires conceptual as well as manual skills. Will Velcro ever replace knotting?
Top left: Metal Knot sign for bakery shop, CA.

Center: Actual knotted rope.

Lower right:
view of metal sculpture at the DeCordova Museum entitled "Knot" by Cosimo Cavallaro. Permission to use courtesy of Sarah Schussel and the DeCordova Museum, Lincoln, MA.

Knot Classifications

This page: Has five crossings. *Photograph of a group of twisted ropes.*

Sailors have used the information of knotting both practically and artfully giving us their knowledge knowledge in the structure of ornamental knotwork known as macrame. The word is Arabic in origin and refers to the trimming of fringe. The practice of knotting was carried to Spain and Italy as well as islands in the South Pacific. At first it was limited to coarse cord fringes and then later was associated with church altarcloths and vestments.

A single knot, like a single atom, can join with other knots to form clusters, like molecules, which form larger and larger systems. They grow by accretion like natural crystals and it is the sheer quantity that builds the artwork. One or more elements, such as cords in looped form, lock together. With essentially only two basic building knots, the square knot and the clove hitch, also called the double half-hitch, and a flexible material, one can create personal and powerful artworks. Various combinations of these two knots are called bars. Combination of bars create patterns. Another knot, the lark's head starts the process. The traditional sailor's material, which can be purchased at any hardware store, is white cotton seine twine which is smooth with a tight twist. It comes in a variety of weights and plies.

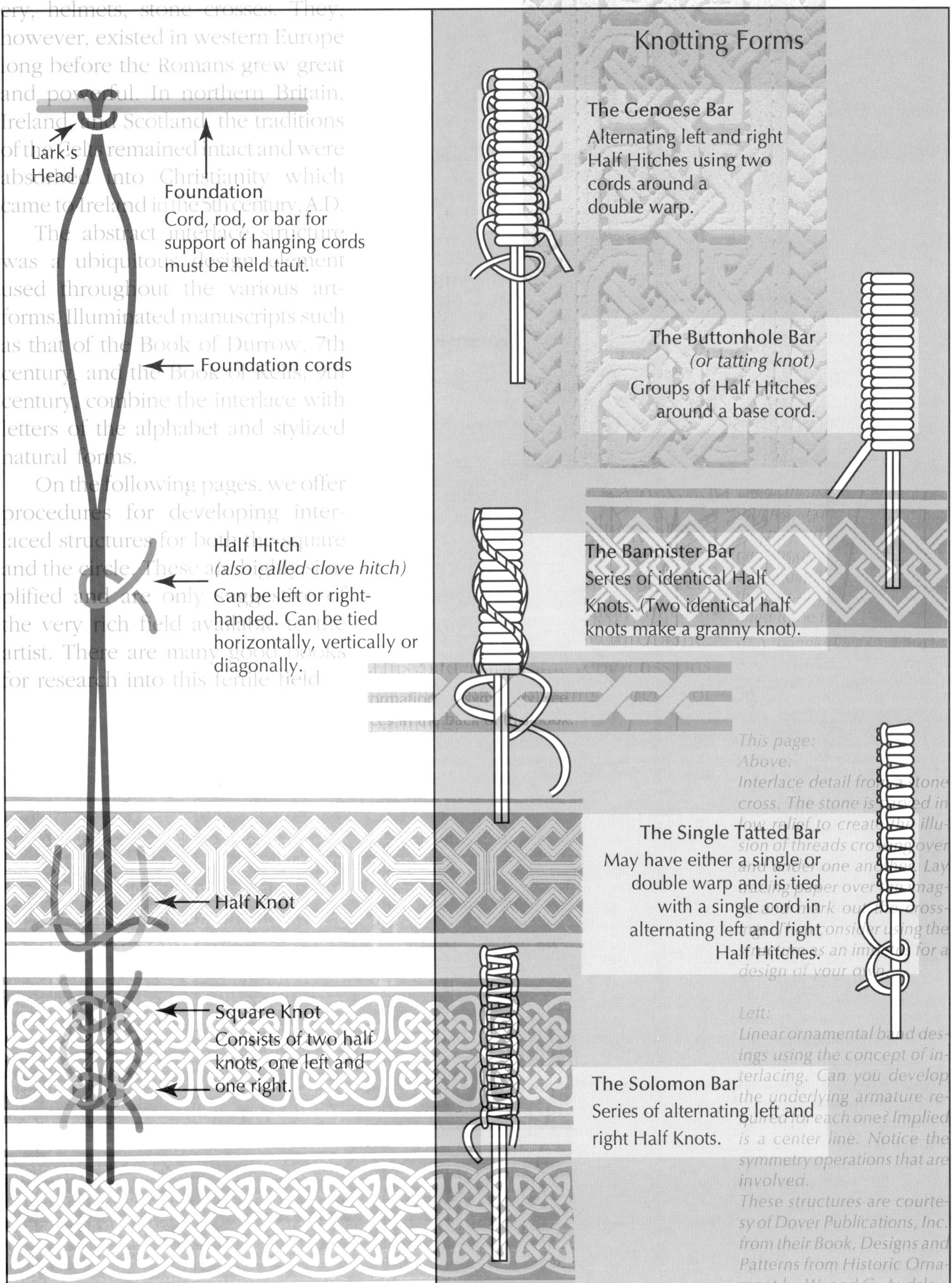

This page:
Above:
Interlace detail from a stone cross. The stone is carved in low relief to create the illusion of threads crossing over and under one another. Layers over layers — imagine working over the image to mark out the cross. Then consider using the design as an impetus for a design of your own.

Left:
Linear ornamental band designs using the concept of interlacing. Can you develop the underlying armature required for each one? Implied is a center line. Notice the symmetry operations that are involved.

These structures are courtesy of Dover Publications, Inc. from their Book, Designs and Patterns from Historic Ornament by W. and C. Audsley.

This page:
Funerary crosses found in cemeteries showing Celtic interlacing.

The three-dimensional fiber arts of plaiting, basketry, macrame, and weaving have given rise to two-dimensional surface embellishment. The interlacing work of the Celts found in stonework, metalwork, and illuminated manuscripts are superior examples of the fusion of geometry, mathematics, and the natural world. Author George Bain in his book, 1973, gives ample evidence of the methods and results of this genre of art.

The tribes that came under the heading of Celtic were described by the Greeks and Romans as barbarians who had no large scale architecture, painting, or sculpture. They did

246

produce portable objects of jewelery, helmets, stone crosses. They, however, existed in western Europe long before the Romans grew great and powerful. In northern Britain, Ireland, and Scotland, the traditions of the Celts remained intact and were absorbed into Christianity which came to Ireland in the 5th century, A.D.

The abstract interlace structure was a ubiquitous design element used throughout the various artforms. Illuminated manuscripts such as that of the Book of Durrow, 7th century, and the Book of Kells, 9th century, combine the interlace with letters of the alphabet and stylized natural forms.

On the following pages, we offer procedures for developing interlaced structures for both the square and the circle. These are highly simplified and are only suggestive of the very rich field available to the artist. There are many good books for research into this fertile field.

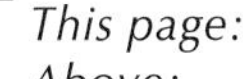

This page:
Above:
Interlace detail from a stone cross. The stone is carved in low relief to create the illusion of threads crossing over and under one another. Lay tracing paper over the images and mark out the crossings. Then consider using the structure as an impetus for a design of your own.

Left:
Linear ornamental band desings using the concept of interlacing. Can you develop the underlying armature required for each one? Implied is a center line. Notice the symmetry operations that are involved.
These structures are courtesy of Dover Publications, Inc. from their Book, Designs and Patterns from Historic Ornament by W. and G. Audsley.

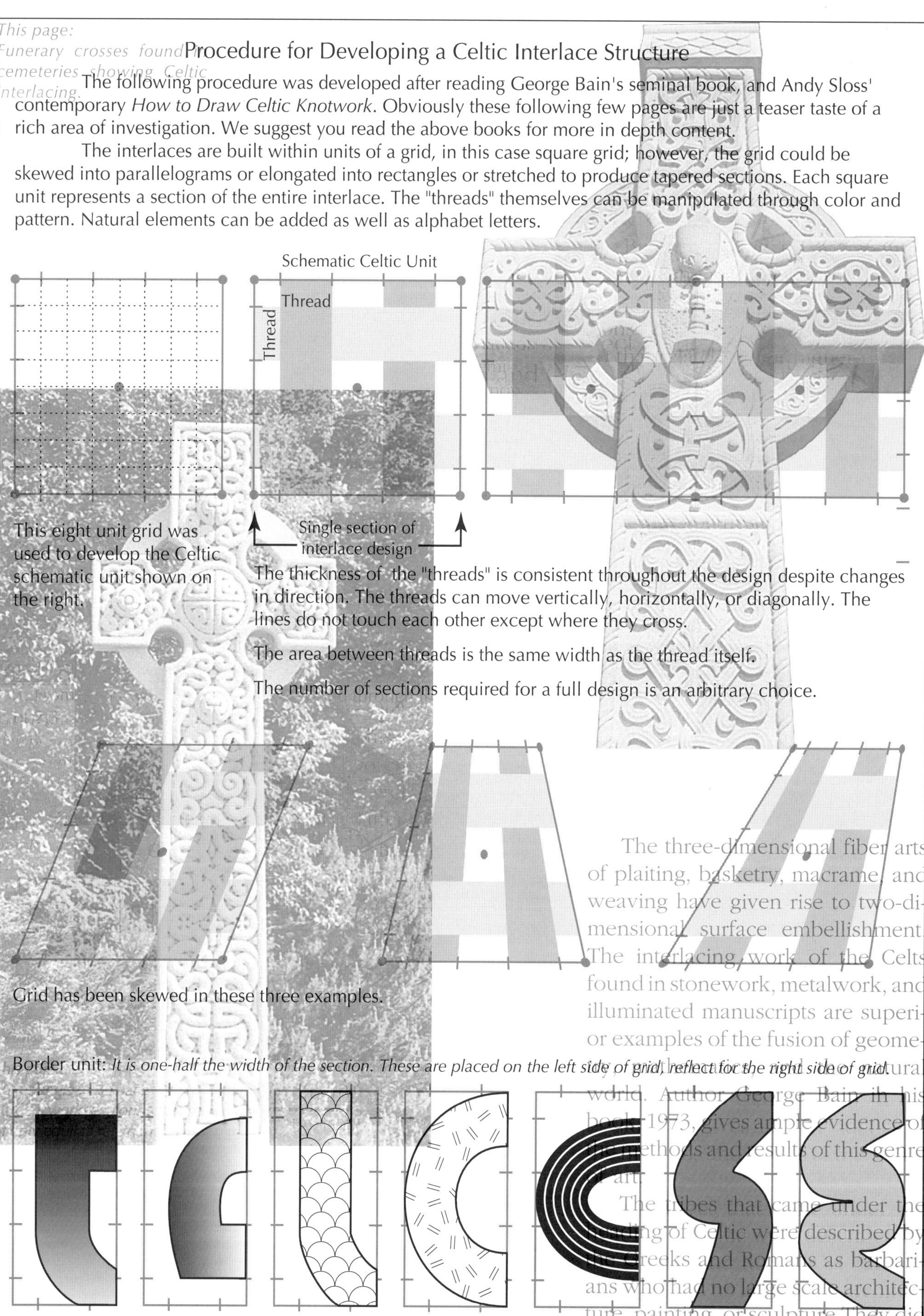

Procedure for Developing a Celtic Interlace Structure

The following procedure was developed after reading George Bain's seminal book, and Andy Sloss' contemporary *How to Draw Celtic Knotwork*. Obviously these following few pages are just a teaser taste of a rich area of investigation. We suggest you read the above books for more in depth content.

The interlaces are built within units of a grid, in this case square grid; however, the grid could be skewed into parallelograms or elongated into rectangles or stretched to produce tapered sections. Each square unit represents a section of the entire interlace. The "threads" themselves can be manipulated through color and pattern. Natural elements can be added as well as alphabet letters.

This eight unit grid was used to develop the Celtic schematic unit shown on the right.

The thickness of the "threads" is consistent throughout the design despite changes in direction. The threads can move vertically, horizontally, or diagonally. The lines do not touch each other except where they cross.

The area between threads is the same width as the thread itself.

The number of sections required for a full design is an arbitrary choice.

Grid has been skewed in these three examples.

Border unit: *It is one-half the width of the section. These are placed on the left side of grid, reflect for the right side of grid.*

The three-dimensional fiber arts of plaiting, basketry, macrame, and weaving have given rise to two-dimensional surface embellishment. The interlacing work of the Celts found in stonework, metalwork, and illuminated manuscripts are superior examples of the fusion of geometry with images from the natural world. Author George Bain, in his book, 1973, gives ample evidence of the methods and results of this genre of art.

The tribes that came under the heading of Celtic were described by the Greeks and Romans as barbarians who had no large scale architecture, painting, or sculpture. They did

Celtic Line Group

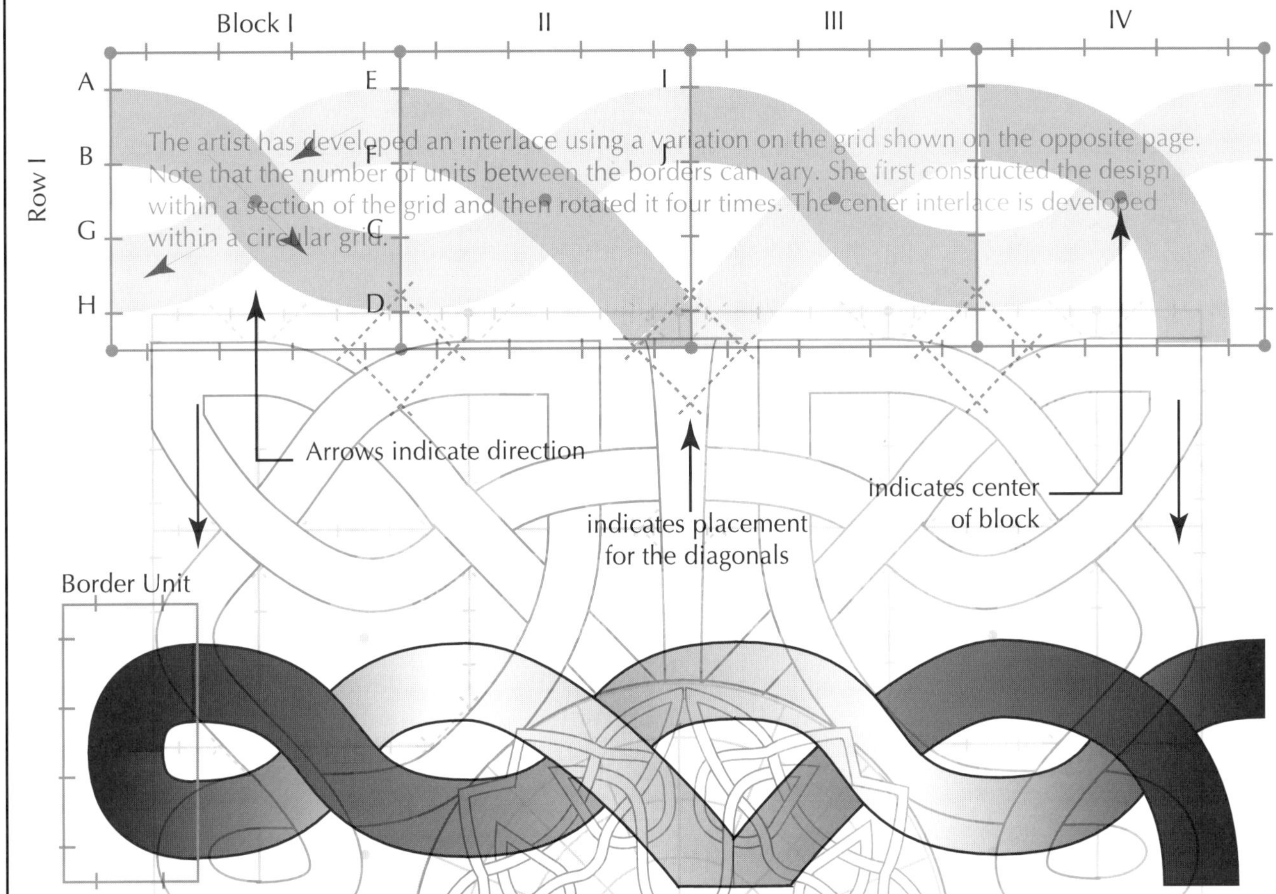

Block I II III IV

Row I

A E I

B F J

G C

H D

The artist has developed an interlace using a variation on the grid shown on the opposite page. Note that the number of units between the borders can vary. She first constructed the design within a section of the grid and then rotated it four times. The center interlace is developed within a circular grid.

Arrows indicate direction

indicates placement for the diagonals

indicates center of block

Border Unit

1. Decide upon the number of sections required for your design. Add border units to both sides.

2. In Row 1, Block 1, start at the upper left side of the unit at points A and B. Choose one of the three directions in which to move (horizontal, vertical or diagonal). In this block we chose horizontal. All moves from left to right become the over-thread. It has been given a darker tone to distinguish it from the next move. Draw the thread so that A goes to C and B goes to D. Now the thread has a thickness.

3. Start at the upper right hand side of the unit at points E and F, move in a horizontal direction again and draw the thread so that E goes to G and F goes to H. This thread becomes the under-thread. The direction of the under-over movement stays constant throughout the design.

4. On Block II, again determine which of the three directions the thread will take. Remember the thread has to be continuous or turn and move back in the direction it came from. Notice what happens in blocks III and IV.

These are nine possible <u>left-to-right</u> variations for placement of interlace threads. Reflect for right-to-left direction.

Celtic Interlace Grid

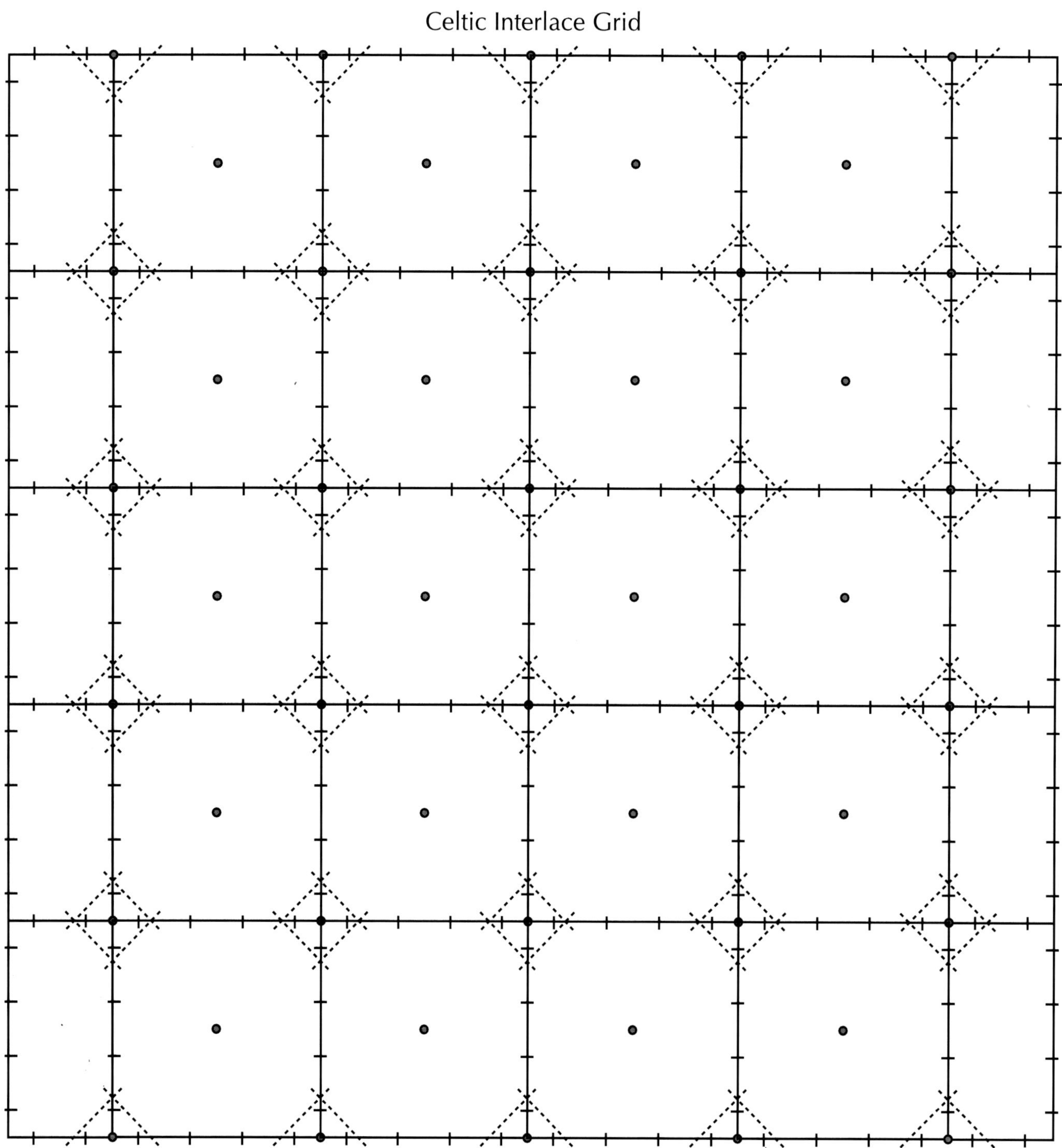

This is a grid for your use. Photocopy it or use a tracing paper overlay and develop an interlace design. Consider which "rules" you can alter and still maintain the interlace.

Hint: Develop the border first in order to establish the direction of the interlace.

The artist has developed an interlace using a variation on the grid shown on the opposite page. Note that the number of units between the borders can vary. She first constructed the design within a section of the grid and then rotated it four times. The center interlace is developed within a circular grid.

This is a grid for your use. Photocopy it or use a tracing paper overlay and develop an interlace design. Consider which "rules" you can alter and still maintain the interlace.
Hint: Develop the border first in order to establish the direction of the interlace.

Notice in this version, the underlying grid is gone. The artist has used a blend tool in the outer interlacing and has developed another structure within the central circle. This will be explained further on.

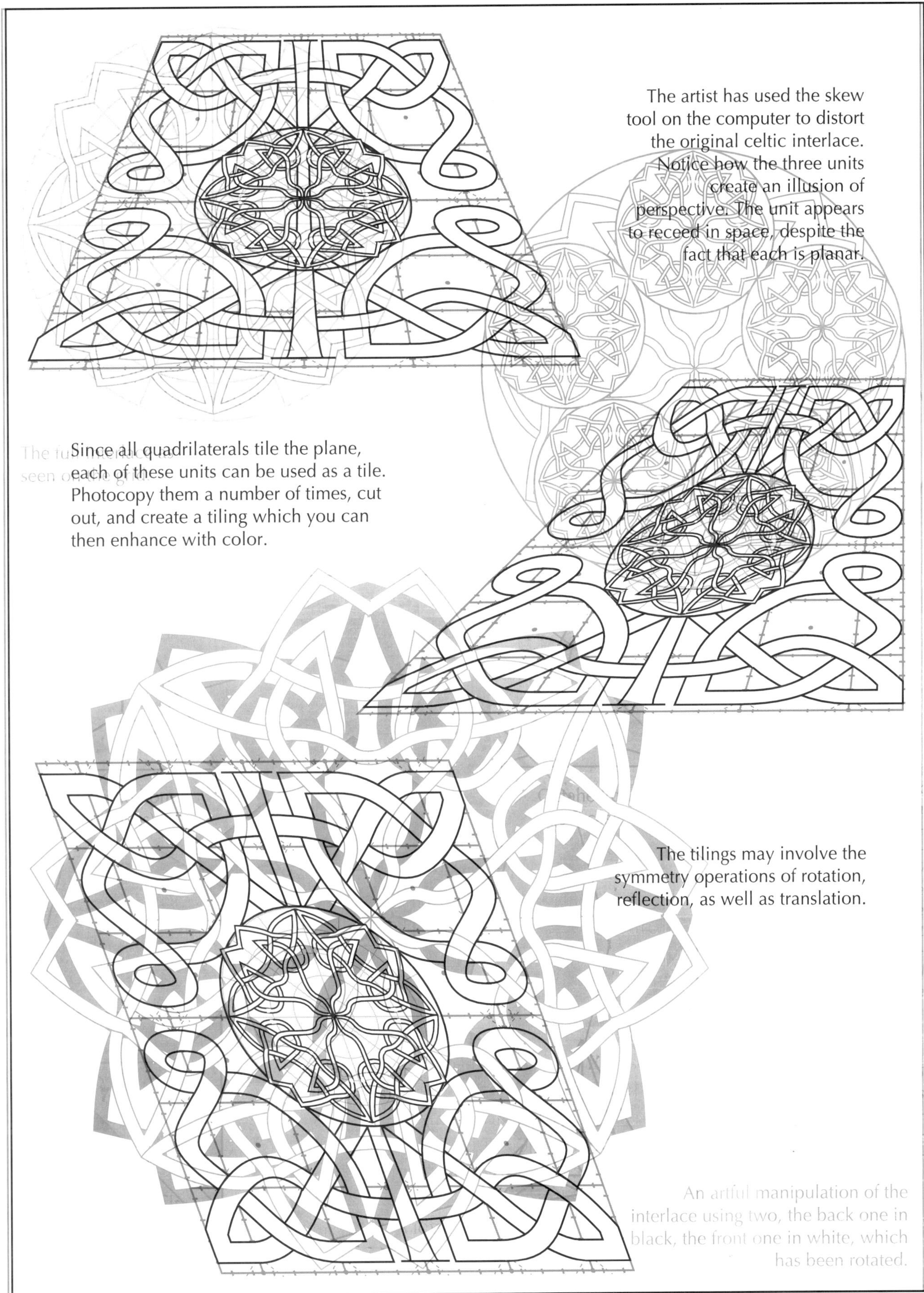

The artist has used the skew tool on the computer to distort the original celtic interlace. Notice how the three units create an illusion of perspective. The unit appears to receed in space, despite the fact that each is planar.

Since all quadrilaterals tile the plane, each of these units can be used as a tile. Photocopy them a number of times, cut out, and create a tiling which you can then enhance with color.

The tilings may involve the symmetry operations of rotation, reflection, as well as translation.

An artful manipulation of the interlace using two, the back one in black, the front one in white, which has been rotated.

Procedure For Constructing a Celtic Interlace Within a Circle

Given circle O, with the center marked

1. On line segment OA mark off five units of constant length (this is an arbitrary choice). See chapter 3. Using each as a radius from center O construct concentric circles.

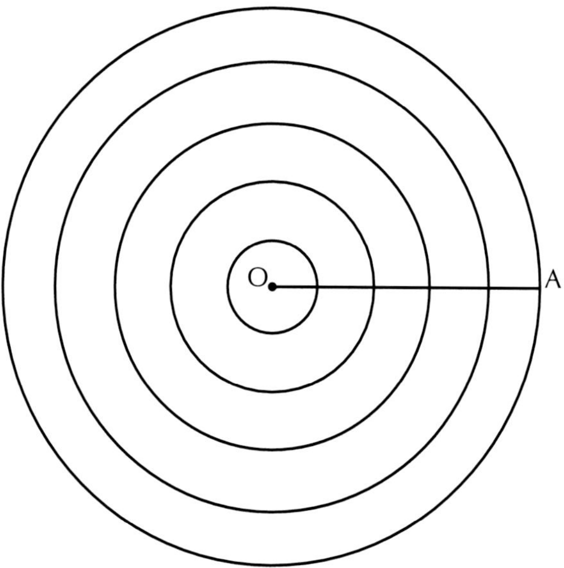

2. Divide circle into eight congruent sectors. See chapter 3. Each sector is then bisected for a total of sixteen sectors, however in order to fit the interlacing within a sector some lines need to be removed. The two center concentric circles remain in eighths.

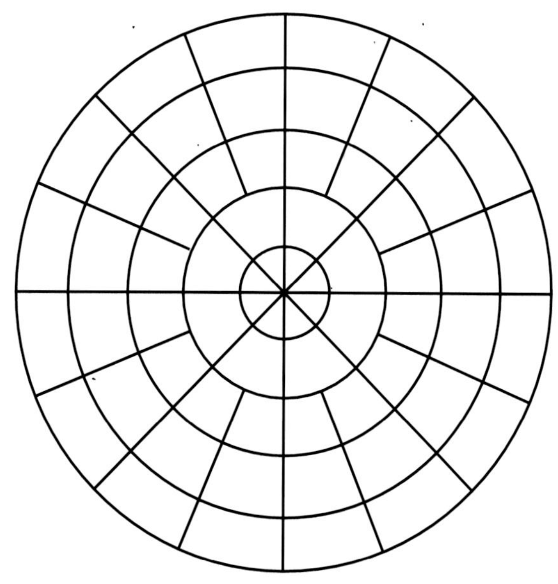

3. The interlace was developed within an eighth of the circle following the "rules" of under and over.

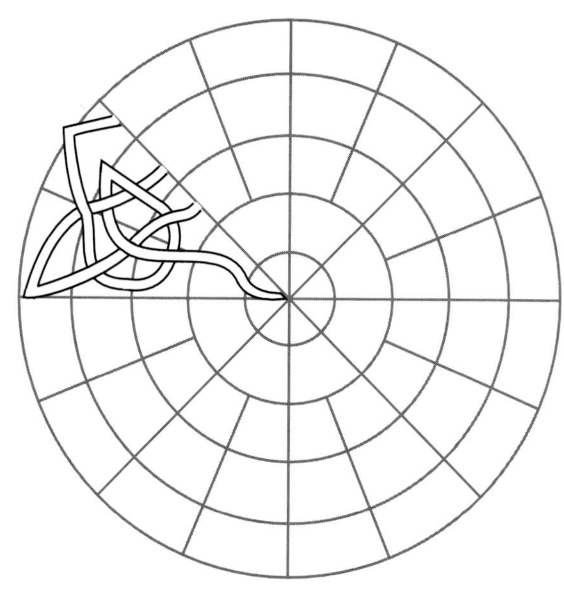

4. The motif was reflected, then the unit was rotated around the circle.

The full interlace as
seen on the grid.

An artful manipulation of the
interlace using two, the back one in
black, the front one in white, which
has been rotated.

Procedure For Constructing Two Circles Within a Single Circle

Given circle with diameter AB and midpoint O.

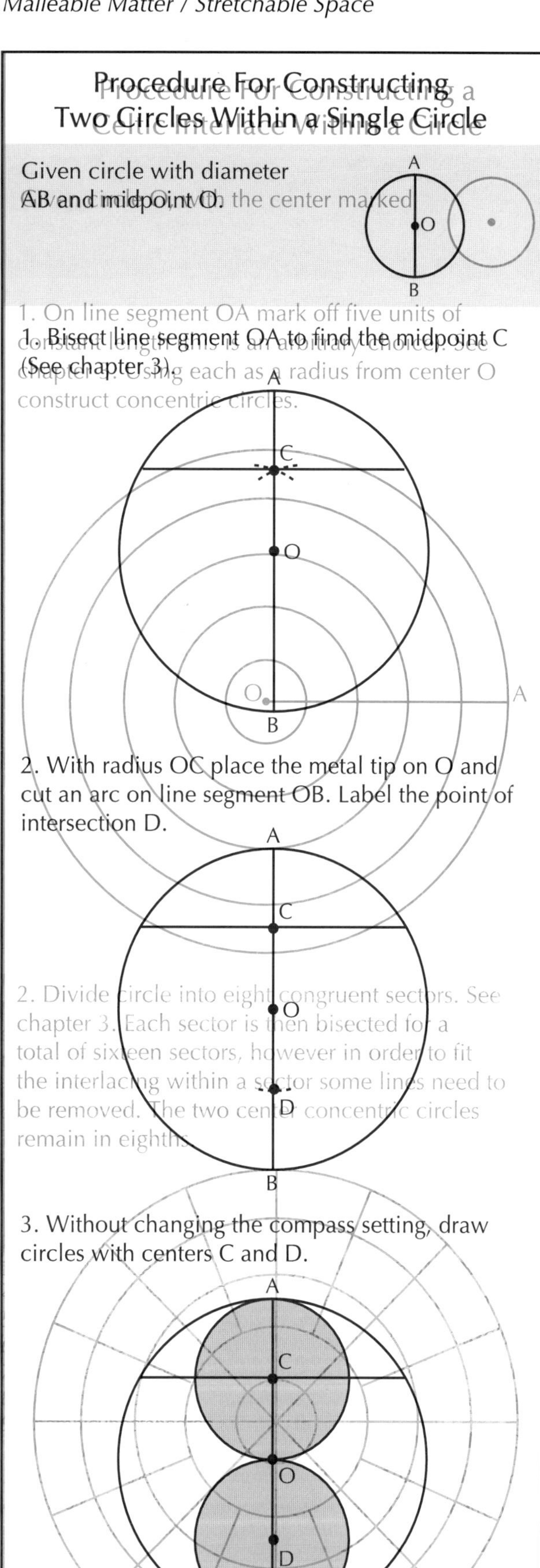

1. Bisect line segment OA to find the midpoint C (See chapter 3).

2. With radius OC place the metal tip on O and cut an arc on line segment OB. Label the point of intersection D.

3. Without changing the compass setting, draw circles with centers C and D.

Now circle O contains two circles within it.

Procedure For Constructing Three Circles Within a Single Circle

Given circle with diameter AB that extends beyond circle and midpoint O.

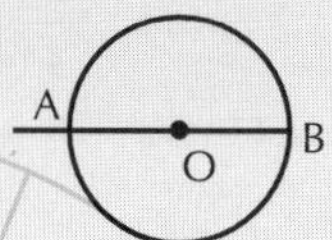

1. With the compass open to the radius of OA, place the metal tip on A and mark point C on the circle. Repeat at B to obtain point D.

2. Draw rays CO and DO and extend them to intersect the opposite side of the circle.

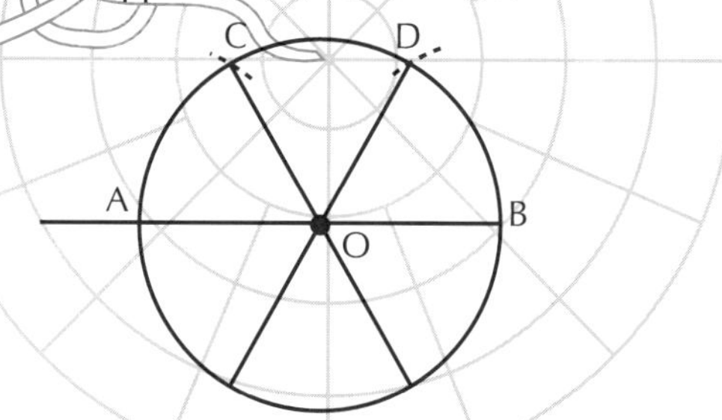

3. Construct a perpendicular to line AB at A, draw ray OC, and extend both until they intersect at E.

4. Bisect angle AEC and extend the bisector to intersect line segment AB at a point, F.

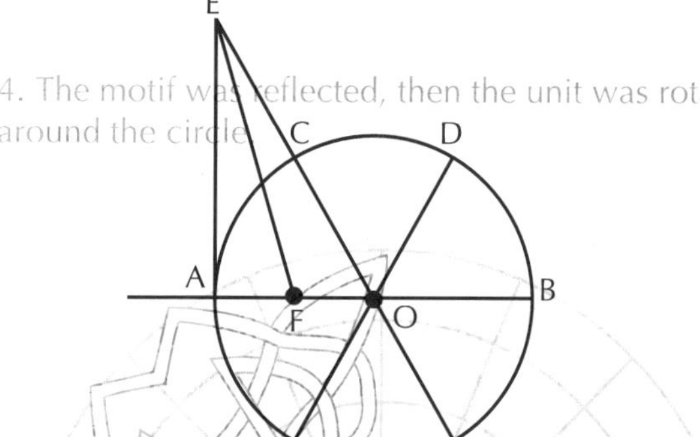

5. Open the compass to measure OF. Place the metal tip on O and cut points G and H as shown. F, G and H will be the centers of the three circles required.

6. Using FA as the measure of the radius, draw circles F, G and H.

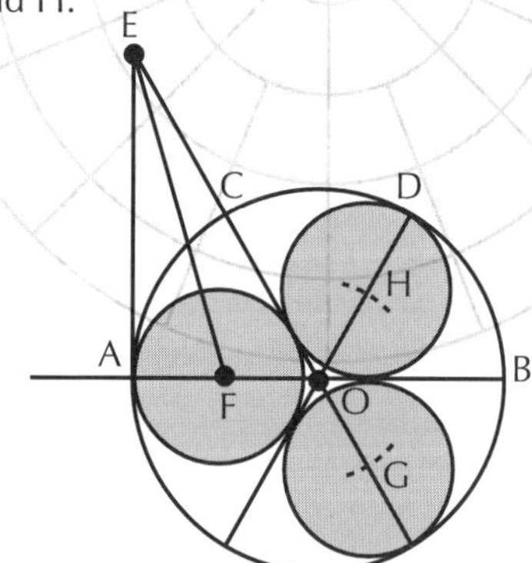

Now circle O contains three circles within it.

Procedure For Constructing *n* Circles Within a Single Circle

Given a Polygon with *n* sides and a circle inscribed within it.
The number of circles desired will determine the choice of polygon.

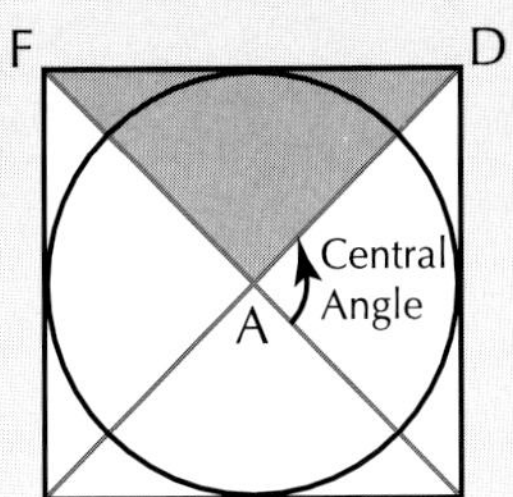

Square with shaded sector. Central Angle:

$$\frac{360°}{4}=90°$$

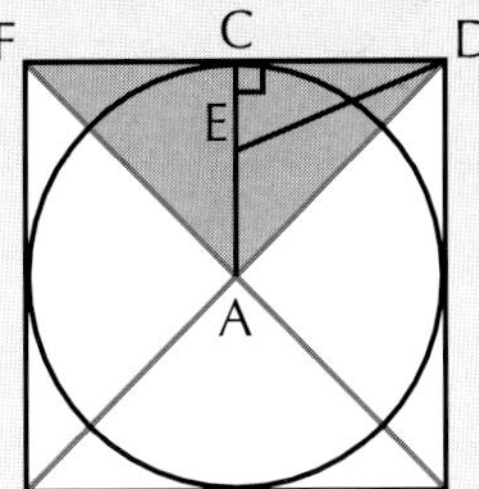

The central angle FAD is bisected. CA is perpendicular to line segment FD. Angle CDA has been bisected.

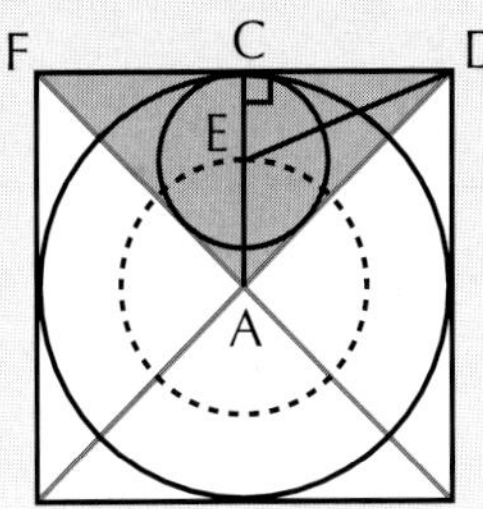

EC is the radius of the small circle used within the bigger circle.

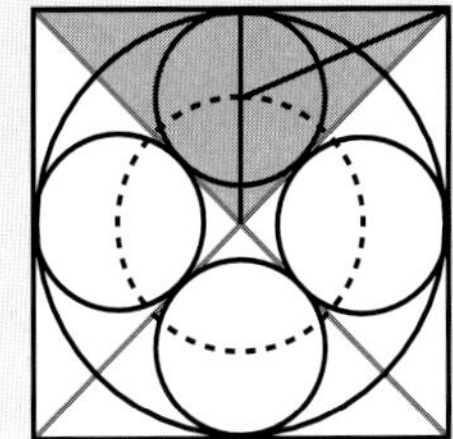

Four copies of the circle fit within a single circle.

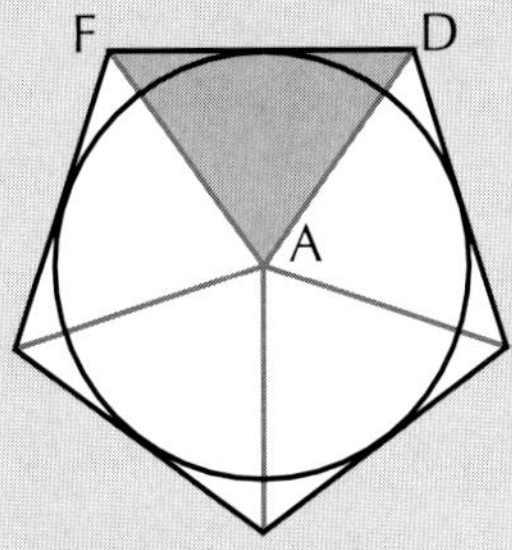

Pentagon: $\frac{360°}{5}=72°$

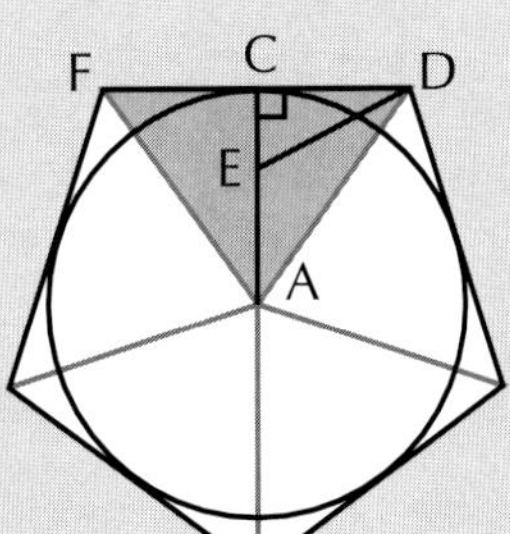

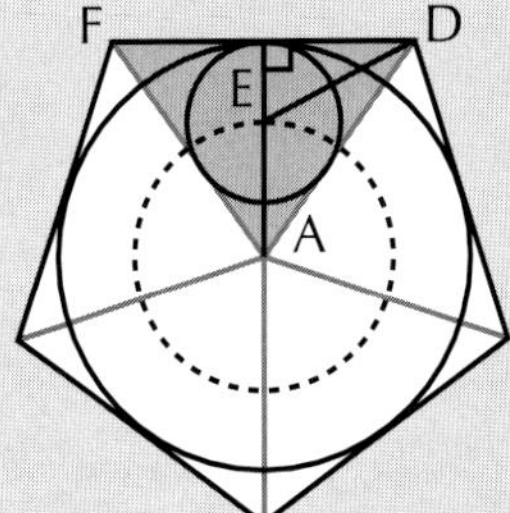

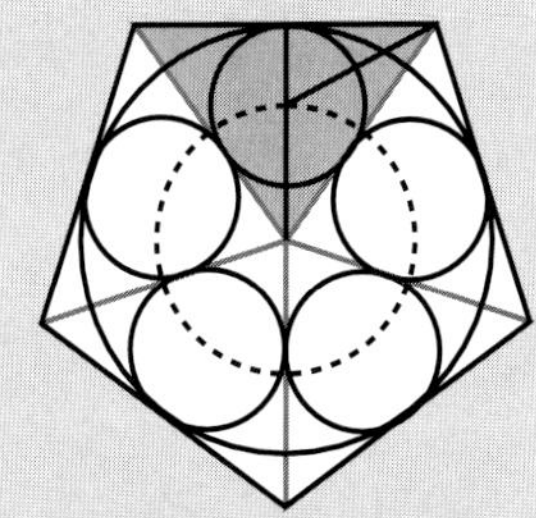

Five circles within a single circle.

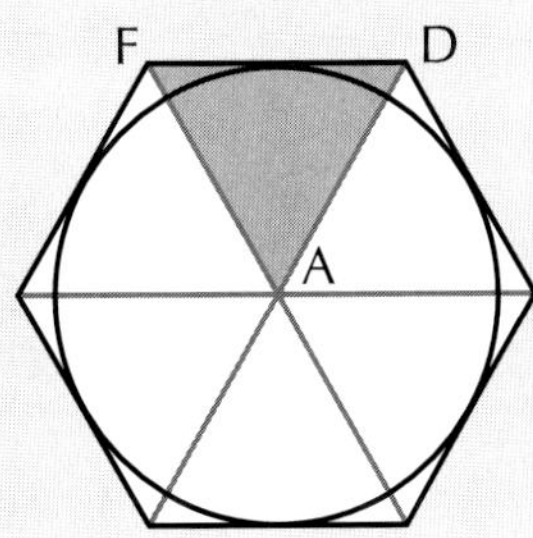

Hexagon: $\frac{360°}{6}=60°$

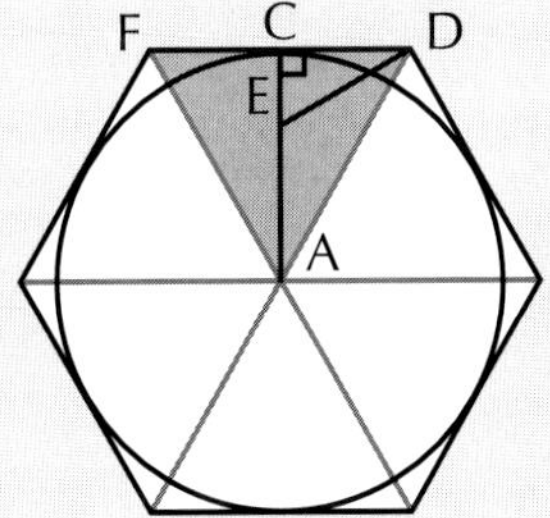

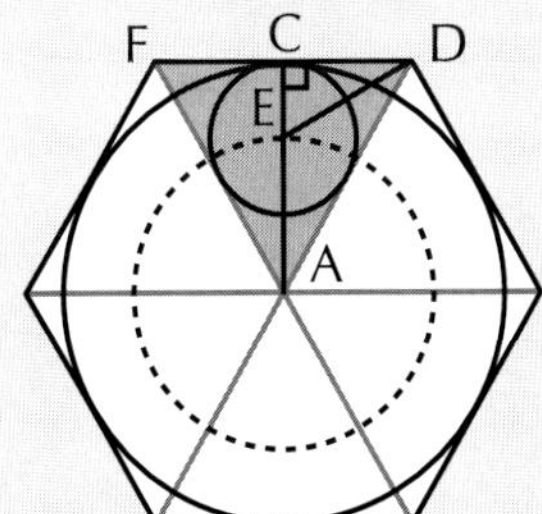

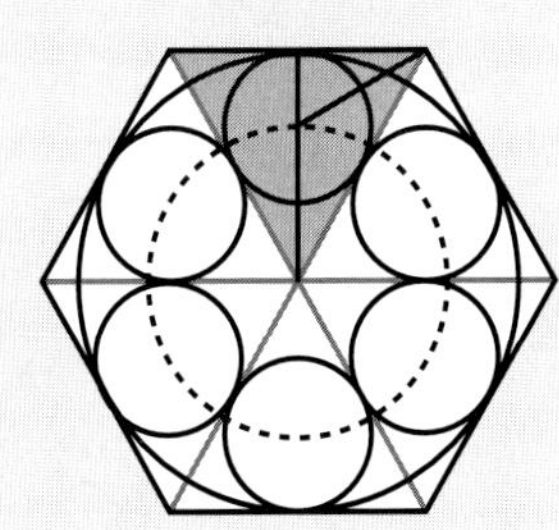

Six circles within a single circle.

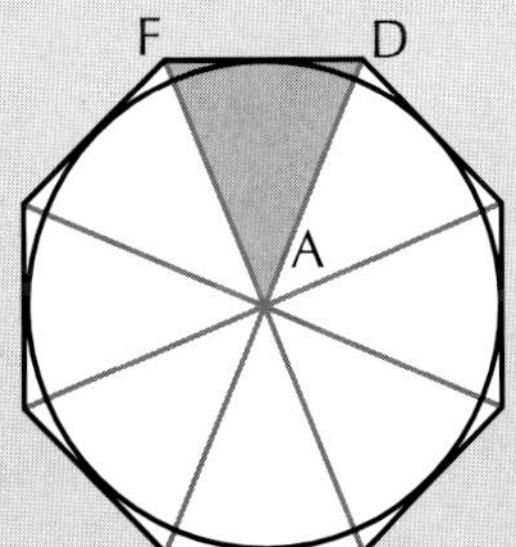

Octagon: $\frac{360°}{8}=45°$

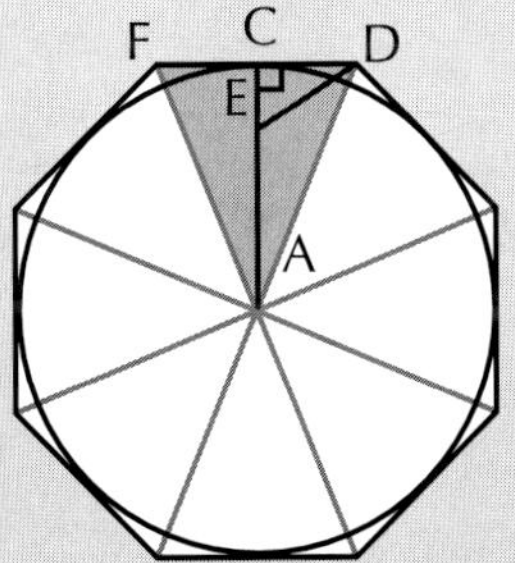

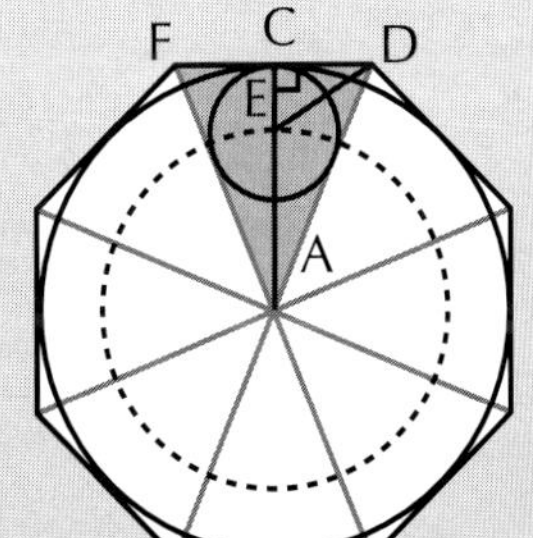

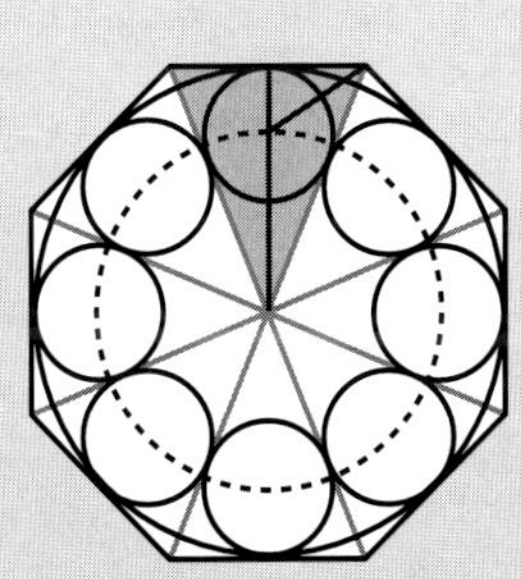

Eight circles within a single circle.

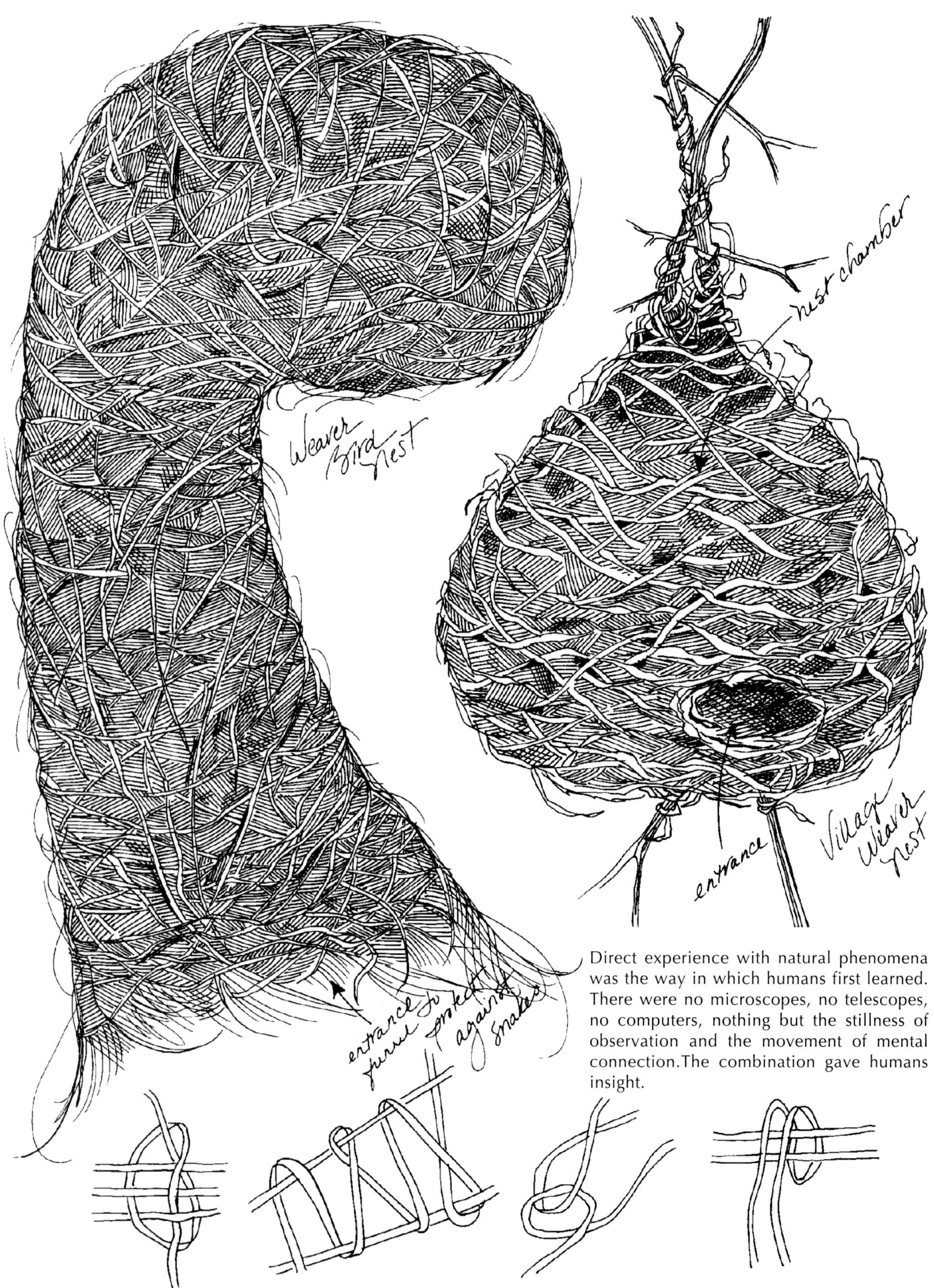

Direct experience with natural phenomena was the way in which humans first learned. There were no microscopes, no telescopes, no computers, nothing but the stillness of observation and the movement of mental connection. The combination gave humans insight.

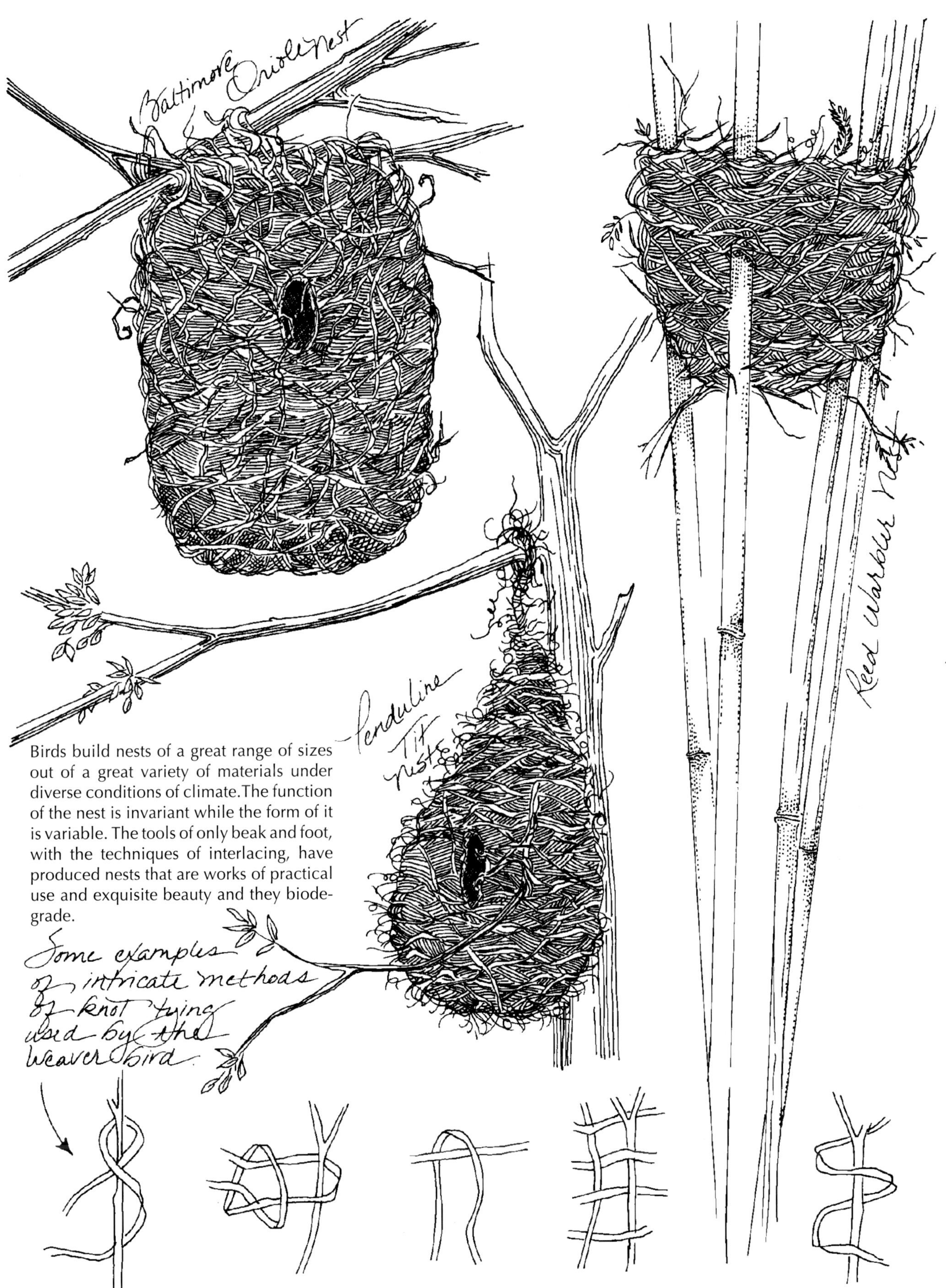

Birds build nests of a great range of sizes out of a great variety of materials under diverse conditions of climate. The function of the nest is invariant while the form of it is variable. The tools of only beak and foot, with the techniques of interlacing, have produced nests that are works of practical use and exquisite beauty and they biodegrade.

The Labyrinth

Once upon a time there was a Greek hero, named Theseus, who was the son of Aegeus, the king of Athens. On one of his many adventures, he was pledged to slay a mighty beast. Every nine years, King Aegeus was forced to send seven young men and seven maidens as sacrifices to King Minos of Crete. Minos was forced to send these hapless creatures down to the bowels of the Labyrinth, built under the palace by the architect, engineer, and inventor Daedalus. There, the lost victims would wander until found and eaten by the Minotaur whose body was that of a man while his head was that of a bull. He was the offspring of Pasiphae, the wife of Minos, who had secretly mated with a White Bull.

Theseus, as heroes do, volunteered to go amongst the victims so that he might slay the beast and put and end to the sacrifices. When he arrived at the island of Crete, he was noticed by the princess/priestess Ariadne, daughter of King Minos. She fell in love with him, and not wanting him to die, but to marry, she gave him a magic sword and a ball of golden twine, symbol of the sphere of being, which he could unravel as he was led into the heart of the Labyrinth. This twine path would then guide him safely out after he had killed the Minotuar with the magic sword.

He successfully completed his mission. He then carried Ariadne off to Naxos, another island, where he left her....but that is another story for another time....King Minos in his anger about this abandonment of his daughter imprisons both Daedalus and his son, Icarus, in the Labyrinth.

This is a very old story and a very old structure. Labyrinths are found in church naves and country gardens. They are found large and they are found small. They are found then and they are found now. The winding path of the labyrinth is symbolic of the life process unfolding in space and time. This labyrinth form is also connected to mazes, meanders, spirals, knots, and the mathematical subject of network theory, a subcategory of topology.

Celtic culture, in pre-christian times, made much use of the labyrinth. Some scholars believe that the form may be as old as 25,000 BC.

In the dictionary, the word maze and labyrinth are used interchangeably. According to the American Heritage Dictionary, a maze is "a graphic puzzle, the solution of which is an uninterrupted path through an intricate pattern of line segments from a starting point to a goal." For some scholars there is a difference between these two forms. A maze is considered a puzzle with points at which choices must be made as to which direction one should move in. It contains twists and turns and angles. A labyrinth, on the other hand, implies a single path, unicursal, with ritual connections. These have no junctions but consist of a single path which leads from the entrance to the goal. There are three types which are called Classical, Roman, and Medieval Christian. All three types have internal rotational symmetry.

Both maze and labyrinth are archetypal symbols of a quest. Both are human constructs suggestive of natural forms such as the intestines and the human brain. Both maze and labyrinth are intended for movement through a journey from a beginning point to an endpoint. Inward in an anti-clockwise direction leads to darkness and death, as with the tale of the Minotaur. The outward clockwise path symbolizes the journey into the light, life, and spiritual enlightenment.

The labyrinth structure involves a combination of two opposing spirals (one centripetal and the other centrifugal). It is also symbolically related to a knot which both binds by hindering and brings together something into unit. Knots are related to power. Strands of DNA are knotted and the double helix is itself a knot. It has the power of life and death.

The Spiral and the Labyrinth are intimately connected. The former is a symbol that suggests the latter.

261

Procedure for Constructing an Archimedean Spiral

(An Archimedean spiral is one in which the spiral lengthens at a constant rate per constant turn).

Given a circle with a set of concentric circles using a constant ratio.

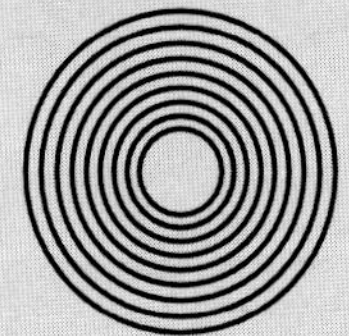

1. Determine the number of degrees required, or desired, that the spiral will move outward from the center. Then determine the number of units between turns. In this example there will be one unit for each turn of 60°.

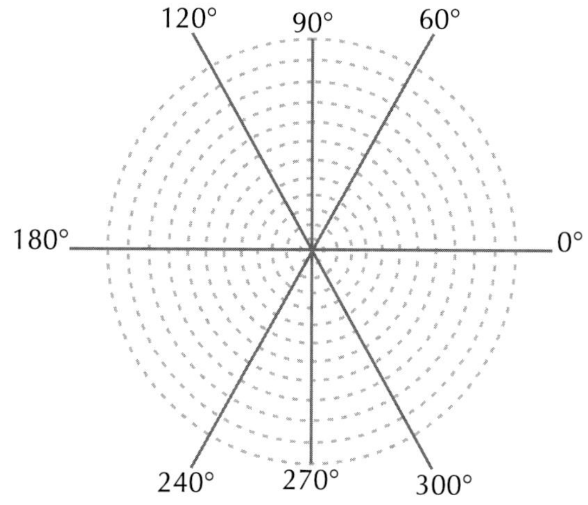

2. Locate point B at the intersection of the first circle and the radius vector corresponding to 60°.

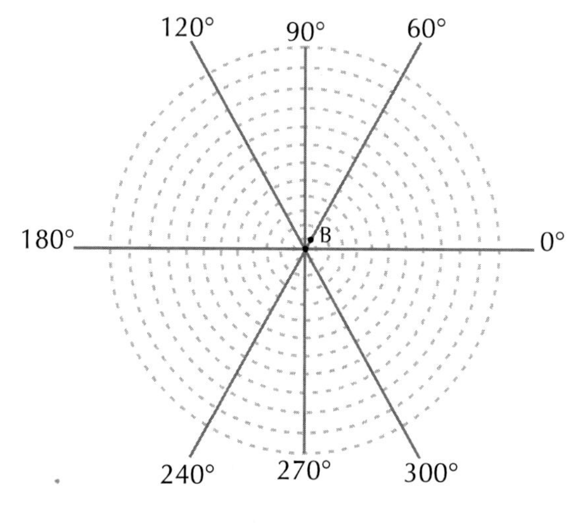

3. Locate point C at the intersection of the next circle and the radius vector corresponding to 120° (60°+60°).

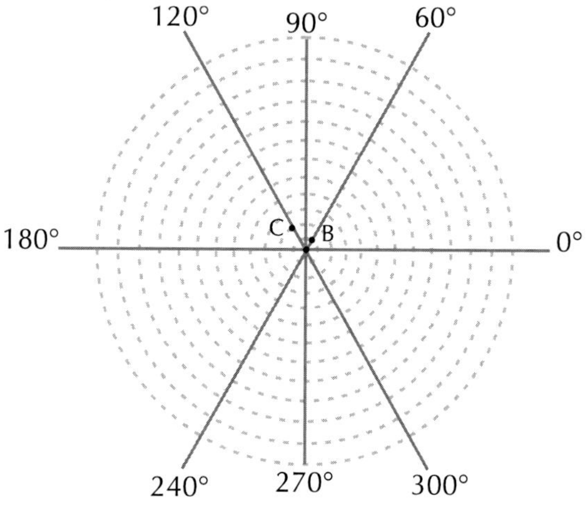

4. Continue this process, each time moving to the next larger circle for each subsequent turn of 60° to obtain the points D,E,F,G,H,I,J,K,L, and M.

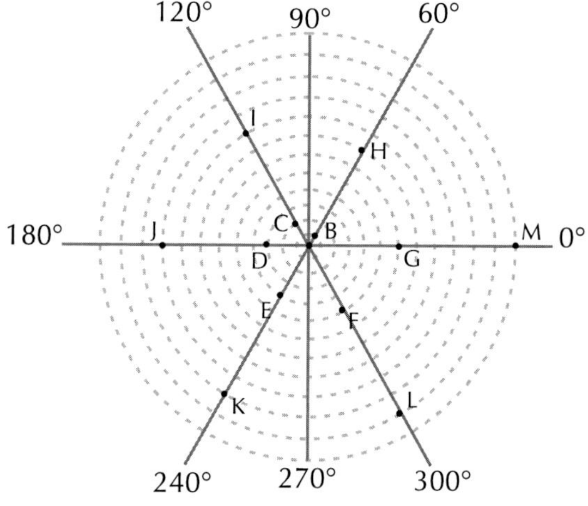

5. Join the consecutive points with a smooth curve. (If working by hand a French curve is a helpful tool here).

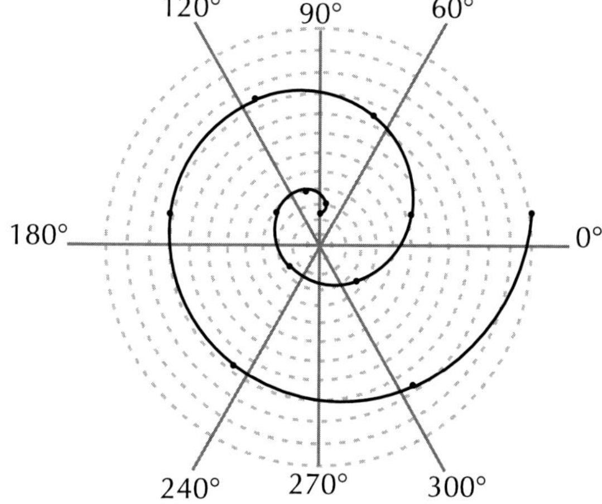

Now the curve is an Archimedean spiral.

Spirals that do not increase by a constant rate but get larger with each turn as with the growth of a nautilus shell. The spiral maintains a constant angle of intersection with the radius vectors.

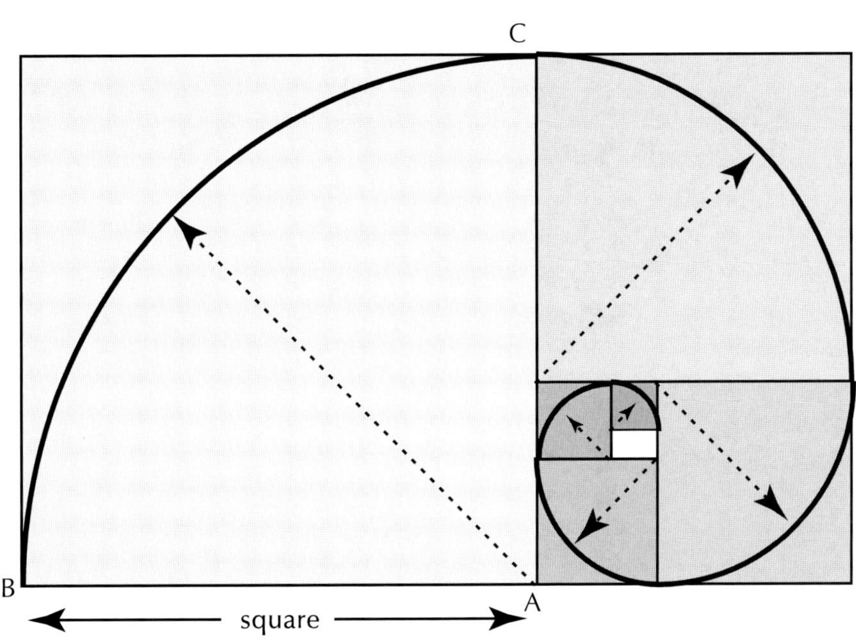

The Golden Spiral within the Golden Rectangle
The spiral is generated by placing the metal tip at vertex A of the
square and swinging an arc from B to C.
The process is continued within each of the subsequent smaller squares.

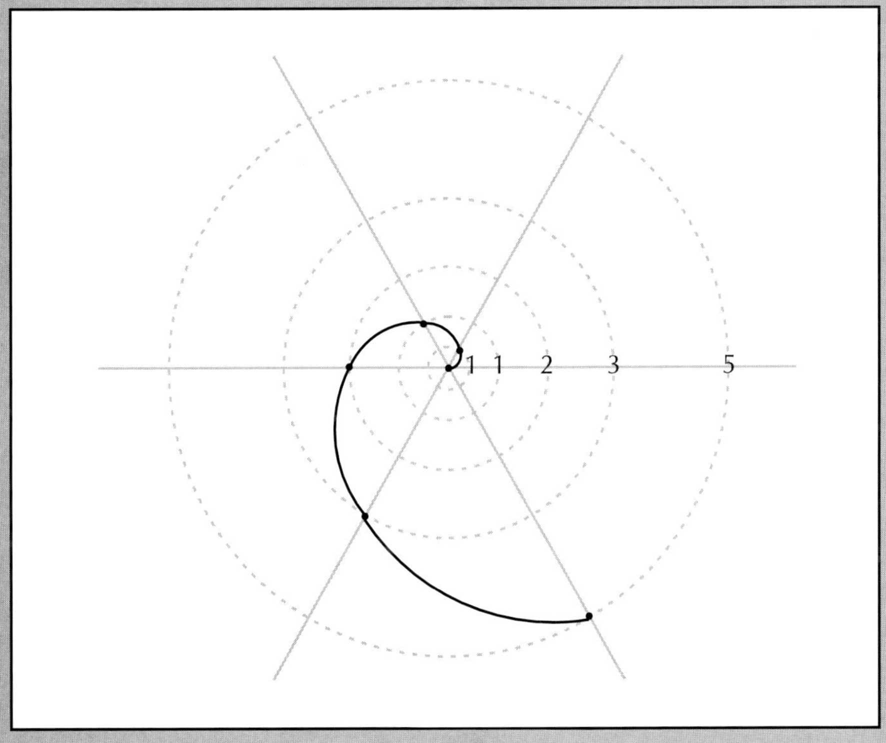

The spiral is generated by using numbers from the Fibonacci sequence (1,1,2,3,5...). The concentric circles increase by these numbers.
In this example for each turn of 60° a point is marked on the circles.

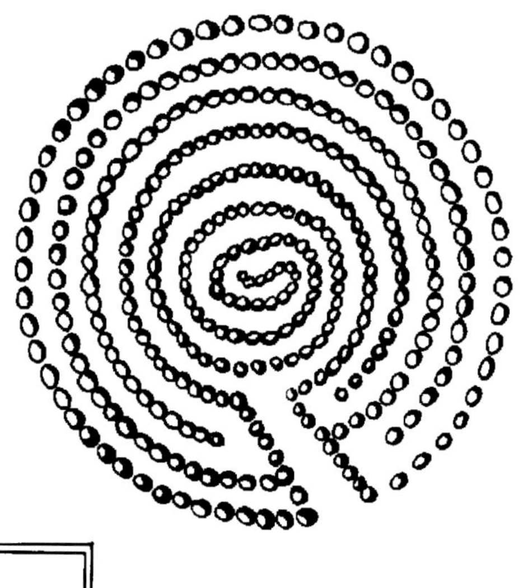

While the maze has dead ends and false passages, the labyrinth has a single path to the center. There is a basic form called the Classic seven-ring labyrinth. It has a single path which never crosses itself, one entrance, and one exit. It requires no measuring equipment and can be constructed with pegs and strings. It is extremely satisfying to construct. It has been around for over one thousand years and, although abstract in form, it is suggestive of natural structures. It is also associated with the knot and the concept of entanglement and entrapment.

This labyrinth exhibits internal rotational symmetry and can be cut along the axis from the entrance to its goal in order to fold out into a structure that is a rectangle. The highest horizontal line represents the first outer path ring, and the lowest line represents the seventh inner path ring.

The earliest surviving full size example of a medieval Christian labyrinth design which is built on this basic form is found in the pavement of the nave of Chartres Cathedral in France. It is circular and named "Chemin de Jerusalem" (the road to Jerusalem). It represents the journey to salvation and the path of life. It relates to the mandala form that psychologist Carl Jung discusses in relation to the unity of the psyche of the self. It also relates to the Mother Earth symbol of Amerindian cultures both in a curved and a straightline version.

The Chartres labyrinth is laid out in blue and white marble stone. It is about 40 feet in diameter and has a relationship with the great west rose window above the west door of the cathedral. They both have the same shape, the same size, and the labyrinth is exactly as far from the west door as the rose window is above it. If the window were brought down over, it would exactly cover the labyrinth. In the center of it, there is rosette of six petals, an affirmation of the form of the rose window above it. The pilgrims to the cathedral would symbolically reenact the journey to Jerusalem by crawling the labyrinth on their hands and knees. The 32 turnings within the labyrinth are believed to bring balance to the two halves of the brain leading to physical and emotional health.

We offer you a procedure for drawing one for yourself. This could very well be used as a plan for building a very large labyrinth using stones, or hedges, or bricks, etc. as well as being the basis for drawings. There are eight turf labyrinths that still exist in England today.

The Labyrinth of Chartres

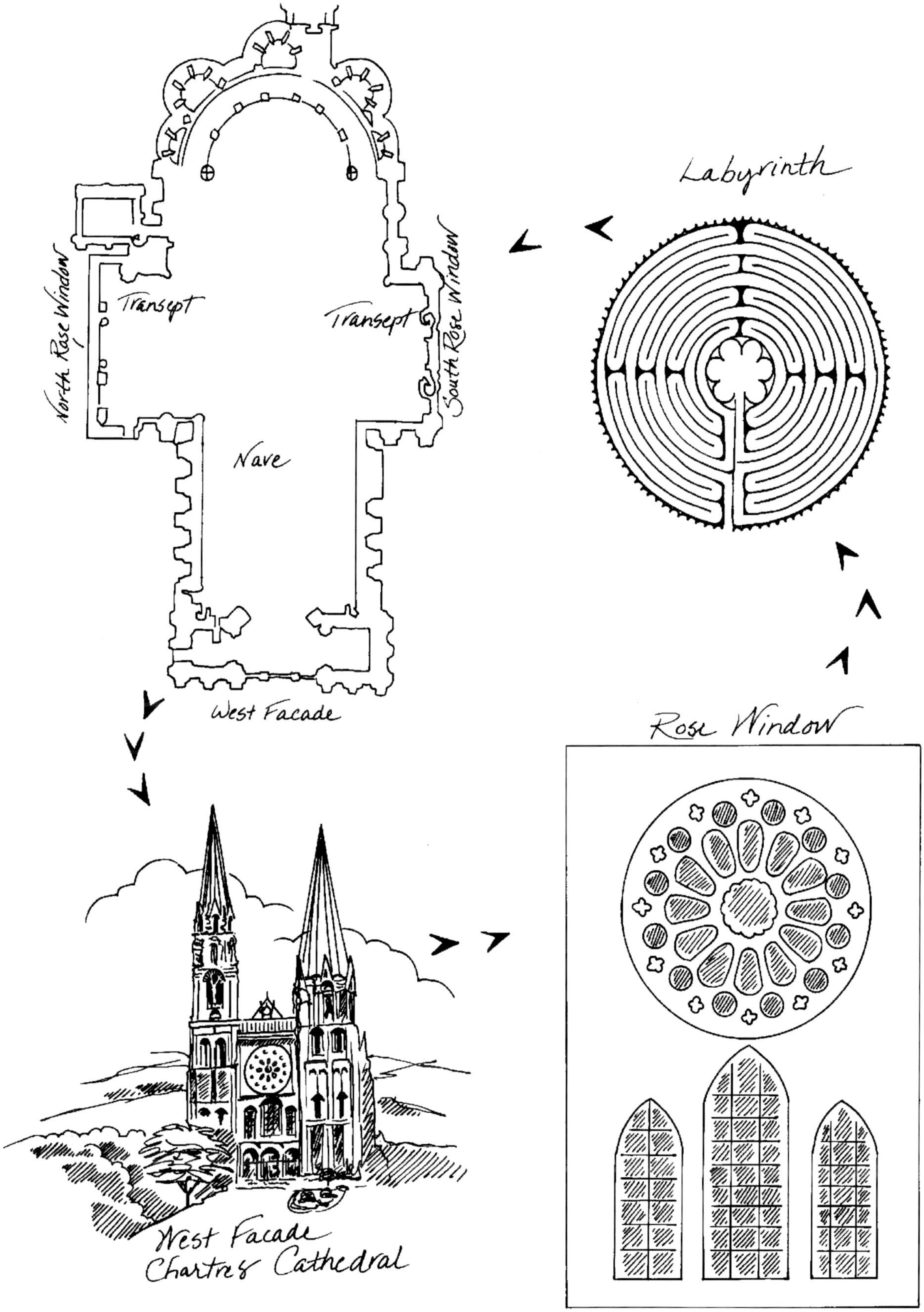
Labyrinth
North Rose Window
Transept
Transept
South Rose Window
Nave
West Facade
Rose Window
West Facade
Chartres Cathedral

This page:
The Labyrinth of Chartres. Built circa 1220 AD. The influence for this particular labyrinth may have come from Roman mosaics. By walking the labyrinth, The Path of Jerusalem, each individual can quiet the mind and slough off the petty details of life; find illumination in the act of meditation at the center; attain unity with the forces of the world by retracing the steps back. The distance from the perimeter of the labyrinth and the center is approximately 286 yds. The labyrinth is a unversal geometric figure and can be found in other places using a wide variety of materials. A single path starts at the outer edge and finishes in the center. The path never crosses itself despite the fact that it makes 32 turns.

Procedure For Constructing an Archetypal Seven Ring Labyrinth

Given line segments AB, CD set perpendicular to each other and with center marked with a dot, O.

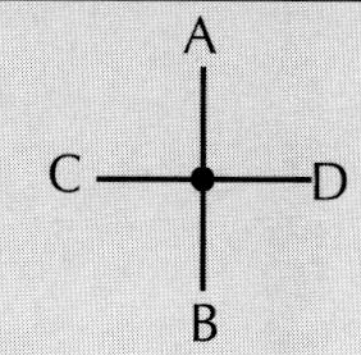

The seven ring design is found in Classical, Roman, Swedish, Finnish, and Medieval Christian cultures. It can be built with no measuring equipment using pegs and strings. It contains a single path, called the unicursal, with one entrance and one exit. It exhibits internal rotational symmetry. It is also known as the Cretan Labyrinth, which is approximately 3,000 years old. The function of a labyrinth is essentially ceremonially.

1. Working from center out on OA, OB, OD, OC, respectively mark eight equidistant points and give them numbers. The number of units depends upon the total number of paths required for the finished labyrinth. Here we use a seven path course.

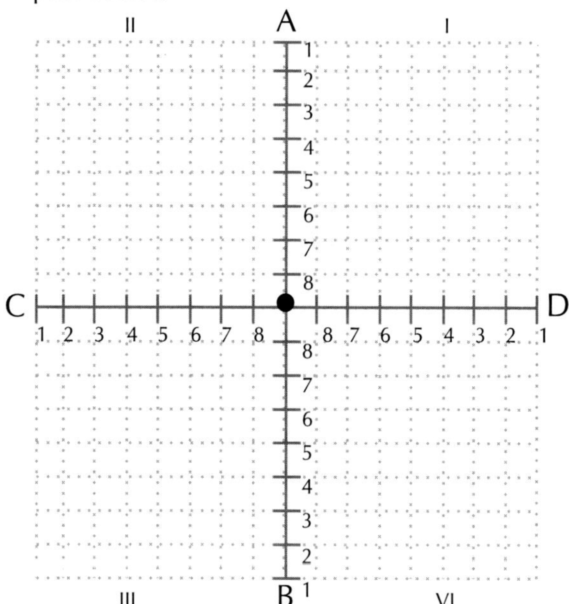

2. There are two boundary lines that move in opposite directions, clockwise and counterclockwise. The areas between the boundary lines is the path for the journey through the labyrinth. The first boundary line will be marked in black, and the second will be marked in gray. The first starts in Quadrant II beginning at unit seven. Notice that the line sometimes parallels itself, but in other places it's separated by the second line. Though we are developing this labyrinth on paper, it can be used as the basis for a physical three-dimensional structure.

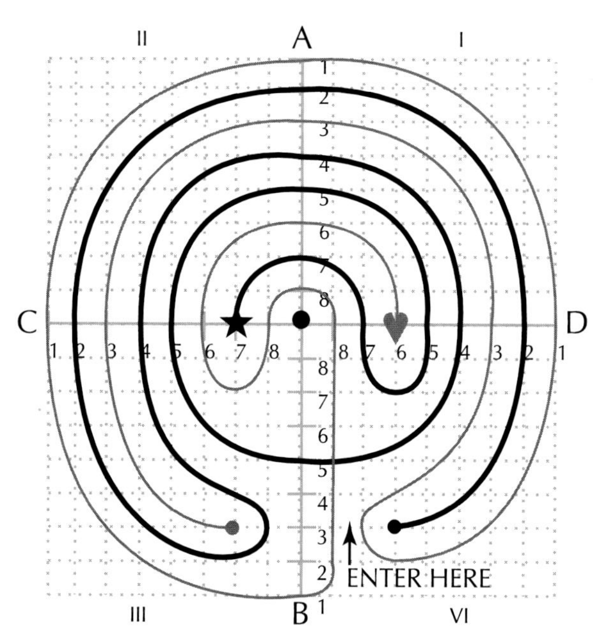

3. The second line starts in Quadrant I beginning at unit six. Follow the direction of the arrows until the endpoint.

The underlying grid is not absolutely necessary, but is helpful for the placement of the curved boundary lines.

Now you can "walk" the path between the boundary lines. Start at the entrance/exit point. In this example, we have used a right hand turn. You can reverse and begin with a left hand version.

A Healing Experience

Now you can take this concept into the landscape and build a version that a person, or group, can walk. In order to work with the energies of the earth, people who dowse are brought into the landscape site to determine where is the best place for the center, where the boundaries of the labyrinth are, where the best place for the entrance/exit should be. Each individual who walks the path poses a question that she or he needs to solve personally and that question is brought into consciousness and contemplated during the walk .

At the center, the individual meditates upon the question and then proceeds to walk back out. The labyrinth is given a conscious thank you for its participation in the process.

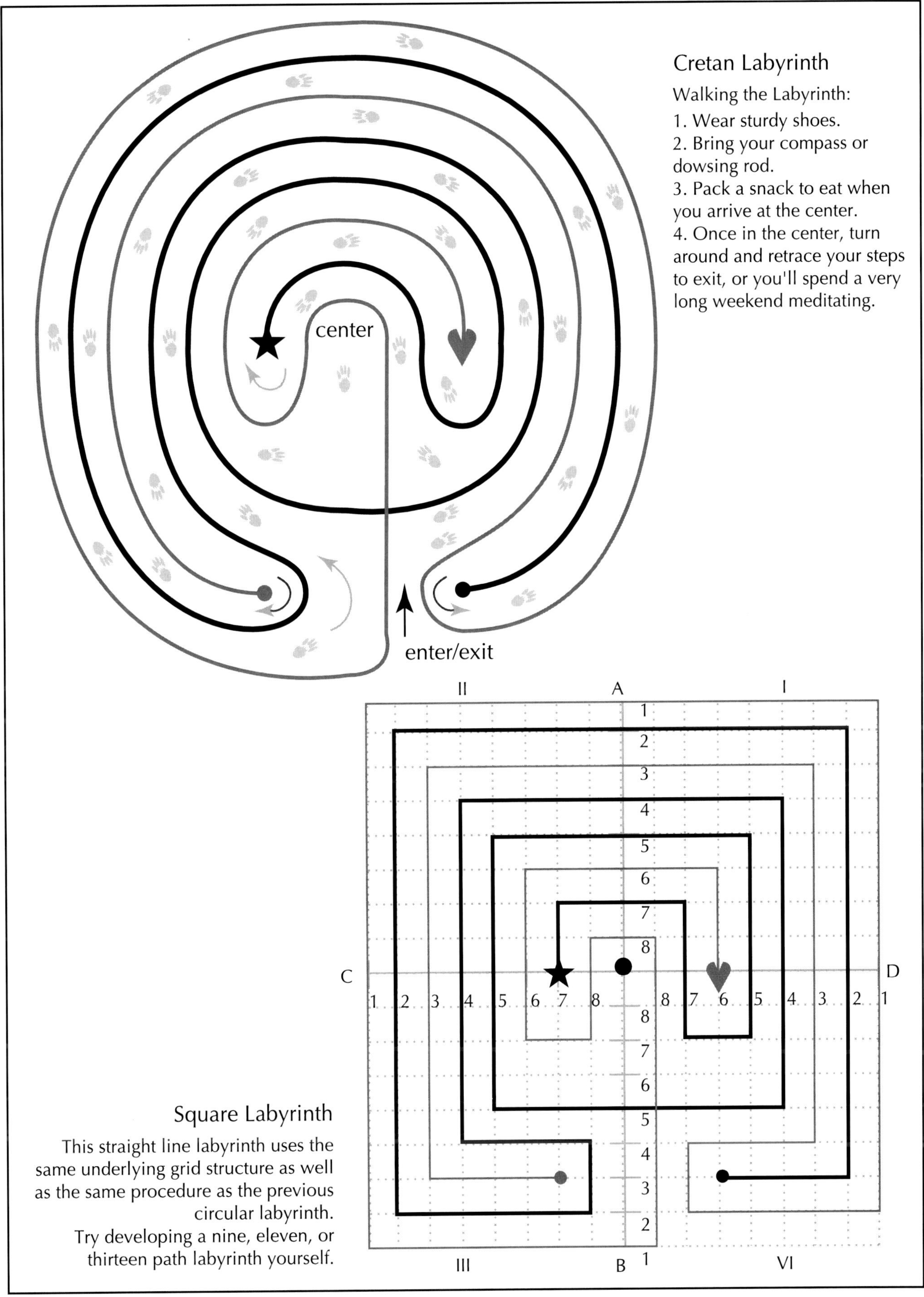

Square Labyrinth

This straight line labyrinth uses the same underlying grid structure as well as the same procedure as the previous circular labyrinth.
Try developing a nine, eleven, or thirteen path labyrinth yourself.

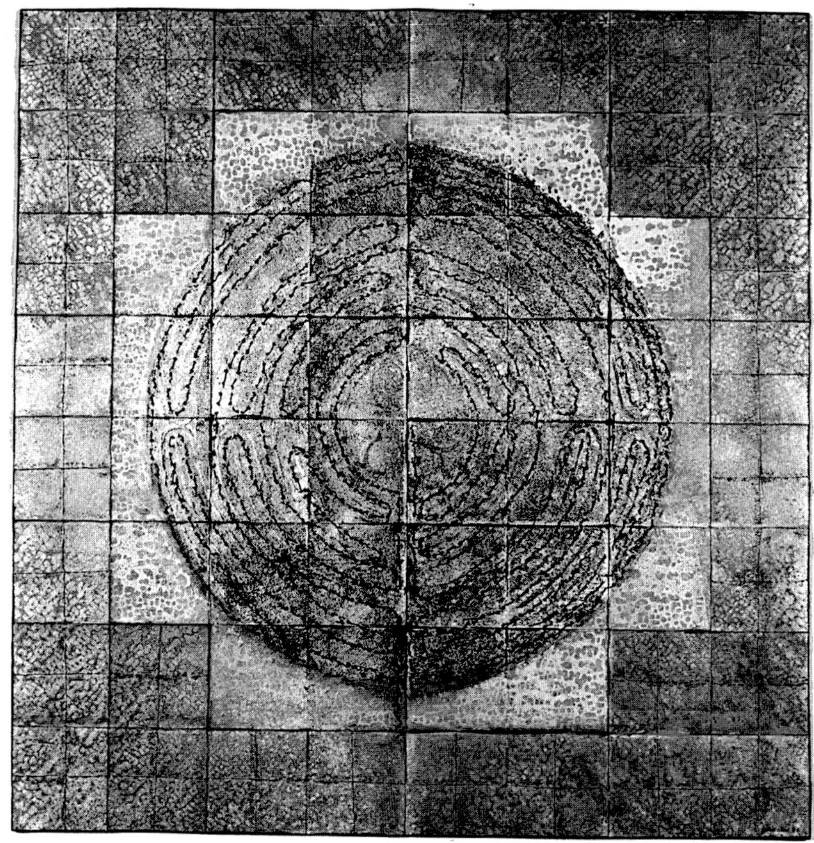

This page:
Juliet Wood. Ariadne's Log-os. *42" x 42". Mixed media of paper, inks, pigments, printing, painting, and stitching. 1995.*

The labyrinth has many images and archetypical layers. For me it serves as an intuitive and meditative aid. It can help the self focus on the health of the soul. It can center the self on our physical journey, for it focuses on issues of perception and intention. It is a fine exercise to "walk the labyrinth" visually.

Since the flow of intelligence is subatomic we often experience its insight as spiritual and soulful. It informs, however, all that we ultimately know of our physical world. This information becomes part of our physical nature and consciously affects our health, our aging process, our relationships, and the attitudes that determine the ecology of the planet.

Since the universal building block is now perceived as multi-dimensional fields of energy, in profound ways, we manifest our world, lives, experiences. So expansive are the fields of possibility that we "cut up" our knowledge, our time, our experiences, into categories that are quickly digestible. However our need to be "whole" or "full" beings speaks forcefully and reminds us on every level of the need for balance.

If you want to change something in your life change your intention and perception first. Most of us feel an intensity now as a new millenium emerges. The lalbyrinth is a sleeping archetypical and spiritual giant; an ancient watercourse and tool continually unfolding with possibility and creation.

California Artist Juliet Wood, San Anselmo, 1994

268

The labyrinth is a symbol of the journey. It can be as long as a lifetime, a history, a story, or as short as a momentary recognition as in "coming back to the place as knowing it for the first time." Because this symbol is archetypal and of the collective unconscious it spans all ages and all cultures. It is timeless. It refers inward to each soul's unique biography. The image is ambiguous enough to question each soul's sense of beauty, myth, vision and calling.

Juliet Wood, 1997

This page:
Juliet Wood.

Above: "Big Games". 48" x 48". Construct. Mixed media of paper, inks, pigments, printing, painting, and playing cards.

Problems, Projects, and Play

1. Decide whether or not the following pairs of figures are topologically equivalent. Give a written explanation for your answers
a. basketball/football
    basketball/icosahedron
b. drinking straw/cylinder
    drinking straw/artery
c. pullover sweater/torus
    pullover sweater/sugar bowl
d. pretzel/ trefoil knot
    pretzel/torus

2. Develop a board game prototype that deals with topological equivalents. Write the rules, construct the board, try it out on your friends.

3. Research the evolution of labyrinths and mazes. Choose one that interests you and develop a maze puzzle, complete with solution. Prepare rough visual sketches of same.

4. Research the subject of knot theory and do a written and visual chart showing the connections between the natural world, fiber forms, and network theory.

5. Research the scullptural work of the English artist Henry Moore. Photocopy examples of his work and arrange them according to a Genus chart.

6. Research the life and times of mathematician August Mobious. Develop a monologue for a radio show that presents the public with information in an entertaining ways about famous mathematicians.

7. Research other ways to construct a torus otherthan baking a bagel. Develop a hand-out that teachers could use with students in elementary school.

8. Find out more about the Peruvian knot (quipu) keepers and how they used knots in accounting procedures.

9. Many contemporary fiber artists used the techniques of knotting, braiding, and plaiting in their artwork. Do research on this subject and try one of these techniques using newspaper.

10. The Book of Kells is an important Celtic illuminated manuscript. Photocopy of an example of a page, analyze the process of interlacing involved, then write a sheet of directions on the procedure.

11. Write a contemporary version of the myth of the Minotaur updating the monster, the hero, the heroine, the golden threads, etc.

12. This project is in several parts.
Take a sheet of paper long enough to be joined into a cylinder and then into a torus. You may have to add folds or pleats in order to accomplish this.

a. Before pleating, develop a pattern for the surface using one of the seven line groups of symmetry. Do it in black and white. Make it artful.

b. Roll the sheet into a cylinder and then join ends into a torus.

c. In writing, answer the following question:

What ratio of length to width is needed in order to achieve a torus? Can it only be done with paper when it is pleated? How is the pattern on the surface changed by the pleatings?

13. Using materials of your choice, create a sculptural transforming using a teacup, saucer, and donut as your inspiration.

14. Find two very dissimilar objects. Take the first object, and through a ten step series, do a drawing that transforms it into the second object. Use black and white, gray tones, or color.

15. Use photographs of of two different faces and transform one into the other through the process of drawing, or collage, or assemblage, or sculpture. Do it in an eight step transformation.

16. Take a two-dimensional object and transform it into a three-dimensional object tthrough the process of sculpture.

17. This recipe can be used for constructing knots without ends, mobious models, toroids, etc.
Inedible Malleable Modelling Material
2 cups of table salt
2/3 cup water
1 cup cornstarch
1/2 cup cold water
Mix the first two ingredients together thoroughly in a saucepan. Stir until well heated. Remove from the heat. Mix the cornstarch and cold water together and add to the saucepan. The mixture should be a stiff dough. If it is not, place over low heat and stir until the desirable density.
If not used immediately, store wrapped in foil or a plastic baggie.
Model objects with imagination.
Place objects on a wire rack or screen so that air can circulate around them. Let dry to harden for about 36 hours depending upon the thickness.
Paint them. Shellac with three coats allowing drying time between them.

18. Using the information on labyrinths, develop a drawing, a painting, acollage, a sculpture, or a model for an outdoor walking experience

19. Tasty Toroids

Using the following recipe, create a batch of tasty toroids. Invite a group of friends over to contemplate the meaning of truth, beauty, and form while you munch away.

4-5 cups unsifted flour
3 TBS. sugar
1 TBS. salt
1 pkg. active dry yeast
1 and 1/2 cups of very warm tap water
1 beaten egg white
1 TBS. cold water

Mix 1 and 1/2 cups of flour, sugar, salt, and undissolved yeast in a large mixing bowl.

Gradually add tap water to the dry ingredients. Beat 2 minutes at medium speed. Scrape the bowl occasionally. Add 1/2 cup flour. Beat high speed for two minutes. Stir in additional flour to make a soft, but not sticky, dough.

Turn out onto a lightly floured surface and knead until smooth and elastic. This takes about 8-10 minutes. Place in an ungreased bowl and cover.
Let rise for 20 minutes in a warm place which is free from draft. Dough will not double in bulk as is usual.

Punch dough down. Turn out onto a lightly floured board. Roll the dough into a 10 x 12" rectangle. Cut dough into twelve 1" x 10" strips. Pinch the ends of the strips together to form a ring.

Place on an ungreased baking sheet. Cover and let rise in a warm place for 20 minutes.

In a large shallow pan, boil water that has a depth of about 2". Lower the temperature and add a few toroids at a time. Simmer for 7 minutes. Remove from the water and place each toroid on a towel to cool for 5 minutes.

Place the toroids on ungreased baking sheets and then bake for 10 minutes at 375°. Remove them from the oven.

Brush them with a combination of beaten egg white and cold water.

Return them to the oven and bake until done, about 20 minutes.

Remove from the baking sheets and cool on wire racks.

Butter and bite!

Further Reading

Adams, Colin C. *The Knot Book.* New York:W. H. Freeman and Co. 1994.

Barr, Stephen. *Experiments in Topology.* New York: Dover Publications, Inc. 1989.

Bain, George. *Celtic Art; the methods of construction.* New York: Dover Publications, Inc. 1973.

Chamberlain, Marcia and Candace Crockett. *Beyond Weaving.* New York: Watson-Guptill Publications. 1974.

Gemignani, Michael. *Elementary Topology,* 2nd Edition. New York: Dover Publications, Inc. 1990.

Hilbert, David and S. Cohn-Vossen. *Geometry and the Imagination.* New York:Chelsea Publishing Co. 1952.

Holden, Alan. *Orderly Tangles.* New York: Columbia University Press. 1983.

Meilach, Dona Z. *Creative Designs in Knotting.* New York: Crown Publishers. 1983.

Newman, Rochelle and Donna Fowler. *Space, Structure and Form.* Bradford: Pythagorean Press/Brown & Benchmark. 1996.

Peterson, Ivars. *Islands of Truth.* New York:W. H. Freeman and Co. 1990.

6 *The Glorious Garden*

Many people believe that once upon a time there was a place called Eden, pleasantness or delight, which was located somewhere in the east where the sun rises. In that place, there was a walled garden which was fruitful and peaceful. All manner of creatures found sanctuary and repose within its boundaries. This Paradise was the very essence of goodness. The word paradise comes from the Greek word paradeisos, for garden, which relates back to the old Persian word pairidaeza, meaning enclosure or park.

Regardless of the names for this place or whether in actuality it existed, the human spirit calls out for this experience. Because it does, pleasure gardens have been created all over this planet and throughout recorded time. Obviously, there have been agricultural sites planted and harvested for the specific needs of food and medicines, but the pleasure garden is a domain of beauty as defined by the persons or cultures that create them.

Long before there were gardens, there was Mother Earth-benign, belligerent, bountiful, barren, always in flux and minimally predictable. The garden is, therefore, the interface between humans and the landscape. It is the edge where the cultivated and the wild conjoin. It is the understanding of the natural by the articulation of the human. It is an enclosure, the separation from what is out there to what is in here. By defining it, we, in a certain measure,

possess it. Enclosed it is tame and subject to our handling while we, at the same time, are bound by its limits, both spatial and temporal. Wind, air, heat, and water play their parts in controlling the happenings within. This garden is both a necessity and a luxury at the same time. Without it we become less than human. In its absence we forget our roots and our beginnings. We see only ourselves mirrored in the glass and concrete surrounding us, arrogantly assuming that we are the makers of all. The garden through all its subtleties reminds us that we are part of nature and subject to the cycles of birth, maturity, and death. We are not above these.

The need for gardens at all scales is great indeed. The personal space provides and atmosphere of introspection and quietude. It allows for experimentation of design and experimentation with materials. The public place is an arena of pleasure and community, but it is also an oasis of research and protection in a time when habitats and species are being devastated at an alarming rate. The public zoo, the national arboretum, the botanical garden all provide what the individual cannot protect.

A garden is a location, a climate, a terrain, a view, an ecosystem, a microcosm, a way of life. It is fact, fantasy, form, function, and philosophy all in one. It will not exist unless one watches through days and seasons the workings of nature. It is an experience of space as well as time.

These facing pages:
Two contemporary interpreta-
tions of the Garden of Eden.
This page:
Pencil drawing by Gail
Maciejewski of The Gar-
den based on the Renais-
sance paintings of Botticelli.

Opposite page:
Pen and ink drawying by
Susan Libby of Adam and
Eve working together in The
Garden.
How would you represent
the Garden of Eden in your
own artwork?

The cycle of seasons. Winter, the time of death and dormancy; spring, the time of bursts of birthings; summer, the time of heating sun and shading clouds, of warmth and color, of storm and shower; and fall, the incredible intensity of final epiphanies. The garden is an experience of patience since all growing things have their own time and their own duration. One cannot will an oak into being. Because of all this, the garden ties a person to a place if one wishes to watch that garden begin and become. And when a garden is done well, with love and wisdom, it becomes an act of optimism, a work of art, and a gesture of the soul.

The garden is an aspect of the visual arts since it is an organized, harmonious, bounded, finite idealized site/space that is partially perceived through the sense of sight. Of course, in a great garden, all the other senses of taste, touch, smell, sound, and movement are activated as well. The garden is related to the visual art of architecture as a transitional space and can be suggestive of an outdoor room or rooms.

But time is also a component of garden space since it physically exists in real time. In order to experience it there must be time of movement through a space, the time of seasons, and the time of growth.

There are essentially two complementary archetypal systems for the designing of gardens. One is designated as formal, or Classical, and the other as informal, or Romantic. The former is conceived and designed by rational universal mathematical principles. It is non-intuitive, usually symmetric, with a single privileged viewpoint from which the overall plan makes the most visual sense. It is time static so that change is de-emphasized by the precise manicuring of all the elements. It moves the participant in particular directions and speed. There are fountains, grottos, statuary, and ponds emphasizing the artificially constructed. Areas are subdivided by parths with the spaces between filled with flowering trees and plants which are laid out in patterns and rows re-affirming the human-made

In contrast, the informal garden is intuitively designed suggesting poetic and picturesque principles. Space is not bounded but limitless and immeasurable. Great vistas abound. Geometry is de-emphasized and the plan is imitative of the natural environment. It encourages meandering through time and place at one's own tempo. It is time dynamic emphasizing change, impermanence, the haphazard, and mystery. The Far East was the birth of this type of garden but it eventually reached western culture and with difference was developed by the English in the eighteenth century. It ultimately, however, is just as tightly designed as the formal garden but the hand of the designer appears less obvious to the untutored.

The great formal garden of Versailles of Louis XIV was begun by his father as a hunting park. The French garden became a political statement of a united nation under the aegis of one man. There is a sense of vast space and the fundamental geometric shapes of the circle and the square as they relate to two dimensions rather than a sense of physical mass. The French temperament found natural wilderness threatening and unpredictable. The garden, therefore, became the place where the irrational forest was outlawed and replaced by extreme order and perfection, building upon the values of rationality, military expertise, and engineering techniques. This work was created by an ensemble of craftsmen working under the vision of Le Notre. It took 30 years to transform the original 250 acres into more than 20,000 acres surrounded by a 26 mile long stone wall. The garden also contained a labyrinth which had 39 groups of fountain sculptures which used water pumped from the Seine.

This page: The formal environment of Versailles by Louis XIV of France.

Above:
The grand palace of Versailles.

Below:
The gardens of the Orangerie. In the summer palm and orange trees were put into the garden to flower and bear fruit. After the picking, they would be brought back inside for the winter.

This page: Gardens in the United States and Canada.

Above:
Bridge from the Dow Gardens, Midland, Michigan.

Center:
Formal flower beds. Botanical Garden, Montreal, Canada.

Below:
Formal plant bed in Naumkeag Gardens, Stockbridge, MA.

This page: Gardens in the United States.
Above:
Longwood Gardens, DuPont Family, Wilmington, DE.
Below:
Municipal Garden, downtown, Indianapolis, IN.

The Persian Garden

The pleasure garden is the quintessential form of the union of art, geometry, and the natural world. Out of the seeming chaos of the primordial landscape, a space is bounded, marked, and designed using the malleable materials of earth, plants, water, and stone. All this is worked together by the wishes of the client, the mind of the landscape architect, and the hands of the gardener. It can be enclosed by the constraints of a window box or be as expansive as a park preserve. But, of course, the form can only come to fruition when joined to the forces of the sun, wind, and rain. Ignoring the limits of a particular ecosystem, the garden can lay barren, a mute, but visible, testimony to the oversight of humans.

Persia (now called Iran) is commonly accepted as the birthplace of the garden as a human construct. Found within their gardens was a connection between geometric perfection and symbolism. Throughout civilization much of Persia has been desert. The lands between the Tigris and Uphrates Rivers had to be irrigated in order to make them flourish since the surrounding mountains absorbed most of the rainfall which left little for the plains.

These gardens exemplify the concept of the "oasis in the desert" and are the precursors to the gardens of Islam which date from around 730 AD and are found in Spain, Iran, and India. Persia was not a hospitable land or climate for the easy development of a garden, but perhaps it was a psychological necessity in order to make life bearable under difficult conditions.

Walls enclosed the garden space protecting it from the hostile environment just outside the enclosure. The design of the garden was highly geometric and included irrigation channels set at right angles to each other. Water was the main focal point, and for good reason, since without it humans die very quickly.

In the pursuit of privacy, Persians walled in their gardens. Within these enclosed spaces, the development of the design included four streams in the shape of a cruciform which signified the four rives of Paradise and their coming together at a node produced the River of Life. Four rectangular planting areas contained the fruit and shade trees which provided both nourishment and escape from the ever pervasive strong light. The type of trees were chosen for their symbolic content as well.

According to Jungian psychology, the number four is a symbol of wholeness for the individual. Groups of four, such as eight, sixteen, thirty-two, are universally recurring religious symbols. There are the four seasons; the four cardinal directions; the four stages of life; and the four functions of human consciousness which are thought, feeling, intuition, and sensation.

One of the wonders of the Ancient World was the hanging gardens of Babylon built by King Cyrus, 6th century BC, for his favorite courtesan who was Persian. It was 100 feet long by 100 feet wide built in terraces. There was also Isfahan, the City of Gardens, which were described in the records of Sir Thomas Herbert in the 17th Century.

The Persians were particularly fond of roses which are one of the oldest plant species available to humans. They have been bred over the centuries to produce innumerable varieties. Some are raised for color, others for fragrance, others for size. The Persians were also familiar with tulips, jasmine, lilies, and violets.

While we cannot experience the ancient gardens firsthand, a great many of the woven rugs of Persia are two-dimensional analogues of the three-dimensional garden space. They indicate a rectangle, garden enclosure, with a fourfold division suggestive of water, with a central unit symbolic of a pavilion or fountain. In the center, there may also be a Tree of Life. In the borders, one finds stylized trees and flowers.

In our diagram, we offer an abstracted version of a generic Persian rug for you to use as a jumping off point for a design of your own. Consider the ratio of the width to the length of the rectangle. This choice has a strong influence on the subsequent divisions found within the rug. Here, we have chosen to use a Golden Rectangle because the subdivisions provide harmony through similarity while providing la contrast to the symmetry within the design structure.

Intricate detail is the essence of all Persian art whether it be painted

miniatures, calligraphic manuscripts, ceramic tilework, or woven carpets. Designs on the carpets had to be abstracted and stylized in part due to the limits of weaving in which horizontal flexible wool weft elements weave at right angles to taut vertical wool or cotton warp elements. The image is painstakingly built colored knot by color knot, moving either right or left, row by row, inch by inch. Large shapes are built up incrementally. The work is accomplished by human mind, eye, and hand only.

This page:
A schematic diagram for the development of a Persian rug design. A central pavillion, or Tree of Life, is at the crossroads of the four rivers of Eden. Flowers, plants, and trees are found within the rectangular subdivisions and around the borders.

The Japanese Garden

Nature is:
> unpredictable
> without scale--exists at
> micro-macro-scopic level
> not necessarily beautiful--
> can be fearsome, awesome, ugly
> indifferent to human life as unique
> concerned with protecting all life
> forms, not just human life
> amoral (ethics & morality are
> culturally determined qualities)

All Japanese art is an example of studied simplicity; not the simplicity of innocence but that of great sophistication. The Japanese garden is intimately connected to the house, not a space apart, while in western culture the garden is usually perceived as separate from the residence. There are notable exceptions such as the famous "Fallingwater" residence designed by Frank Lloyd Wright. Designing gardens is a fine art equal to painting and sculpture. For the Japanese its importance is parallel to cathedral building in the medieval period in western art.

A Japanese house can only become a home by having a garden. Yet, there is a clear separation as to what is house and what is garden. The scale of the garden is small and intimate like that of the interior of the house and its design is as carefully adjusted and controlled. The garden is contained by an enclosure the height of which is far above eye level. This guarantees a separation from the outside communal world and the inside individual world. Yet within the confines of a finite space, the garden designer creates a sense of the infinity of nature within human comprehension. The Japanese garden is both a fact and a symbol. It speaks of itself and yet of things beyond itself. It is of the moment and it is timeless. It is concrete and it is abstract.

The small residential garden is built to be seen only from the house interior. It is not meant to be walked through, or played in. It is a place of meditation. Thus, it becomes architecture in that it is a purposely designed space that establishes human scale in which to reflect upon Nature, Life, the Universe, and human attitudes toward these. A heavy burden for the garden, indeed. Zen Buddhist philosophy helped shape the form of the garden. Zen is derived from the Chinese world "Ch'an" which means meditation. Its philosophy emphasizes an immediate personal ,direct, intuitive, path in order to attain spiritual Enlightenment.

Zen believes that all the various physical forms that exist upon this planet, be they animate or inanimate, are the outward material expression of a single force, the oneness of all things. Zen stresses simplicity and sincerity of feeling through direct contact with objects and events. No complex dogma or ritual is involved with the experience of the ineffable. The aim is to calm the mind so that it can concentrate on spiritual awakening.

All the various components of a garden, such as the plants, the stones, the water, need the same respect as that given to humans. The latter are not placed in a higher position in the scheme of things. The garden, therefore, is no longer wild nature or even a miniature copy, but rather a "symbolic abstraction". Openings, terraces, and balconies are all placed with an eye toward the landscape, near and far scenery and vistas. It has a casual asymmetric appearance and yet is carefully planned to be an experience for all seasons.

The standardization of the features of the garden allows even the humblest of persons to partake of the symbolism. This forces the maker of the garden to find expression only in the art of composition rather than the free choice of objects and materials. There are the basic elements of plants, stones, and water. In Buddhist philosophy an object is not a thing or a substance but an event that opens an individual to contemplation. The color of the plant material as well as the other garden objects is limited to browns, grays, and greens in order to prevent the viewer from being attracted to the superficial colors and forms instead of the deeper symbolic essentials. Pine trees especially, play a most important part in the composition.

They are evergreen all year round and do not change color conspicuously.

Stones in the garden perform a function similar to the column in the house. They provide a design module and a scale for all the objects found within. They provide a stability and strength as well as drama. They are eternal and appear never changing. The particular garden stones are chosen to suggest a mood by contrasting round with jagged, smooth with rough, green areas again gray stone. They act as guides for the sense of the viewer. They can hint at deep ravines or mountain peaks. They can suggest the flow of the river and the movement of a stream. Usually the stones are arranged asymmetrically and with an odd number group. Stones become the gravel for ground cover, basins for washing hands, and lanterns for both illumination and decoration.

Water is found in many forms and can be suggestive of rain, snow, fog, clouds, and ice. It can be actual or symbolized in waterfalls, streams, fountains, and pools. Raked gravel and sand in dry landscape gardens suggest water images.

This page:
A section of the Japanese Garden at the Denver Botanical Gardens, Denver, Colorado.

Elements in the Japanese Garden

Japanese Concept of Beauty

acceptance of what is of the impermanence of all things

belief in the union of opposites

celebration of the ordinary and the commonplace. All the objects of human use are considered design worthy

direct experience with an object or event

freedom from attachment to things

humility by not impressing the individual personality on things

mistrust of language and reason

naturalness so that one sees the effect of time on objects

reverence for all aspects of nature

silence by searching for serenity and away from the noise of the world

simplicity by the elimination of the unnecessary

subtlety by attention to the intrinsic quality of things

suggestion by leaving something out which must be completed by the perception of the viewer

textural awareness through the interest in irregularities and imperfections

Haiku Poem

Fallen petals rise
back to the branch--
I watch:
oh...butterflies!

Moritake, 15th century

Views of Japanese Gardens

Proportional relationships play an essential role in the composition of all Japanese gardens.
David Slawson, Scholar

This page:
Elements within a Japanese Garden.

Above left: Stone lantern from Denver Botanical Gardens.

Center right: Stone bridge from Brooklyn Botanical Gardens.

Left: Stone and gravel path from Brooklyn Botanical Gardens.

The Dry Landscape Garden at Ryoan-ji

Zen Buddhist priests have created some of the most beautiful and austere gardens using the concept of the dry landscape. The gardens express the essence of pure abstraction of thought through the elimination of all but the most essential elements. It is meant for contemplation which would lead to the enlightenment of the spirit in which the viewer and viewed become one. Physical access to the garden is forbidden except to the monks that must tend it. The garden is meant to be seen from an elevated mobile viewpoint with no sense of horizon lines.

The most famous of these gardens is at the Ryoanji monastery laid out at the city of Kyoto, Japan, in about 1488, the Muromachi period. The creation of the garden is attributed to So-ami (1472-1523) who may have designed it as an abstraction of an ocean experience of islands, sea, and waves. It is the essence of understatement and restraint. In a little more than an acre of land, the designer created a feeling for the concept of a large vista. Stones, sand, and quartz are the only elements employed in the composition. In an enclosed area of approximately 75 feet long by 31 feet wide (less than four hundred square yards), fifteen stones are arranged in five groups of: five stones, two stones, three stones, two stones, and three stones, set within an area of level raked white quartz sand. This material is essentially a crushed stone that western gardeners would call fine gravel. An earth wall surrounds the area of the garden on three sides and the fourth side is bordered by a veranda. There are no plants within this garden.

According to researcher Gyorgi Doczi in his book, *The Power of Limits*, the garden is filled with harmonic proportions. The area filled by the raked sand is built of two overlapping Golden Rectangles. Golden Cuts and diagonals within this rectangle establish key points for the placement of the five groups of stones. Each of these groups represents another connection to the harmony of the symbolic geometrical and numerical relationships. The number five is a combination of the male positive force of yang and the female negative force of yin. The number three stands for heaven, earth, and man. These numbers are also related to the Fibonacci sequence which in turn relates to the Golden Ratio. See Design Appendix. The ratio of height to width of each stone within a group is also important to the overall proportional arrangements.

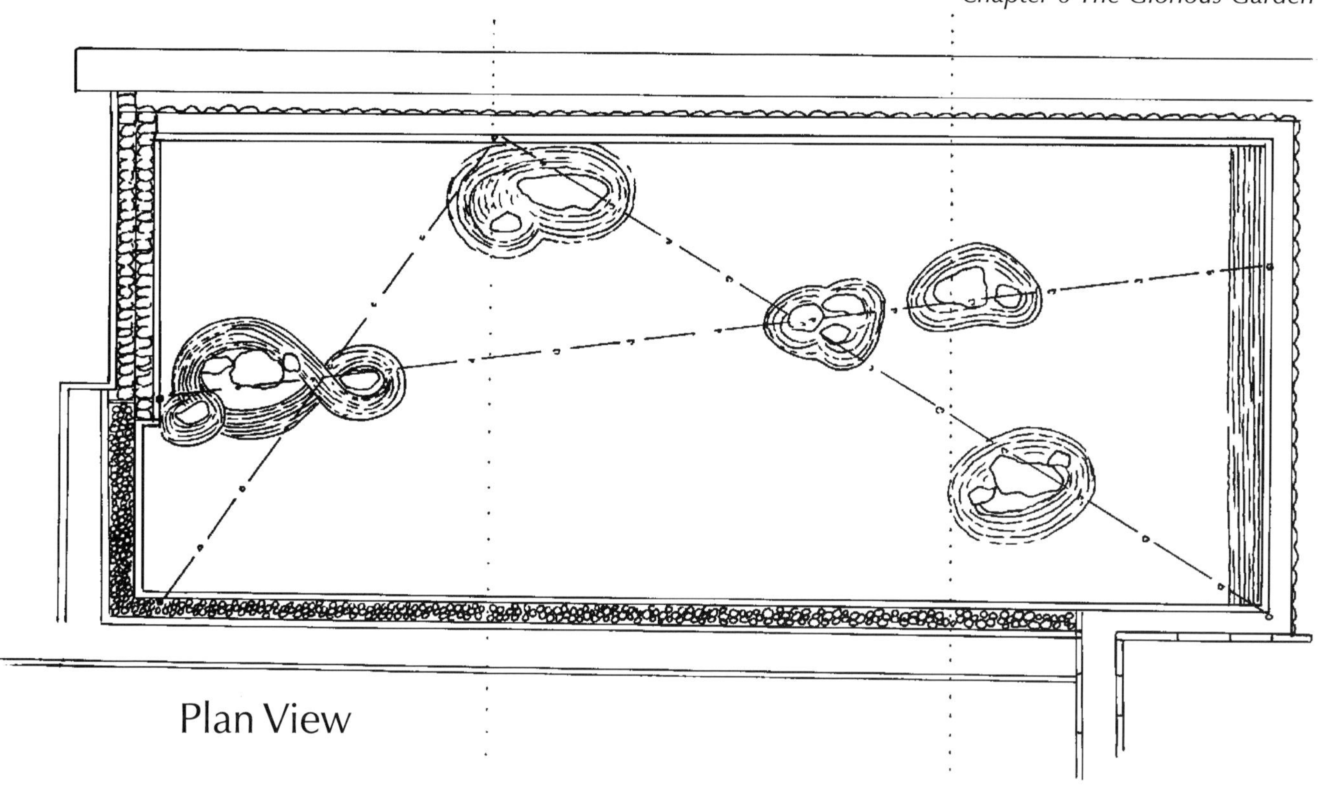

Plan View

Elevation View

This rectangular space is 30 meters from east to west and 10 meters from north to south. Fifteen rocks, of various shapes and sizes, are arranged so that viewers can only see fourteen of them at once regardless of the vantage point from which the garden is viewed. Only when one has attained spiritual enlightenment can the missing stone be seen in the mind's eye.

This page:
Drawing of dry landscape garden of Ryoanji as seen from both above and from a side.

The landscape can also be seen in abstract form and expression in the design of the kimono. In a single garment, geometric and organic forms combine along with the idea of the natural seasons.

All classes of Japanese society enjoy a sense of aesthetic expression through the art of clothing. The Edo period, 1615-1868, was the high point in the development of dress as an art form. The precursor of the modern kimono was the Kosode, an outer garment. Beginning with the 16th century, all persons wore such a piece of apparel. The shape of the garment provides comfort and freedom of movement and airiness to counteract the humidity of the climate.

Kimonos are usually reserved for formal occasions such as weddings and important holidays. The designs vary according to age, social station, and season. Children and young women wear brightly colored kimonos with long sleeves while married women wear darked, more subdued shades in garments with shorter sleeves. The sleeve indicates the age and marital status of its wearer. Men wear solid dark colors.

The obi is the sash which ties around the garment. It may be as wide as one foot, as long as 13 feet, and is tied in a number of ways.

On this page we offer a schematic diagram so that you can design a paper or fabric kimono for your own pleasure. Choose a particular season, or month, as inspiration for your design. What flowers, trees, plants, etc. are found within that time period in your particular geographic location? What colors do you associate with that season? What geometric elements can you use that would remind the viewer of a trellis, fence, pond, waterfall, etc. Think of designing with a sense of restraint and elegance. First do some more research on Japanese culture and aesthetics.

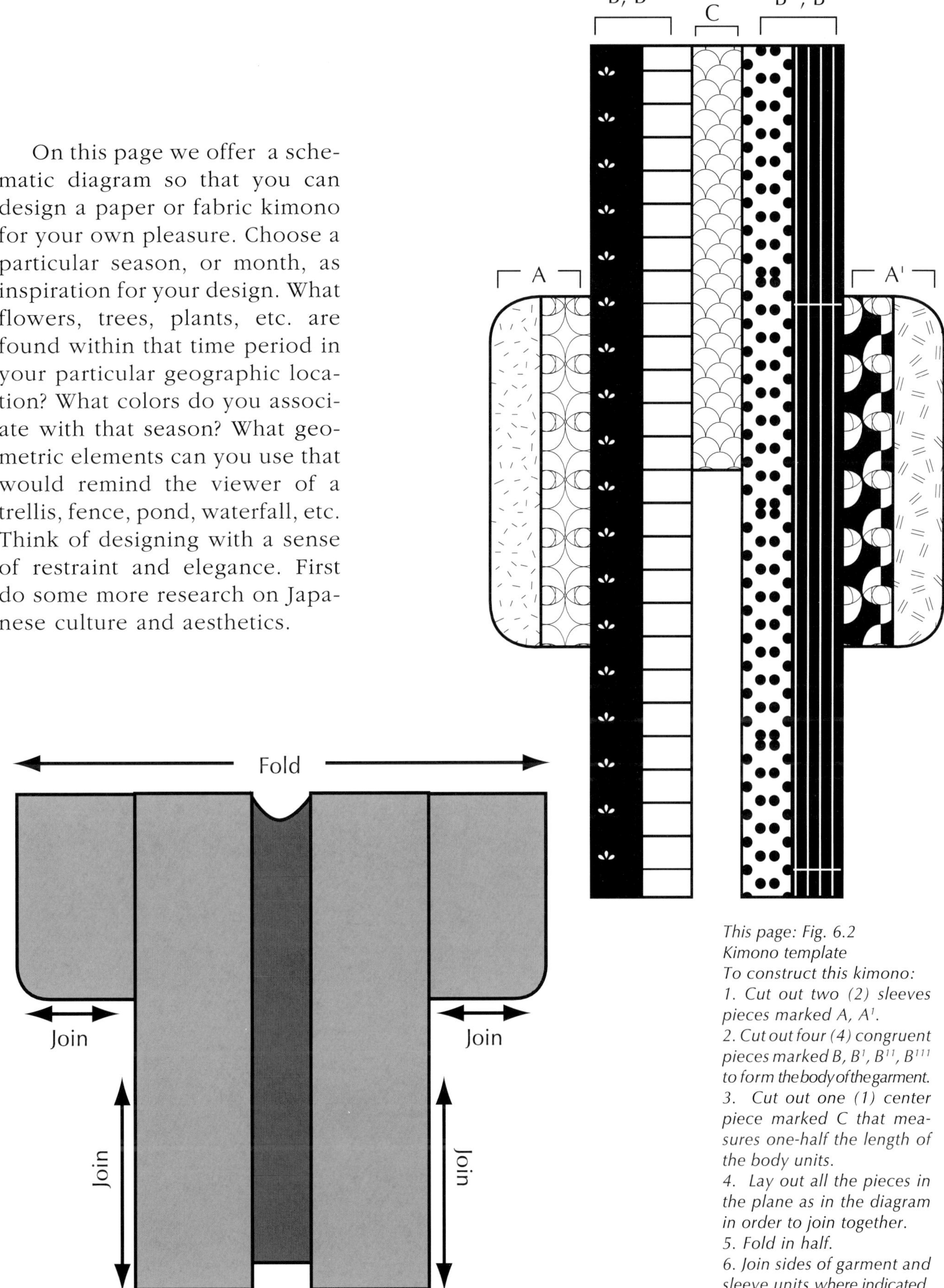

This page: Fig. 6.2
Kimono template
To construct this kimono:
1. Cut out two (2) sleeves pieces marked A, A^1.
2. Cut out four (4) congruent pieces marked B, B^1, B^{11}, B^{111} to form the body of the garment.
3. Cut out one (1) center piece marked C that measures one-half the length of the body units.
4. Lay out all the pieces in the plane as in the diagram in order to join together.
5. Fold in half.
6. Join sides of garment and sleeve units where indicated.

The Golden Garden

Though there are specialists whose profession it is to design landscape environments and gardens, each of us can also engage in the process of garden creation knowing that perfection is ever-elusive, nature is full of surprises, green things can die from a variety of causes, and ideal concepts of the mind can be flawed realities of the physical. In fact, the expression of creativity in all forms should be an inalienable right of all human beings with or without specialized training. Creation, when approached with an open mind, a loving heart, and spiritual integrity is always meaningful.

So dig right in! Begin with a garden the size of a window box. Think of it as a miniature outdoor room having length, width, and depth. Consider its proportions and the proportions of its inhabitants, the plants. If your desire is to work on a large scale, bring together the design of the garden and the design of your house, assuming that a sensitive architect developed proportional relationships within the structures.

Elements of the facade of the house, such as windows, doors, roofline, can be the starting point for developing the modules for design. A garden, at its very best, is created to provide a complex set of related but varied sensations involving sound, sight, smell, touch, movement, and seasons. Each section of a garden encourages one to stop and contemplate a particular aspect of nature. Some cultures prefer to accentuate the scent and color of a particular plant or flower. Others emphasize the enduring forms of shrubs and trees. Start your garden with a trip to the library, or a surf through the Internet, or a walk through an arboretum or local garden center.

FACT

Seasons with climate, temperature range, wind conditions,amount of rainfall
Site size, shape, and situation of east, west, north, and south
Sunlight hours and shadowed sections
Soil conditions of acidity, alkalinity, porosity

FUNCTION

Place of meditation
Playground
Parties and celebrations
Pet participation
Practical concerns of food growing or flowers

FANTASY AND PHILOSOPHY

Cultural connections: Persian, Japanese, Chinese, Italian, English, French
Classical or Romantic
Contemporary or Historical

FORM

Asymmetric or symmetric
Architectonic or natural
Archetypal elements:

>A tree, group of trees, sacred groves, forests
>A gazebo, garden house, or hut
>A terrace with seating areas of bench, chairs, tables, swings,
>An enclosure fence, wall or screen, natural, or artificial
>A threshold, actual or symbolic such as an arch or gateway
>A bridge that connects two areas
>A water feature as a birdbath, foutain, pool, pond, lake, ocean
>A sacred mound which provides a high vantage point
>A green area of meadow, field, or lawn
>A single stone, a grouping of stones, or a stone wall
>A pathway of gravel or bricks or stones

These facing pages:
Garden details

Left: Fountain in Vizcaya Gardens, Miami, Fla.

Above right: Wooden bridge in Japanese Garden, Brooklyn Botanical Gardens.

Facing page:
Stones in stream, Dow Gardens, Midlands, Michigna.

The garden tames the natural order of things. It grows in the three dimensions of space and the one-dimension of time. Design organizes all the variables into a coherent whole so that all the elements play their part in creating unity. Since a garden occupies physical space, it is a piece of geometry. Unless the land is flattened, or covered with concrete, the plane is no longer Euclidean but has aspects of other kinds of surfaces.

The layout of a garden is a spatial design problem for humans. The internal and external structures of the plant material within the garden is a design problem. Natural forms have growth patterns that emerge from their chemical and biological interactions. Cells, which are composed of molelules, less than 0.1mm across,

are the building blocks of organic life. Inside each cell is the necessary genetic information that must be pASSed onto the offspring cells . Each unit cell grows internally but can only expand so far, thus the cell has to split into two cells, each carrying the molecular information of the parent cell. In order to grow, cells need energy which ultimately comes from the sun.

The links between the sun and the earth are the green plants because plants use the process of photosynthesis to convert ligsunht into energy. The plants take in carbon dioxide and hydrogen and give off oxygen which is required by other living creatures.

Though the variations within the garden are legion, the fundamental forming structures such as branching, spiraling, exploding, meandering, cracking are few. So while mother nature is profligate in species, she is frugal with structures.

There is a mathematical number sequence that relates to the physical growth patterns of plants, phyllotaxis. It is called the Fibonacci Sequence, named for the mathematician of the Middle Ages that first speculated about this group. This sequence has a relationship to the Golden Ratio. After the 15th term, the ratio between a pair of consecutive numbers is 1: 1.61803.... which is the Golden Ratio. See the Design Appendix for more information.

Plants need light, moisture, and air to frow as they lengthen and spiral about a branch or a stem or a trunk. As new leaves are formed about a stem of a plant, they do not grow one over the other, but place themselves at intervals around the stem that are ratios from the Fibonacci Sequence, ie. 2:5; 3:8; 8:13, etc. Fig. 6.2 shows an example of such growth.

Above:The garden elements of rocks and water in a stream.

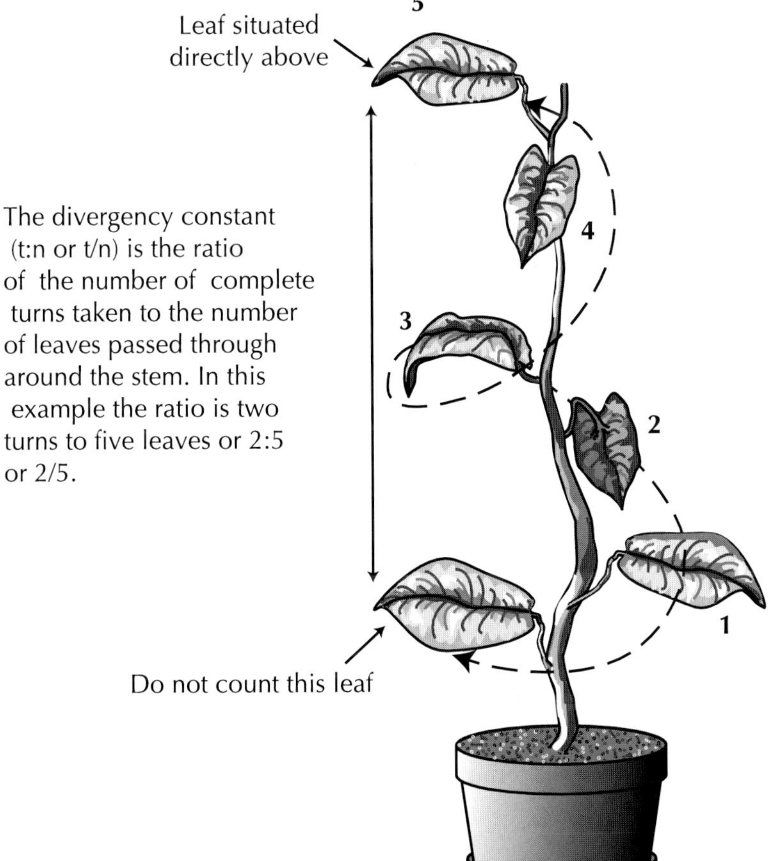

Below left: Fig. 6.2
An example of Fibonacci numbers in plant growth. Fibonacci ratios can be seen by looking at leaves on plants. First locate a leaf on a stem and then find another leaf directly above it. Count the number of leaves and turns required to move around the stem between them.
In the diagram at the left, leaf number 5 is situated directly above the unnumbered leaf. When moving from the first leaf to the other directly above it, five leaves are passed through in two complete turns around the plant stem.
See the Design Appendix for the relationship between the Fibonacci numbers and the Golden Ratio.

Though Euclidean geometry is useful for the armature of the garden, it is not the geometrical tool that can explore the dynamical growth systems of the plants or the variables involved in the weather that affects the growth. For this investigation, there is a new geometry discovered and named by the 20th century mathematician, Benoit Mandelbrot. It is called fractal geometry, derived from the latin verb frangere.

The triangle, square and pentagon are ideal figures that do not exist in the natural world except as implied structures. Look closely at any natural object, and while they can be set within an idealized form, they are more complex than any of these. What a fractal form exhibits is the characteristic of self-similarity. A tree twig, a tree branch exhibit a branching structure that is like that of its trunk and they are all subject to the same set of rules. When any one of these objects is magnified, the same kind of structure results. In contrast the increasing magnification of a section square simplifies down to a line segment and gives no clue as to the overall shape of the square.

A square has a dimension of 2 since it is a plane figure, but a fractal object can have a the fractional number that lies between two dimensions. Every surface can be assigned a fractal dimension. It is the ratio between the altitude of the larger surface projections and the smaller ones. Think of the difference between the surface of an orange and that of a gourd. These spherical shapes have dimensions between the smoothest which is designated as 2 and the roughest which is designated as 2.9.

The advent of the computer and the discovery of fractal geometry go hand-in-hand. In order to develop a fractal, it takes a rule and many, many repetitions of the rule to produce a visual able to be seen. Fortunately , the computer does not get bored. On the facing page is a computer generated fractal fern using only a drawing program. Natural branching patterns, such as in the fern, follow rules that can be modelled by the mathematics of fractals. Remember, though, that a computer model is only an approximation of the real thing existing in nature. There are myriad influences upon the growth of a real fern. The geometric "fern" in Fig. 6.4 is generated by first constructing a simple branching unit, which exhibits the symmetry operation of glide reflection. Scaled down copies of it are placed along the branches as seen in the second diagram. Another set of scaled down copies are placed on these branches.

This page:
The garden element of a Chinese dragon sculpture made of stone.

This page:
Left: Fig. 6.4
A computer generated fractal fern using an illustration program.

Above: Looking down on a water garden in the city of San Antonio, Texas.

Below: A flower and shrubs border path in the Numinous Gardens, Haverhill, MA.

The essence of fractal geometry involves a rule that is repeated, iterated, over and over again, too many times for the hand and eye to manage, but comfortable for the computer to produce.

While technically time consuming, the process can be understood easily by looking at a geometric figure, such as a triangle, and seeing how the iterations occur. Look at the two examples in Fig. 6. 3. The original figure is called the initiator and is the seed figure to which the rule, called an algorithm, is applied. The first round of the rule is called a "second generation" figure. This result is then used as the input for the third generation and so on and so on.......for many generations.

This page: Fig. 6.3
Fractal development using a geometric figure.
In both cases, the initiator figure is an equilateral triangle. The development of the second, third, and fourth generations are shown.
In the first group, the sides of the triangle are divided into thirds and the middle third is replaced with the sides of another equilateral triangle. This process could be carried to another generation, then another, etc.

In the second generation of the second group, the interior of the equilateral triangle is subdivided at the midpoints of its three sides. An equilateral triangle is set within the interior with its vertices at the midpoints.
In the third generation, the same operation is performed within the three light triangles.
Using the equilateral triangle, develop some other rules for this figure and using the computer, or constructing by hand, show four generations of the figures.

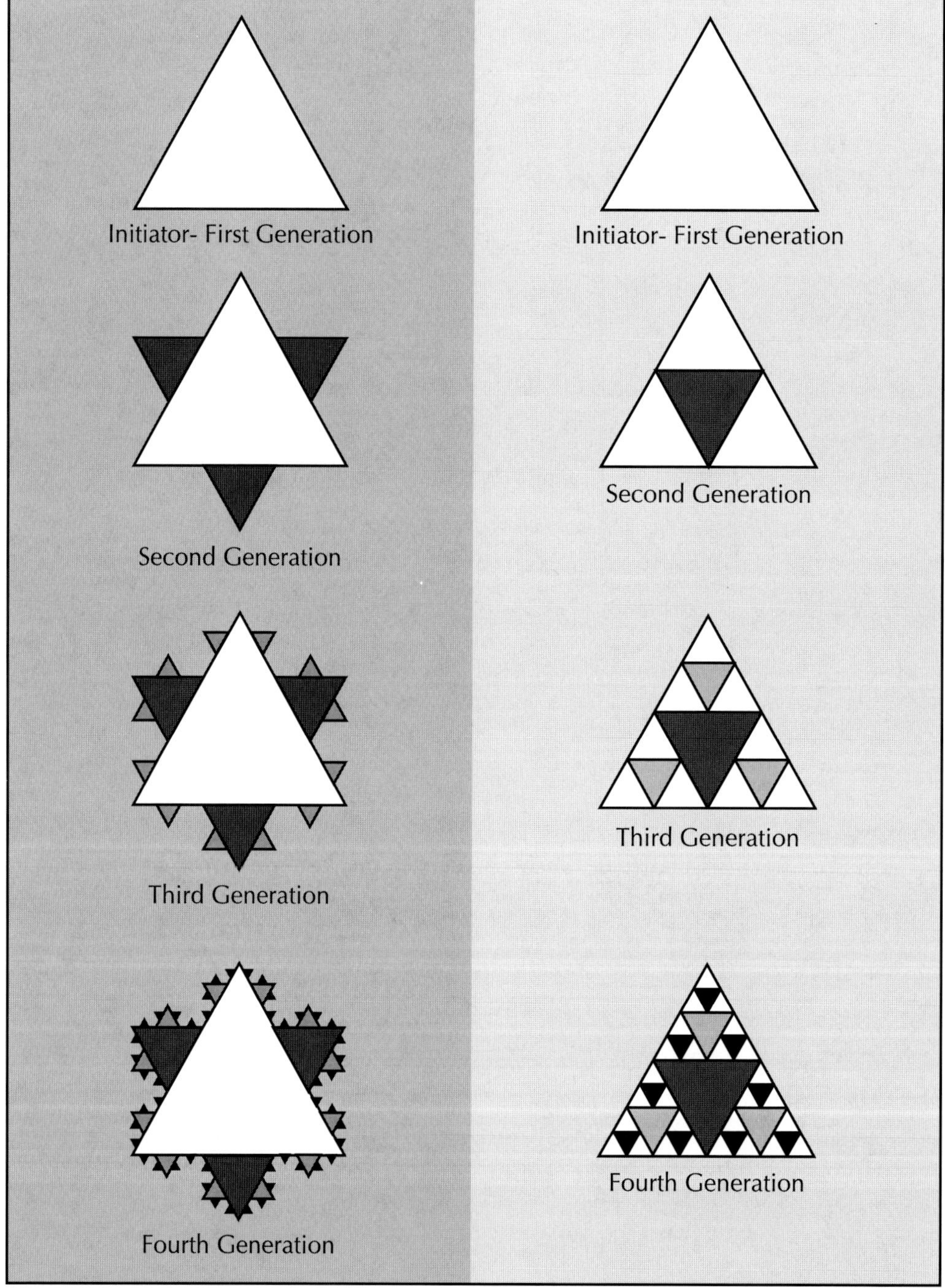

298

Mathematically, this process can also be explained through the use of an algebraic equation which makes operations easier to perform, albeit more abstract conceptually. We can use the concept of a tree and its branchings to illustrate. The branches are shortened in a consistent manner and the angle between branches stays the same. On the four following pages we give a procedure for growing an artful tree based on this fractal branching.

Ideal mathematical fractals and natural fractals are variations on a theme. In nature, actual objects do not grow infinitely large nor have infinitely small parts. Physical trees are constrained by a number of other physical systems. But mathematical fractals can act as models for behavior. Let us give an example here.

Let the initiator, which is the "trunk" of the tree, have a unit length of 1.

 Let r stand for the length of each new branch compared to the previous line segment (branch). In this case it is 3/4 the length of the previous branch.

Let the mathematical symbol theta stand for half the branching angle. In this case it will be 45°. The subsequent lengths of the new branches can be written as an iterative function (f(x) = 3/4 (or.75) where x stands for the length of the segment in any generation.

We have built two symmetrical highly simplified geometric trees but asymmetry could become part of the equation to make them look more natural.

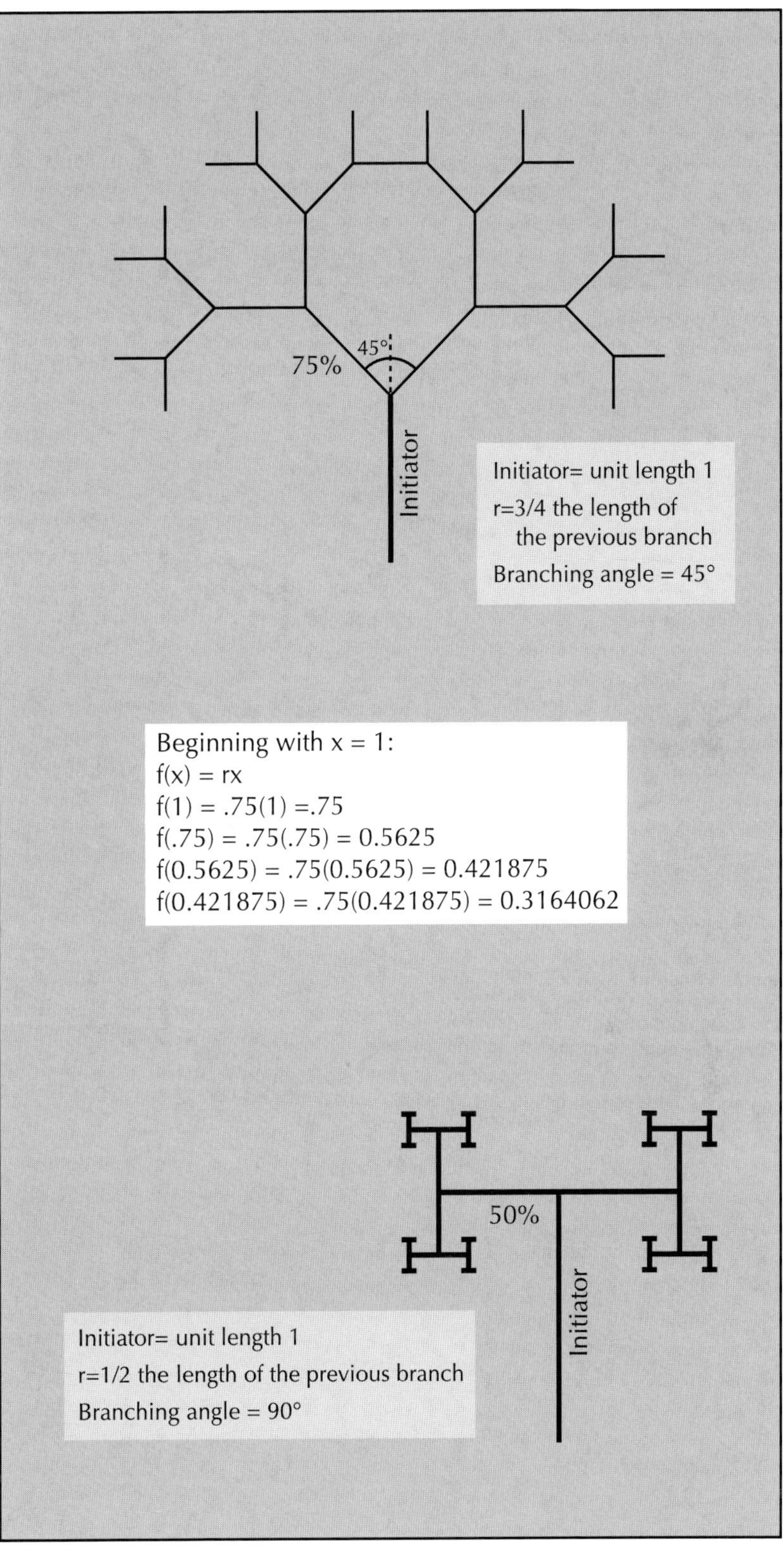

This page: Fig. 6.4
A tree and its branchings using an initiator of 1, and different lengths for r. The same branching angle is used. Notice that the length of each new branch gets smaller with each generation.

Procedure For Constructing a Branching Tree

Given line segment c

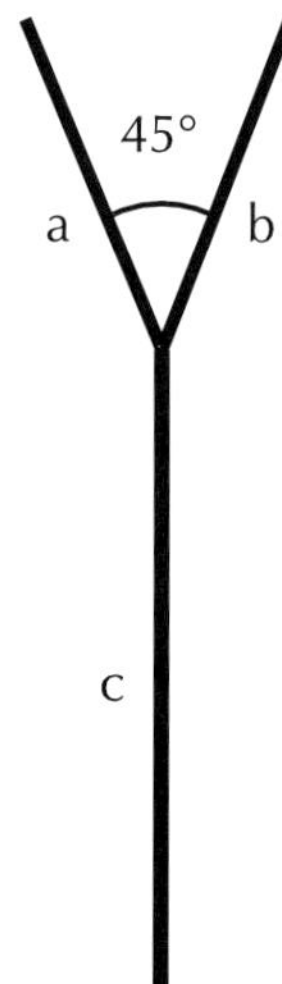

1. Add "branches" line segments a and b, that have an arbitrary angle, to line segment c. In this example it is 45°.

2. To the available "tips" of branches a and b, add branches d, e, respectively, making their lengths 1/2 of a and b.

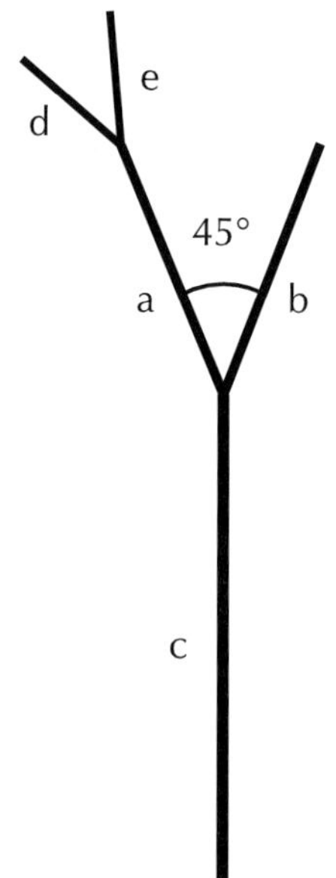

3. Repeat this process as many times as you like.

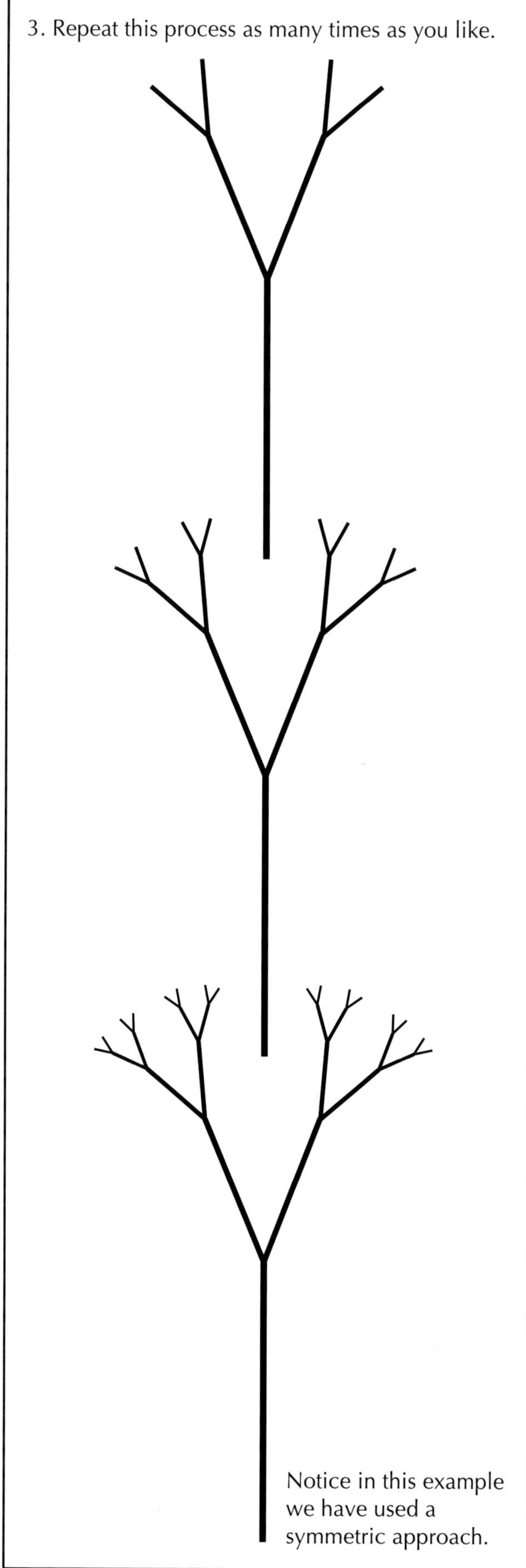

Notice in this example we have used a symmetric approach.

Notice here that this tree is developed from an asymmetric branching of elements.

To the right is a real tree, notice the similarities and differences.

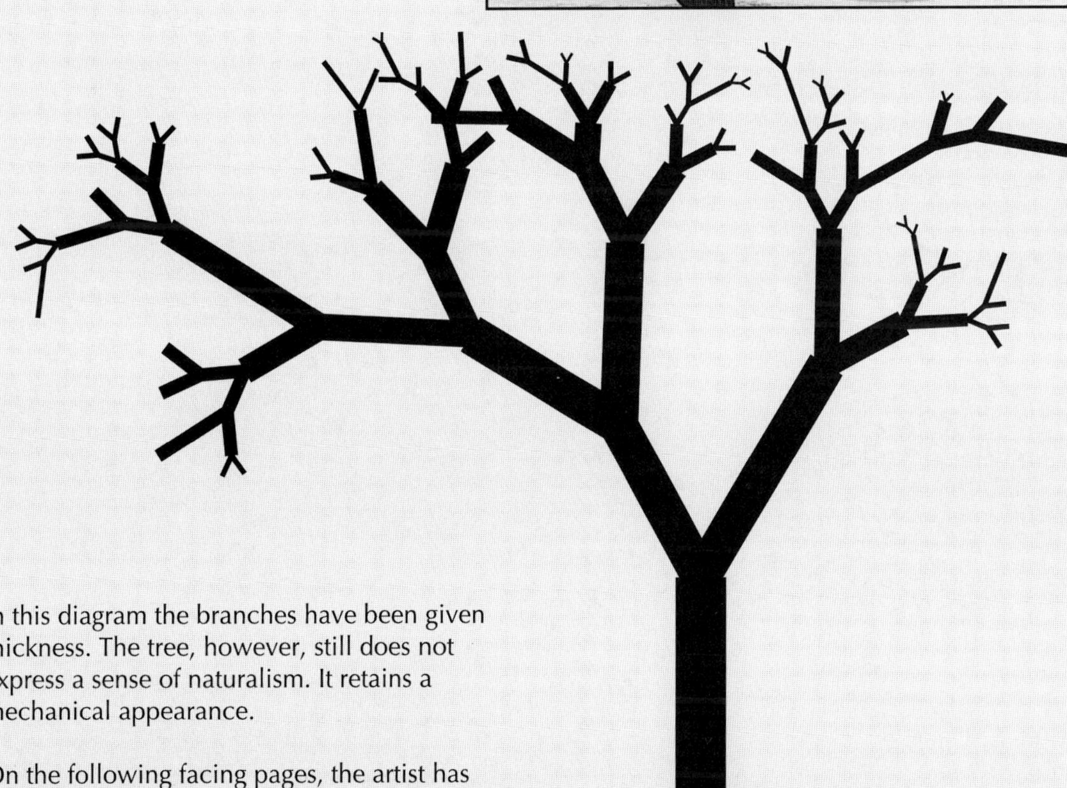

In this diagram the branches have been given thickness. The tree, however, still does not express a sense of naturalism. It retains a mechanical appearance.

On the following facing pages, the artist has "stretched" the tree structure into a more personal expressive vision. Other landscape elements have been added to create a composition.

Let us create a garden on paper in order to explore the design process. Imagine a garden area as if it were a contained in a volume of space of an implied rectangular solid which has the harmonic proportions of the Golden Ratio. Here we first show a plan view of a garden contained within an approximate Golden Rectangle of 10 units by 16 units (1: 1.6). Then, on the facing page, we give you a perspective view of the same garden.

Using colored pencils or watercolor on a photocopy of the plan given here, create a colored variation of this garden.

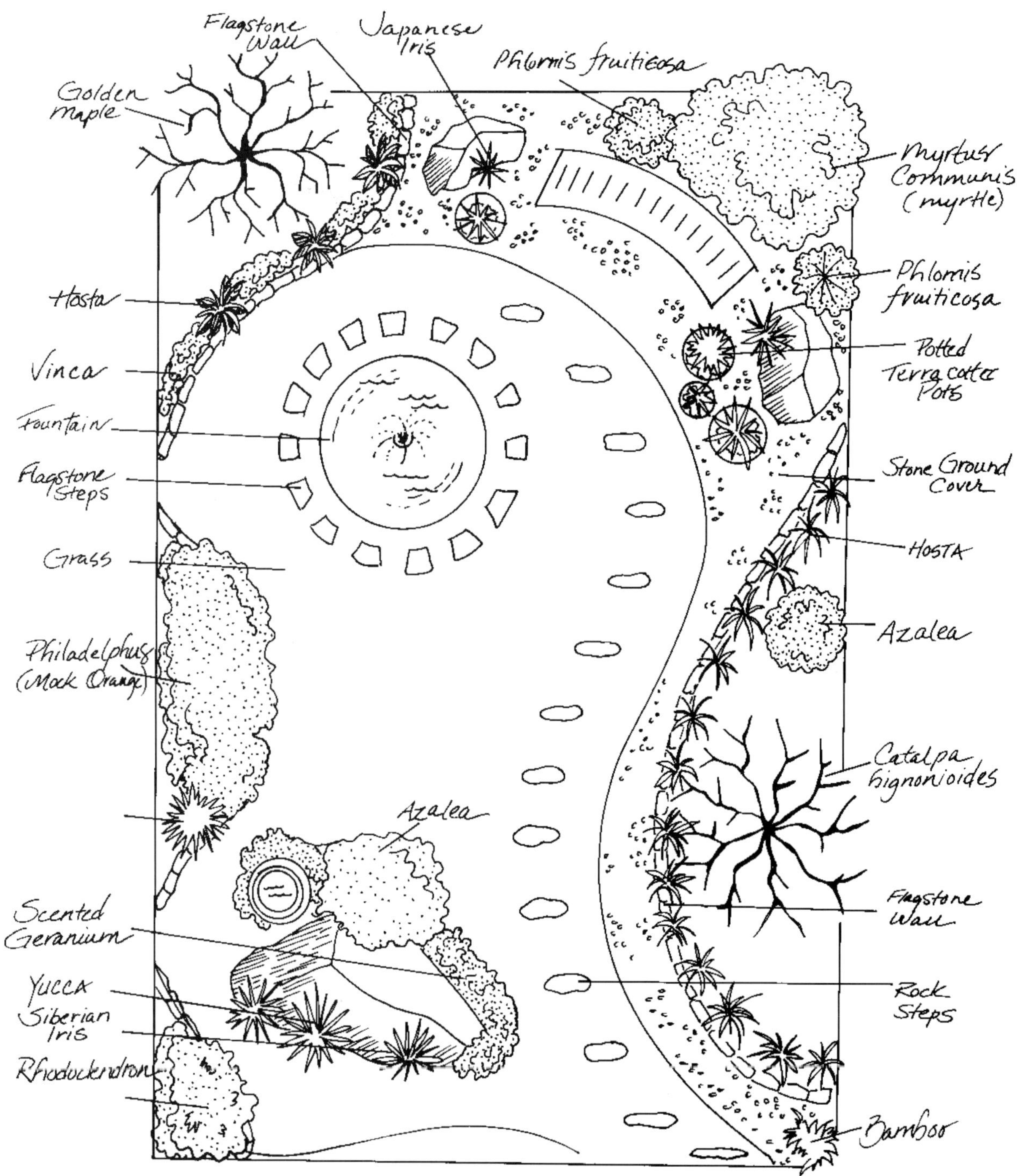

Try transferring the information given on these pages to an oblique and/or an isometric projection seeing how the different systems of spatial representation change your response to the garden.

Look at paintings, garden magazines, catalogs, and internet sites to learn more about gardens, plants, and the environment. Go to arboretums, zoos, botanical gardens, to see actual spaces in all their physicality of smell, movement, touch, space.

Opus 40

> Work is love made visible.
> The Prophet
> Khalil Gibran, poet

These following pages: Thanks to Tad Richards of the Opus 40 Foundation for allowing us conversation and permission to photograph this very special sculptural and environmental experience.

We encourage our readers to rejoice in the spirit of creativity that brings forth such works.

Opus 40, 1939-1976
7480 Fite Road
Saugerties, New York
6.5 acres in a bluestone quarry.

At the end of Chapter 1, we showed the work of an individual's two-dimensional images of the landscape. We close this chapter, and book, with a look at a three-dimensional work that is created of the landscape by a single individual. We celebrate the uniqueness of vision that each brings to our perceptions of this world. The multiplicity of perceptions brings richness to the tapestry of Life.

The journey to the center of the self never follows a straight path but involves the curve of the spiral. Artist Harvey Fite, born 1903 in the United States, in his search for the center travelled through the areas of law, theology, and acting until a chance encounter with a whittling knife and an empty three spool touched the vital never of sculpture-making. In his attempt to define sculpture for himself, a turn in the road took him to Honduras, Central America, where he encountered the dry laying of stone masonry and ancient Mayan sculpture.

Back in his home territory of New York State, he bought a 12 acre abandoned bluestone quarry that he anticipated would provide him with the material to build bases for his figurative sculpture. The quarrymen had left a legacy of flat stone in mounded piles. He first used the rubble to build pedestals for his works and for ramps to join pedestal area to pedestal area.

A twenty-five year bend in the road brought him face-to-face with a nine ton monolith that required him to again rethink his journey. Realizing that the entire quarry landscape was his sculptural milieu forced the artist to let go of his figurative subject matter, but not his materials, in favor of an abstract composition that was a fusion of architecture, sculpture, earthwork, garden, and personal philosophy. He brought together mountain background, full grown trees, pools of water, and contoured land into a single statement. Using only his own instincts, his own physical strength, simple hand tools, and humble winches, he stacked, carved, moved and re-moved the inarticulate raw bluestone. With patience and inner sight, he shaped the landscape to his own vision joining geometry to intuition. Sculptural elements rise and fall. Circles connect by curvings walls to other circles. The single erect male yang dominant monolith high on a platform is set in a context of the multi-curved concave feminine yin. Sky high and earth low embrace each other.

While architectural marvels such as Stonehenge in England, the Cathedral of Chartres in France, and the Great Pyramids in Egypt were communal efforts, Opus 40 is the passion and persistence of a single individual and stands as a tribute to creative courage. Harvey Fite had intended to work the land for 40 years, but in his 37th year, he accidentally fell from a ladder onto the rocks of his sculpture. He died making his life and art one.

beauty is life,
When life unveils her holy face.
But you are life and you are the veil.
The Prophet
Khalil Gibran

When you work you are a flute through whose heart the whispering of the hours turns to music.

The Prophet
Khalil Gibran

Problems, Projects, and Play

1. Research proportion and number symbolism in the gardens of Islam. Document your work with writing, drawings, collage, etc

2. Research and document the public gardens, parks, conservation areas in your city or town. Do a comprehensive survey. Locate areas that you think could use more public gardens and open natural space. Write a proposal to your city government about such a project.

3. Research the public garden projects of Frederick Law Olmstead. In an essay, written, oral, visual, discuss the effects of such large-scale planning on the environment.

4. In an essay with photos, drawings, etc, compare and contrast some of the formal gardens of France and Italy. Show how each uses the elements of water, sculpture, plant materials, etc

5. Research the development of the total environmental complex in Arizona called "Arcosanti" by Paolo Soleri. Could lsuch a project be developed elsewhere? Develop in writing, an argument forl such a viewpoint for your own geographic area.

6. Research the concept and desing of the private residence called "Fallingwater" by architect Frank Lloyd Wright. In an essay, compare and contrast this work with theresidence of Alden B. Dow, a disciple of Wright, in Midllands, Michigan.

7. In writing, compare and contrast attitudes toward Nature of the Judaic-Chiristian heritage in western civilization and the Zen Buddhist heritage of the Japanese in eastern civilization.

8. Discuss, in a tape recording, the attitudes toward Nature of several Native American cultures of the United States.

9. Using a still, movie, or video camera, document a walk in the woods.

10. Through photocopies or reproductions, do an analysis of landscape painting in: Western art compared to Japan, China, or Persia.

11. On paper, develop a garden plan for your own particular livingenvironment. It could be a window box, a riased garden bed, a yard, etc. Use information on the Golden Ratio to set key subdivisions within your plan. Write a statement about the philosophy of your garden.

12. Research Persian rugs and analyze the geometry found within. the compositions.

13. Research the subject of the Japanese art of Bonseki. Construct a shallow tray based on the ratios of harmonic rectangles (Newman and Boles). Subdivide the rectangle using the Golden Ratio. Develop a miniature Japanese rock garden using stones, sand, and balsa wood.

14. Group Project. In conjunction with a local garden organization, develop and execute a pcket part for a neighborhood within your city or town. You will have to do a lot of research in preparation for such a project.

15. Locate a public wall in your community that is in need of enhancement. Design and execute a wall mural on the them of the four seasons. This could be an individuall or a group project.

16. Research and develop through written statements, photographs, drawings, paintings, collage, sculpture, etc. a low maintenance, envonmentally hardy, diverse, and highly visible garden for a highway median strip. Make sure you consider the speed at which vehicles will travel by this garden.
Contact your city officials, garden clubs, etc. to see if this garden can be executed. Find some hardworking and compatible friends and then do it as a gift to your community.

17. Develop and plant a group of window box fragrance gardens for the visually impaired.

18. Develop and execute a colored drawing, painting, collage, assemblage, scullpture of your impression of The Garden of Eden. You must include at least the following elements in your artwork:
a. a tree
b. a snake
c. a water element
d. a butterfly
e. a fish
f. an animal

19. If you have skills in any of the fiber arts, translate the above project into a fabric quilt, a woven tampestry, a knotted macrame, a knitted sweater, a crochet garment, an embroidered hanging, a beaded image.

20. Using found images from magazines and garden catalogues, create a Garden of Eden collage.

21. With typing paper and black crayon, make "rubbings" of natural objects you find in the course of a nature walk. Create an outwork from your results.

22. Find a garden in your neighborhood and attempt to capture a modest aspect of it by drawing, photography, painting, or sculpture.

23. Research the Persian garden and rug. In painting, drawing, collage, needlework, etc. conceive and execute a variation on this theme. Be modest in your choice of size for this project.

24. Using the pattern given in this chapter, execute four small kimonos in paper using the theme of the four seasons.

25. Research the artwork of Italina artist Giuseppe Archimboldo. Use mgazine components of vegetable, fruits, flowers, etc and do a self-portrait in the style of this artist.

26. Write and lay out a rough mock-up story for the very young child on an aspect of nature that is very special to you.

27. Contact your local cable TV station. Using their equipment and help from their staff, create a videoprogram that shows the positive aspects of gardens and gardening.

28. Listen to the musical expression. "The Rites of Spring" by Russian composer Stravinsky and create a colored drawing or painting in response to the music.

29. Find several poems about nature and do a collage in response to your reading of them

30. Read about Haiku poetry. Then do a black and white drawing of something in nature that has the same spare essence as the poetry. (Remember simple is not necessarily easy!)

31. Design and execute a poster that could be used for celebrating "Earth Day", Earth Week, etc.

32. Locate a nature protection group such as the Humane Society, Greenpeace, The Audubon Society, The Sierra Club, etc. do a series of drawings related to the content of their particular cause, and donate these images to that organization for their use.

33. Find out information on the Gorilla Foundation so that you could develop a series of drawings for an ad campaign, brochure, billboard, poster that would help their cause. Donate your results.

34. "Think globally, act locally" in the preservation of this your only home planet. Design and execute a TV commercial or a Website.

Further Reading

Anderson,William. *Green Man*. San Francisco: HarperCollins. 1990.

Barrie,Thomas. *Spiritual Path, Sacred Place*. Boston: Shambala Press. 1996.

Boles, Martha and Rochelle Newman. *The Surface Plane*. Bradford: Pythagorean Press/Brown & Benchmark. 1996.

Doczi, Gyorgy. *The Power of Limits*. Boulder: Shambala Press. 1981.

Francis, Mark and Randolph T. Hester, Jr., Editors. *The Meaning of Gardens*. Cambridge:MIT Press. 1993.

King, Ronald. *Great Gardens of the World*. London: Peerage Books. 1985.

Messervy, Julie Moir. *The Inward Garden*. Boston: Little, Brown and Company. 1995.

Miller, Mara. *The Garden as an Art*. Albany: State Univesity of New York Press. 1993.

Newman, Rochelle and Martha Boles. *Universal Patterns*. Second Revised Edition. Bradford: Pythagorean Press. 1992.

Peterson, Ivars. *The Mathematical Tourist*. New York: w.H. Freeman and Co. 1988.

Slawson, David. *Secret Teachings in the Art of Japanese Gardens*. New York: Kodansha International. 1987.

Stevens, Peter. *Patterns in Nature*. Boston: Little, Brown and Company. 1974.

Conclusion

This is the end of this book. I have come to realize that my role as fool who jumped in where angels feared to tread, has allowed me to joyfully experience the mystery and beauty of this world, and to share my connections to it. Had I been a specialist who "knew" things, I would not have jumped so innocently into the abyss of my not knowing. I hope that my vulnerability will encourage others to take their own journeys down little trod paths.

For to be foolish is not really to be stupid.

Rochelle Newman

The Greenman, archetypal figure, and male counterpart to the Great Goddess figure has been found throughout time and re-appears when civilization is in need of him.

This page:
"Gaia in the Springtime of
Her Life."
Artist Gail Maciejewski.
Drawing done in the man-
ner of the Italian artist
Archimboldo.

Selected References

Atkins, P.W. *Creation Revisited*. Oxford and New York: W. H. Freeman & Co., 1992.

Baggott, Jim. *Perfect Symmetry*. Oxford, New York, and Tokyo: Oxford University Press, 1994.

Banchoff, Thomas. F. *Beyond The Third Dimension*. New York: Scientific American Library, 1990.

Bartusiak, Marcia. *Through a Universe Darkly*. New York: HarperCollins Publishers, 1993.

Billings, Robert. *Power of Form Applied to Geometric Tracery*. London: W. Blackwood and Sons, 1851.

Boles/Newman. *The Surface Plane*. Bradford: Pythagorean Press, 1992.

Bouleau, George. *The Painter's Secret Geometry*. New York: Harcourt Brace and World, 1963.

Bruni, James. *Experiencing Geometry*. Belmont: Wadsworth, 1977.

Capra, Fritjof. *The Tao of Physics*. New York: Bantam Books, 1965.

Capra, Fritjof. *The Turning Point*. New York: Bantam Books, 1983.

Cole, Rex Vicat. *Perspective for Artists*. New York: Dover, 1976.

Colman, Samuel. *Natures's Harmonic Unity*. New York Putnam's and Sons, 1911.

Cotterill, Rodney. *The Cambridge Guide to the Material World*. Cambridge: Cambridge University Press, 1985.

Cohn-Vossen. *Geometry and the Imagination.* New York: Chelsea Publishing Co, 1962.

Cooper, Doublas. *Drawing and Perceiving.* Wecond Edition. New York: Van Nostrand Reinhold, 1992.

Courant, Richard and Herbert Robbins. *What is Mathematics?* New York: Oxford University Press, 1941.

Coxeter, H.S.M. *Introduction to Geometry.* New York: John Wiley and Son, 1961.

Critchlow, Keith. *Order in Space.* New York: The Viking Press, 1970.

Cundy and Rollett. *Mathematical Models.* New York: Oxford University Press, 1961.

Doczi, Gyorgy. *The Power of Limits.* Boulder: Shambala Press, 1981.

Edwards, Edward B. *Patterns and Design with Dynamic Symmetry.* New York: Dover, 1967

Feininger, Andreas. *The Anatomy of Nature.* New York: Dover, 1956.

Field, J. V. *The Invention of Infinity.* Oxford, New York, Tokyo: Oxford university Press, 1997.

Fomenkp, Anatolii T. *Mathematical Impressions.* Rhode Island: American Mathematical Society. 1990.

Friday, Adrian and David S. Ingram, General Editors. *The Cambridge Encyclopedia of Life Sciences.* New York: Cambridge University Press, 1985.

Fuse, Tomoko. *Unit Origami.* Tokyo: Japan Publications Inc., 1990.

Gardner, Martin. *The New Ambidextrous Universe,* Third Revised Edition. New York: W.H. Freeman and company, 1990.

Ghyka, Matila. *The Geometry of Art and Life.* New York: Dover, 1977.

Grunbaum/Shephard. *Tilings and Patterns.* New York: W.H. Freeman and Company.1987.

Hambidge, Jay. *The Elements of Dynamic Symmetry.* New York: Dover, 1967.

Hambidge, Jay. *Practical Applications of Dynamic Symmetry.* New Haven: Yale University Press, 1932.

Holden, Alan. *Shapes, Space and Symmetry.* New York: Columbia University Press, 1971.

Holden, Alan. *Orderly Tangles.* New York; Columbia University Press, 1983.

Huntley, H.E. *The Divine Proportion.* New York: Dover, 1970.

Kaku, Michio. *Hyperspace.* New York and Oxford: Oxford University Press, 1994.

Kline, Morris. *Mathematics in Western Culture.* New York: Oxford University Press, 1953.

Lord, E.A. and C. B. Wilson. *The Mathematical Description of Shape and Form.* New York. John Wilcy & Sons, 1984.

Mandelbrot, Benoit. *The Fractal Geometry of Nature.* New York: W. H. Freeman and Co., 1983.

Montague, John. *Basic Perspective Drawing, Second Edition.* New York: Van Nostrand Reinhold, 1993.

Newman/Boles. *Universal Patterns: The Golden Relationship: Art, Math & Nature.* Bradford: Pythagorean Press, 1992.

Newman, Rochelle and Donna Fowler. *Space, Structure and Form.* Bradford: Pythagorean Press/Brown & Benchmark, 1996.

Pearce, Peter and Susan Pearce. *Polyhedra Primer.* New York: Van Nostrand Reinhold, 1978.

Peterson, Ivars. *Island of Truth.* New York: W.H. Freeman and Company, 1990.

Peterson, Ivars. *The Mathematical Tourist.* New York: W.H. Freemand and Company, 1988.

Schattschneider, Doris. *Visions of Symmetry.* New York: W.H. Freeman and Company, 1990.

Stevens, Peter. *Patterns in Nature.* Boston, Little, Brown and Company, 1974.

Trudeau, Richard, *The Non-Euclidean Revolution.* Boston: Birkhauser, 1987.

Washburn, Dorothy K. and Donald W. Crowe. *Symmetries of Culture.* Seattle: University of Washington Press, 1988.

Wenninger, Magnus J. *Polyhedron Models.* Cambridge: Univesity Press, 1971.

Wenninger, Magnus J. *Spherical Models.* Cambridge: University Press, 1979.

Weyl, Hermann. *Symmetry.* Princeton: University Press, 1952.

Williams, Robert. *The Geometrical Foundation of Natural Structure: A Source Book for Design.* New York: Dover, 1979.

Zee, A. *Fearful Symmetry.* New York: MacMillan Publishing Co., 1986.

Zeier, Franz. *Paper Constructions.* New York: Charles Scribner's Sons, 1980.

Design Appendix

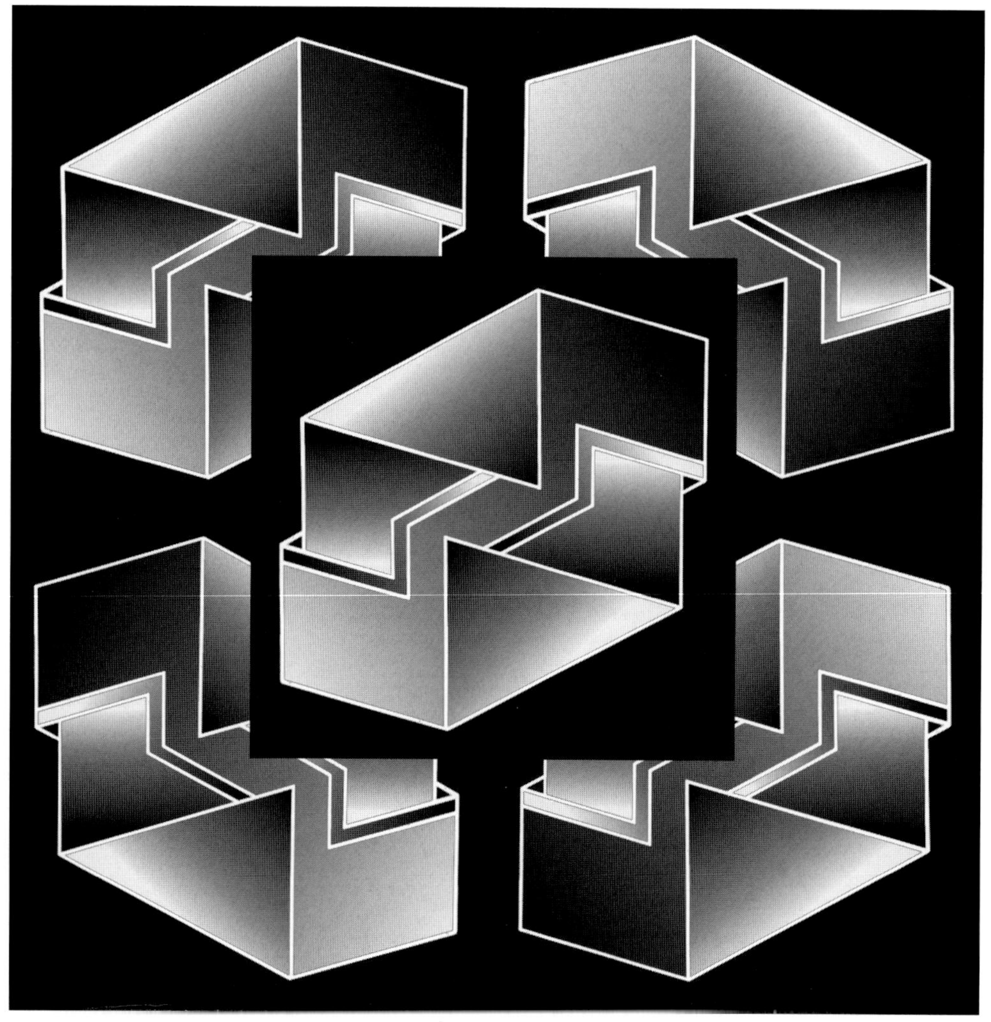

The charts on the following pages are synopses of concepts developed throughout the six chapters of this book. For detailed information, return to the body of the text.

The information found in the above mentioned charts is related to the design issue of composition which involves surface, space, and spatial sub-divisions. The source of these is the structure of the natural world with its efficiency, economy, organization, function and form. Geometry is used to define and describe these spatial relationships in a precise manner.

A composition is a system of relationships within a given field of space that has unity through the balancing of opposing elements as well as the repetition of them. This is essentially accomplished through the control of proportion, whether it be shapes, colors, textures, patterns, lines, tones, or figure and ground.

The design process is a problem-solving activity which involves a person, a concept, materials, techniques, and a form structure that is created by the imaginative manipulation of the design principles. Design is not about unlimited freedom but freedom within limits.

Any quality work of art is a self-contained organism built from diverse elements but having the common goal of unity. The whole and the parts partake of the same order. To achieve this wholeness, full attention must be given to all of the designalphabet of point, line, shape, texture, pattern, color which are manipulated by the basic design grammar of symmetry and asymmetry; repetition and similarity; ratio and proportion; harmony and contrast.

Chart A is devoted to the universal concept of symmetry which involves a motif and its position and orientation in space. Symmetry involves the spatial operations of translation, rotation, and reflection. These make use of repetition around a point, along a line, across a plane.

Chart B refers to a set of rectangles that has special properties of similarity which make them especially useful for design in both two and three dimensions.

Chart C is an extended look at the concept of the Golden Ratio and how it provides a special kind of harmony. It is embedded within a great many forms in the natural world and provides unity in design.

In addition to these three charts, we provide a fourth, D, that is concerned with color and tone. Both of these are very important elements of the design vocabulary. In the design of this book, tone plays a key role.

The Concept of Repetition:
Symmetry Groups in Two Dimensions *(relates to position and orientation of objects in the plane)*

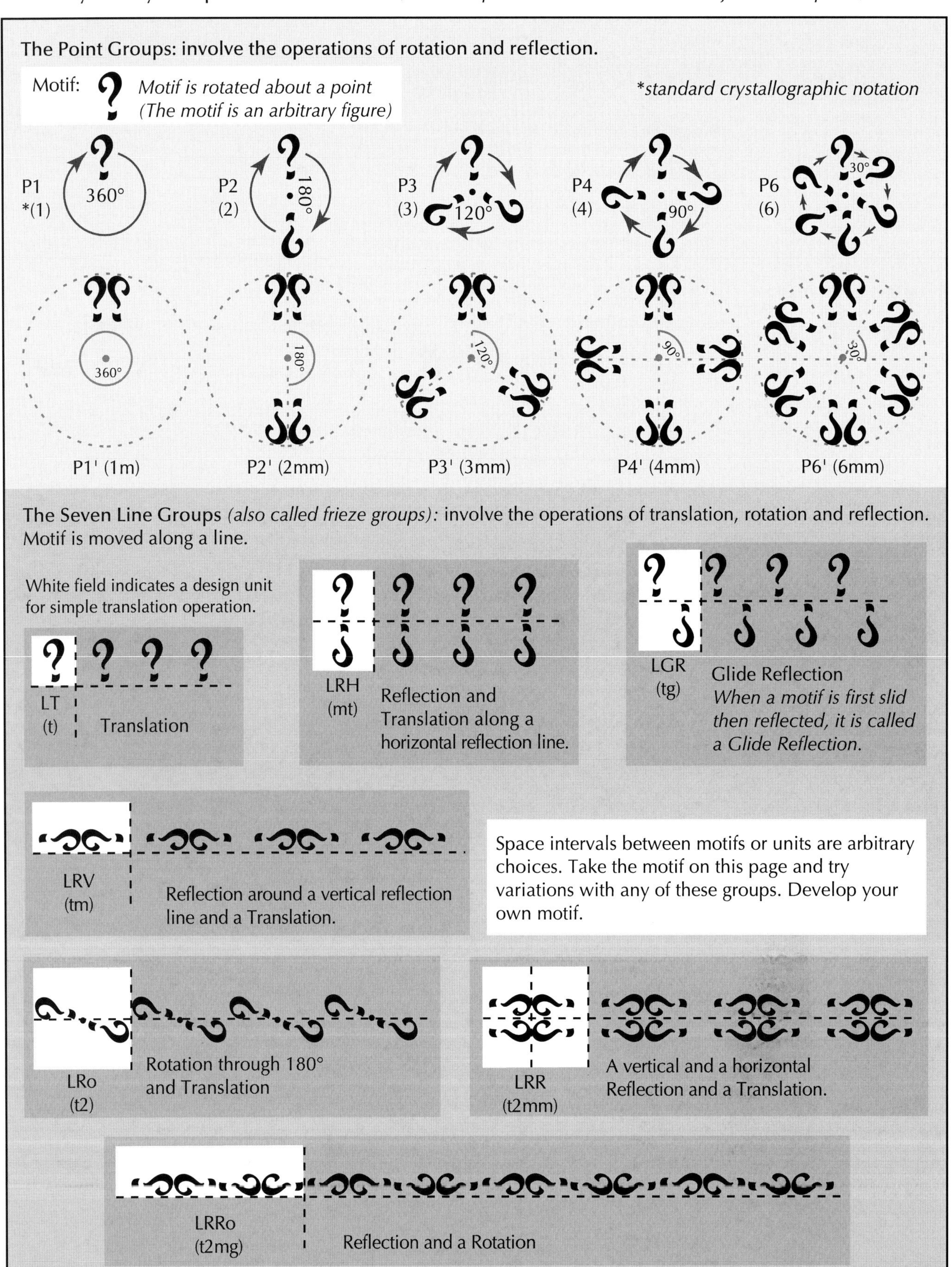

The Concept of Repetition:

The Seventeen Plane Groups *(also called the wall paper groups)*

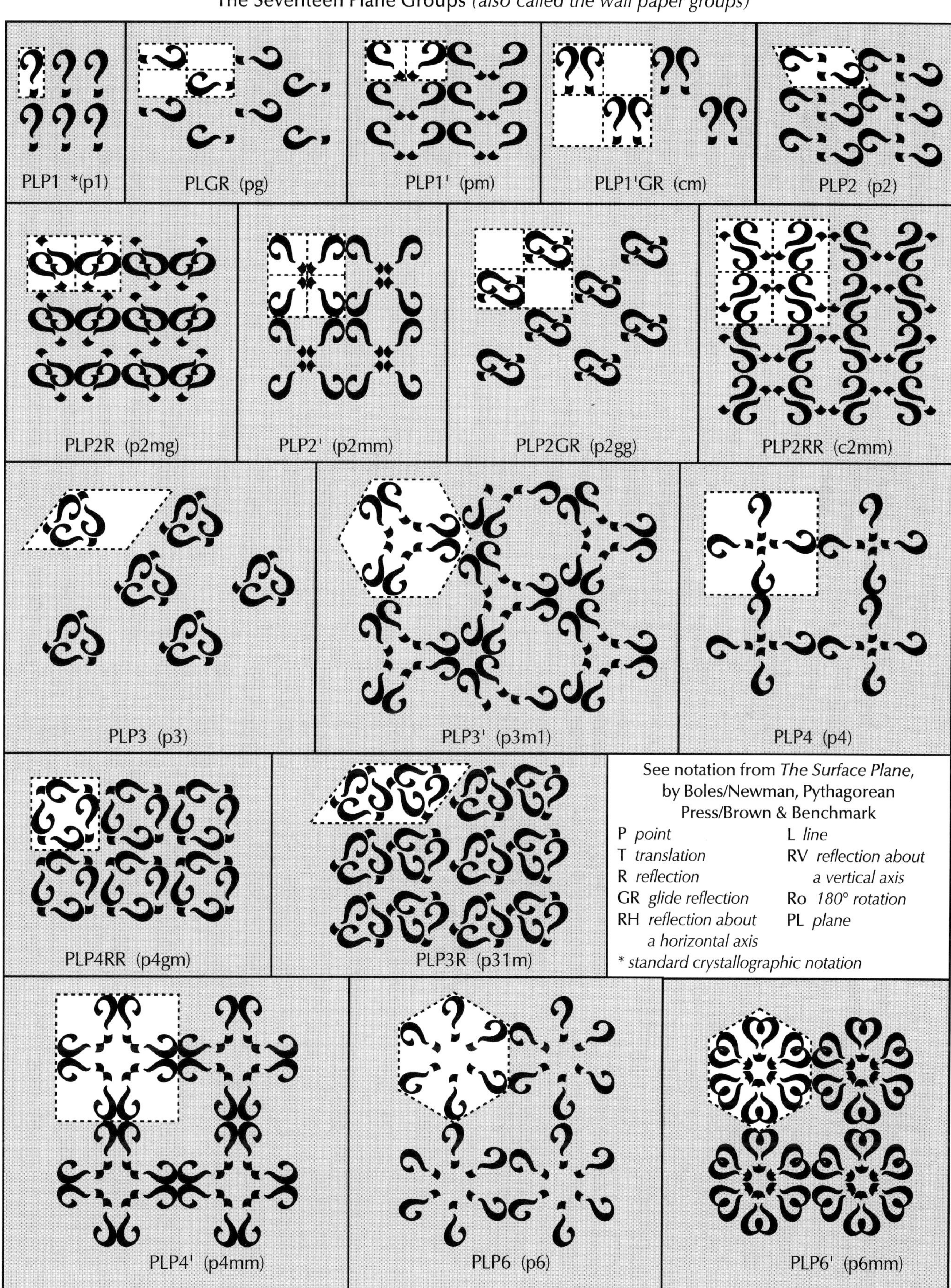

The Concept of Similarity:
The Dynamic Rectangles

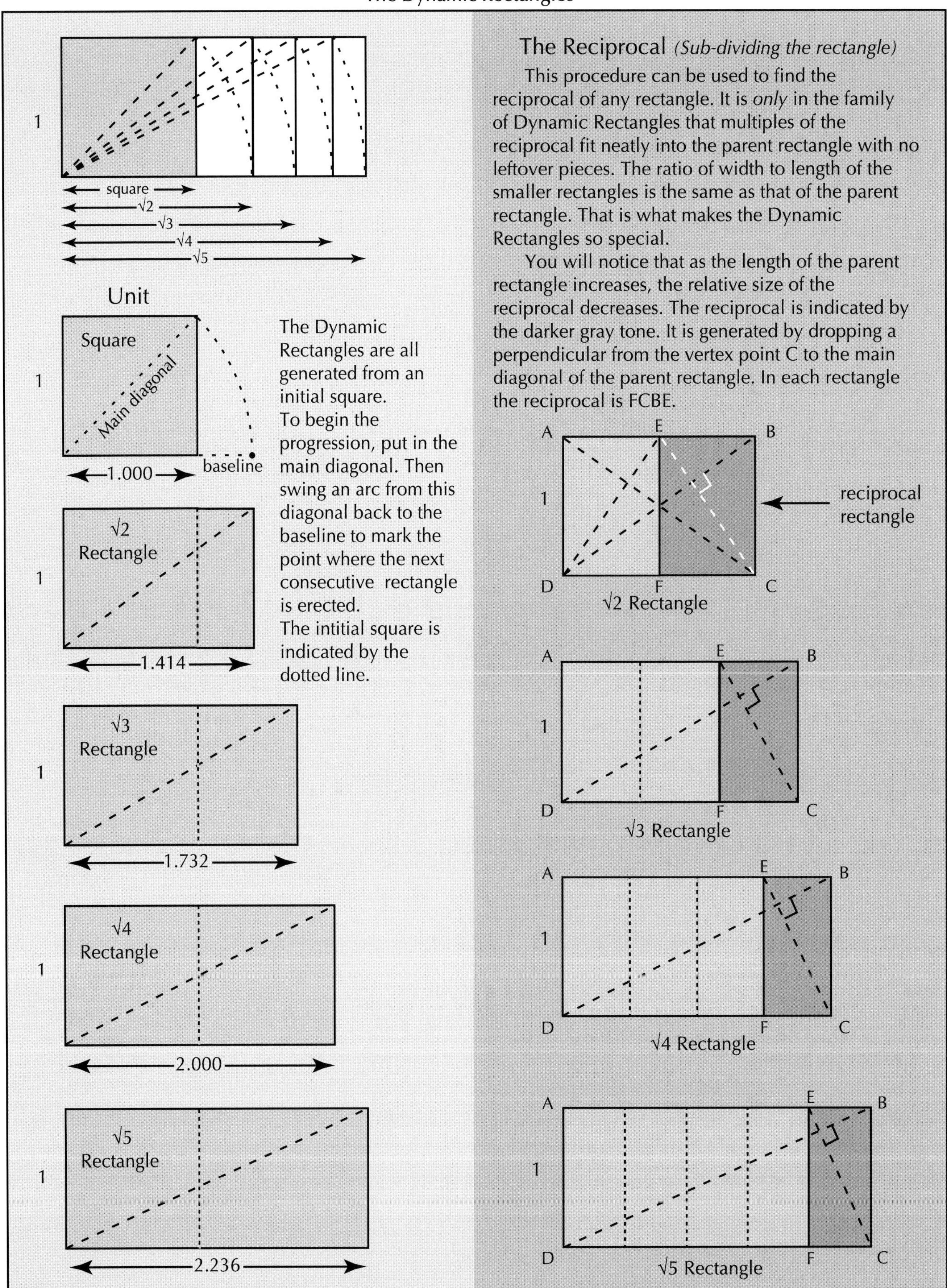

The Dynamic Rectangles are all generated from an initial square.
To begin the progression, put in the main diagonal. Then swing an arc from this diagonal back to the baseline to mark the point where the next consecutive rectangle is erected.
The intitial square is indicated by the dotted line.

The Reciprocal *(Sub-dividing the rectangle)*

This procedure can be used to find the reciprocal of any rectangle. It is *only* in the family of Dynamic Rectangles that multiples of the reciprocal fit neatly into the parent rectangle with no leftover pieces. The ratio of width to length of the smaller rectangles is the same as that of the parent rectangle. That is what makes the Dynamic Rectangles so special.

You will notice that as the length of the parent rectangle increases, the relative size of the reciprocal decreases. The reciprocal is indicated by the darker gray tone. It is generated by dropping a perpendicular from the vertex point C to the main diagonal of the parent rectangle. In each rectangle the reciprocal is FCBE.

The Dynamic Rectangles cont....

A spiral can be generated by using the Dynamic Rectangles consecutively. Starting with the square place the next rectangle along the main diagonal. Continue this process until you've used as many rectangles as you need to produce the desired spiral.

To produce a skeletal structure on which to build the body of a composition, construct an armature. To subdivide the area of each rectangle, use the main diagonals, the diagonals of reciprocals, midpoints, etc.

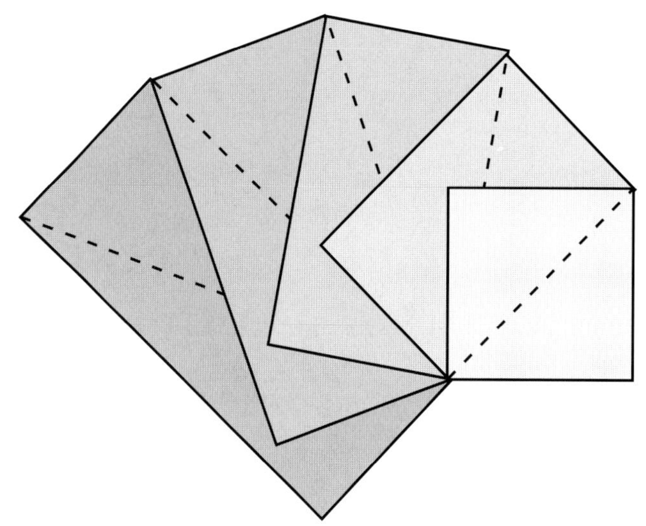

Harmonic Subdivisions *(armatures of design)*	Angular Spirals	Parallelograms
√2 Rectangle	√2	√2
√3 Rectangle	√3	√3
√4 Rectangle	√4	√4
√5 Rectangle	√5	√5

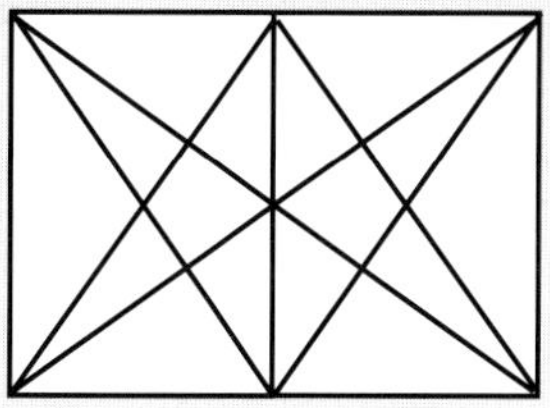

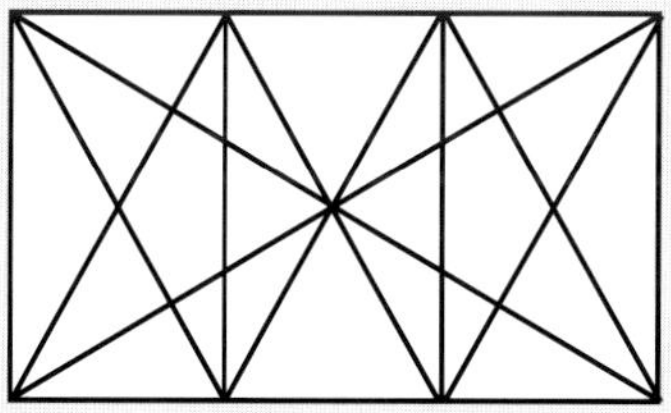

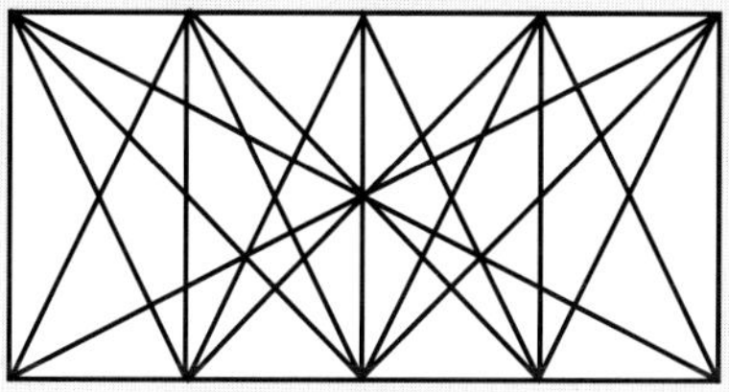

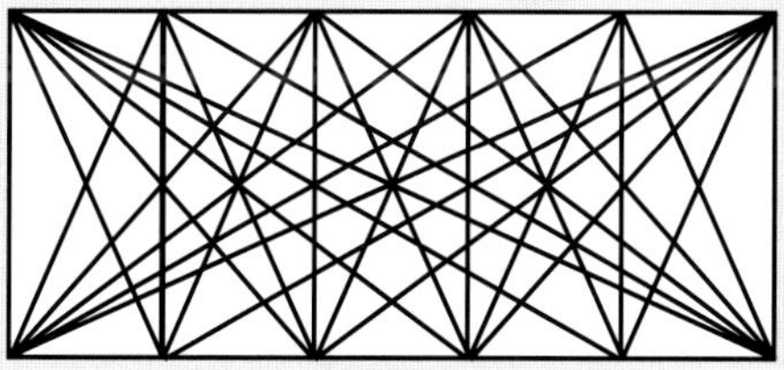

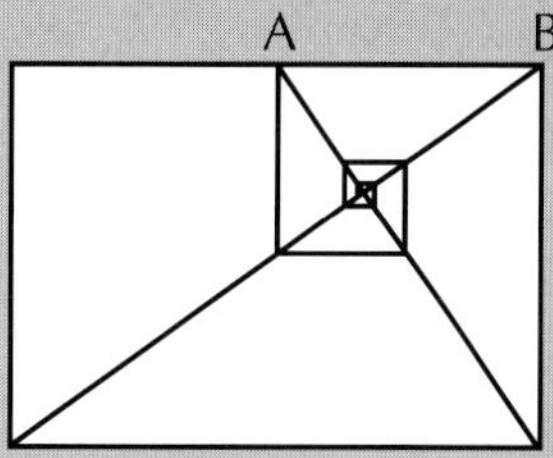

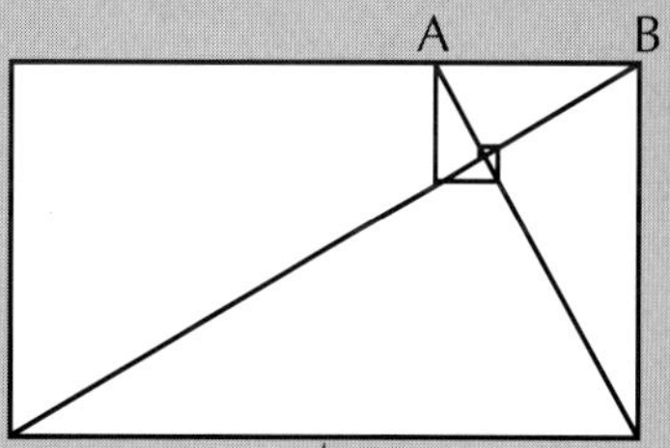

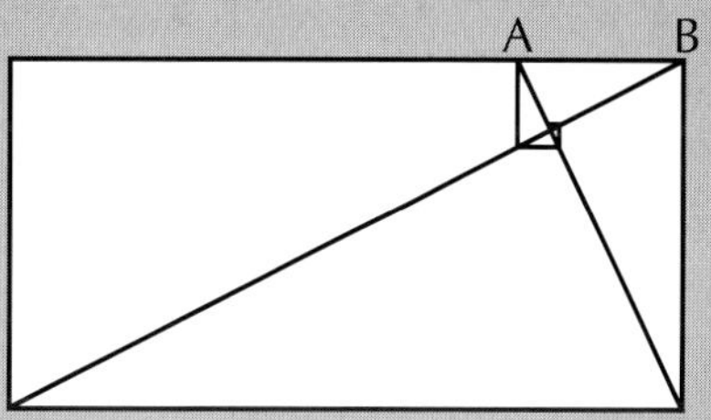

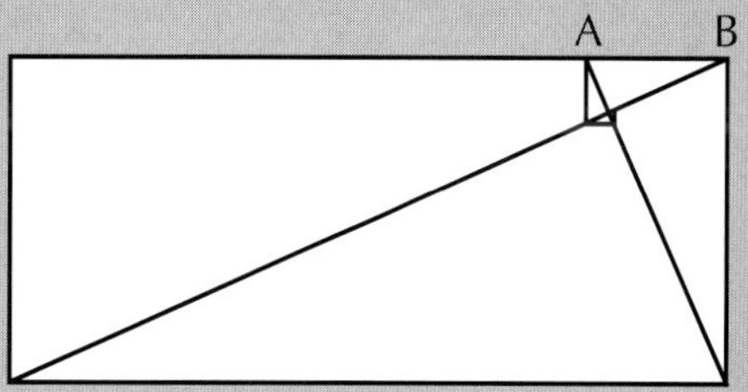

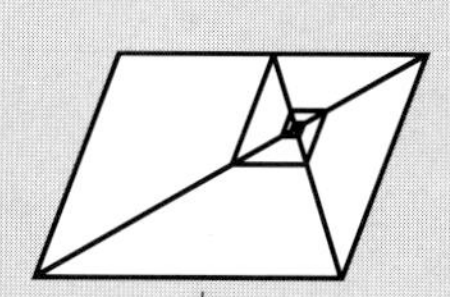

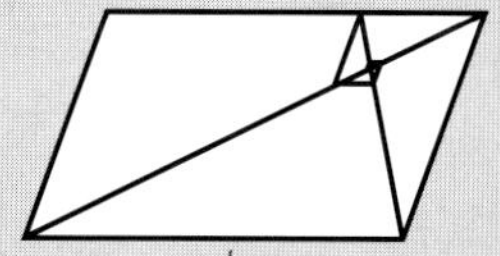

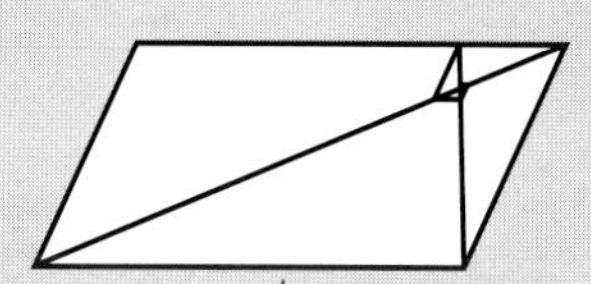

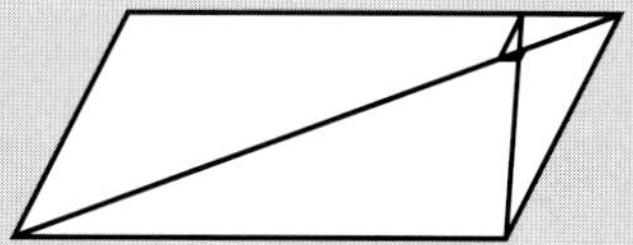

The Dynamic Rectangles cont....

Procedure For Constructing the Dynamic Rectangles Within a Square

Given square ABCD

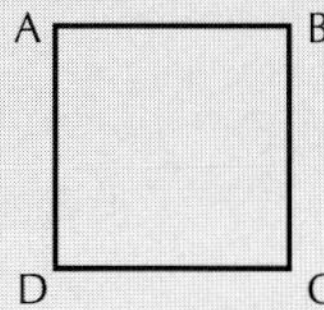

I

1. Place the metal tip on D and the pencil tip on A and cut arc AC. Draw line segment DB, and label the point of intersection E of arc AC and line segment DB.

2. From E construct a perpendicular to line segment BC and extend it so that it intersects line segment AD. Label the points of intersection F and G.
Now FDCG is a √2 rectangle.

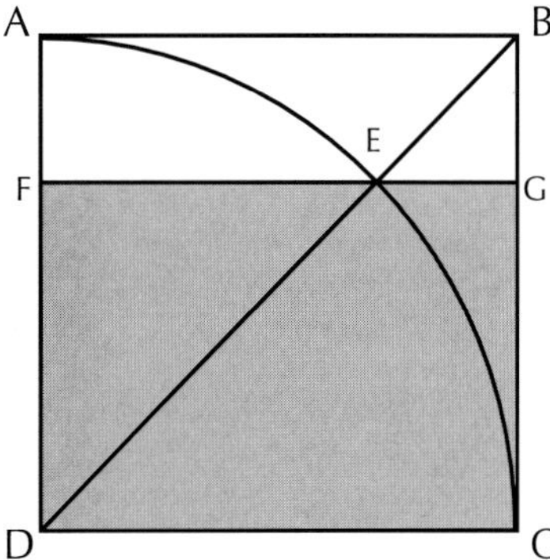

II

1. Draw line segment DG, and label H the intersection of arc AC and DG.

2. From H construct a perpendicular to line segment BC and extend it so that it intersects line segment AD. Label the points of intersection I and Q
Now IDCQ is a √3 rectangle.

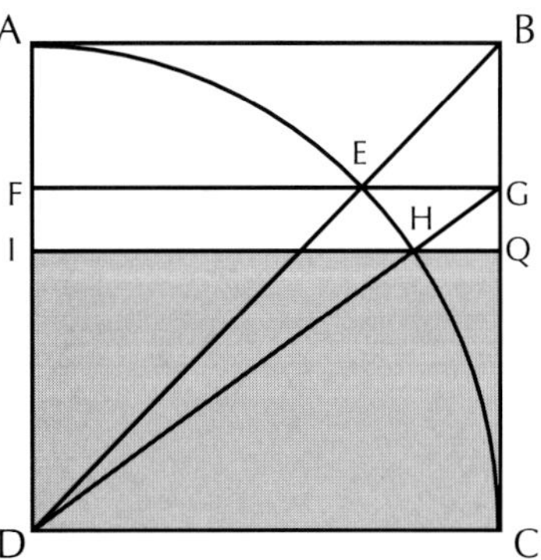

III

1. Draw line segment DQ, and label K the intersection of arc AC and DQ.

2. From K construct a perpendicular to line segment BC and extend it so that it intersects line segment AD. Label the points of intersection L and M.
Now LDCM is a √4 rectangle.

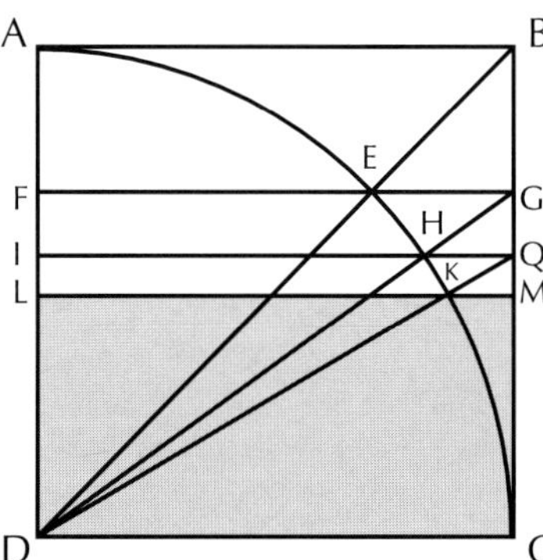

IV

1. Draw line segment DM, and label N the intersection of arc AC and DM.

2. From N construct a perpendicular to line segment BC and extend it so that it intersects line segment AD. Label the points of intersection O and P.
Now ODCP is a √5 rectangle.

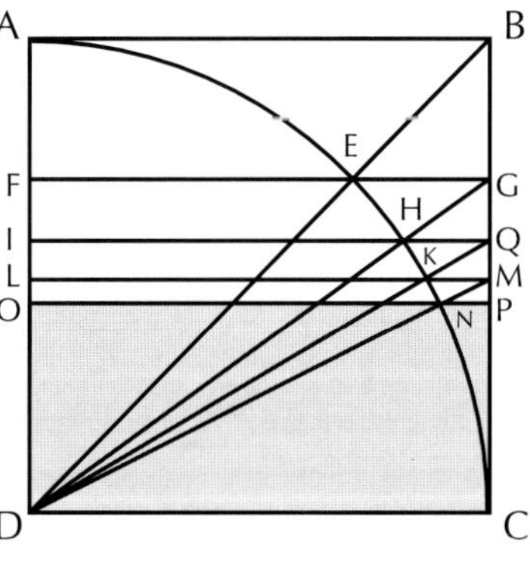

The Dynamic Rectangles cont....

Dynablocks

The Dynamic Rectangles generate a set of three dimensional solids called the Dynablocks. Two of the faces are squares (unit 1), and two of the faces are any of the dynamic rectangles. The series can go beyond the ones given on this page.

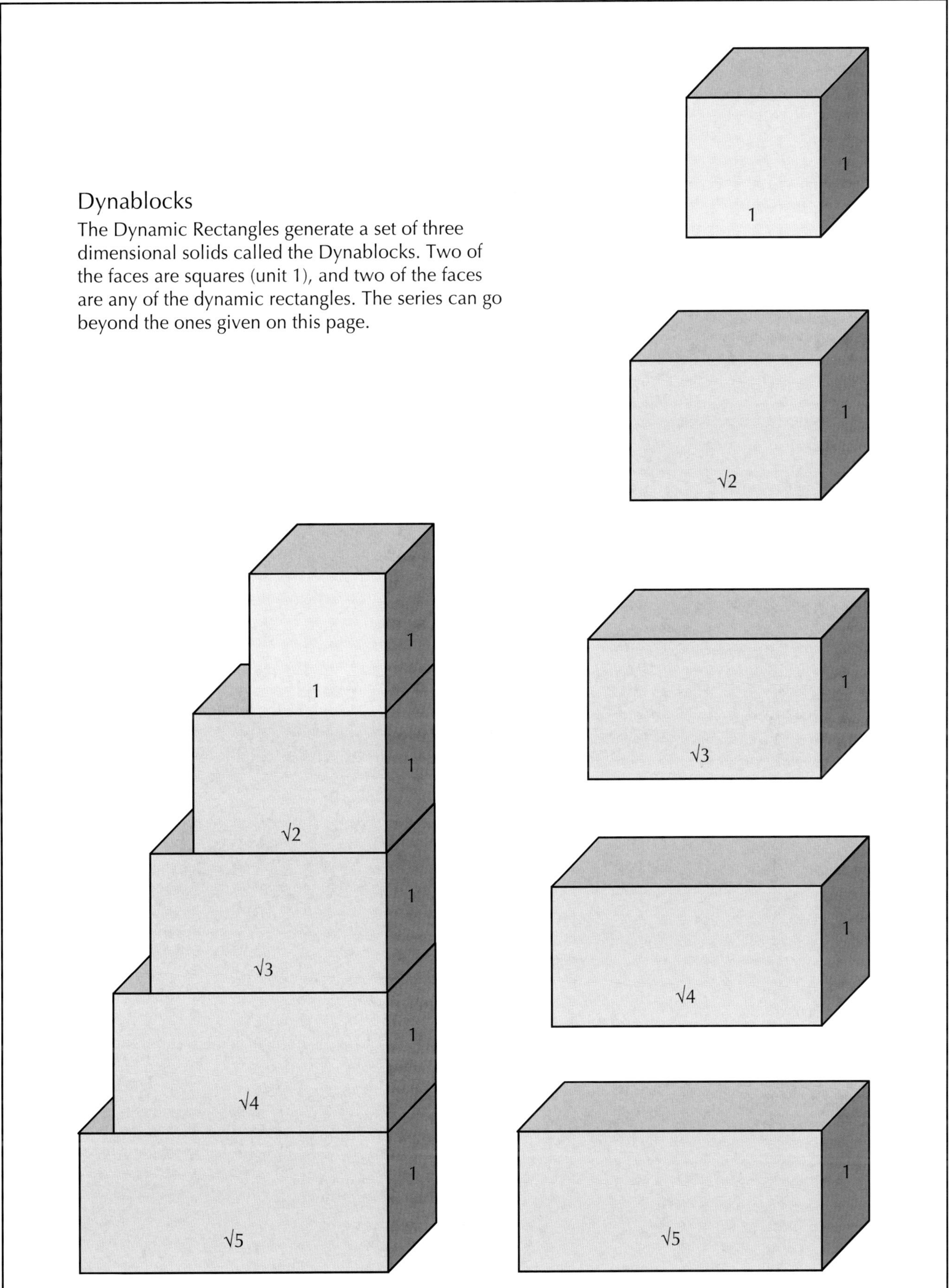

The Golden Thread that weaves Art, Math, and Nature Together

The practice of design involves working with the concepts of repetition (symmetry), contrast (difference), and similarity (ratio and proportion). The latter concept is what we will focus on here. Similarity involves the idea that two figures are the same shape but not the same size.

There is a special relationship called the "Golden Ratio" which is embedded in the natural world. It is growth by similarity. This ratio generates a special proportion called the "Divine Proportion".

1:1
1:2
1:3
1:4

A ratio is a comparison of numbers only, as in the illustrations below. These comparisons can be expressed algebraically as a:b where a and b stand for any number.

Any length can be designated as the number *1*. Once this "unit length" is determined for a given situation, all other lengths are related to it. The relationship is that of quantity only. There is no consideration of the distinct qualities that a particular object has. Notice that we are comparing hands to faces.

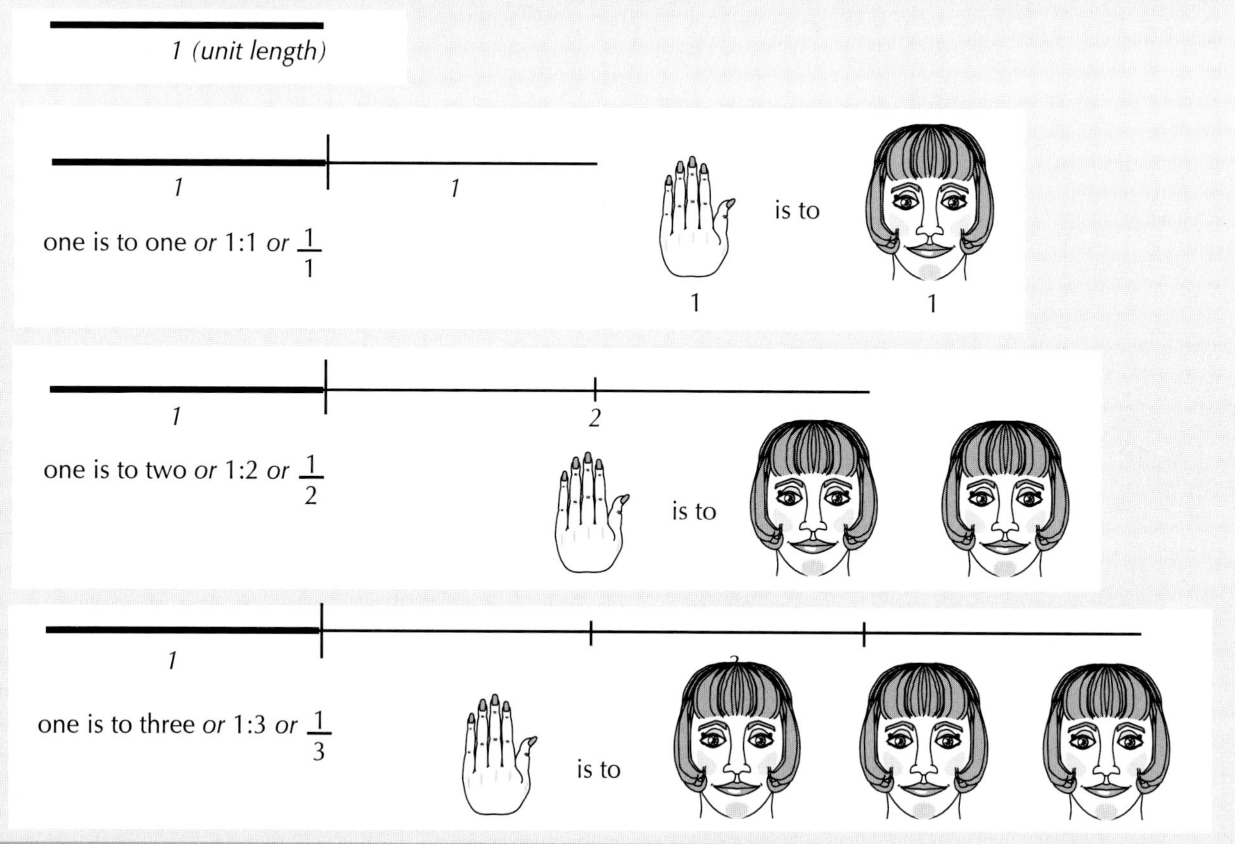

A proportion is a comparison of ratios.

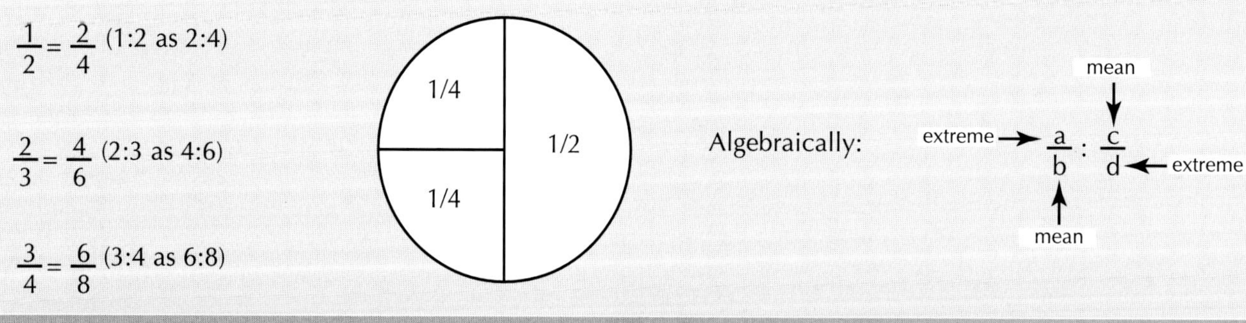

A mean proportion is one in which the means are equal: $\frac{a}{b} = \frac{b}{c}$

The Divine Proportion is a mean proportion: $\frac{AB}{AC} = \frac{AC}{CB}$

Procedure For Finding the
Golden Cut of a Given Line Segment AB

Given *any* line segment, there is a unique place on it that can be marked (with a point) called the Golden Cut that divides that line segment into two pieces such that the ratio of the smaller piece to the larger piece is the same as the ratio of the larger piece to the whole. Each piece can be given a numerical value.

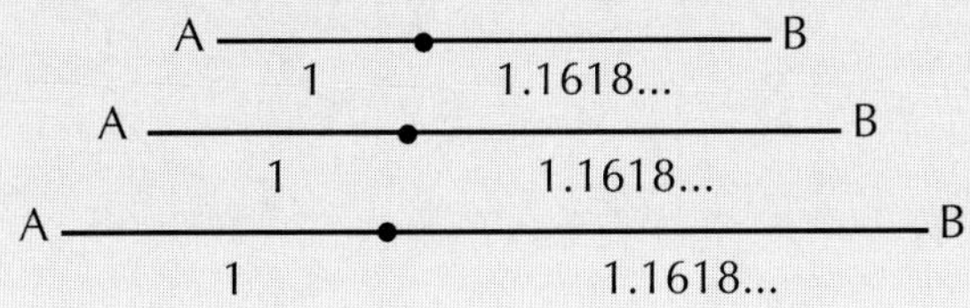

1. Choose a unit length *a*. Place a copy of *a, a'*, contiguous to it. Label the entire line segment AB.

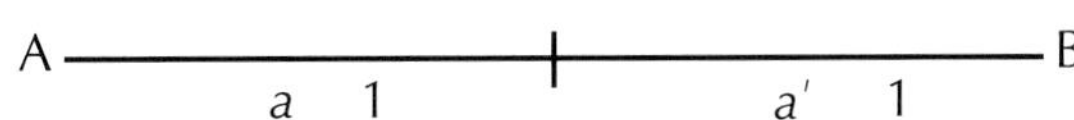

2. Set line segment *a"*, which is the same length as *a*, perpendicular to the other two segments. Label that BC.

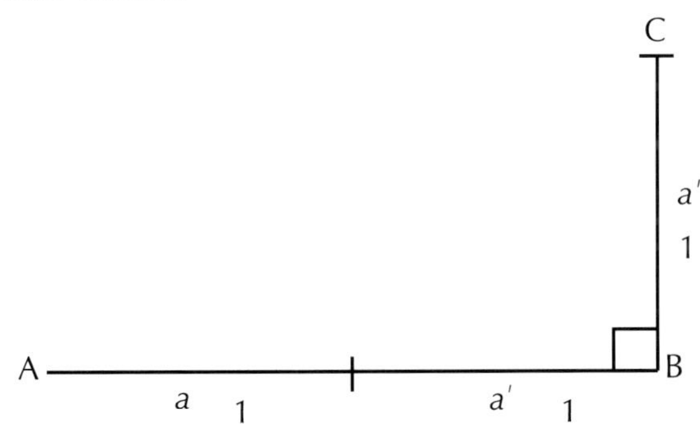

3. Connect AC with dotted line. This is for procedural purposes only. With the metal tip of your compass at C, measure from C to B, and cut an arc that intersects line segment AC. Mark the point of intersection E.

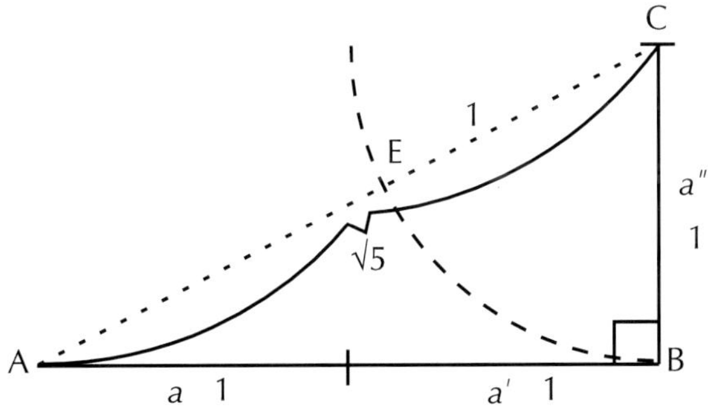

4. With the metal tip of your compass at A, measure from A to E, and cut arc ED. D becomes the Golden Cut of line segment AB.

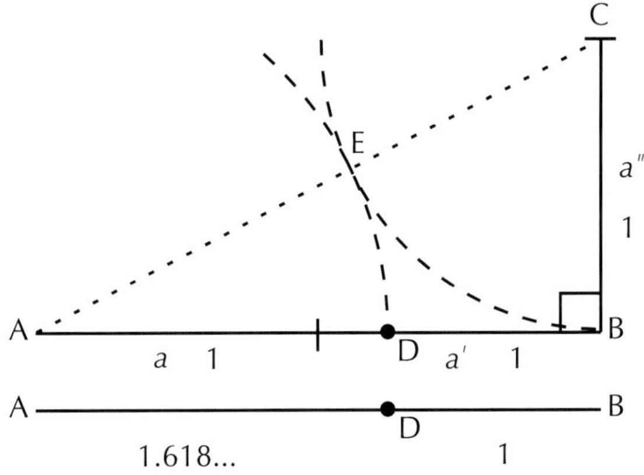

This can be proved algebraically using numbers.

Now you have two pieces that are design elements that can be manipulated playfully.

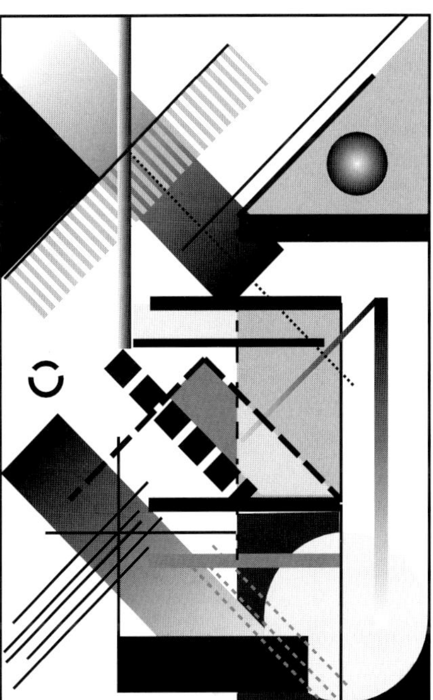

This artful image has been constructed primarily by using the two line segments of 1 and 1.618.

The Golden Thread that weaves Art, Math, and Nature Together

329

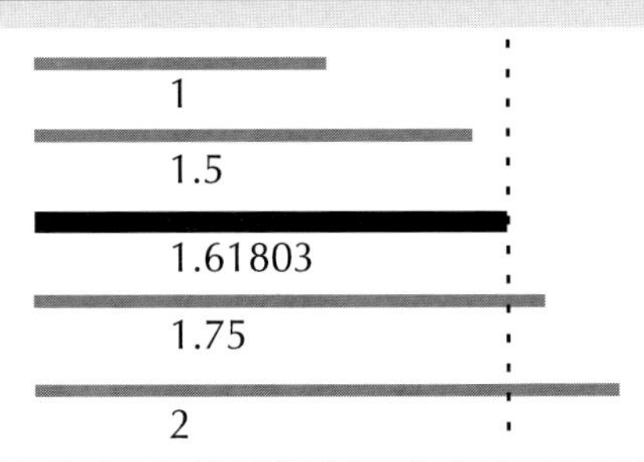

The Golden Ratio Comparison

As you have seen, the Golden Cut can be found on any given line segment. The Golden Cut is located at a point between 1.5 and 1.75. In order to visualize the subtleties of placement, the visual to the left shows the relationship.

A compass is not always necessary in order to find the Golden Cut of a line segment for a design project. A calculator can be very useful here.

Example: Given a line segment of 8 cm, punch in the number 8 into the calculator and use the divide button to divide 8 by 1.61803 which will find the number that designates the Golden Cut. The answer is 4.9443, which is approximately 5. For a mathematician 5 is not accurate enough but for an artist it is adequate.

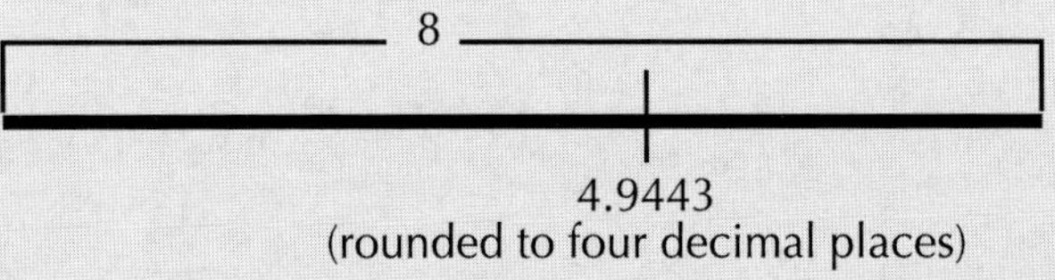

(rounded to four decimal places)

The Golden Cut is the simplest asymmetric division of a line segment that has a proportional relationship.

Now try dividing 13 cms by 1.61803 . The answer is 8.0334.

Now try it with 21 cms. The answer is 12.9789.

The particular whole numbers that we have used here belong to the Fibonacci Sequence discussed later on in this chart. The Golden Ratio was discovered by the Greeks way back in 5th century BC and was the basis for much of the art produced at the time. You might like to research Classical Greek sculpture , particularly that of the human figure, and architecture. You might also like to compare the proportions of the figure found in chapter two shown here, with examples from art history. On an ideal body the Golden Cut is found at the navel. The Golden Ratio is found in the proportions of the hand.

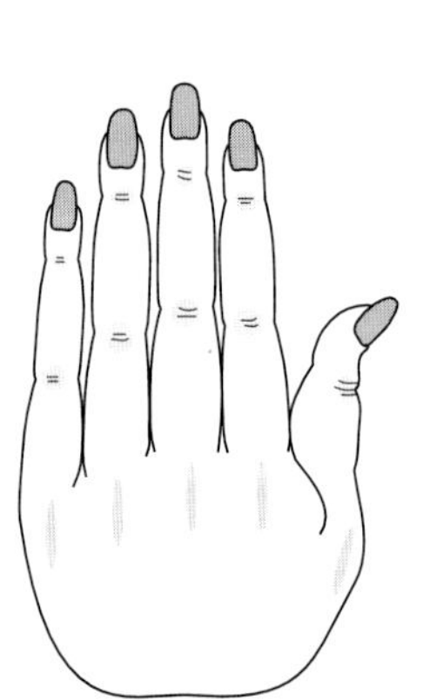
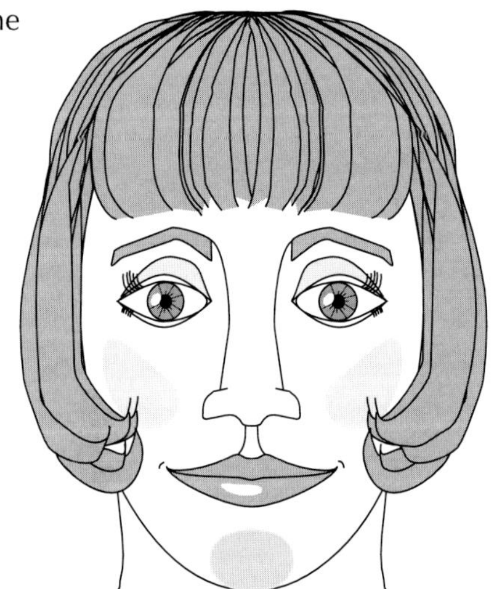
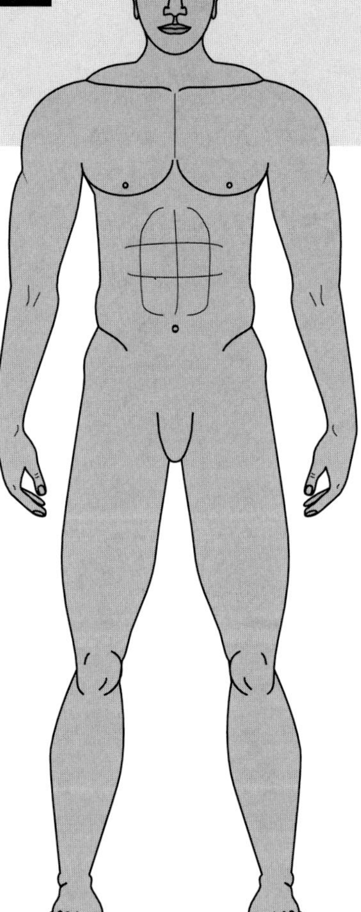

The Golden Thread that weaves Art, Math, and Nature Together

The Divine Proportion (*or Golden Proportion*)

As you have seen, a ratio is a comparison of quantities. In our example, we use a comparison of lengths which represent quantities. A proportion is a comparison of ratios. The Divine Proportion is special. It divides a line segment into two parts such that the shorter length (CB) is related to the longer length (AC) in the same way that the longer length (AC) relates to the entire length (AB). The Golden Ratio and the Divine Proportion offer the most elegant example of similarity. Examples abound in the natural world. When we looked at a line segment we were able to find the Golden Ratio. In that same line segment is found the Divine Proportion. The parts relate to each other.

The Divine Proportion: $\dfrac{AB}{AC} = \dfrac{AC}{CB}$ or $\dfrac{a+b}{a} = \dfrac{a}{b}$

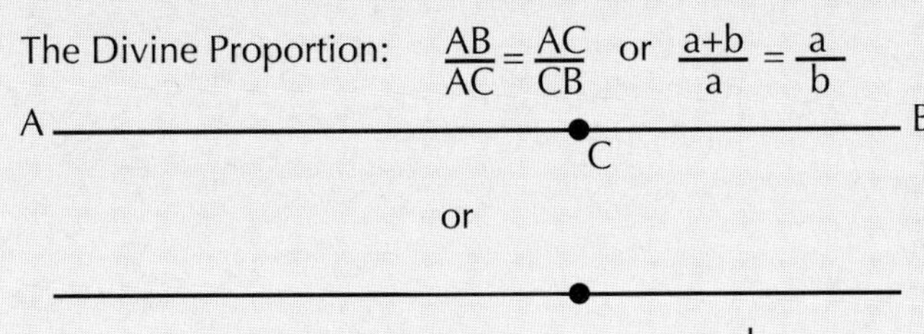

The concept of proportion carries with it the idea that there is some kind of permanent quality which is transmitted from one ratio to another.

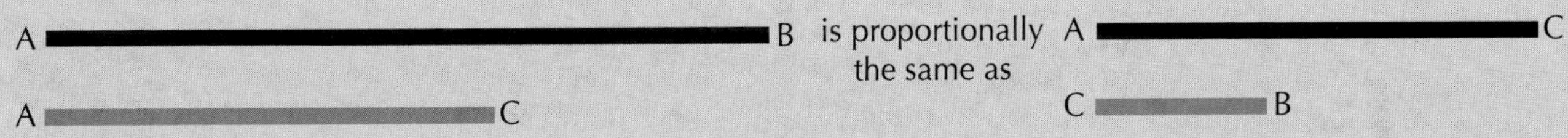

The concept of beauty is related to the concept of proportion. When we say that something looks good we are making value judgements about how the parts relate to the whole. We talk of a beautiful woman, or man; a beautiful painting; or a beautiful piece of music. Think of the things that you consider beautiful and what makes them so. Grotesque can be thought of as the opposite of beautiful. Think of the things you consider grotesque. Below we have taken liberties with the proportions of the male figure presented on the previous page. What kinds of things do you see? Notice both symmetry and proportion are involved.

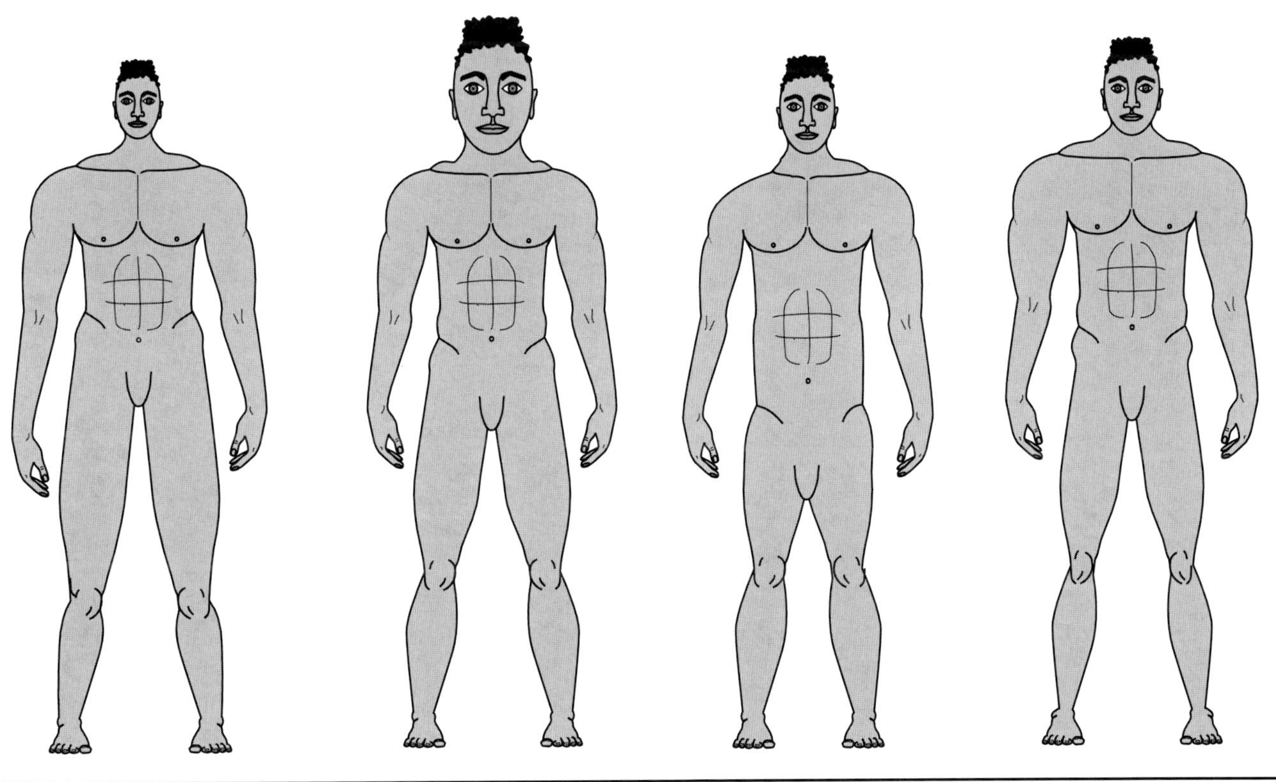

Procedure For Finding the Golden Cut of a Square

Given Square ABCD

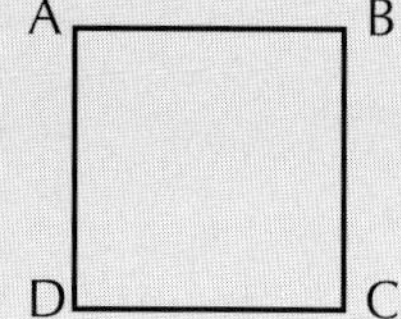

1. Find the midpoint of line segment DC, labeling it E. Place the metal tip of the compass on E, and cut an arc that connects points D and C.

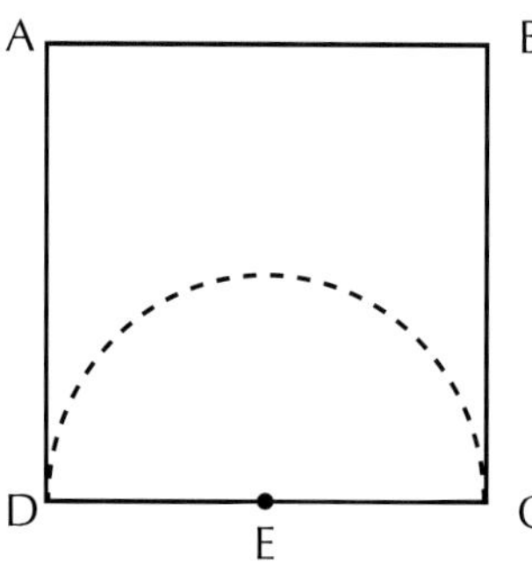

2. Draw a line segment from point E to the vertex A. Label the point of intersection F with the arc just drawn. With the metal tip on A and the pencil tip on F cut an arc that intersects line segments AD and AB. Label these points of intersection G and H.

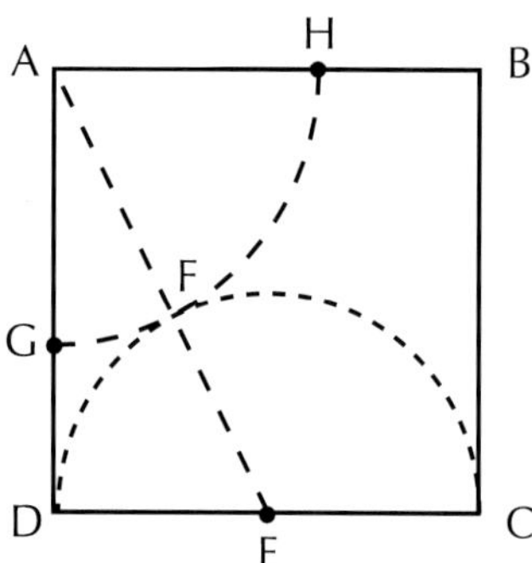

2. Draw a line from H that runs perpendicular to line segment DC. Cut an arc from G perpendicular to BC. *Now the square is divided into Golden Rectangles and squares.*

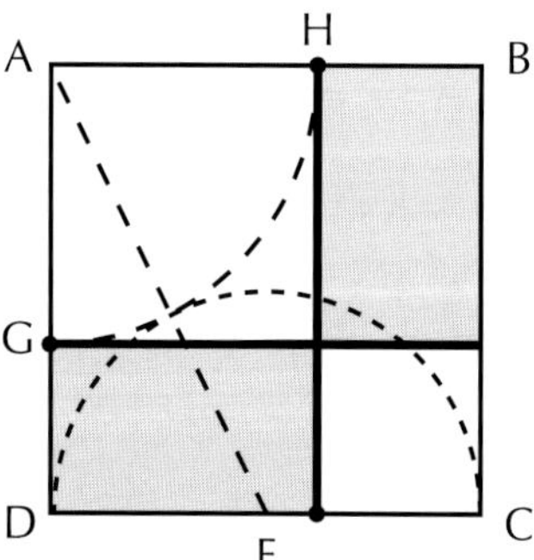

Procedure For Dividing the Radius of a Circle into Ø Proportions

Given a regular decagon with side AB inscribed in circle O

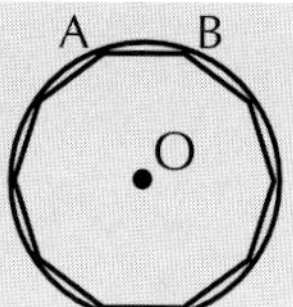

1. Draw a circle with the radius equal to line segment AB. Connect points OA and OB. Where OA and OB intersect the shaded circle label the points of intersection C and D. Connect points C and D to form line segment CD.

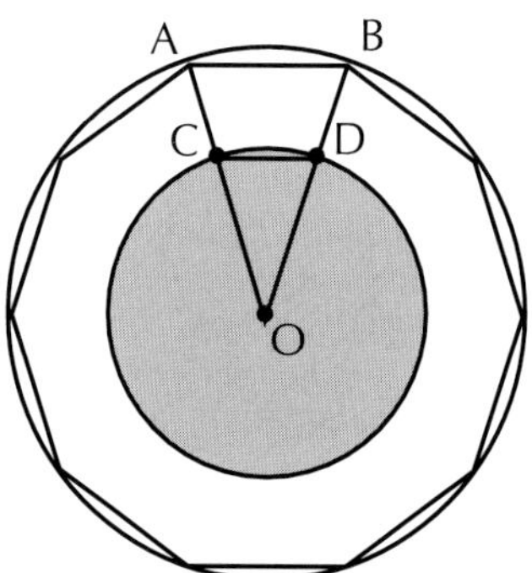

2. Draw a circle with the radius equal to line segment CD. Label the points of intersection E and F. Connect points E and F to form line segment EF.

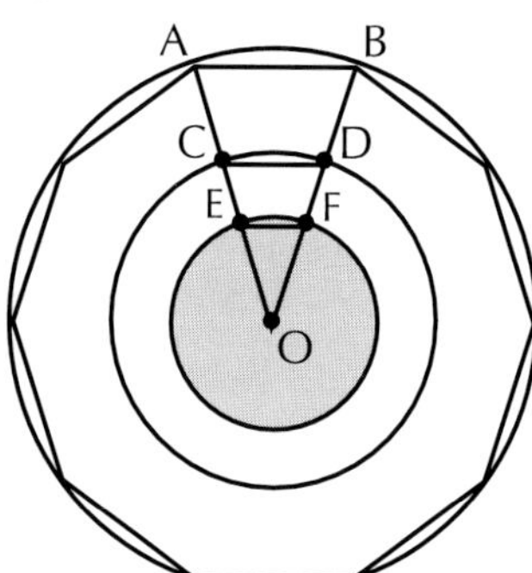

2. Draw a circle with the radius equal to line segment EF. Label the points of intersection G and H. Connect points G and H to form line segment GH. *Now both radii OA and OB are divided into Ø proportions.*

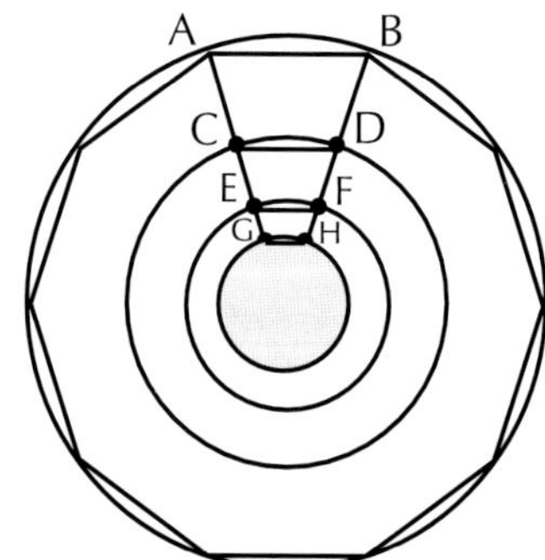

The Golden Thread that weaves Art, Math, and Nature Together

The Golden Thread that weaves Art, Math, and Nature Together

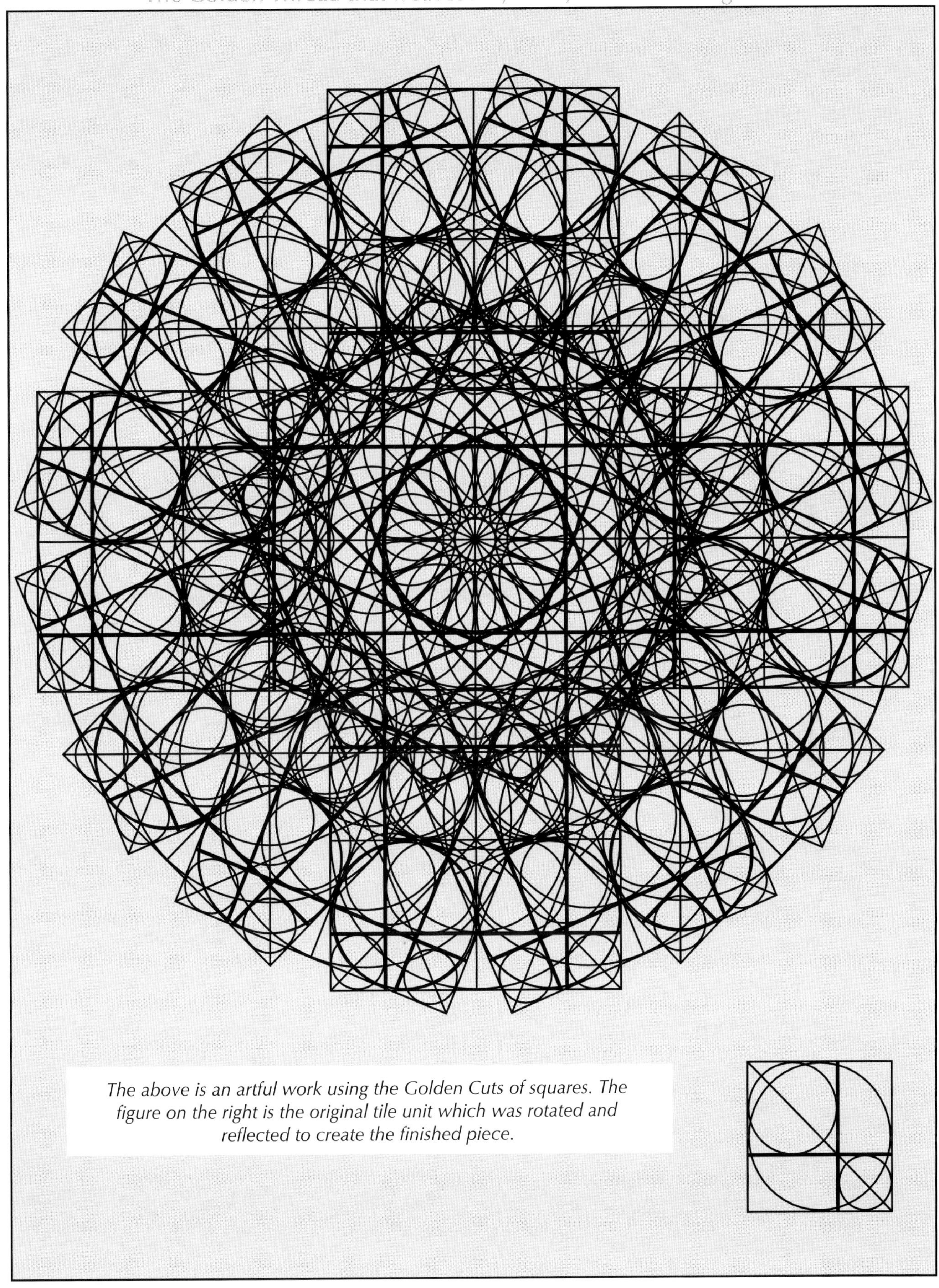

The above is an artful work using the Golden Cuts of squares. The figure on the right is the original tile unit which was rotated and reflected to create the finished piece.

The Golden Triangle

The Golden Ratio generates many figures that carry the initial similarity which then results in harmony of design. The simplest Golden figure to construct is the Golden Triangle. It requires three line segments, one line segment that has unit length *1* and two line segments that have length 1.618.

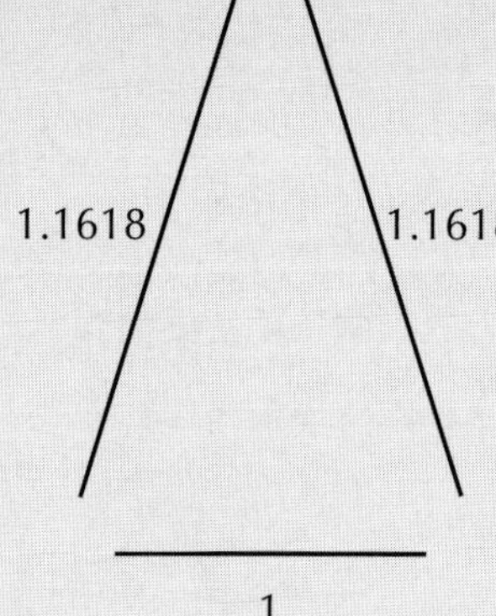

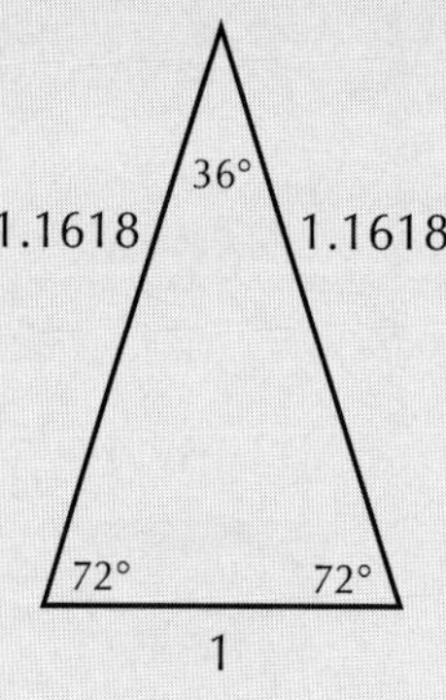

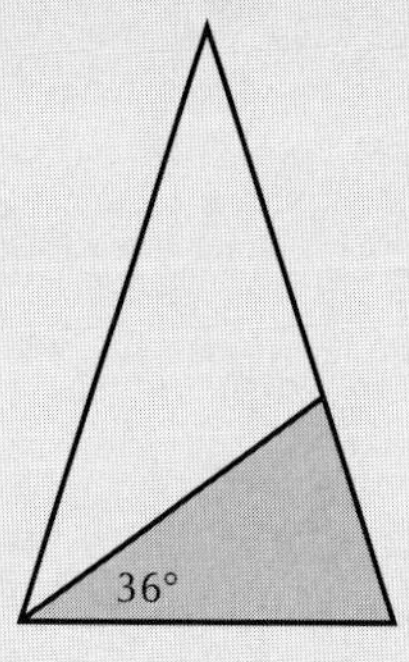

By bisecting a base angle of the parent triangle, another Golden Triangle is generated.

Continued bisection of the Golden Triangle gives smaller and smaller triangles. Use arcs to construct a spiral. Can you locate the placement points for the compass?

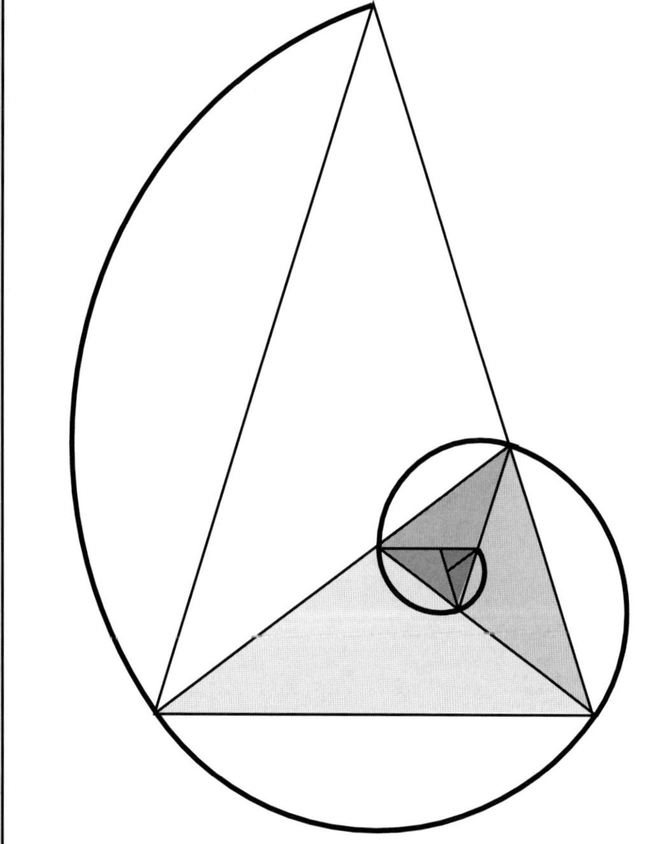

Notice the concepts of similarity and symmetry coming together in the two artful manipulations of the spiral. The border design below is a symmetry line group while the other design above is a symmetry plane group.

Procedure For Generating the Lute of Pythagoras

Given Golden Triangle ABC

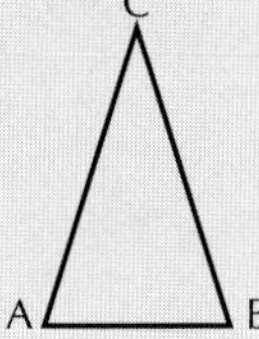

1. Open the compass to measure line segment AB. Without changing the compass setting, place the metal tip first on A. Cut an arc on the side AC of the triangle. Place metal tip on B and repeat the process. Label the points of intersection D and E.

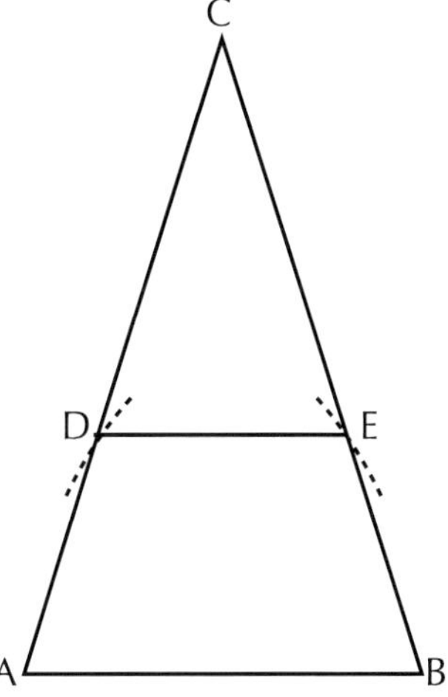

2. Continue this process upward through the triangle, each time, setting the compass by the previously cut pair of arcs (eg. open the compass to measure DE to cut the arcs that intersect at the points you will label F and G). Join the points of intersection by line segments.

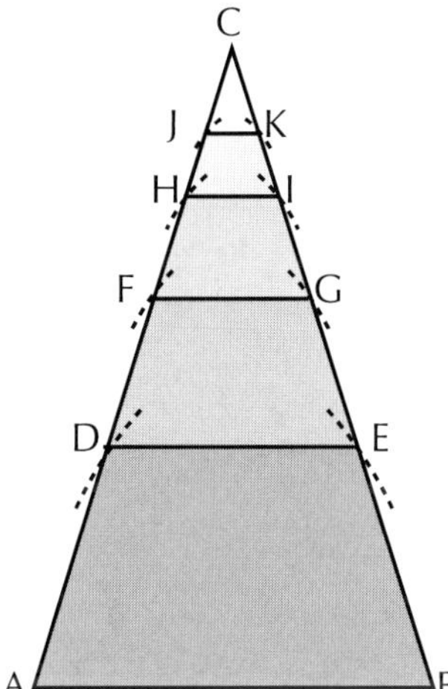

The rungs of the ladder also divide the sides of the larger Golden Triangle into Phi proportions. That is:

$$\frac{BE}{EG} = \frac{EG}{GI} = \frac{GI}{IK}$$

Above is an artpiece that uses the Golden Triangle and its Golden Cuts.

The Golden Rectangle & The Ø Family Rectangles

The second simplest Golden figure to construct is the Golden Rectangle. It requires four line segments, two line segments that have unit length *1* and two line segments that have length 1.618.

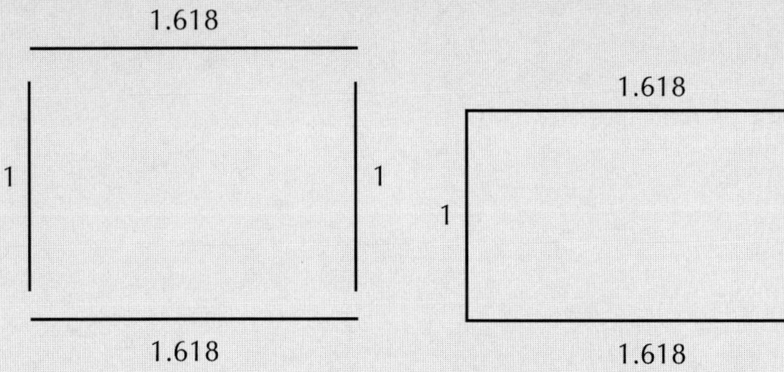

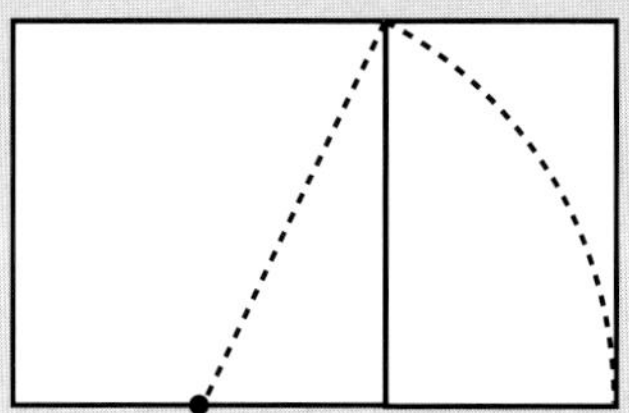

Another way to generate the Golden Rectangle is to begin with a square and the midpoint of a side. An arc is dropped to the baseline from a vertex.

A spiral can be generated within the Golden Rectangle by using arcs within the squares. Because of this it is also called the Rectangle of Whirling Squares. It is the only rectangle that has a square for a reciprocal. Each time a square is constructed a smaller Golden Rectangle results.

The Phi-Family Rectangles
This Family of rectangles relates to the square and the Golden Rectangle. They provide a compatible design system.

Below is an artful piece using the Whirling Squares and straight line spirals.

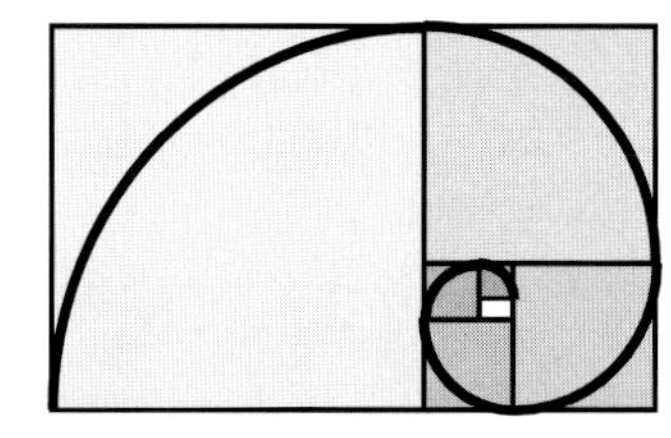

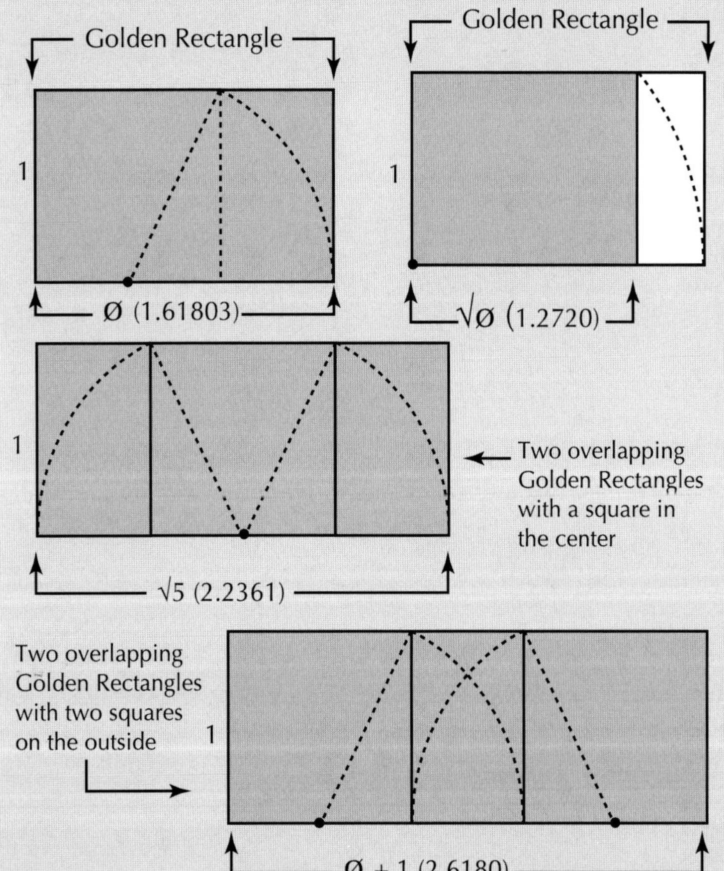

Procedure For Constructing a Golden Triangle Using a Golden Rectangle

Given Golden Rectangle ABCD 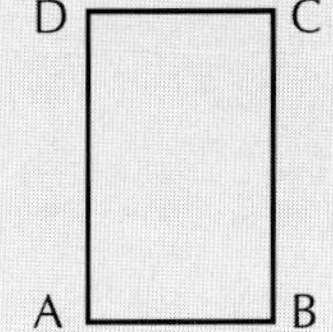

1. Place the metal tip of compass on B and the pencil tip on C. Cut an arc in the interior of ABCD. Repeat the process placing metal tip on A and pencil tip on D. Label point of intersection E.

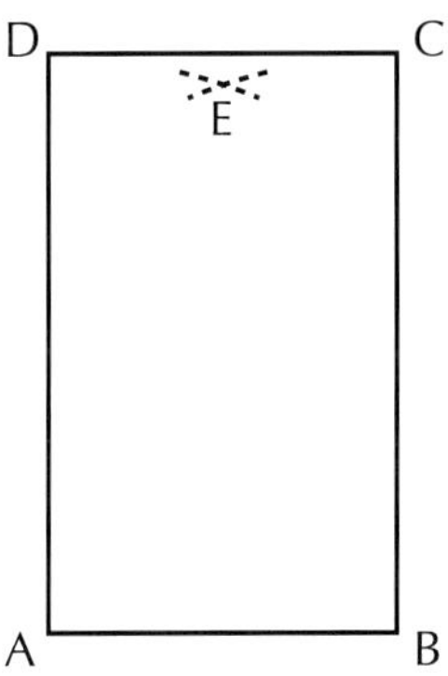

2. Draw a line segment from A to E, then from E to B.

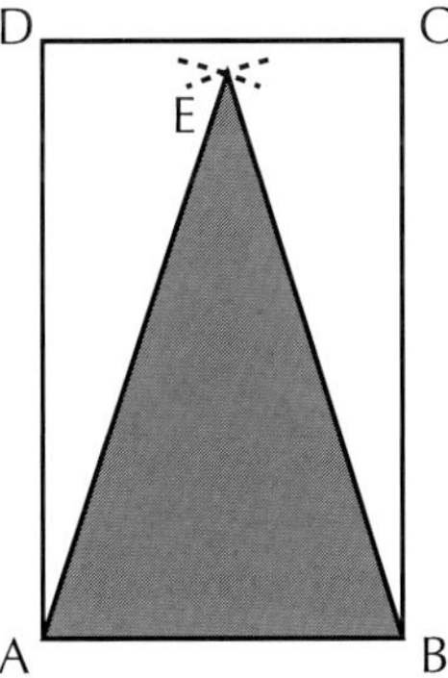

Now you have a Golden Triangle.

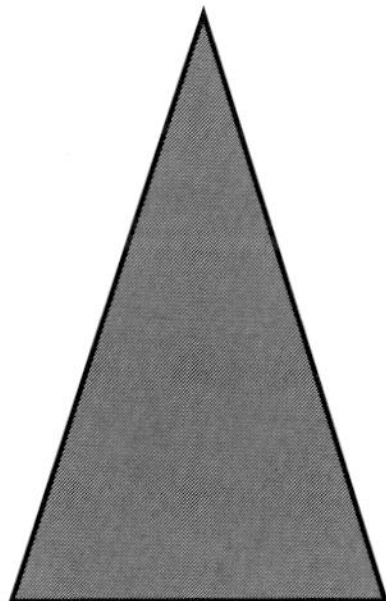

Procedure For Constructing a Rectangle Similar to a Golden Rectangle

Given Golden Rectangle ABCD 

To Construct a similar smaller rectangle:

1. Draw a diagonal from A to C. Choose any point along line segment AC and label it E.

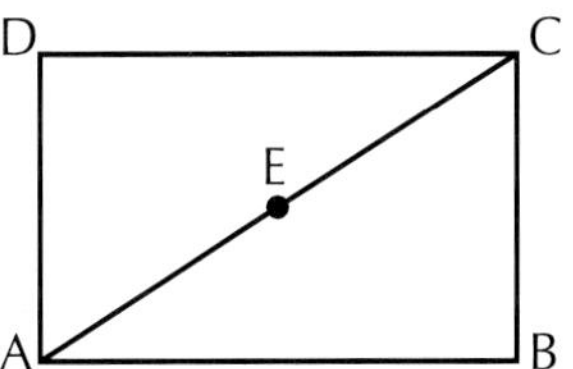

2. Through E, construct perpendiculars to line segments AB and AD. Label points of intersection G and H.

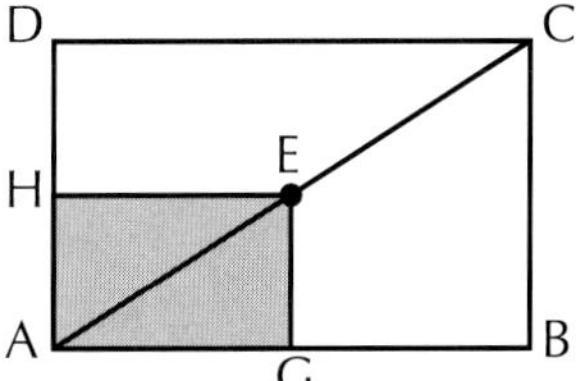

To Construct a similar larger rectangle:

1. Draw a diagonal from A that extends beyond C. Extend line segments AD and AB. Choose any point beyond the original rectangle and label it F.

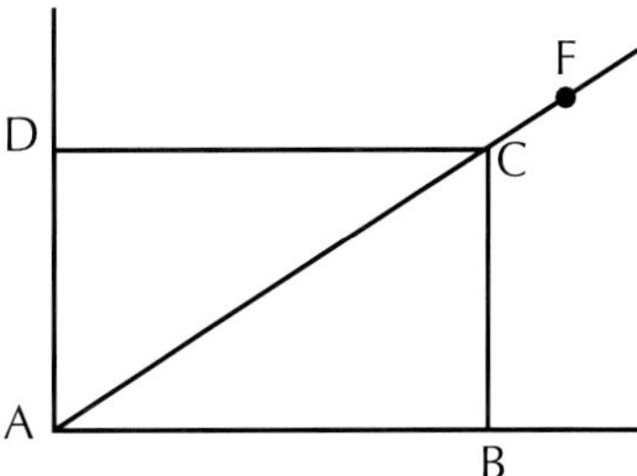

2. Construct perpendiculars through F to line segments AB and AD. Label points of intersection I and J.

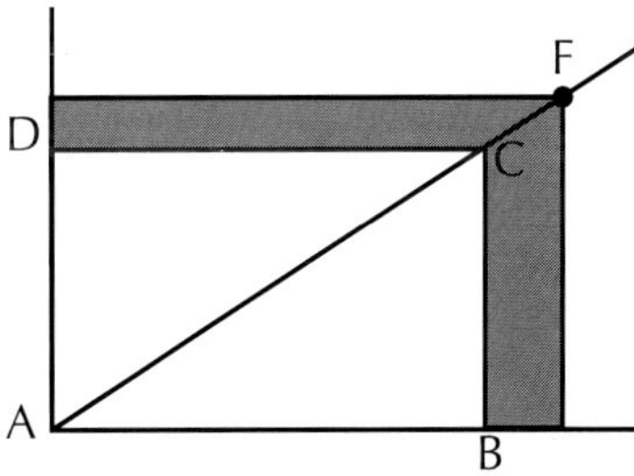

Procedure For Constructing a Golden Rule

Given Golden Rectangle ABCD
with its unit square ADEF

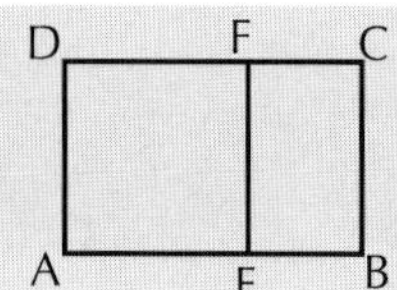

1. Extend line DA any distance, and choose a
point O on it. That length determines the length of
the Rule.

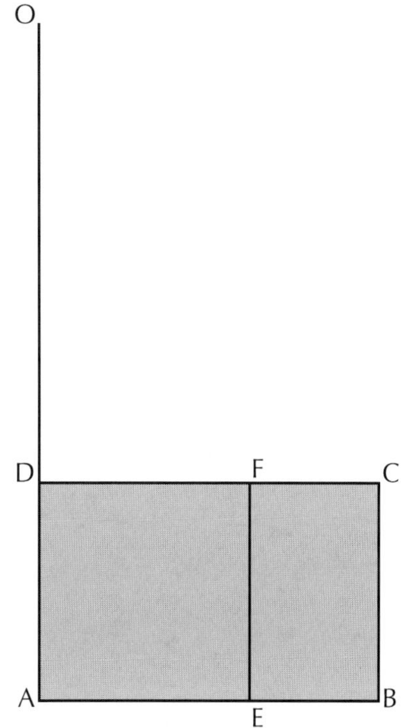

2. Draw line segments OF and OB.

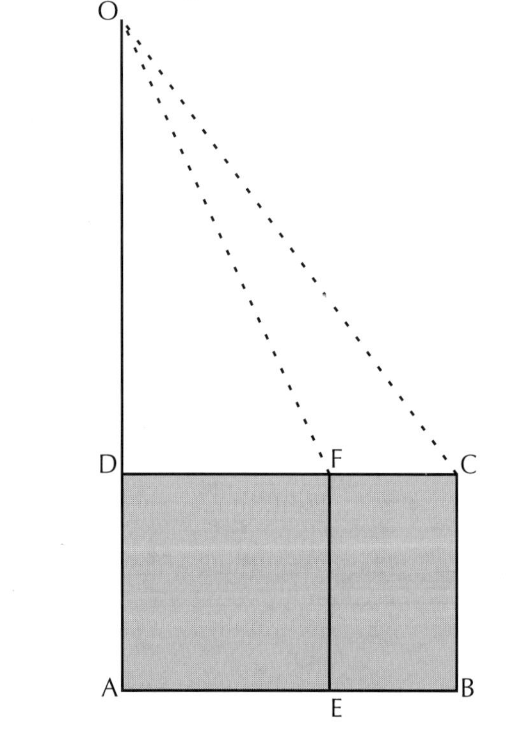

Any line segment drawn parallel to line segment
AB in triangle ODC will be divided at the Golden
Cut by line segment OF.

$$\text{ex.} \quad \frac{GH}{HI} = \frac{JK}{KL} = \frac{MN}{NP} = \frac{QR}{RS} = \emptyset$$

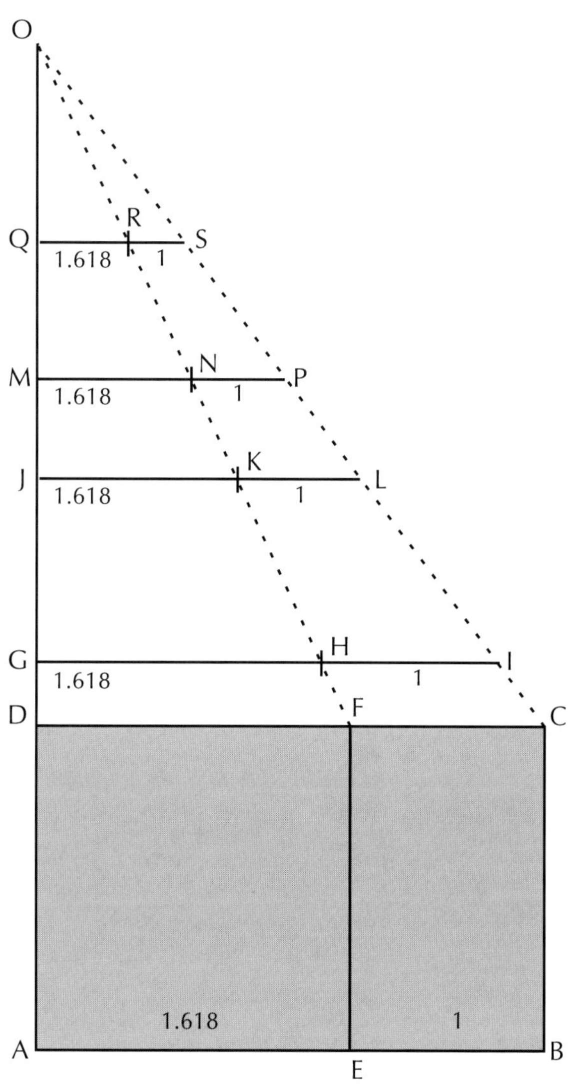

Now a Golden Rule has been constructed.

Hint:
You can use this ruler alongside
your traditional inch and centimeter
measuring tools. The latter two give you
constant units while the Golden Rule
gives you units that are similar.

The Golden Thread that weaves Art, Math, and Nature Together

The Golden Thread that weaves Art, Math, and Nature Together

Golden Grids

Here is a sampling of sections of grids developed from variations on the Golden Ratio. You might want to develop others.

Golden Rectangle

Golden Parallelogram

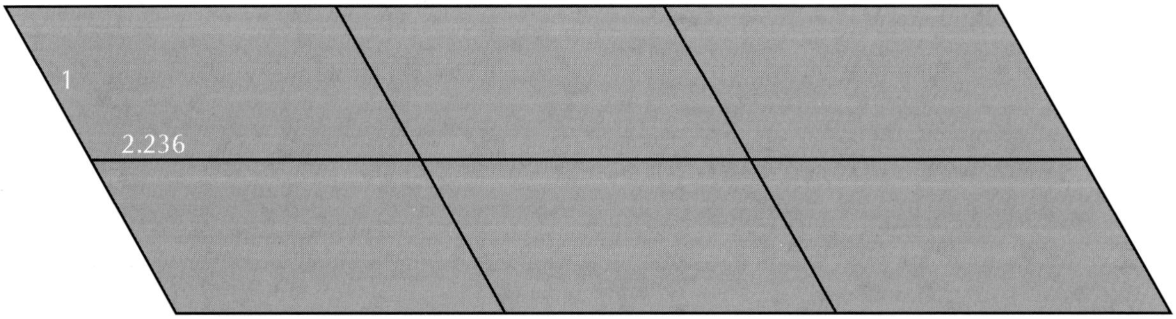

The Golden Thread that weaves Art, Math, and Nature Together

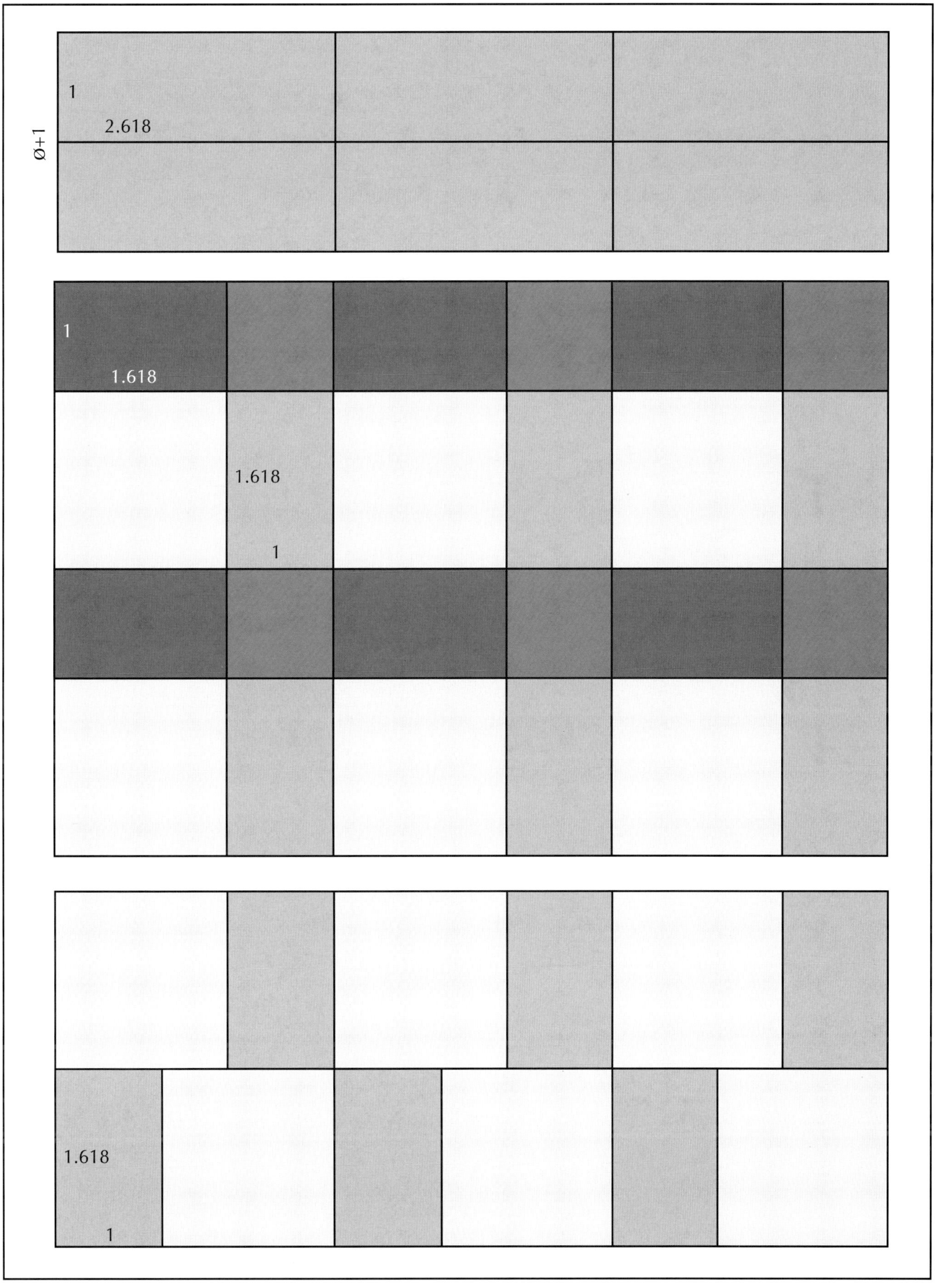

Ø+1

1
2.618

1
1.618

1.618
1

1.618
1

The Golden Polygons: The Pentagon and Pentagram

The regular pentagon and pentagram are filled with Golden Ratios.

In Each Case

$$\frac{AB + BC}{BC} = \frac{BC}{AB} = \emptyset$$

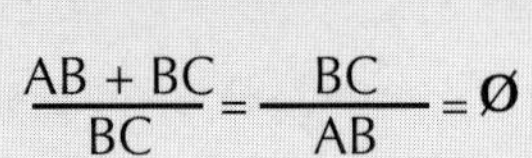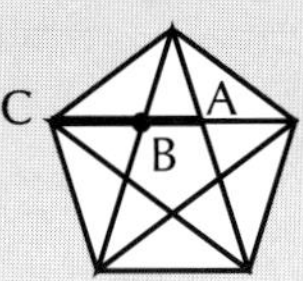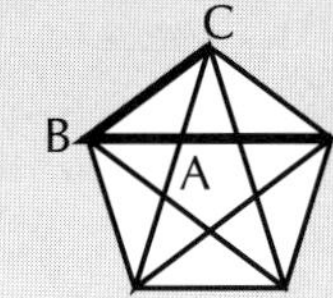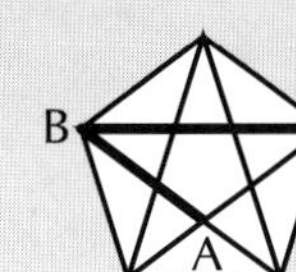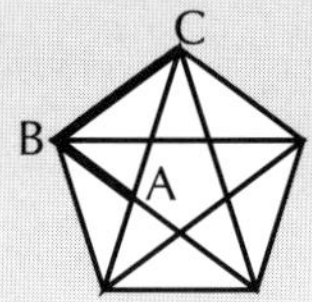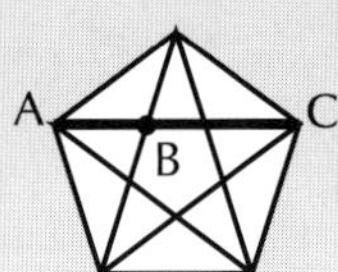

Procedure For Constructing a Regular Pentagon Given the Golden Cut of a Line Segment

Given the line segment AB with C as the Golden Cut

1. Open the compass to measure CB. With metal tip on C, cut an arc above line segment AC. Without changing the setting, place the metal tip on A and cut an arc that intersects the other. Label the point of intersection D.

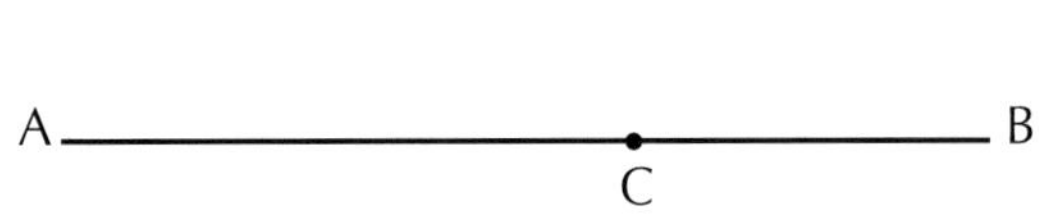

2. Draw a line segment from A to D, then from D to C. A,C and D will be three vertices of the regular pentagon.

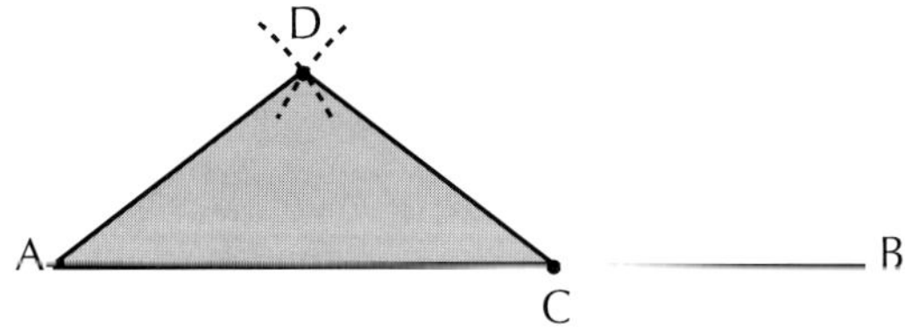

3. Open the compass to measure AC. With the metal tip on A, cut an arc below C in approximately the position of the fourth vertex. Without changing the setting, and with the metal tip on C, cut an arc below A in approximately the position of the fifth vertex.

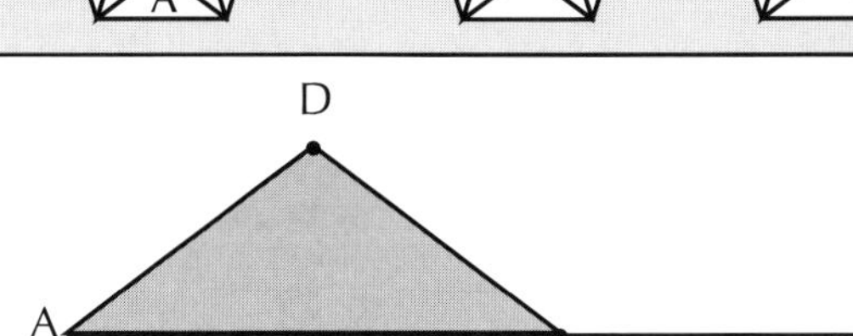

2. Open the compass to measure line segment AD. With the metal tip on A first and then with the metal tip on C, cut arcs that intersect those drawn above. Label the points of intersection E and F.

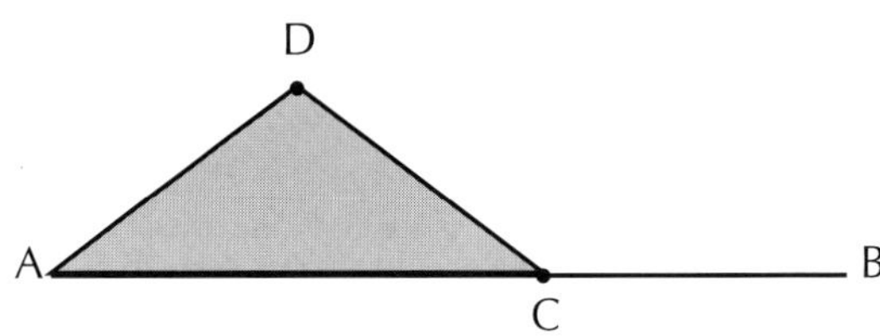

2. Draw a line segments AE, EF, and FC. *Now you have a regular pentagon labeled AEFCD.*

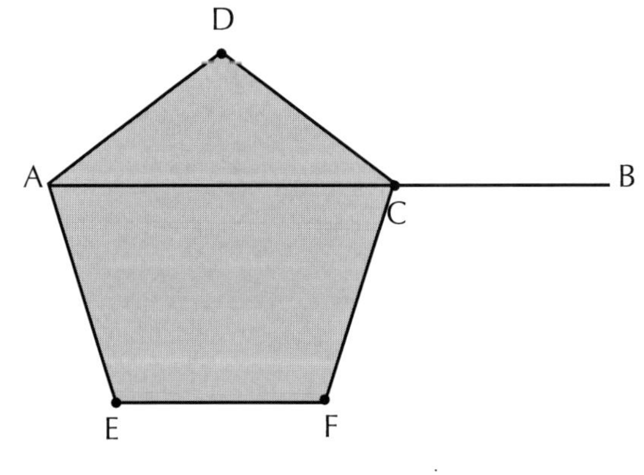

Lute of Pythagoras

The rungs of the ladder divide the sides of the larger Golden Triangle into Phi proportions. Within the lute are found pentagons and pentagrams with their contained Golden Ratios.

The above is an artful work using the Lute of Pythagoras as a tiling.

The Golden Tiles

Subdivisions of the pentagon also give two isosceles triangle that are used to form tiles that have Golden Ratios. Each tile alone can be used to cover the plane periodically. These tiles are called Penrose Tiles after the 20th century mathematician Roger Penrose.

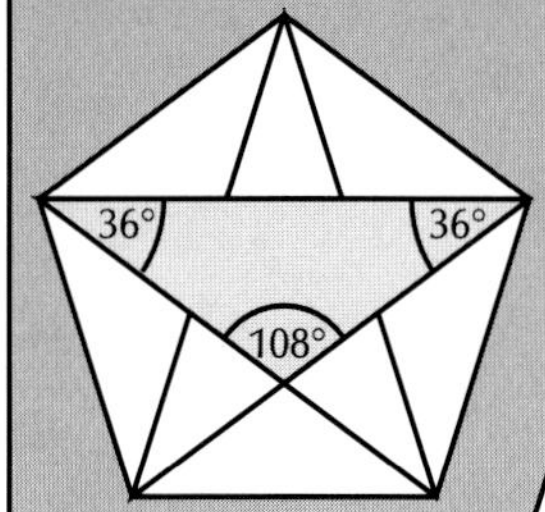

The "fat" Penrose tile
(which is a quadrilateral)

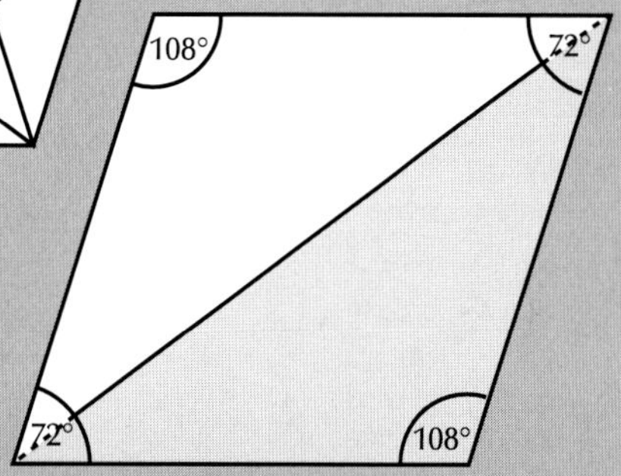

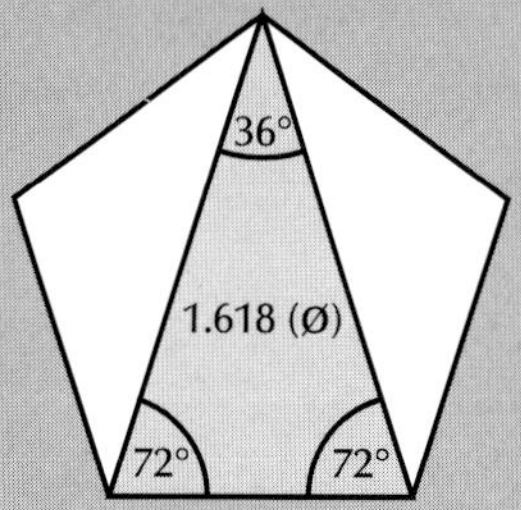

The "skinny" Penrose tile
(which is a quadrilateral)

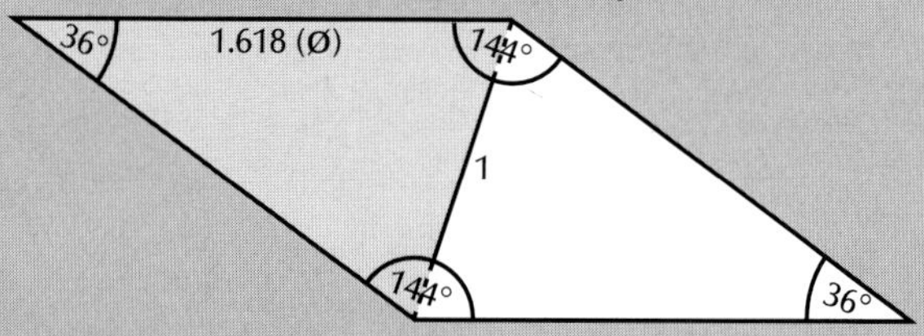

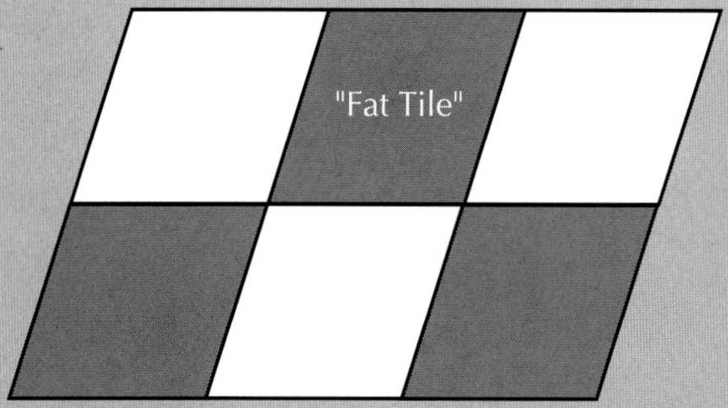

Both these tilings are periodic which means they repeat infinitely by translation.

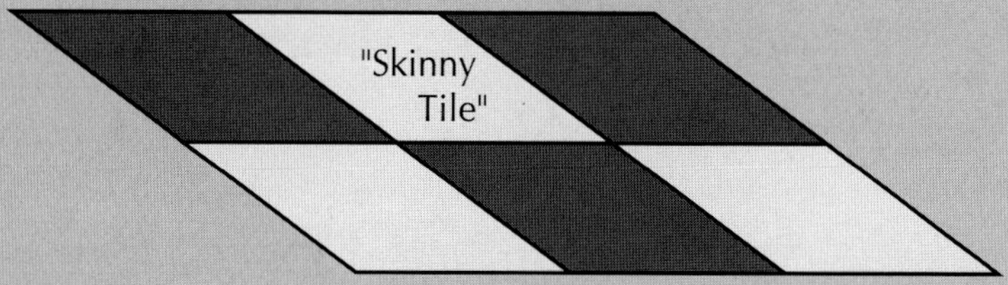

"Fat Tile"

"Skinny Tile"

If, however, these two tiles are used together, they form a tiling that is called aperiodic which means that they will have some elements of repetition and some of difference. Aperiodicity is dependent on following a set of matching rules.

There is an interesting relationship between the number of fat tiles and skinny tiles and it involves the Golden Ratio. Using combinations of these units try a tiling for yourself.

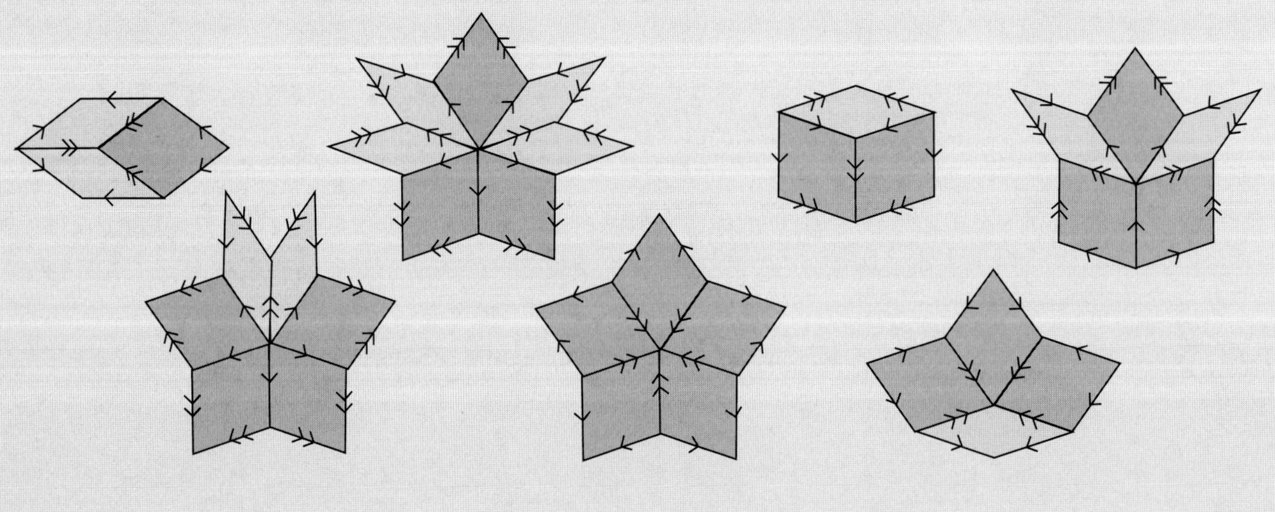

The Golden Solids

The Penrose tiles can also be used to build two three-dimensional figures called the Golden Rhombohedra.

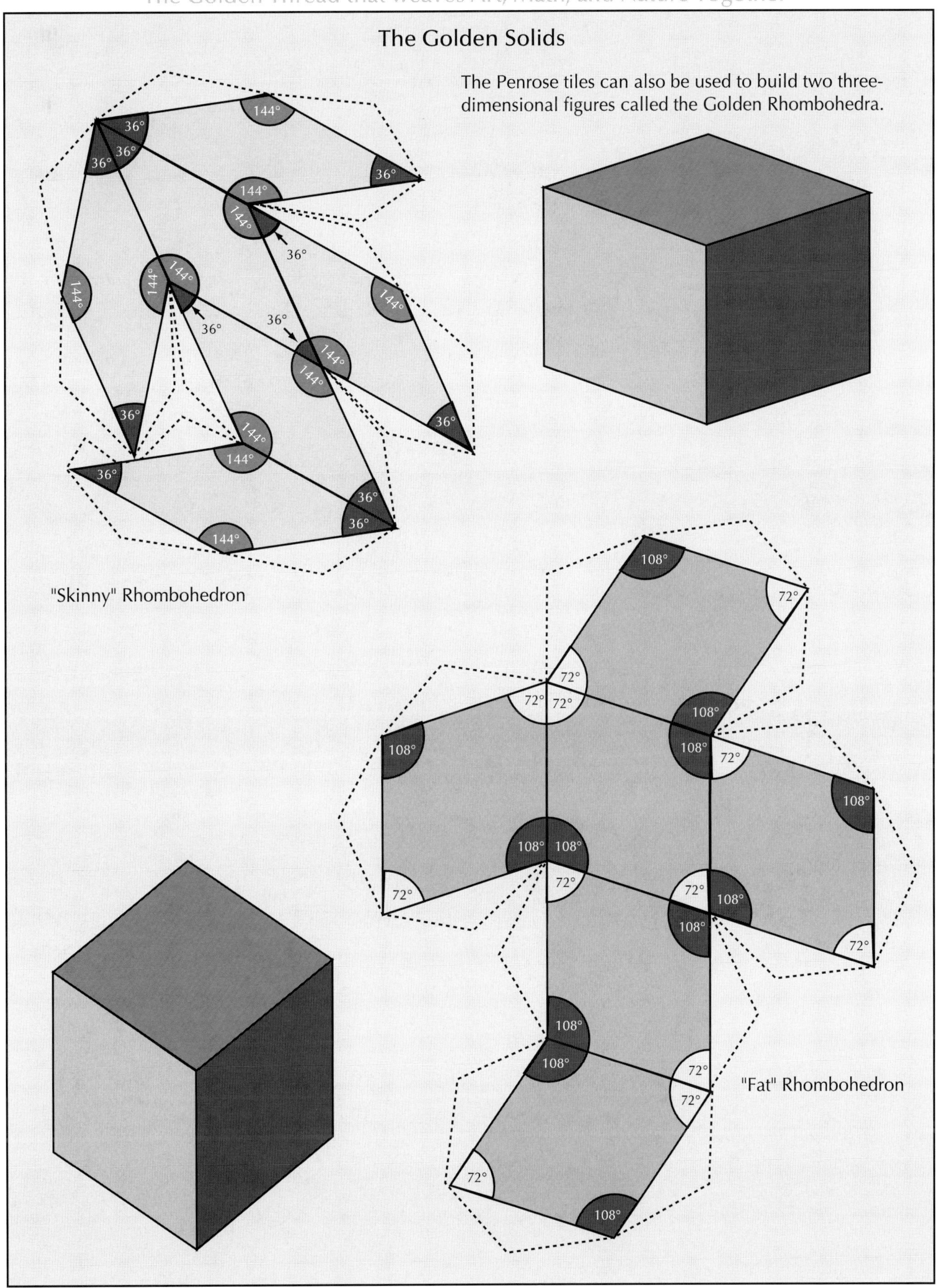

The Fibonacci Sequence

There is an interesting relationship between a certain sequence of numbers and the Golden Ratio (1.61803...). If the number 987 is divided by the number 610 the result is approximately 1.6180327; if the number 1597 is divided by 987 the result is approximately 1.6180344; if the number 2,584 is divided by 1,597 the result is approximately 1.6180338. The difference between the pairs of numbers described occurs after the fifth decimal place. This process can be continued indefinitely. You will have noticed that we require a pair of numbers. Let us go back to the beginning terms of this special sequence which is called the Fibonacci Sequence named after a mathematician of the Middle Ages.

The Rule for the Fibonacci Sequence

Start with 1 and add 1. The result is 2. Take this result and add the previous number 1. The result is 3.

$$2+3=5; \ 3+5=8; \ 5+8=13; \ 8+13=21...$$

Notice that each term in the sequence is the sum of the two preceding terms:

Fibonacci Sequence: 1, 1, 2, 3, 5, 8, 13, 21, 34, 55, 89, 144, 233, 377, 610, 987...

If you divide a number in the sequence by the previous number you will not get as accurate an approximation until you reach the fifteenth term (which is 610). For example, 5 divided by 3 equals 1.5. For your own understanding, do the divisions of the other terms so that you can see the progression towards the Golden Ratio.

These whole numbers can be translated into lengths of line segments. Let 1 be the unit length.

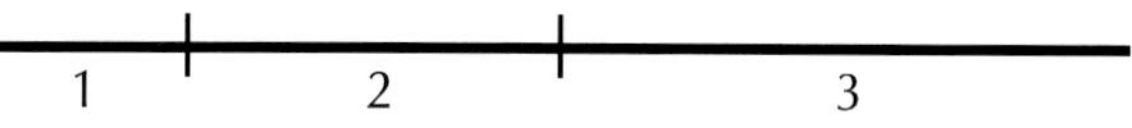

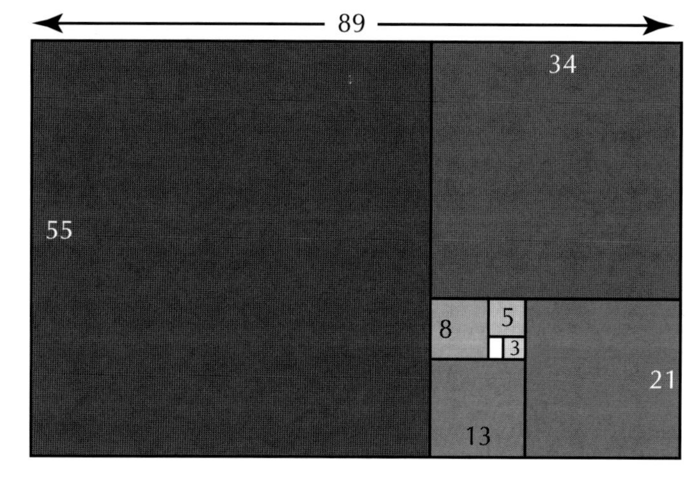

These lengths can be used as the length and width of rectangles. The Golden Rectangle can be approximated by using multiple squares that use different Fibonacci numbers.

The Golden Thread that weaves Art, Math, and Nature Together

The above is an artfull work using the Fibonacci numbers 55, 34, 21, 13, 8, 5, 3, as lengths for a set of squares. The work also uses the symmetry operations of translation, rotation and reflection. The unit to the right was the generator for the artwork. Take the unit and skew it, and develop your own artful variation. Try a color or texture composition.

The Fibonacci Sequence in Plant Growth (Spiral Phyllotaxis)

The growth patterns of natural forms such as sunflowers and daisy heads; pineapples and pinecones; petals, buds, and leaves have a relationship to Fibonacci numbers. The way in which leaves spiral around the stem of a plant, as seen in the diagram to the right, shows a Fibonacci connection between the number of leaves and number of turns taken around the stem. The fifth leaf is found directly above the leaf which is not counted. We pass through five leaves in two complete turns around the stem. Examine a group of plants and see if you can determine which leaves are directly above one another and how many turns it takes to get from one leaf to another. This relationship is represented by a fraction such as 2/5 as in the plant at the right.

Pineapples have three sets of spirals because of the hexagonal shape of their scales. They exhibit Fibonacci numbers of 5, 8, 13, 21, 34. Pine cones have numbers of 2, 3, 5, 8, 13.

In a sunflower head, seed pockets grow from a central point, increase in size and spiral out. The size of the sunflower head determines which pair of Fibonacci numbers it has. See facing page.

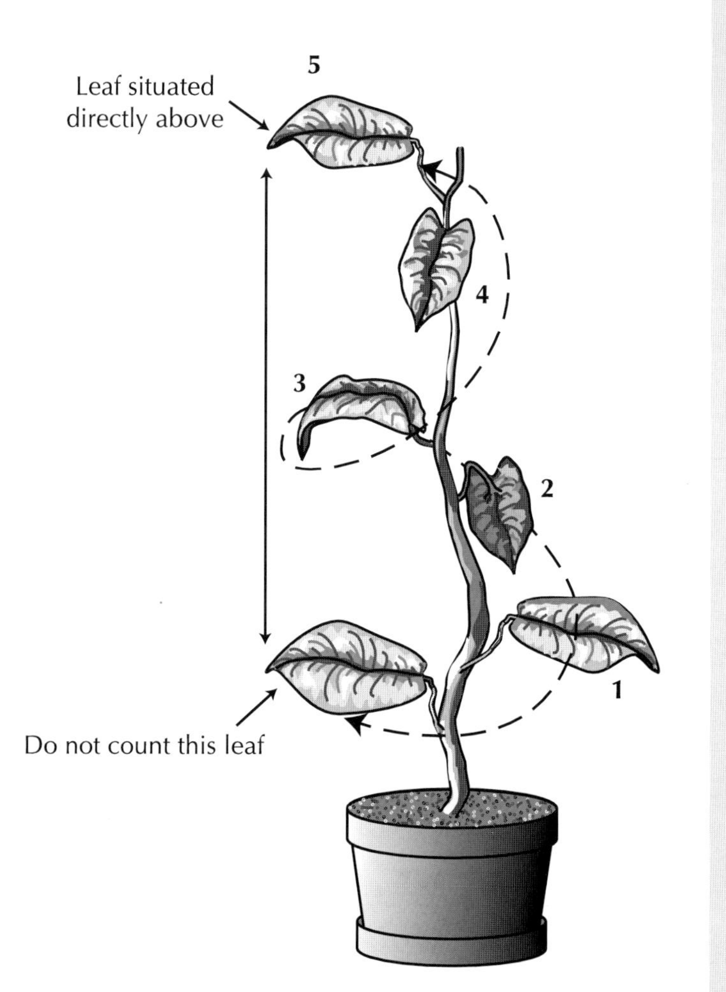

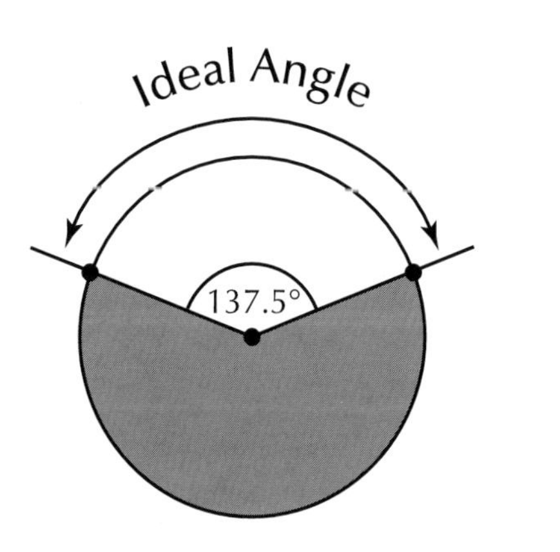

"Plant-Petal Numerology"

The ideal angle is that which allows for efficient packing of primordia, the tiny lumps which eventually develop into leaves and petals. To obtain this angle take a pair of consecutive Fibonacci numbers, 34/55 (0.61818) x 360° to get 222.5°- subtract from the 360° of a circle and the angle that remains is 137.5°.

The ideal angle is the circular arc between any two consecutive points.
137.5° is an approximation of 137° 30' 28".
It is $360° \times Ø^2 (360 \times .618^2) \longrightarrow 360 \times .381924 = 137.49264$

This page:
Plants in the natural world exhibiting Fibonacci connections.

The Fibonacci Sequence in Plant Growth (Spiral Phyllotaxis)

The head of a sunflower can be represented as a figure in the plane so we can schematically explore what happens in nature and can use this information as a basis for design. At the beginning stages of the plant's development, there are undifferentiated pieces of matter called primordia, shown in the diagram as dots. These become the specific florets, leaves, petals. As these pieces grow larger and move away from the apex, they grow out, in a spiral formation. The earliest primordia appear the furthest out from the center. Each new primordia is spaced sparsely along a tightly wound spiral. It is the tightness that effects the final form. The essential feature is the angle between successive primordia which are equally spaced.

The most efficient packing comes from the angle of 137.5°. This spiral is developed on a polar graph, which is essentially a circle with inner concentric circles, divided into 360° marked with 10° intervals. In the first diagram, the concentric circles decrease in size 80% starting from the outer circle.

Starting with the inner circle successive points have been placed 137.5° apart. Their points are then connected to indicate the spiral. It is the underlying natural growth armature. However, this spiral is not what you see when you look at the sunflower head. In the outward appearance you will see two sets of opposing spirals forming a pair of Fibonacci numbers, such as 8 and 13.

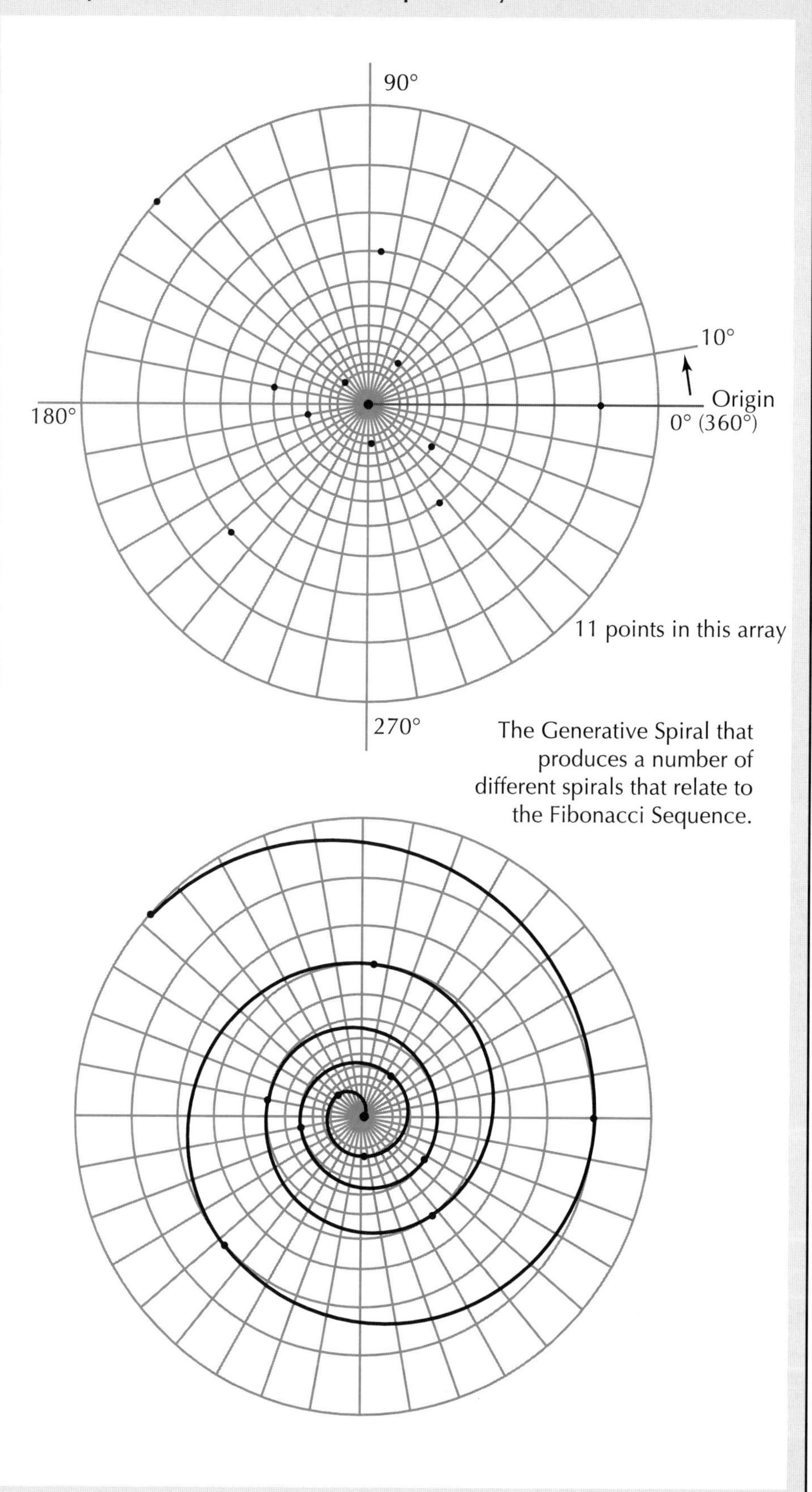

The Generative Spiral that produces a number of different spirals that relate to the Fibonacci Sequence.

The Fibonacci Sequence in Plant Growth (Spiral Phyllotaxis)

The appearance to the eye of two distinct opposing spirals is the result of points from the generative spiral. In this example, there are 8 spiral arms in one direction and thirteen in the other, suggestive of a sunflower head. A larger head might have 21 and 34 spirals.

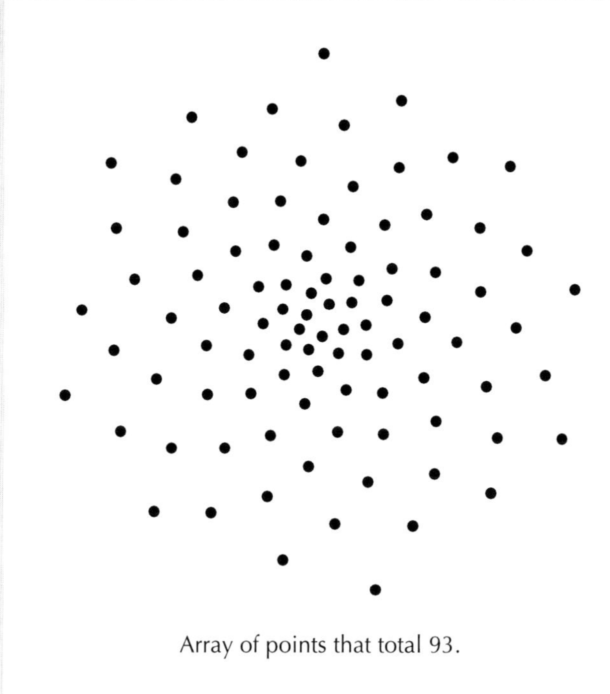

Array of points that total 93.

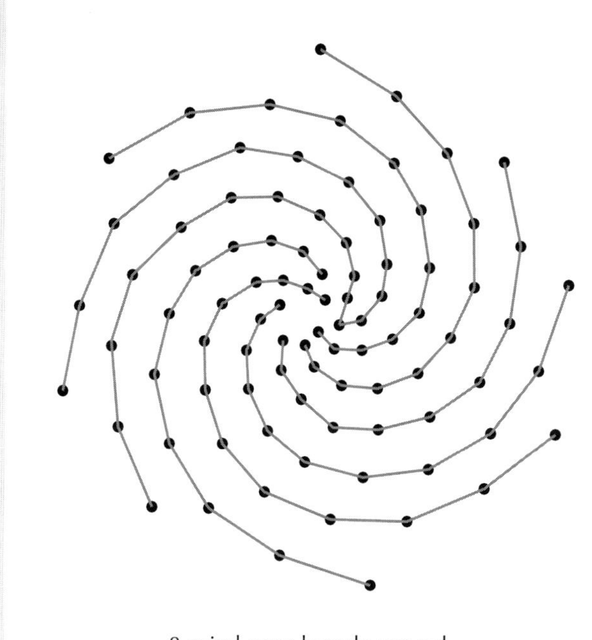

8 spiral arms loosely wound.

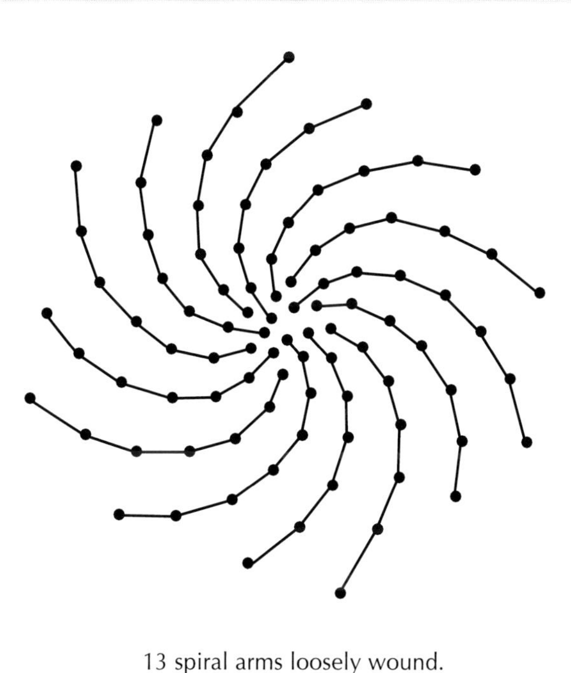

13 spiral arms loosely wound.

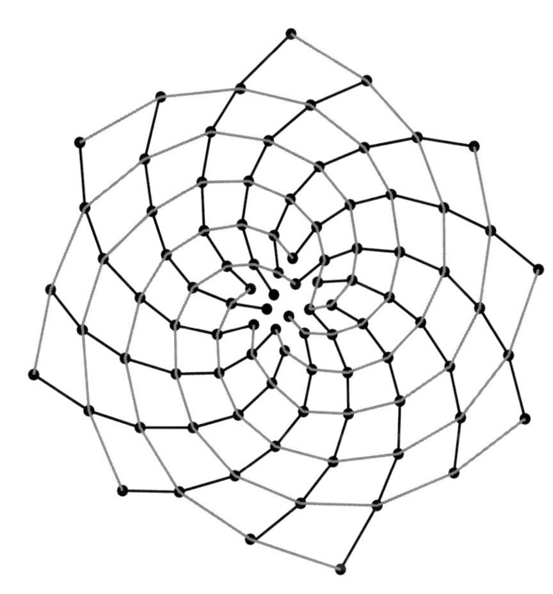

The result of overlapping the two spirals.

To develop a spiral design in one direction, either clockwise or counterclockwise, with n number of arms (in this case 8 and 13) we need: a polar grid, a decision about which Fibonacci number is desired, and the use of diagonals within subsections. On the facing page the artist used the asymetric motif of the spiral with 8 arms and the symmetry operation of translation to create the artwork.

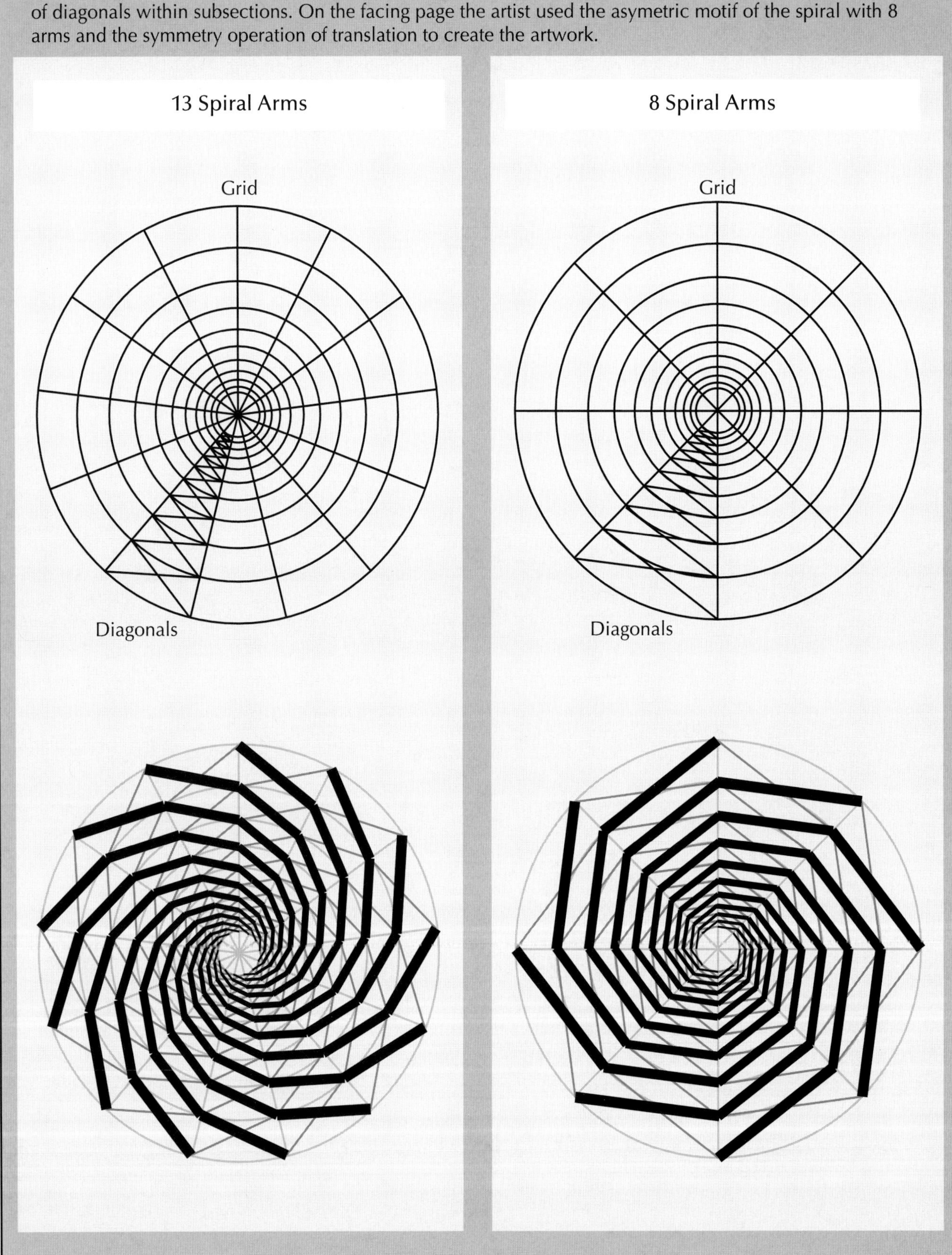

The Golden Thread that weaves Art, Math, and Nature Together

The Golden Thread that weaves Art, Math, and Nature Together

These two facing pages are the last connections to be made to the Golden Thread. If you recall the construction for finding the Golden Cut of any line segment at the very beginning of this chart, it involved constructing a right triangle, thus, the proportional relationship of the Golden Ratio is embedded within this triangle. X marks the Golden Cut of line segment AB.

The proof of this construction involves the Pythagorean Theorem as indicated in the diagram.

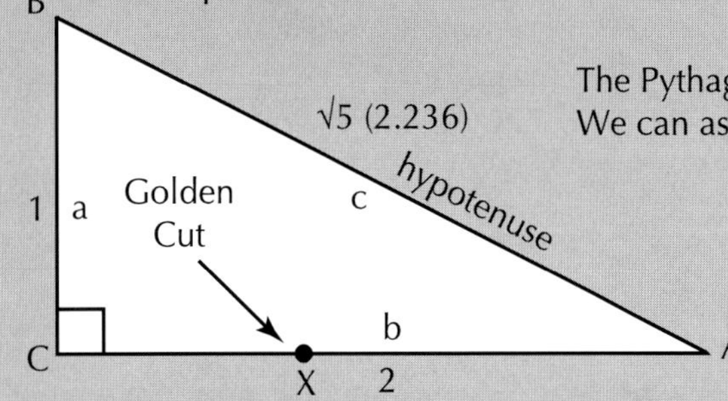

The Pythagorean Theorem states: $a^2 + b^2 = c^2$ (the relationship of the sides)
We can assign numerical lengths to the sides of the triangle:

$$1^2 + 2^2 = c^2 \;\rightarrow\; 1 + 4 = c^2 \;\rightarrow\; 5 = c^2 \;\rightarrow\; \sqrt{5} = c$$

Notice in this particular trangle the length of the hypotenuse is an irrational number.

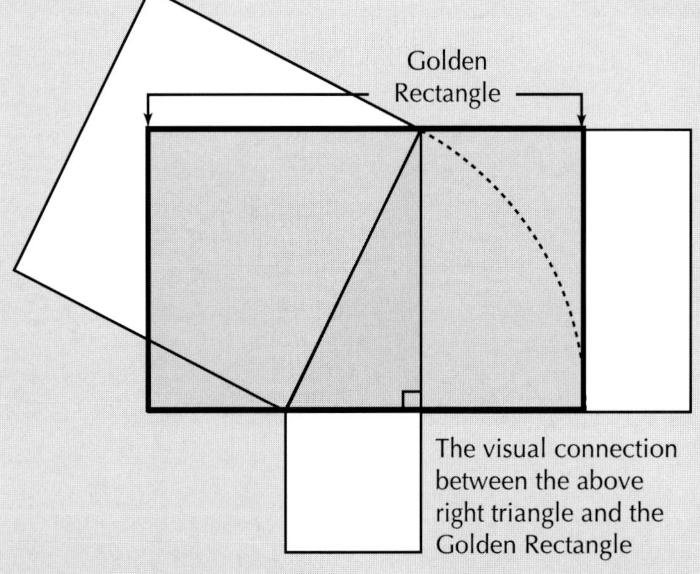

The visual connection between the above right triangle and the Golden Rectangle

The 3-4-5 triangle is the first in the series of right triangles to have the lengths of the three sides in whole numbers.

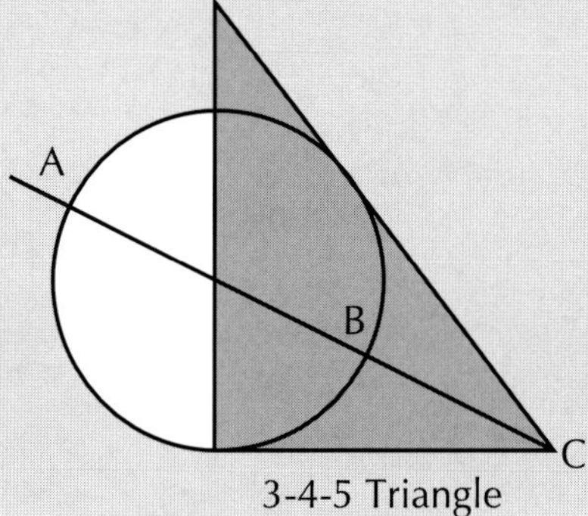

3-4-5 Triangle

The connection between the 3-4-5 triangle, the Golden Cut of a line segment and a circle. The ratio of AB to BC is Ø.

Visually the difference between these two rectangles is indiscernible; however, numbers reveal the difference.
5 divided by 3 equals 1.6666....Compare this to 1.618. The difference between the two is .0486. In nature these small differences can make huge changes. For design purposes you may use either one.

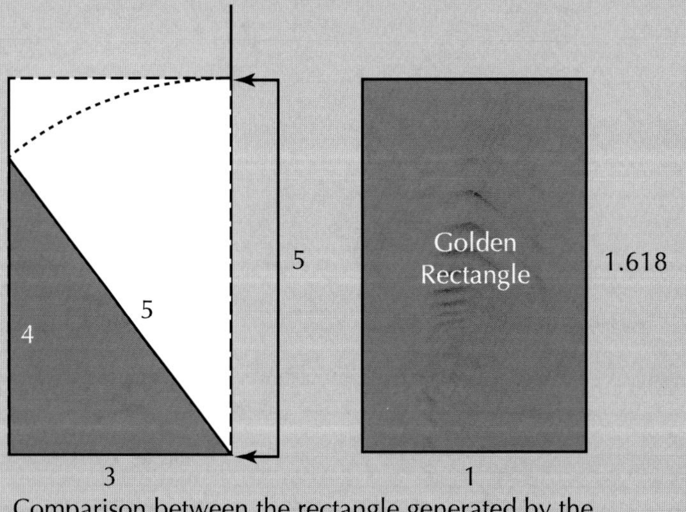

Comparison between the rectangle generated by the 3-4-5 triangle and the Golden Rectangle

The Golden Thread that weaves Art, Math, and Nature Together

Color Structures

In order for colors to be harmonious, a few basic guidelines can be followed in limiting the number of colors used. This can be done by using one of the following:

1. Color to a tone to a gray.
2. Color to a tint to a white.
3. Color to a shade to a black.
4. Tint to a tone to a shade.
5. Tint to a shade to a tone to a gray.
6. Shade to a tone to a white.
7. Color to a tint to a white to a gray to a black.

Monochromatic
Uses one color and the light and dark variations therof.

Complementary
Uses a pair of colors directly opposite one another with light and dark variations thereof.

Analogous
Uses three adjacent colors with light and dark variations thereof.

Triad
Uses three colors that form the vertices of either an isosceles or equilateral triangle with the light and dark variations thereof.

Tetrad
Uses four colors that form the vertices of a rectangle (including the square) with the light and dark variations thereof.

Color Wheel

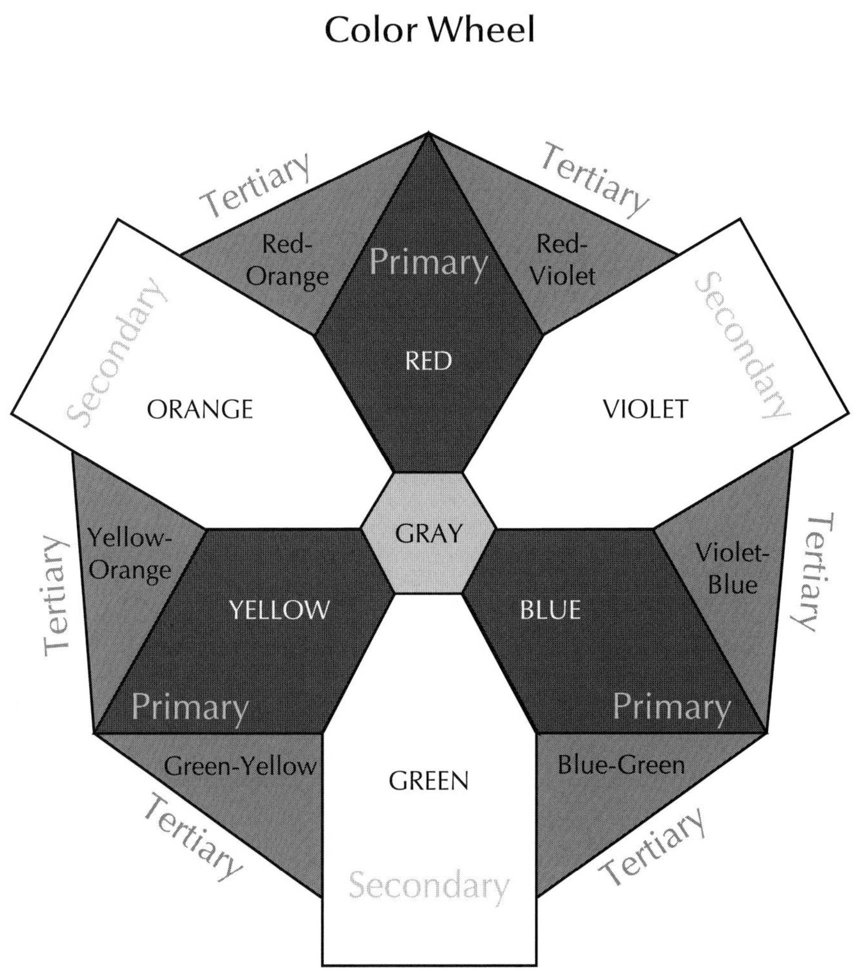

The color wheel is based on a twelve tone structure. Yellow, blue and red are the three *primary* colors. When they are mixed together a gray will obtained. Theoretically, all the other colors can be mixed from only these three. It is best, however, when purchasing pigments to include white, black, and violets to gain the maximum amount of flexibility in mixing color.

A *secondary* color is obtained by mixing a pair of primary colors. The secondary colors are orange, green and violet.

A *tertiary* color is obtained by mixing a primary and secondary color together. The tertiary colors are blue-green, yellow-green, yellow-orange, red-orange, red-violet, and blue-violet.

Tints are obtained by mixing the color with white, while *tones* are gotten by mixing gray into a color and shades are obtained by mixing in black.

The above information is used primarily for mixing color with pigments. The computer, however, uses two systems: cyan, magenta, yellow and black; and red, green and blue.

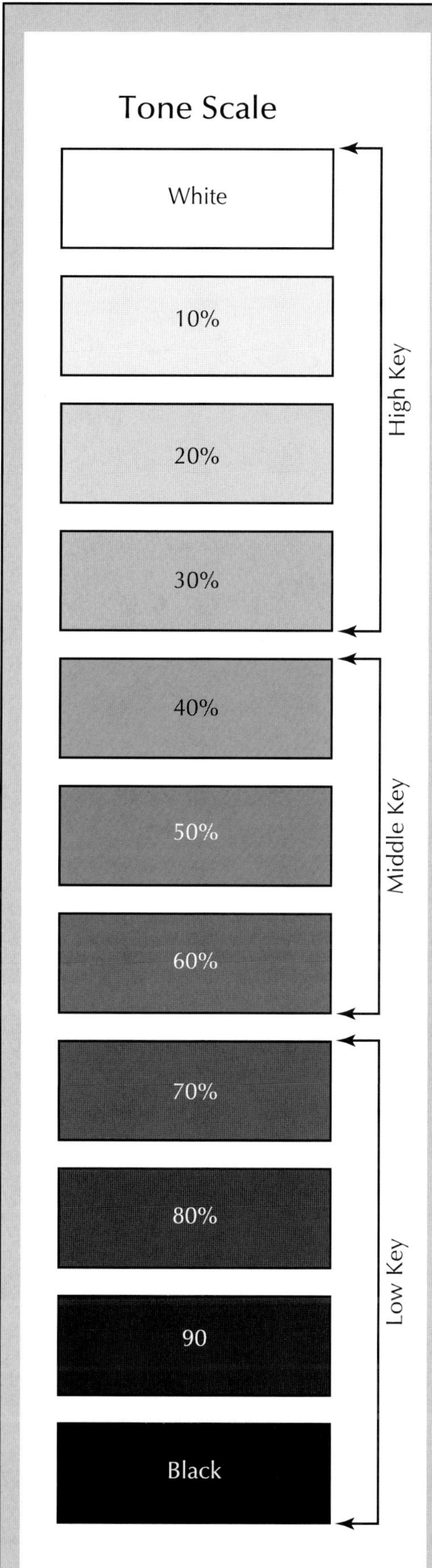

Value refers to the light to dark range of any hue.
High key refers to the lightest values.
Low key refers to the darkest values.
Neutrals are called achromatic.
Hues are called chromatic.

Terms Related to Tone

Tint- when white is added to a color.

Tone- when gray is added to a color.

Shade- when black is added to a color.

Warm or Cool grays- when particular hues are added to grays. Orange will warm a gray while blue will cool the color.

Chiaroscuro- subtle modulation of tone from light to dark in order to create the illlusion of three-dimensional volume on a two-dimensional surface plane.

Grisaille- an underpainting of tone that sets the light to dark relationships before color is applied.

Tenebrism- the use of extreme contrasting chiaroscuro to create highly dramatic effects.

1. High Contrast Range
One that uses the extremes of black and white only.
2. Neutral range
One that uses black, white, and only middle gray.
3. Light Tone Range
One that uses the grays at the light end of the value scale where middle gray functions as the darkest value.
4. Dark Tone Range
One that uses the grays at the dark end of the value scale where middle gray functions as the lightest value.

Achromatic		Chromatic
White		
10% (B)	90% (W)	Lemon Yellow
20% (B)	80% (W)	Yellow
30% (B)	70% (W)	Golden Yellow, Orange Yellow
40% (B)	60% (W)	Orane, Yellow-Green, Magenta
50% (B)	50% (W)	Red, Red-Orange, Green, Cyan-Blue
60% (B)	40% (W)	Green-Blue, Cobalt-Blue, Torquoise
70% (B)	30% (W)	Purple, Ultramarine, Violet
80% (B)	20% (W)	Blue-Purple, Indigo, Prssian
90% (B)	10% (W)	None
Black		

Glossary

acute angle
An angle that measures between 0° and 90°.

acute triangle
A triangle having three acute angles.

algorithm
A mathematical rule.

altitude
In a triangle: the line segment from a vertex perpendicular to the opposite side. Every triangle has three altitudes which might not all be in the interior of the triangle.
In an acute triangle: all three altitudes lie in the interior of the triangle.
In an obtuse triangle: two of the altitudes lie in the exterior of the triangle.
In a parallelogram: the altitude is the distance between two parallel sides. Any line segment joining two sides and perpendicular to both is *an* altitude.

analogous color scheme
A color scheme that uses three to four hues that are next to each other on the color wheel. For example, you can use yellow, yellow-green, and green. Light and dark values of the hues may be included to extend the scheme.

angle
The figure formed by two rays having a common endpoint. The common endpoint is called the vertex of the angle and the rays are called the sides of the angle.

arc
The figure formed by two points and the portion of a circle between those two points. The two points are called the endpoints of the arch. An arc is named by its endpoints.

Archimedean Solids
A family of 13 polyhedra composed of two or more different regular polygons.

Archimedean Spiral
It is named for the Greek mathematician Archimedes, 3rd century B.C.
It is a curve traced by a point continually moving around and away from a fixed point called the pole.

area
The measure in square units of any bounded portion of a plane.

armature (of a rectangle)
The subdivisions of a rectangle that provide a scaffolding for a composition.

axiom
An undemonstrated proposition

concerning an undefined set of elements, properties, functions and relationships.

axis
A line used as a reference. In the Cartesian coordinate plane, all points are located with reference to two perpendicular axes: the x-axis and the y-axis.
An axis of symmetry: a line which divides the plane into two halves which are mirror images of each other (also called an axis of reflection or a mirror).

axis of reflection
See axis.

axis of rotation
In three-dimensional space, an axis serves as the center of rotation.

axis of symmetry
See axis.

base
The plane regions on either end of a cylinder.

base angles
In an isosceles triangle: the two congruent angles.
In a trapezoid: a pair of angles with a base as a common side. Each trapezoid has two pairs of base angles.

base group
Any one of the point groups developed from the symmetry operation of rotation only.

bisect
To divide into two congruent figures.

bisector
The line or ray that divides a segment or angle into congruent figures.

bisector, perpendicular
A line that passes through the midpoint of a segment and is at right angles to that segment.

Buckyball
The carbon molecule, C_{60} ,that has been named after the geometer/designer Buckminster Fuller.

Cartesian coordinate plane
A rectangular system for locating points in a plane by using a pair of perpendicular axes: the horizontal line is the x-axis and the vertical line is the y-axis.

The Catalan Solids
The family of 13 polyhedra composed of non-regular polygons. They are the duals of the Archimedean family.

center
A point from which all points on a circle are equidistant.

center of rotation
In two-dimensional space, a point serves as the center of rotation.

center of vision
 In linear perspective graphic representation, the focal point of the viewer is called the center of vision. Where the line of vision intersects the horizon line is the center of vision.

central angle
The angle formed by two radii of a circle.

chord
A line segment whose endpoints lie on a circle.

circle
The set of all points in a plane equidistant from a given point called the center. Although it is not part of the circle, a circle is named by its center. Circle O is a circle whose center is O. The circle is a special case of the logarithmic spiral. It's growth rate is 0.

circumference
The distance around a circle. The circumference of a circle is given by the formula C=d, where d is the diameter of the circle. It is also given by the formula C=2 where r is the radius of the circle.

collinear
Lying on the same line.

compass
An instrument used for drawing circles and arcs.

complementary angles
Two angles whose sum is 90°.

complementary color scheme
A color scheme that uses two hues that are directly opposite each other on the color wheel. Red is the opposite of green, violet is the opposite of yellow, blue is the opposite of orange, etc.

concave polygon
A polygon that has at least one interior angle that measures more than 180°.

concentric
Having the same center. Two circles with the same center are considered concentric.

cone
A three-dimensional figure with a circular base and a surface generated by the lines passing through a point outside the plane of the cirlce and the points on the circle.

cone of vision
It is the angle of vision of the viewer. The cone converges at the eye of the viewer and the angle varies in relation to the distance from the viewer to an object. An angle between 45° and 60° prevents great distortion in the perspective view.

congruent
Having the same size and shape.

consecutive
In order, or in sequence. In a polygon, consecutive sides or consecutive vertices are those that appear in order as you travel around the polygon in either direction starting at any vertex.

convex deltahedra
A group of 8 polyhedra composed of only equilateral triangles.

convex polygon
A polygon whose interior angles measure less than 180°.

cross-cap
In topological terminology, it is a mobius band which attaches to one hole on the surface of a sphere.

cube
    As a noun: A polyhedron

having six square faces, sometimes referred to as an hexahedron.

As a verb: To raise to the third power. 3^3 = 3 cubed = 3 x 3 x 3 (27).

curve
A continuous set of points having only the one dimension of length. By this definition, a straight line is a curve.

cylinder
A three-dimensional figure generated by a circle whose center moves a long a line segment.

decagon
a polygon having ten sides. It may be either regular or non-regular.

deduction
See deductive reasoning

deductive reasoning
A method of reasoning that allows specifid conclusions to be drawn from general truths. This is the form of reasoning used in mathematical proof.

degree (°)
The unit of measure in angles. There are 360° in a circle. Thre are 180° in an Euclidean triangle.

denominator
In a fraction, this is the number that is written below the line. In the fraction two-thirds, 2/3, 3 is the denominator.

Désargues Triangle
this refers to the proof by the mathematician of the same name of perspectivity.

diagonal
A line segment that joins the non-consecutive vertices of a polygon. The number of diagonals that a polygon has is dependent upon the number of sides of the polygon.

diameter
A chord that passes through the center of a circle is called the diameter. A circle has infinitely many diameters all having the same length. The measure of any one of them is called the diameter of the circle. A diameter cuts the circle into two semicircles.

distance
Between points: the length of the line segment joining the points.
From a point to a line: the length of the line segment from the point to the line and perpendicular to the line.
Between parallel lines: the length of any segment having one endpoint on each line and perpendicular to both.

Divine Proportion
The proportion derived from the division of a line segment into two segments such that the ratio of the whole segment to the longer part is the same as the ratio of the longer part to the shorter part.

dodecahedron
Any polyhedron having twelve faces which may or may not be regular. There is a pentagonal dodecahedron which has twelve congruent pentagonal faces. There is a rhombic dodecahedron that has twelve congruent rhombic faces.

Dynamic Parallelogram
Any parallelogram whose sides are in the same ratio as one of the Dynamic Rectangles.

Dynamic Rectangles
A family of rectangles derived from the square through the use of diagonals. If the width is *1*, then the lengths of the rectangles are given by the square roots of the natural numbers. The Dynamic Rectangles are sometimes called the Root Rectangles and sometimes the Euclidean Series.

Dynamic Spiral
Any logarithmic spiral which uses one of the Dynamic Rectangles as its framework.

Dynablocks
A family of rectangular solids which are derived from the Dynamic Rectangles.

edge
In a polyhedron, the common side between two faces. In a cube there are twelve edges joining the eight vertices.

elements of symmetry
In three-dimensional space, the elements of symmetry are: the axes of rotation and the planes of reflection.

elevation
In architectural graphic rendering the view that is the front view when the viewer is standing on the ground line parallel to the view.

equiangular
Having angles of the same measure. In an equilateral triangle, all the sides and all the angles have the same measure.

equilateral
Having sides of the same measure. In a regular polygon, all the sides have the same measure. In an equilateral triangle, all the three sides have the same measu re.

even number
A whole number that is divisible by two. 2, 4, 6, 8....are numbers that are divisible by two. If *n* is any whole number, *2 n* will always designate an even number.

exponent
A number that indicates the number of times a quantity is to be used as a factor. In the expression b^3, 3 is the exponent. $b^3 =(b)(b)(b)$.

exterior
Of an angle or a closed curve: The set of all points in the plane that do not lie on the figure or in the interior of the figure.
exterior angle of a polygon: An angle formed by a side and the extension of an adjacent side in a polygon. There are two exterior angles at each vertex.

extremes
See proportion

face
A polygonal shape, together with its interior, that forms part of the surface of a polyhedron. A square is the face of a cube. A triangle is the face of a tetrahedron.

factor
A number that divides another number without leaving a

remainder. 3 is a factor of 9 since 3 divides 9 three times without leaving a remainder.

Fibonacci Numbers
The summation sequence whose terms are: 1, 1, 2, 3, 5, 8, 13....

Fibonacci Sequence
The summation sequence whose terms are: 1, 1, 2,3,5,8,13,21,34,555...

figure/ground
In any system of graphic representation, it is the relationship of the object(s) to its background.

foreshortening
In a drawing, the representation of the long axis of an object by contracting its lines so as to produce an illusion of an extension into space.

fractal
A curve that exhibits self-similarity at all levels of magnification.

function
A set of ordered pairs in which no first element is paired with more than one second element.

generation
In fractal geometry, the figure determined by a particular number of iterations of the rule.

generator
In fractal geometry, the figure that results from one application of the rule.

genus
In the subject of topology, objectsa are classified according to their genus.

a. all closed, orientable surfaces can be continuously transformed into a sphere with a certain number of handles.

b. All closed non-orientable surfaces can be continuously transformed into a sphere with a certain number of cross-caps.

geodesics
The shortest line between two points on any mathematically derived surface.

glide reflection
The symmetry operation that combines moves of translation and reflection.

Goldbricks
A set of rectangular solids that are derived from the Phi-Family Rectangles.

Golden Cut
The point on a line segment that divides the segment into two segments whose lengths are in the Golden Ratio.

Golden Mean
the Golden Mean is the same as the Golden Cut.

Golden Ratio
The Golden Ratio is the same as the Golden Cut and the Golden Mean. The Golden ratio is the ratio of approximately *1.61803* to *1,* derived from the Divine Proportion. The Golden Ratio is sometimes known by the Greek

letter ø (phi).

Golden Rectangle
Any rectangle whose sides are in the ratio of ø to *1* where ø is approximately equal to 1.61803.

Golden Section
The point on a line segment that divides the segment into two segments whose lengths are in the Golden Ratio.

Golden Triangle
Any isosceles triangle in which the ratio of a leg to the base is ø to 1, where to five decimal places ø = 1.61803. The triangle is also called the Triangle of the Pentalpha or the Sublime Triangle.

Golden Zonohedra
A family of five polyhedra in three-dimensional space that is related to the cube of higher dimensions.

grid
A repeating pattern superimposed on the plane consisting of lines and/or circle.
　　polar grid: A grid consisting of concentric circles with radius vectors emanating from the center, or pole.
　　rectangular grid: A grid whose units are rectangles.
　　triangular grid: A grid whose units are triangles.

ground line
It is the intersection of the drawing surface (picture plane) and the ground plane.

handedness
Referring to the orientation of right or left-handedness.

height
　　In a cylinder: The distance between the bases.

hexagon
A polygon having six sides which may be regular or non-regular.

helix
A spiral-like curve formed by wrapping a cylinder uniformly in a single layer. The thread of a metal screw is a helix.

hexahedron
A figure having six faces which may or may not be regular. A cube is a hexahedron. Plural is hexahedra.

horizon line
The intersection of the plane of vision of the viewer and the drawing surface. It is at the eye level of the viewer.

hue
The name of a particular color. Red is a hue.

hyperbolic geometry
A non-euclidean geometry that does not include Postulate # 5 of Euclid's geometry. Through any point not on a straight line, infinitely many straight lines can be drawn having no point in common with the given line.

hyperbolic plane
In the Poincaré model, the Euclidean plane is replaced by the interior of a circle.

hypercube
The cube in the fourth dimension

hypotenuse
In a right triangle, the side opposite the right angle is the hypotenuse.

hypothesis
A conjecture that requires testing for verification.

icosahedron
A polyhedron having 20 faces which may be regular or non-regular.

included angle
The angle formed by two specified adjacent sides of a polygon.

included side
In a polygon, the common side between two specified angles.

induction
See inductive reasoning.

inductive reasoning
A method of reasoning that allows a generalization to be made from observing specific cases. Inductive reasoning can be likened to making educated guesses based on experience. Unlike deductive reasoning, it can lead to a false conclusion.This is the form of reasoning used in the Scientific Method.

initiator
In fractal geometry, the initiator is the original figure to which the rule is applied (the first generation curve).

inscribed angle
An angle whose vertex likes on a circle and each of whose sides intersects the circle.

inscribed circle
A circle lying in the interior of a polygon such that each side of the polygon is tangent to the circle.

inscribed polygon
A polygon whose vertices like on a circle.

integer
Any member of the following infinite set of numbers:....-3, -2, -1, 0, 1, 2, 3....

interior
Of an angle: The set of all points not on any angle having the property that if any two are joined by a line segment, the segment will not intersect the angle.
Of a closed curve: The set of all points in the plane that are enclosed by the curve.

interior angle
Of a polygon: An angle formed by adjacent sides of a polygon.

intersect
To have at least one point in common.

invariant
Something which is not affected by a designated mathematical operation.

irrational number
Any number that cannot be

expressed as the ratio of two integers, 2:3. The Golden Ratio is an irrational number, 1:1.618...

Isosceles Right Triangle
A triangle having a pair of congruent sides and one right angle (90°).

Isosceles trapezoid
A trapezoid in which the nonparallel sides are congruent.

isosceles triangle
A triangle having a pair of congruent sides.

isometric projection
This is a drawing system for representing three dimensions on the two-dimensional surface. This form of projection presents three faces visible to the viewer and the receding angles are congruent.

iteration
A repetitive procedure generally linked to a mathematical function wherein the prevous output becomes the input for the next term.

labyrinth
A single ritual path which leads from an entrance to a goal.

lattice
See space lattice.

lateral surface
The portion of the surface of a cylinder that lies between the bases.

leg(s)
In an isosceles trapezoid: the nonparallel sides.
In an isosceles triangle: the congruent sides.
In a right triangle: the sides that form the right angle.

line
An undefined term in geometry. It has the properties of infinite length, continuity, and no width. It is a straight curve. A line is named by any two of its points or by a lower case letter.

line group
A line group is a linear pattern resulting from translating a unit in a single direction. The unit is developed by manipulating a two-dimensional motif with the symmetry operations of reflection or rotation or both. There are seven two-dimensional line groups.

line segment
The figure formed by two points on a line and all the points in between those two points. The two points are called the endpoints of the segment. A line segment is named by its endpoints.

linear perspective
(See Renaissance perspective)

logarithmic spiral
(Equiangular spiral)
Each full rotation of the spiral increases its distance, from the pole of the spiral, by a fixed ratio. Thsi determines its rate of growth. Every straight line through the pole intersects the

spiral at the same fixed angle.

macrame
A form of lacework that is made by knotting cords into a patterned structure.

mapping
A rule of correspondence established between two mathematical sets that associates each member of the first set with a single member of the second set.

mean proportion
A proportion in which the means are equal.

means
See proportion

midpoint
The point that divides a line segment into two congruent segments.

minute
One-sixtieth of a degree in angle measure. $60' = 1°$. $35°12'$ reads 35 degrees and 12 minutes.

Möbius band
Named for the mathematician August Möbius. This type of band has only one surface and one bounding edge; therefore, it has no inside nor outside.

monochromatic color scheme
A color scheme that uses one hue and the light and dark values thereof.

mirror
See axis of reflection.

motif
A two-dimensional design that is asymmetric.

mountain fold
In the techniques of paper folding, the fold in which the peak points upward as a mountain is called a mountain fold. It is the reverse of a valley fold.

non-collinear
Not lying in the same line.

non-Euclidean geometry
Any geometry which differs from Euclidean geometry.

obtuse angle
An angle that is greater than 90° and less than 180°.

obtuse triangle
A triangle having one obtuse angle.

octant
One of the eight sptail subdivisions created by the x, y, z axes in 3-space.

odd number
A whole number that is not divisible by two.

order of (vertex net)
The mathematical numerical notation that indicates what kind and how many polygons are found at a single vertex.

ordered triple
A trio of numbers used to locate a point in 3-space. In order, they give positions relative to the x,y, z axes, respectively.

Parablock
A set of rectangular solids that are derived from parallelograms.

parallel
Lines: two or more lines that lie in the same plane and have not points in common.
Segments: noncollinear segments which lie on parallel lines.

parallel projection
This is a drawing system for representing three dimensions on the two-dimensional surface. A projection that sets a face parallel to the picture plane while the lines that recede are at angles to that picture plane.

parallelepiped
A three-dimensional solid that is composed of 6 faces with three pairs of parallel ones.

parallelogram
A quadrilateral in which opposite sides are parallel.

pattern
A repetition of units in an ordered sequence.

Penrose rhombohedra
the two rhombohedra that are derived from the two Penrose tiles named the "fat" and "skinny" rhombuses.

Penrose tiles
The tiling units discovered by the British mathematician Roger Penrose. These units can be used in pairs to form infinite nonperiodic tilings. Each is derived from the pentagram within a pentagon.

pentagon
A polygon having five sides which may or may not be regular.

perimeter
The distance around a figure.

periodic
A tiling is periodic if any region of adjacent poygons can be moved by the symmetry operation of translation, without the use of rotation, to another position where it can coincide with a congruent region.

perpendicular
Lines: lines that meet to form a right angle.
Segments: noncollinear segments that lie on perpendicular lines.

perpendicular bisector
See bisector, perpendicular.

phi
The number naming the ratio that appears in the Divine Proportion (approximately 1.61803). The Greek letter phi (ø) is used in this text, but the letter tau, O, is also used by mathematicians to name this ratio.

Phi-Family Rectangles
The group of rectangles related to the Golden Rectangle or generated by phi.

phyllotaxis
The principles governing the arrangment of leaves in plant growth.

pi
The irrational number 3.14... that

represents the ratio of the circumference of a circle to its diameter. C:D

picture plane
Refers to the surface, usually rectangular, on which an artist constructs an image. It is usually parallel to the viewer.

plan view
An architectural drawing made to scale, showing the single view of an object when looked down upon.

plane
An undefined term in geometry. A plane may be thought of as "a slice of space" having infinite length and infinite width but no thickness.

plane group
A pattern resulting from the repetition of a unit in two non-parallel directions in order to fill the plane. The unit is developed by manipulating a motif using one or more of the symmetry operations of rotation, reflection, translation. There are 17 plane groups.

plane of reflection
In three-dimensional space, the plane of reflection divides the object into mirror halves.

Platonic Solids
The five regular polyhedra that are associated with the Greek philosopher Plato. They are: the tetrahedron, the cube, the octahedron, the dodecahedron, and the icosahedron.

point
An undefined term in geometry. A point is often thought of as a location in space having no dimension.

point group
A pattern that results from the rotation of a unit about a pole.

pole
In a polar graph, it is the point corresponding to the origin.

polygon
A simple closed curve composed of line segments each intersecting exactly two others, one at each endpoint. A polygon is named by its vertices read consecutively in either a clockwise or counter- clockwise direction. The line segments are called the sides of the polygon. The points of intersection are called the vertices of the polygon.

polyhedron
A three-dimensional figure formed by polygons which intersect in such a way that vertices and sides are common. The polygonal figures are called the faces of the polyhedron. The common sides are called the edges of the polyhedron. The common vertices are called the vertices of the polyhedron.

polyplex
A polytope in 4 dimensions composed of polyhedral cells.

polytope
This term refers to the category of geometric figures bounded by portions of lines, planes or hyperplanes. The families within this category include the polygons, the polyhedra, the polyplexes, and n-polytopes.

primary colors
In pigments, the primary colors are yellow, red, and blue. These are the essential hues from which all other colors can be obtained.

projection
The collection of all light rays coming from an object and converging at the eye of the viewer, or artist, or the lens of a camera.

projection systems
The various systems os drawing three-dimensional objects on the two-dimensional plane.

projective geometry
The study of geometric properties that are invariant under projection (those figures that are left unchanged by a projection.

proportion
A pair of ratios set equal to each other. a/b =c/d. In this proportion, *a* and *d* are called the extremes of the proportion while *b* and *c* are called the means. These words refer only to the position of the terms.

protractor
A tool used for measuring the number of degrees in an angle. It is shaped like a semicircle.

Pythagorean Theorem
It is the theorem named after Pythagoras in which the sum of the squares on the sides of a right triangle equal the sum of the squares on the hypotenuse.

Pythagorean triple or triad
The integers that have the property that the sum of the squares of two of them equals the square of the third. 3, 4, 5 is a Pythagorean triple since 32 = 42 = 52, or 9 + 16 = 25.

quadrilateral
A four-sided polygon which may or may not be regular. A square is a quadrilateral. A rhombus is a quadrilateral.

radicand
The quantity under the radical sign.

radius
    Of a circle: a line segment that joins the center of a circle to any point on the circle is *a* radius of the circle. The measure of any of these segments which are all congruent is *the* radius of the circle.
    of a regular polygon: If a regular polygon is inscribed in a circle, *the* radius of the polygon is the radius of the circumscribed circle. A radius is a line segment from the "center" of the polygon to a vertex.

radius vector
Any ray emanating from the pole of a spiral and intersecting the spiral curve.

ratio
A comparison of numbers to each

other. The ratio of 4 to 5 may be expressed as 4:5 or simply 4 to 5 or as a fraction 4/5.

ray
The figure formed by a point on a line and all the points on the same side of that point. The point is called the endpoint of the ray. The ray is named by its endpoint and any other point on the ray.

reciprocal (of a rectangle)
A similar rectangle cut from a parent rectangle the length of which is the width of the original rectangle.

reflection
The symmetry operation that flips a motif changing both its position and its handedness.

Renaissance perspective
A pictoral drawing system widely used in the Renaissance. it is also called linear perspective. It involves lines that converge at a point on a horizon line.

rhombus
A parallelogram in which all sides are congruent.

rhombohedron
A solid composed of 6 rhombuses.

right angle
An angle that measures 90°.

right circular cone
A cone whose base is at right angles to its axis of rotation.

right circular cylinder
A circular cylinder whose bases are at right angles (90°) to its axis of rotation.

right triangle
A triangle having one right angle.

Rope Knotter's Triangle
The right triangle whose legs measure 3 and 4 respectively, and whose hypotenuse measures 5. Also known as the 3-4-5 Right Triangle. It was so named because it was discovered by the Ancient Egyptian rope knotters who prepared ropes for the purpose of measuring .

Root Rectangles
The rectangles derived from the square such that the width measures one (1), ands the length measures the square root of a natural number. These rectangles are the members of the family of Dynamic Rectangles.

rotation
The symmetry operation that involves the turning of a motif about a point or an axis.

rotocenter
A point about which a motif is rotated to create the two-dimensional point groups of symmetry.

scalene triangle
A triangle having no congruent sides.

secant
A line that intersects a circle in exactly two points.

second
One-sixtieth of a minute in angle measure. 60" = 1' or 3600" =1 °. 20 ° 40' 13" reads 20 degrees, 40 minutes, 13 seconds.

section
A portion of the plane that intersects rays of light coming from an object to the viewer.

sector
A region in the interior of a circle bounded by two radii and the circle itself.

self similarity
The property of a figure whose shape remains constant regardless of the level of magnification.

sequence
An ordered progression of terms. A sequence may have a finite number of terms, eg. 1, 2, 3, 4, 5, 6, 7; or an infinite number of terms, eg, 1, 1, 2, 3, 5, 8, 13, 21.....(This is known as the Fibonacci Sequence) .
If the notation S_1, s_2, s_3, ...S_n is used to denote a sequence, the subscript names the position of the term, eg. S_{12} would be the 12th term.

set
A collection of distinct elements.

side
> *Of an angle:* see angle.
> *Of a line:* either of the half-planes formed by a line in the plane.
> *Of a polygon:* see polygon.

similar
Having the same shape. Two polygons are similar if corresponding angles are congruent and corresponding sides are in proportion.

simple closed curve
A plane curve that can be traced without lifting pencil from paper in such a way that the starting point and the ending point are the same. No point is traced more than once.

space frame
In three dimensions a repeating array of points or lines. Also known as space lattice.

space lattice
See space frame.

spiral
A curve which moves around a central point, called the pole, while continually increasing its distance from the pole.
> *Archimedean spiral*: A spiral whose successive wholrs are spaced at equal intervals along any radius vector.
> *logarithmic spiral:* A spiral whose successive wholrs form congruent angles with any radius vector and are spaced in geometric progression along the radius vector.

square
> *Noun:* a rectangle in which all of the sides are congruent.
> *Verb:* To multiply a quantity by itself. The number 2 used as an exponent indicates the

operation of squaring, eg. $a^2 =$ (a)(a).

square root
a number that when multiplied by itself produces a given number. The operation of taking the positive square root of a number is indicated by a radical sign. The given numer, under the sign, is called the radicand, eg. $\sqrt{16} = 4$ since 4 x 4 = 16.

Square Root Phi Rectangle
Any rectangle in which the ratio of the length to the width is phi to 1, or approximately 1.271.

straight angle
An angle measuring 180°.

station point (SP)
An element of a Renaissance perspective drawing. It is the location of the eye of the viewer.

stereographic projection
The technique of depicting three-dimensional solids on a plane surface.

symmetry operations
The symmetry operations are those of rotation, reflection, glide reflection, and translation.

tangent
A line lying in the same plane as a circle that intersects the circle in exactly one point.

template (for a polyhedron)
A two-dimensional prototype that has glue tabs added.

tessellation
a tiling.

tetrahedron
a polyhedronhaving four triangular faces which may or may not be regular.

theorem
A proposition that is provable on the basis of explicit assumptions.

tiling
Noun: the resulting pattern when the plane is tiled.
Verb: the process of covering the plane with closed units so that there are no gaps or overlaps.

topology
The study of the properties of geometric configurations that are invariant under transformation by continuous mappings.

torus
A surface having the shape of a doughnut.

transformation operations
The operations of chamfering, skewing, slicing, stretching, truncating, that alter a three-dimensional figure in a particular way.

translation
The symmetry operation which involves the sliding of a motif in two-dimensions, or a figure in three-dimensions without changing its orientation.

trapezohedron
A three-dimensional object that has trapezia for its faces.

trapezoid

A quadrilateral with exactly one pair of parallel sides. The parallel sides are called the bases of the trapezoid and the angles including either base are called the base angles. There are two pairs of base angles.

trapezium

A quadrilateral having no parallel sides. The plural is trapezia.

triangle

A polygon having three sides which may or may not be congruent.

triangulate

The operation of subdividing a figure into smaller triangular units.

truncate

To cut off either planar sections or three-dimensional sections in a systematic fashion.

uniform tiling

A tiling in which all the vertices are congruent.

unit

In measurement: a length or volume that acts as a basic element of measure.

In symmetry: a pattern element determined by a motif or figure used with one or more symmetry operations.

unit measure

The length taken to represent the number *one (1)* in a given set of circumstances.

valley fold

This is a fold in the technique of paper folding that has the peak pointing downward as a valley. It is the reverse of a mountain fold.

vertex

Of an angle: see angle.
Of a polygon: see polygon.

vertex net

A numerical notation that gives the kind and number of polygons around a particular vertex, eg. 4 - 4 - 4 tells that three squares surround a vertex.

volume

The measure in cubic units of any three-dimensional object.

x-axis

The horizontal axis on the Cartesian coordinate plane.

y-axis

The vertical axis on the Cartesian coordinate plane.

z-axis

The oblique axis on the Cartesian coordinate plane.

zonohedra

A group of solids composed of zonogons. A zonogon is a 2-n sided polygon where pairs of opposite sides are parallel and congruent.

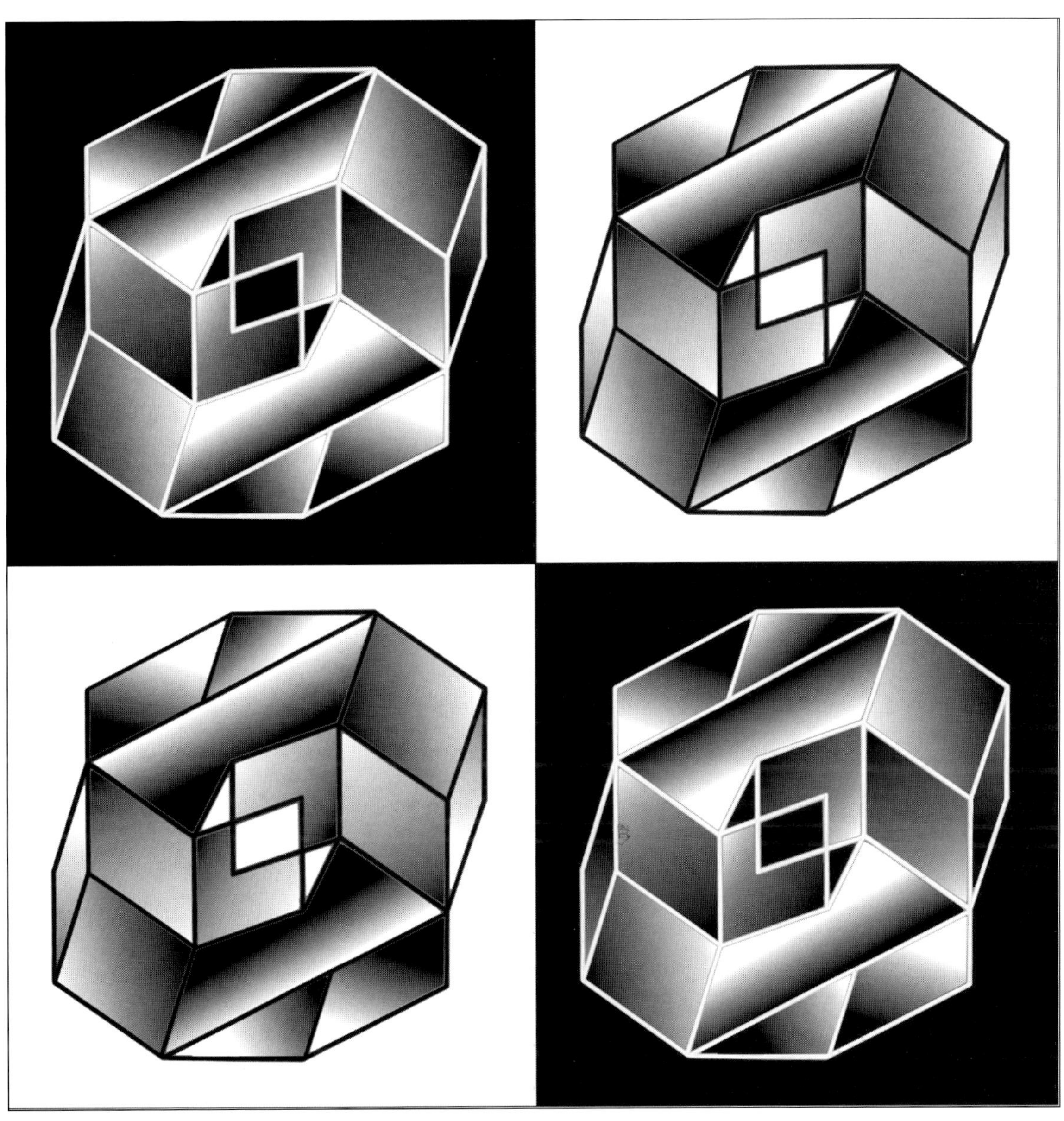

The above artwork by Richard Newman was created using the software program Adobe Photoshop.

Construction Procedures

Charts

Index

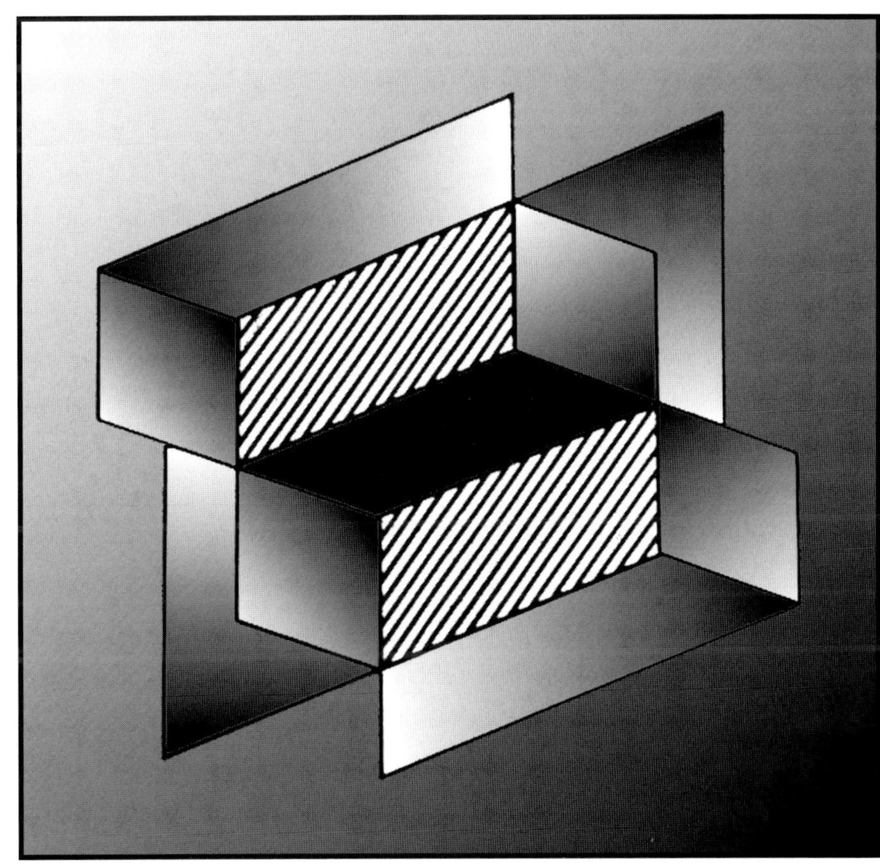

This page:
Artworks created by Richard Newman using the software program, Adobe Photoshop.

Notes

Notes

Pythagorean Press produces books that are interdisciplinary in content.
These bring together the subject areas of Art, Math and Nature.
The first book, *Universal Patterns*, is currently out of print.

The Surface Plane

by Martha Boles and Rochelle Newman
ISBN: 0-697-355578-0

This book explores two dimensions with emphasis on pattern. In the first chapter, readers are introduced to the notion of the grid and its function. The second chapter describes in detail the aspects of point, line, and plane symmetry which become the underlying concept that weaves the chapters on circles, tilings, and fractals together. The last chapter, the Pliable Plane, is the lead into three dimensions.

Space, Structure, and Form

by Rochelle Newman and Donna Fowler
ISBN:0-697-333058-3

This book involves the third dimension. It brings to the emerging field of Design Science , a lively text filled with diagrams, drawings, and photos. Its use is as diverse as its content. Not only will teachers in the disciplines of Art, Math, and Science find the information useful, but artists and designers can learn much for their work in three- space. This volume begins with the manipulation of the pliable plane to create 3-d folded structures. It moves on to show how the eternal triangle is used to build stable forms. The familiar cube generates other families of rectangular solids. Also explored are the properties of the sphere and its connections to solids called polyhedra. The latter are explored in great detail.

To order copies of these books, check your leading bookstore
or contact Pythagorean Press as stated below.
Volume discounts available to schools, libraries, bookstores.

These books may be ordered by contacting:
Pythagorean Press
c/o Corporate Fulfillment Systems
1 Bert Drive
P.O. Box 339
West Bridgewater, MA. 02379-9970
1-800-344-4501 or 508-583-5239
Fax: 508-583-9904